HEAT-RESISTANT POLYMERS

Technologically Useful Materials

HEAT-RESISTANT POLYMERS

Technologically Useful Materials

J. P. Critchley

G. J. Knight

and

W. W. Wright

**Royal Aircraft Establishment
Farnborough, England**

PLENUM PRESS • NEW YORK AND LONDON

Library of Congress Cataloging in Publication Data

Critchley, J. P., 1932–
  Heat-resistant polymers.

  Bibliography: p.
  Includes index.
  1. Polymers and polymerization—Thermal properties. 2. Heat resistant plastics. I.
Knight, G. J., 1937–   . II. Wright, W. W. (Walter William), 1928–   . III. Title.
TA455.P58C74 1983   620.1′9204217   83-3987
ISBN 0-306-41058-3

First Printing—July 1983
Second Printing—June 1986

ACKNOWLEDGMENTS

Some of the figures and tables which appear in this book have previously been published elsewhere and we wish to express our thanks to the following for granting us permission to reproduce the material listed in this work.

Academic Press Inc.: Figures 6.9 and 6.11.

Air Force Wright Aeronautical Laboratory (USA): Figures 6.17, 6.18, 6.19, 6.20, and 6.21.

American Chemical Society: Figures 3.15, 3.16, 5.27, 8.4, 8.5, 8.6, 8.7, 8.10, 9.10, and 9.11; Tables 3.21, 3.23, and 5.5.

Applied Science Publishers: Figures 1.10, 2.30, 2.31, 2.32, and 5.7; Tables 4.1 and 5.9.

Birkhäuser Verlag: Figure 3.4.

Carl Hanser Verlag: Figures 2.23, 2.24, 2.25, and 5.19; Table 2.12.

Communications Channels Inc.: Figure 3.11.

Elsevier Scientific Publishing Co.: Figure 3.3.

Freund Publishing House: Tables 5.27, 5.28, 5.29, 5.30, 5.31, 5.32, 5.33, and 5.34.

Gordon and Breach Science Publishers Ltd.: Figures 2.26 and 2.27.

Hüthig und Wepf Verlag: Figure 4.17.

Institute of Electrical and Electronic Engineers Inc.: Figures 9.1 and 9.2.

IPC Science and Technology Press Ltd.: Figure 4.20.

John Wiley & Sons Inc.: Figures 2.14, 2.34, 4.11, 4.13, 4.14, 5.1, 5.2, 5.3, 5.4, 5.6, 5.36a–e, 6.13, 7.1, 7.2, 7.3, 7.4, 7.5, 7.6, 7.7, 7.9, 7.10, 7.11, 7.12, 8.1, 8.2, 8.3, 8.8, 8.9, and 9.17; Tables 2.3, 2.21, 4.2, 4.9, 4.14, 4.20, 5.37, and 9.8.

Marcel Dekker Inc.: Tables 7.3 and 9.3. Reprinted from *J. Macromol. Sci. Chem. A* **1**(6), 1967, pp. 1116 and 1122, and *J. Macromol. Sci. Rev. C* **6**(1), 1971, by courtesy of Marcel Dekker Inc.

McGraw-Hill Book Co.: Figures 2.16 and 5.18. From *Handbook of Epoxy Resins* by H. Lee and K. Neville. Copyright © 1967 by McGraw-Hill Book Company.

Naval Research Laboratory (USA): Figure 9.9.

Naval Surface Weapons Center (USA): Figures 9.6 and 9.7.

Office National d'Etudes et de Recherches Aérospatiales: Figure 2.33

Penton/IPC: Table 3.22. Reprinted with permission from *Materials Engineering,* copyright Penton/IPC, Cleveland, Ohio.

Pergamon Press Inc.: Figures 2.20, 2.21, 2.28, 6.15, 6.16, 9.12, and 9.16; Tables 5.10, 6.7, and 9.9.

Royal Society of Chemistry: Table 6.1.

Rubber World: Figure 8.11.

Society for the Advancement of Materials and Process Engineering: Figures 2.34, 2.35, and
   9.5; Tables 2.19, 2.21, and 2.22.
Society of Chemical Industry: Figures 2.8 and 2.29; Table 2.13.
Society of Plastics Engineers Inc.: Figure 7.8.
Society of the Plastics Industry: Figure 2.19.
Springer-Verlag: Figure 2.4.
Textile Research Institute: Figure 4.12.
Van Nostrand Reinhold Co.: Table 6.9.
Verlag Chemie: Table 5.47.

PREFACE

Definitions of what is meant by a heat-resistant polymer vary considerably. We have taken the term to mean a polymer which can be *used*, at least for short time periods, at temperatures from 150°C. The greatest problem which arises in writing a monograph on such materials is the tremendous amount of data that is available. More than 2000 references have been published on one heat-resistant polymer system alone over a period of little more than two years. The result is that a very high degree of selectivity must be exercised with respect to the information reproduced. We have chosen to restrict our coverage to polymers that have received at least some degree of commercial exploitation and to details of their methods of preparation, their thermal and thermo-oxidative stabilities and modes of degradation, and their properties at elevated temperatures. It must be emphasized that other properties not cited, e.g., hydrolytic and chemical stability, and resistance to ultraviolet radiation, may be equally important in particular uses of these materials.

The "older" heat-resistant polymers, e.g., the thermosets and some of the fluorine-containing materials, are not dealt with in such depth as are the "newer" polymers with aromatic and/or heterocyclic rings in the chain. This is because books have been available for some time on the well-established commercial polymers and developments in them have not been as marked recently as in the aromatic and heterocyclic macromolecules. It is, nevertheless, salutary to reflect that despite the immense effort put into the synthesis of new thermally-stable polymers based on heterocyclic ring systems, relatively few have reached any stage of commercial development, and were it not for the polyimides (variants of which comprise about forty percent of the named commercially available heat-resistant polymers), the impact would have been negligible.

The crux of the problem is that the type of organic chemical structures that convey good thermal and thermo-oxidative stability are also those which lead to infusible, insoluble materials that are difficult, if not impossible, to fabricate. As a consequence, the history of the development of heat-resistant polymers is, more often than not, one of compromise between stability and fabricability. In recent years, in fact, effort has been devoted more to modifications of existing polymer systems to improve processability without concomitant loss in thermal stability, than to synthesis of novel materials.

There are currently no indications of organic polymers in the "pipe-line" that will have better thermal stability than currently available materials. It is considered that the upper temperature limits of use of organic polymers have been reached and that future advances in heat-resistant polymers must await developments in the inorganic field.

CONTENTS

Chapter 3

FLUORINE-CONTAINING POLYMERS 87

Chapter 4

POLYMERS WITH AROMATIC RINGS IN THE CHAIN . . 125

Chapter 5

POLYMERS WITH HETEROCYCLIC RINGS IN THE CHAIN 185

Chapter 6

SILICON-CONTAINING POLYMERS—SILICONES......323

Chapter 7

BORON-CONTAINING POLYMERS— THE CARBORANESILOXANES

Chapter 8

PHOSPHORUS-CONTAINING POLYMERS— THE PHOSPHAZENES

Chapter 9

FUTURE DEVELOPMENTS

Appendix

SOME GENERAL REVIEW ARTICLES FROM 1970
ONWARDS

1

INTRODUCTION

SCOPE OF REVIEW

There are many applications for which heat-resistant polymers are needed, but there is little doubt that the main impetus for their development has come from the aerospace field. Synthetic activity was at its height during the late fifties and early sixties, but has declined lately for a number of reasons. These include the cutback in aerospace expenditure, the basic difficulties encountered in producing useful materials, and the relatively small markets for and cost of these speciality products. Progress in heat-resistant polymers has been governed by the opposing requirements of thermal stability and processability. The types of chemical structure conveying heat resistance have tended to lead to insolubility and infusibility, and hence structural modifications must often be made so that fabrication is possible. As a consequence, relatively few of the very many polymers that have been synthesized have achieved commercial exploitation. In the past few years no completely new thermally stable organic polymer systems have emerged; effort has been concentrated upon improved methods of synthesis of known structures, or upon modifications to these structures so that easier processing and fabrication are possible without a concomitant loss of stability. There is little evidence at present that any marked increase in thermal stabilities of organic polymers over those currently obtainable will be forthcoming. A survey of those materials available in commercial or development quantities is therefore appropriate, as it is considered probable that such materials are unlikely to be superseded in the near future.

A good case could be made for a detailed critical review of all the polymers that have been synthesized and evaluated with heat resistance in mind, but this would now be a task of considerable magnitude (since 1972 there have been some 4000 references in the literature concerning polyimides alone), and the data would comprise a large volume, much of whose contents would be only of academic interest. This review has therefore been restricted to those polymers that have given indications of being technologically useful, the criteria being that they are or have been commercially available. By restricting the scope in this way the data are of manageable proportions. It is also hoped that this approach will be of more practical value and will serve to show why many research polymers have remained precisely that. Three books have previously been published on the topic of heat-resistant polymers.[1-3] These publications are all somewhat dated and none of them adopted the approach now proposed. Review articles that have appeared since the date of publication of these books are listed at the end of this chapter in the supplementary bibliography and in the Appendix.

(While this manuscript was in preparation, a book entitled *Thermally Stable Polymers* by P. E. Cassidy was published by Marcel Dekker in 1980. Professor Cassidy's book deals with all thermally stable polymer systems and concentrates on synthetic aspects; the coverage is therefore radically different from that of the present volume.)

BACKGROUND TO HEAT-RESISTANT POLYMER DEVELOPMENT

Why has there been this specific need for heat-resistant materials in the aerospace field? The problem is perhaps best illustrated as follows: Figure 1.1 shows the temperature profile for a flight by the supersonic aircraft Concorde from London to New York. For the major part of the flight, somewhat over two hours, the skin temperature is approximately 110°C. Hence the materials of construction must be capable of withstanding exposure to this temperature for the aircraft's flying lifetime, say 30,000 hours. Concorde's flight profile is partly governed by the fact that 110°C is about the maximum temperature of use of currently available aluminum alloys. At higher speeds the problem is magnified because the skin temperature rises with the square of the speed (Figure 1.2). Thus for a Mach 3 aircraft the skin temperature would be about 300°C, and titanium or stainless steel would have to be used in its construction instead of aluminum. Nonmetallic components, e.g., windows, radomes, tires, seals, paints, and insulants would also have to be capable of withstanding such temperatures for long time periods without undue deterioration. This question of time at temperature is a very important consideration in any discussion concerning the application of heat-resistant polymers. The other main aerospace use—guided weapons, reentry vessels, space shuttle service—may involve the imposition of very high temperatures (perhaps thousands of degrees), but for very short time periods. Under such conditions, of course, nonmetallic materials are degrading rapidly, and it is the very breakdown that conveys insulating properties. The ablation problem is, perhaps surprisingly at first sight, simpler to solve than is the problem of protracted use of polymers at much lower temperatures, and conventional phenolic resins have proved very successful.

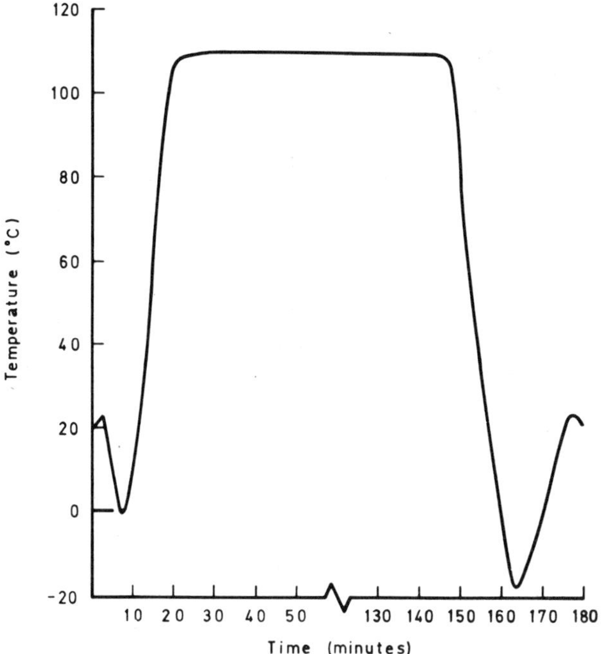

Figure 1.1.   Surface stagnation temperature profile for flight by Concorde from London to New York.

Having established a need for heat-resistant polymers, what approaches have been used to satisfy this need? The requirements for *use* of a polymer at high temperatures may be very simply stated:

1. Retention of mechanical properties—high softening point, high glass transition temperature.
2. High resistance to thermal breakdown.
3. High resistance to chemical attack, i.e., oxidation, hydrolysis.

The softening point and glass transition temperature (and it should be noted that it is the latter that often governs the maximum practical use temperature of a polymer rather than the former) can be raised by increasing the intermolecular forces between chains. This can be done by incorporating polar side groups (the glass transition temperature of atactic polypropylene

$$\left[CH_2 - \underset{\underset{CH_3}{|}}{CH} \right]_n$$

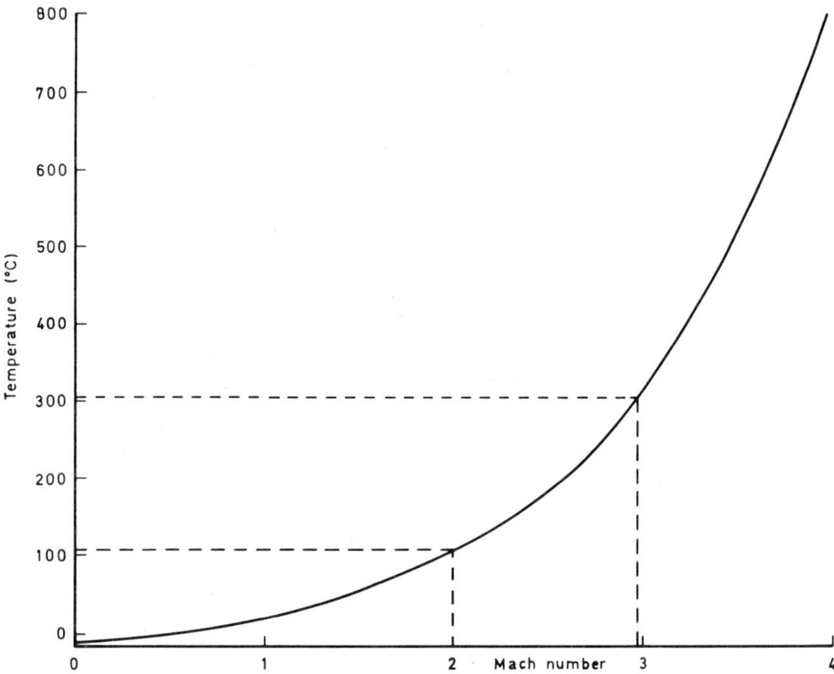

Figure 1.2. Adiabatic wall temperature for flight at constant speed in air at 15,000 m.

is −20°C, whereas that of polyacrylonitrile

$$\left[\begin{array}{c} CH_2 - CH \\ | \\ CN \end{array}\right]_n$$

is about +105°C), by increasing the opportunities for hydrogen bonding, and by actual chemical cross-linking of the chains. Other methods are to increase the regularity of the chain with possible consequent increase in the degree of crystallinity, e.g., by stereospecific polymerization, or by making the chain more rigid by the incorporation of bulky cyclic groups, especially para-linked, in the main backbone.

Pure thermal stability is influenced by the strength of chemical bonds and it is axiomatic that combinations of atoms with known weak bond strengths should not be used. Table 1.1 lists some of the bond strengths for common combinations of atoms found in polymer structures.[4,5]

In the search for thermally stable polymers, undue emphasis upon the bond strength criterion led to an initial concentration of effort on the development of inorganic polymers. The higher bond strengths inherent in these systems are,

TABLE 1.1. Bond Strength Values (kJ/mole) (References 4, 5)

C—S	273	C—N	307	C—H	416	P—O	528	C=C	609	B—O	777
B—H	294	Si—H	319	C—F	428	P—C	580	C≡N	617		
		Si—C	328	Si—N	437			Ti—O	672		
		C—Cl	340	Si—O	445						
		C—C	349								
		C—O	361								
		C—B	374								
		B—N	386								

however, more often than not offset by chemical attack involving processes of relatively low activation energy, e.g., hydrolytic or oxidative cleavage reactions. As a consequence relatively few useful inorganic polymers have yet been developed. The most widely known and used polymers with a completely inorganic backbone are the siloxanes, and these provide an example of the dangers of concentrating upon bond energies to the exclusion of all other considerations. In polysiloxanes the Si—O bond energy is 445 kJ/mole and that of the Si—C bonds 328 kJ/mole. It would therefore be expected that cleavage would occur preferentially at Si—C bonds with the elimination of the alkyl, or aryl, framing groups. In practice Si—O bond breakage occurs with the formation of low-molecular-weight cyclic products because overall this is energetically favored.[6]

Other pointers to thermally stable structures are given by the properties of model compounds. The drawback to this is that the polymer chain environment in which degradation occurs often has effects that would not have been predicted from the behavior of model compounds. It must be emphasized that it is very difficult, if not impossible, to prepare a polymer as a pure compound in the strict chemical sense. Besides the questions of molecular weight and molecular weight distribution, there are the possibilities of different end-groups, different steric configurations, branching, and the presence of impurities such as initiator fragments, occluded solvent, etc. An example may be given from the oxadiazoles.[7] The thermal breakdown was studied of 2,5-diphenyl-1,3,4-oxadiazole as a model compound for the poly-1,3,4-oxadiazoles. The model compound decomposed to give as major products benzonitrile and phenylisocyanate, which are formed by the rearrangement shown:

$$
\begin{array}{c}
N - N \\
\| \quad \| \\
C_6H_5 - C \quad C - C_6H_5 \\
\diagdown_O\diagup
\end{array}
\xrightarrow{500°C}
C_6H_5CN + C_6H_5NCO
$$

In the polymer, however, the main breakdown route involved elimination of molecular nitrogen, the long chain substituents on the oxadiazole rings being presumably too bulky to undergo the rearrangement observed with the model compound.

Thermal breakdown of poly(phenylene-1,3,4-oxadiazole).

Another example of the effect of increasing molecular size is given by the perfluoro-polyphenylenes.[8] With perfluoro-*p*-terphenyl the degradation products are those that would be expected from scission of the inter-ring bonds. As the molecular weight increases, however, silicon tetrafluoride and carbon dioxide become the main volatile products and the carbonized residue contains lower amounts of fluorine. For degrees of polymerization greater than sixty, the residue contains virtually no fluorine at all. The volatile products arise through reaction of fluorine-containing moieties with the glass of the reaction vessel.

Perhaps the most useful data guiding the search for heat-resistant polymers have come from the study of the thermal stabilities and mechanisms of thermal breakdown of sets of polymers of closely related structure. Considering the total data the following generalizations can be made:

To attain high thermal stability:

(a) Only the strongest chemical bonds should be used.
(b) The structure should allow no easy paths for rearrangement reactions.
(c) There should be a maximum use of resonance stabilization.
(d) All ring structures should have normal bond angles.
(e) Polybonding should be utilized as much as possible.

Seldom is condition (a) the limiting one. Most bonds would be strong enough to give adequate stability provided that the only possible degradation mechanism involved bond scission as the first step, and that this was not followed by a chain reaction or elimination of small, simple fragments. Nearly all systems break down because less energetic mechanisms are operative [condition (b)]. Resonance stabilization (c) is advantageous because more energy is needed for bond rupture, and if bond angles are normal (d), then once a bond is broken the natural angles of the structure hold the atoms in close proximity so that bond healing is possible once the excess energy has been dissipated through the molecule. The polybonding condition (e) requires that each skeletal atom is linked in the polymer chain by more than one route, so that chains cannot be broken by the rupture of a single bond, e.g., the concept of the so-called ladder polymers. It must be stressed that all these conditions really relate to a perfect structural unit in isolation. In practice there are interactions within and between molecules, and seldom does a polymer have a perfect, idealized structure.

Despite these reservations the data are sufficiently clear for some conclusions to be drawn relating polymer structure to thermal stability.[9]

1. Chains containing para-linked rings have the highest thermal stabilities. Such structures also have the highest softening points and lowest solubilities. To achieve an acceptable compromise between ease of fabrication and stability, meta-linked units often have to be introduced in place of some or all the para-linked rings.

2. Rings containing only hydrogen substituents give optimum heat resistance. Replacement of hydrogen by any other atom or group normally results in a reduction in stability. At elevated temperatures in oxidizing atmospheres, hydrogen substituents themselves become reactive.

3. Again, to ease fabrication problems flexible linking groups often have to be introduced between rings. These too cause a reduction in stability. Of all the linking groups that have been used, those having the least effect are:

$$-CO-\quad -COO-\,,\quad -CONH-\,,\quad -S-\,,\quad -SO_2-\,,$$

$$-O-\,,\quad -\!\!\left[CF_2\right]_n\!\!-\,.\quad -\!\!\left[C(CF_3)_2\right]\!\!-$$

4. Poly-*p*-phenylene would appear to be a good standard of what can be achieved with respect to a thermally stable organic polymer. Of the very many heterocyclic polymers synthesized, the best have stabilities differing only marginally from this standard. Other factors such as ease of processing, mechanical properties, cost, etc., have therefore governed which particular polymers have been commercially developed.

5. From theoretical considerations, "ladder" or "double-strand" polymers should be somewhat more stable than analogous linear polymers. The synthetic problems in achieving a perfect ladder structure are so great, however, that in practice no such improvement has yet been attained.

The following chapters will provide examples of all the points itemized above, but before proceeding to a detailed review of the preparation, processing, and elevated temperature properties of technologically useful heat-resistant polymers, brief mention should be made of thermal stability criteria.

THERMAL STABILITY CRITERIA

A problem immediately arises because there is neither a clear definition of nor standard for thermal stability of polymers, nor is there a generally accepted method for its assessment. For the user of a material, clearly the principal criterion is that the property, or properties, of interest should be maintained at the desired level at a specified maximum temperature for a certain length of time. This tends to be loosely referred to as "heat resistance." Because of the diversity of requirements, it is not surprising that a host of different tests and test conditions have been employed. There are basically two different mechanisms by which property deterioration occurs, both of which are temperature-dependent. The first is a reversible reaction involving ''softening'' of the material, which is governed by temperature alone. This is of greater practical significance for thermoplastics, as most of these soften and flow at temperatures below that at which decomposition commences. The second mechanism is the irreversible degradation of a polymer by heat, and this is both time- and temperature-dependent. Environment also has an important influence. It is this irreversible process that is most often implied when the thermal stability of a polymer is referred to. It is of most practical importance for thermosetting materials.

The various experimental techniques that have been used to study the reversible and irreversible deterioration of polymers (broadly termed thermo-analytical methods) have been dealt with in detail elsewhere, and it is not intended to reproduce the information here.[10] Mention must be made, however, of one particular technique, thermogravimetry, because of its very wide application and because of the undue reliance that has been placed upon data obtained by it. The question is: Why has thermogravimetry been so popular? A number of reasons can be cited:

(a) The measurement is easy to carry out.
(b) A wide choice of commercial instrumentation is available.
(c) Only small samples (milligrams, or less) are required; hence diffusion problems are minimized.
(d) In principle, measurements may be made in any atmosphere.
(e) No fabrication of the sample is necessary. Insoluble, infusible materials can be examined.
(f) There was at an early stage a considerable theoretical background based on the concept of chain depolymerization, which related the rate of decomposition as expressed by weight loss to the degree of degradation.

The weight loss results obtained are considerably influenced by experimental parameters and also by the following features of polymers themselves:

(a) Molecular weight. The effect depends upon whether initiation of breakdown is preferentially at chain ends, or at some structural irregularity, e.g., a weak link, or whether it occurs randomly along the chain. It also depends on whether initiation is followed by depolymerization (''unzipping'') of the polymer chain. Figure 1.3 shows for some fractionated polystyrenes the effect of molecular weight upon the rate of weight loss as a function of degree of degradation.[11]
(b) Chain branching. Small numbers of branches have little effect, but as the number increases the stability of the polymer *decreases*. Figure 1.4 shows the effect of amount of branching upon the rate of weight loss as a function of degree of degradation for copolymers of ethylene and propylene.[12]

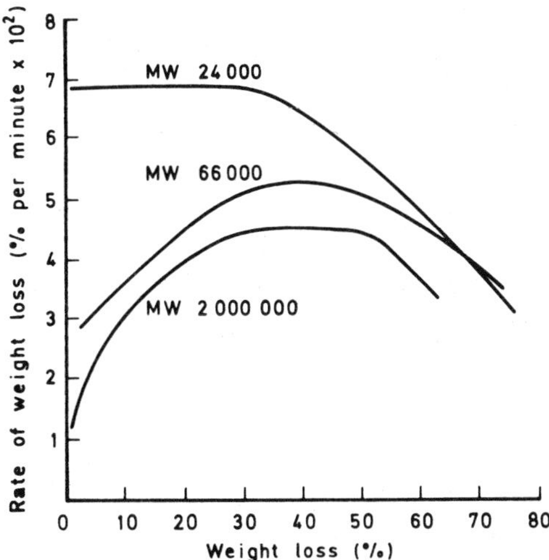

Figure 1.3. *Effect of MW of polystyrene on rate of thermal degradation. Reference 11.*

(c) Cross-linking. Again, small amounts of cross-linking have no measurable influence, but as the amount increases, the stability of the polymer *increases*. This is illustrated in Figure 1.5, which shows the change in thermal stability of copolymers of styrene and trivinylbenzene with degree of cross-linking.[13]

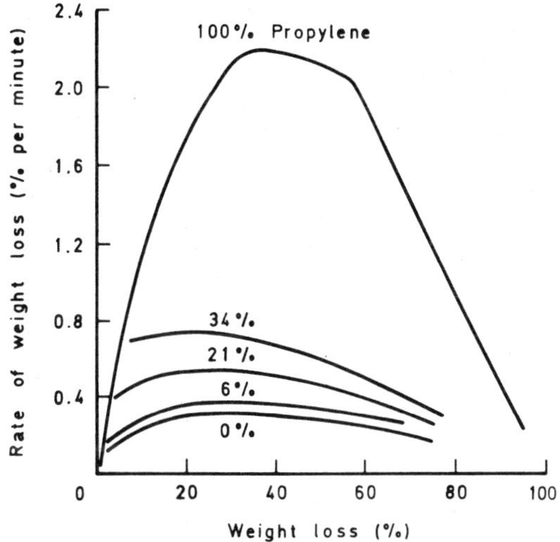

Figure 1.4. *Effect of proportion of propylene (chain-branching) in ethylene-propylene copolymers on rate of thermal degradation. Reference 12.*

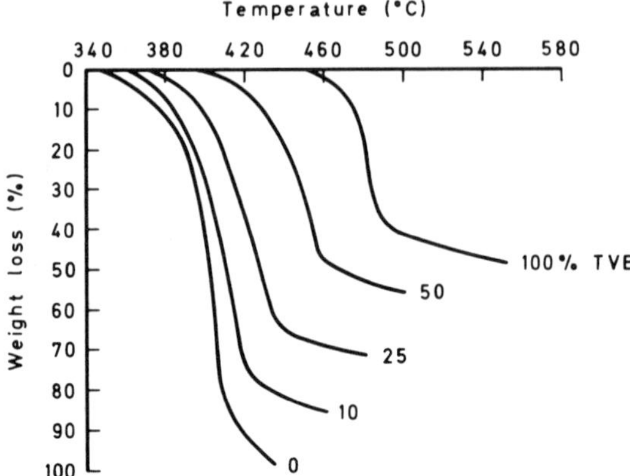

Figure 1.5. *Change in thermal stability of styrene-trivinylbenzene copolymers with proportion of trivinylbenzene (degree of cross-linking). Reference 13.*

(d) Tacticity. This would appear to have little effect upon weight loss data, but the evidence is based upon relatively few results.

(e) Crystallinity. The degree of crystallinity can influence thermo-oxidative stability.

(f) Trace impurities. These can be very deleterious indeed. Figure 1.6 shows the catastrophic change in the thermal stability of a poly(fluorosiloxane) because of the presence of a small amount (less than 1%) of basic catalyst.[14]

(g) Change in chemical structure as degradation proceeds. This can account for changes observed in the apparent overall activation energy for breakdown during the course of a reaction. Figure 1.7 illustrates this point for a polyimide.[15] Treatment of a polyimide with hydrazine hydrate breaks the polymer down to the original diamine and the cyclohydrazide derivative of the dianhydride from which the polymer was made.

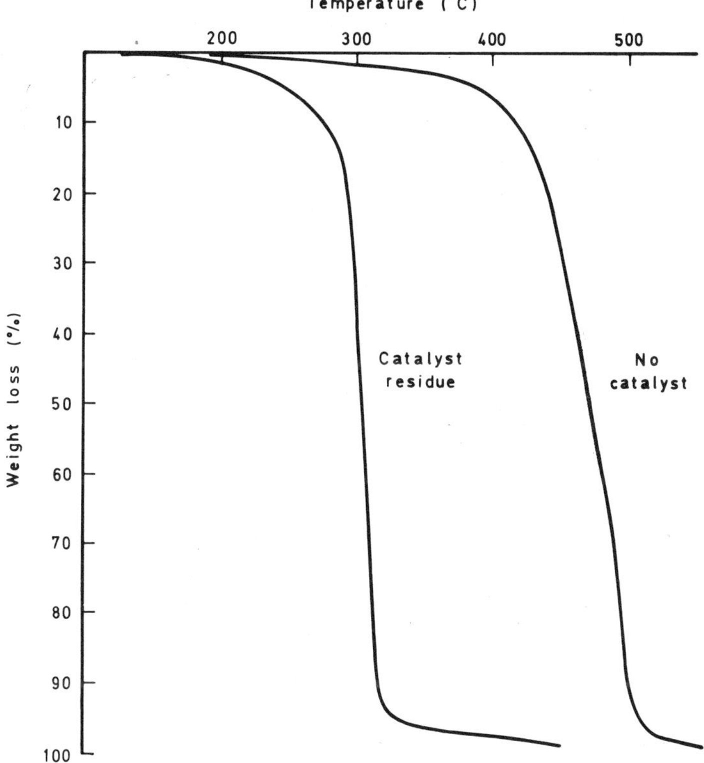

It is therefore possible to take a polyimide film that has been oxidized to different degrees, treat it with hydrazine hydrate and determine the extent to which the chemical structure has changed. Figure 1.7 shows that at about two per cent weight loss, over half the units in the polymer chain have been chemically modified.

Considerations such as these explain why at times there have been wide divergences in the results reported for what are nominally the same systems. Definitive results can only be obtained if the polymer system under examination is very well characterized.

The number of papers in the literature which now reproduce thermogravimetric curves as a measure of thermal stability of a polymer are legion.

Figure 1.6.   Effect of residual catalyst on the thermal stability of a poly(fluorosiloxane). Reference 14.

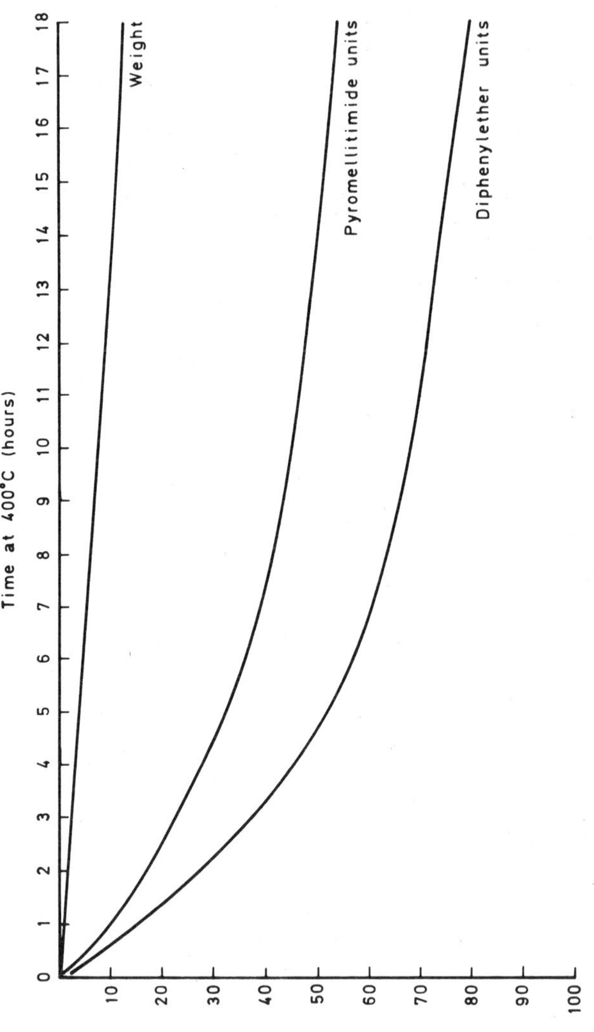

Figure 1.7.   Change in structure of a polyimide film on aging in air at 400°C. Reference 15.

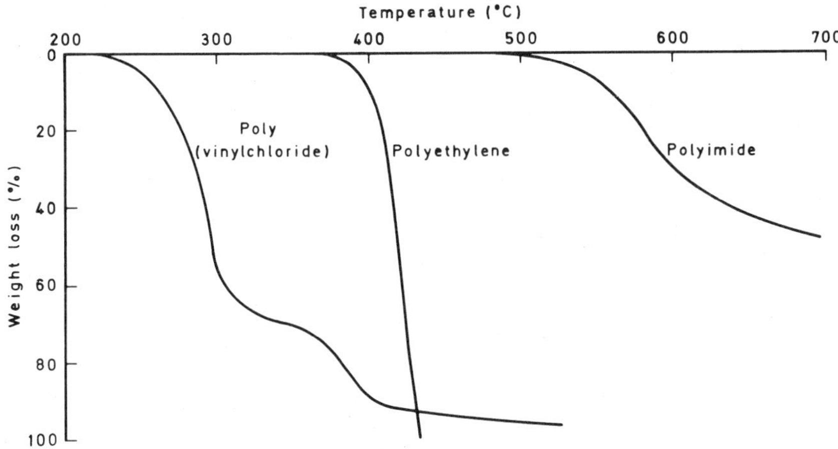

Figure 1.8. Different types of thermogravimetry curves.

Most of the curves fall into one of three main classes and these are illustrated in Figure 1.8.

Class 1. Complete degradation in one step over a narrow temperature range, e.g., polyethylene.

Class 2. Complete degradation in two or more steps over a more extended temperature range, e.g., poly(vinylchloride).

Class 3. Incomplete degradation over a very extended temperature range, e.g., polyimide.

Such plots in themselves obviously give valuable qualitative information on the overall behavior of a polymer as a function of temperature. Attempts have been made to derive a number giving a relative thermal stability on the basis of such curves, e.g., the area under the curve; the temperature for a specified loss in weight; and the temperature for initial weight loss. Such numbers may be reasonably valid if the degradation is of the type shown by polyethylene, but not otherwise. The difficulties in quoting such a single number are well illustrated if the curves of Figure 1.8 are superimposed (Figure 1.9). What should be quoted now?

Similarly, attempts have been made to quote a relative thermal stability number from an isothermal thermogravimetric curve rather than a dynamic one. Quantities used, for example, have been the percentage weight loss in a specific time at a specific temperature; the time required to attain a specific weight loss at a specific temperature; and the calculated rate of weight loss at a specific temperature. Again, such numbers only have reasonable validity if the polymer decomposes completely to volatiles over a narrow temperature range. Where the degradation occurs over an extended temperature range and cross-linking and chain

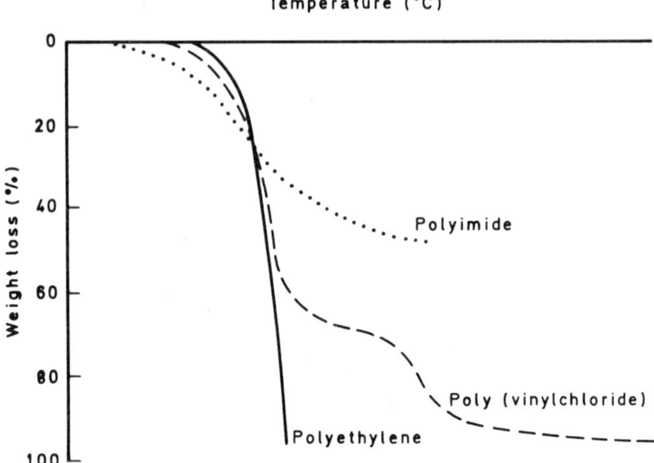

Figure 1.9. Superimposition of curves of Figure 1.8.

scission are in competition, a single figure is not satisfactory. It is better in such circumstances to plot values derived from a *set* of isothermal curves. Two possible methods are shown in Figures 1.10 and 1.11. Figure 1.10 is a comparison of the thermal stability of some polymers based on the weight losses that occur in two hours at different temperatures.[9] Figure 1.11 is a comparison based on

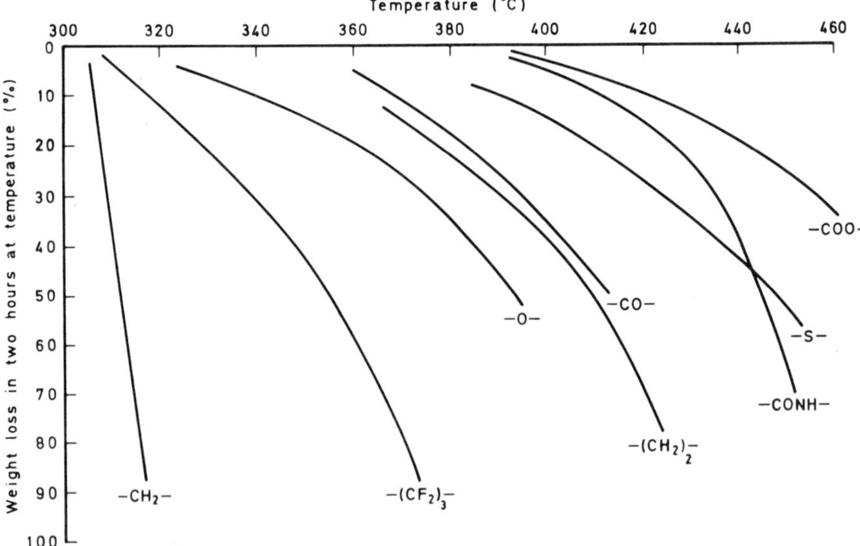

Figure 1.10. Comparison of thermal stability in air of phenylene polymers with different linking groups between the rings. Reference 9.

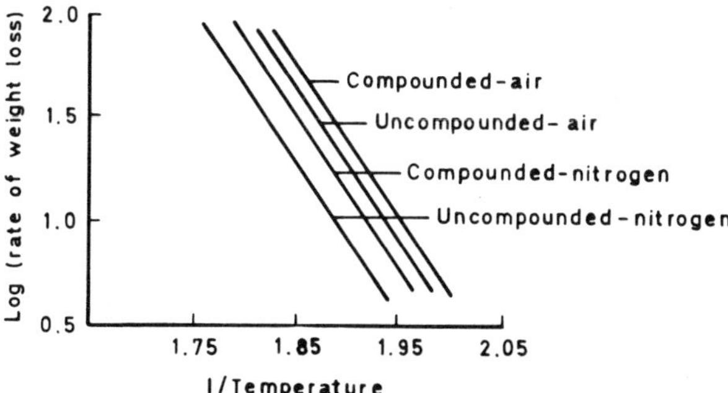

Figure 1.11. Comparison of thermal stability of epichlorhydrin elastomers by means of Arrhenius plots. Reference 16.

Arrhenius plots of the log (rate of weight loss) vs. the reciprocal of the absolute temperature.[16]

Weight loss data have also been used to determine kinetic parameters such as the order of reaction, the overall activation energy for degradation, and the frequency factor in the Arrhenius expression $k = Ae^{-E/RT}$. There are very many published papers giving different mathematical methods of analysis for derivation of these parameters from a single dynamic thermogravimetric trace.[17,18] These methods may be broadly divided into two types, differential or integral, depending upon whether the initial equation set up is differentiated or integrated. It is the authors' opinion that none of these yields reliable kinetic data unless the polymer degrades by a single, simple reaction mechanism and the order of reaction is already known from other experiments. This is rarely the case. The difficulties are illustrated in Figures 1.12 and 1.13, which show the relatively small changes that occur in the shape of a weight loss curve for large changes in the order of reaction,[19] or overall activation energy.[18]

The mathematical treatments also weight different parts of the thermogravimetric curve to different extents, and the values of the activation energy calculated may differ by as much as 20% using the *same* experimental data for calculation purposes. As a consequence there is a growing tendency to recommend that, for determination of kinetic parameters, a set of weight loss experiments at

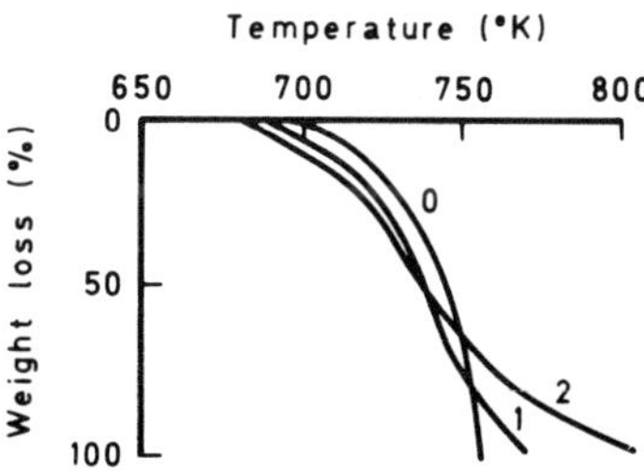

Figure 1.12. Effect of order of reaction on thermogravimetry curves. Reference 19.

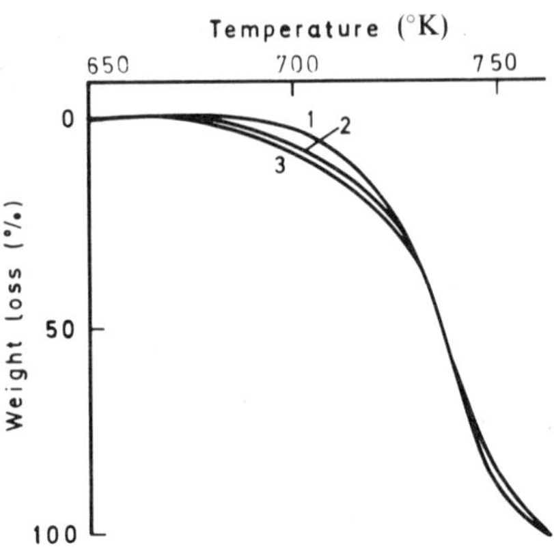

Figure 1.13.   Effect of activation energy on thermogravimetry curves. Reference 18.

different heating rates, or experiments in which the temperature is "cycled" or "jumped" during a run, should be used.[18] This, of course, immediately removes some of the advantages that thermogravimetry has over isothermal mass-change measurements.

It is our belief that the latter is much more appropriate for the detailed study of the kinetics of degradation of polymers. For such studies the Arrhenius relationship is normally applied in which "a rate of weight loss" is plotted against the reciprocal of the absolute temperature (see Figure 1.11). A rate of weight loss is deliberately emphasized, as the rate more often than not varies during the course of an experiment, and the question then arises as to precisely which rate to use in the plot. If the weight loss data fit a standard (apparent) zero-order, first-order, or second-order relationship, then rate coefficients can be derived and used. If this is not the case, initial rates of decomposition can be used. This has much to commend itself, as the start of breakdown is often the most important practical consideration. Unfortunately, initial rates are the most difficult to measure with certainty in an isothermal experiment because of the time needed for the sample to reach temperature and the weight loss which may have already occurred before equilibrium is attained. Two other rates that have been extensively used are the maximum rate of weight loss and the rate of weight loss at a specific percentage decomposition. The former has some theoretical justification from the theory of chain depolymerization.[20] The latter is useful because a number of values of the overall activation energy can be derived for different weight loss percentages. Any changes occurring are indications of changes in the se-

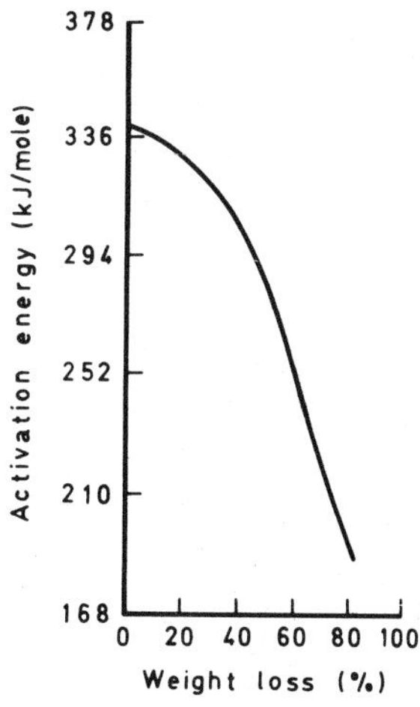

Figure 1.14. Change in activation energy with weight loss for poly(tetrafluoroethylene). Reference 21.

quence of degradation reactions. For example, a change from a lower to a higher activation energy may be caused by the elimination of weak links in the chain, a change from a hydrolytic to a thermal mode of breakdown, or progressive stabilization through cross-linking. The reverse indicates the production of secondary structures that destabilize the polymer. Figure 1.14 shows that this must occur in the thermal degradation of poly(tetrafluoroethylene).[21]

The question must be posed as to the precise meaning of the overall activation energy value derived. It is a ''global'' figure encompassing a number of different reactions, each with its own rate constant and individual activation energy. Pure thermal decomposition involves:

(a) An initiation reaction, which may be of a random or specific nature.
(b) A propagation reaction. Depolymerization may have short or long zip lengths.
(c) A termination reaction. This may involve all or any of first-order, recombination, or disproportionation steps.
(d) Chain transfer reactions.
(e) Volatilization of small chain fragments.
(f) Diffusion, which can affect all the above, (a) to (e).

(Degradation in an oxidizing atmosphere adds further complications.)

It is true, of course, that the initiation, chain breaking, step (a) tends to dominate the overall activation energy. Reactions (a) to (d) and (f) also occur in polymerization, but it is possible by working at low degrees of conversion to eliminate some of them and to isolate others, so that the individual rate coefficients and activation energies can be determined. In degradation studies this is not possible. What then is the point of determining overall activation energies?

We believe that, taken on their own, their value is questionable, but when *coupled with other data*, useful mechanistic information is obtained. The information given by the change in activation energy with degree of degradation has already been mentioned. Another example that can be quoted is the progressive increase in activation energy for thermal breakdown, which occurs on drying of polyamides, indicating that unless suitable precautions are taken, the initial chain scission is a hydrolytic rather than a thermal reaction.[22] With a copolymer of vinylidene fluoride and chlorotrifluoroethylene, the activation energies for yield of hydrogen chloride and hydrogen fluoride can be compared with that for weight loss. This shows that in some circumstances the major breakdown reaction involves elimination of hydrogen fluoride rather than hydrogen chloride—a result contrary to expectation.[23] The constancy of the activation energy for the thermo-oxidative degradation of a vinylidene fluoride/hexafluoropropene/tetrafluoro-ethylene terpolymer before and after cross-linking with a diamine, and before and after addition of such materials as carbon blacks and acid acceptors, shows that the principal mode of decomposition has not been altered by these formulation changes.[24]

This rather lengthy consideration of weight loss studies has been included because the results of such studies are quoted extensively in this volume, and it is important to appreciate precisely what they mean. It cannot be overstressed that the technique only measures processes involving conversion to, and elimination from a system of, volatiles. Since, as we have seen, polymer degradation can be a very complex phenomenon with a number of simultaneous reactions taking place, it is essential for a full understanding of the process to use a variety of thermo-analytical techniques. It is significant that there is still some controversy over the precise mechanism of degradation of relatively simple polymers such as polystyrene and poly(methylmethacrylate)!

In practice we are concerned not only with the effect of heat on a polymer, but possibly with the superimposed influences of heat, light, moisture, stress, and chemicals.

The following chapters deal with the preparation, processing, and properties of heat resistant polymers, the divisions being based upon chemical types.

REFERENCES

1. A. H. Frazer, *High-Temperature Resistant Polymers,* Interscience, New York (1968).
2. E. Behr, *Hoch Temperatur-Bestandige Kunststoffe,* Carl Hanser Verlag, Munich (1969).
3. V. V. Korshak, *Heat-Resistant Polymers,* Israel program for scientific translations, 1971. (Originally published in Russian in 1969.)
4. T. L. Cottrell, *The Strength of Chemical Bonds, 2nd ed.,* Butterworths, London (1958).
5. V. I. Vedeneyev, L. V. Gurvick, V. N. Kondratyev, V. A. Medvedev, Y. L. Frankevich, *Bond Energies, Ionization Potentials and Electron Affinities,* Edward Arnold, London (1958).

6. T. H. Thomas and T. C. Kendrick, *J. Polym. Sci. A 2* **7**, 537 (1969).
7. J. L. Cotter, G. J. Knight, and W. W. Wright, *J. Gas Chromatography* **86** (1967).
8. J. L. Cotter, G. J. Knight, J. M. Lancaster, and W. W. Wright, *J. Appl. Polym. Sci.* **12**, 2481 (1968).
9. W. W. Wright, *Degradation and Stabilization of Polymers,* G. Gueskens, (ed.), Applied Science Publishers, London (1975), pp. 43–75.
10. P. E. Slade and L. T. Jenkins, (eds.), *Techniques and Methods of Polymer Evaluation,* Vol. 1, *Thermal Analysis,* Edward Arnold, London (1966); Vol. 2, *Thermal Characterization Techniques,* M. Dekker, New York (1970).
11. S. L. Madorsky, D. McIntyre, J. H. O'Mara, and S. Straus, *J. Research Natl. Bur. Standards* **66A**, 307 (1962).
12. L. A. Wall, *High Temperature Resistance and Thermal Degradation of Polymers,* SCI Monograph No. **13**, 145 (1961).
13. F. H. Winslow and W. Matreyek, *J. Polym. Sci.* **23**, 315 (1956).
14. J. M. Cox, B. A. Wright, and W. W. Wright, *J. Appl. Polym. Sci.* **8**, 2935 (1964).
15. R. A. Dine-Hart, D. B. V. Parker, and W. W. Wright, *Br. Polym. J.* **3**, 235 (1971).
16. J. Day and W. W. Wright, *Br. Polym. J.* **9**, 66 (1977).
17. J. H. Flynn and L. A. Wall, *J. Research Natl. Bur. Standards* **70A**, 487 (1966).
18. J. H. Flynn, in *Aspects of Degradation and Stabilization of Polymers,* H. H. G. Jellinek, (ed.), Elsevier, New York (1978), pp. 574–603.
19. L. Reich and S. Stivala, *Elements of Polymer Degradation,* McGraw-Hill Book Co., New York (1971), p. 100.
20. R. Simha and L. A. Wall, *J. Phys. Chem.* **56**, 707 (1952).
21. B. Carroll and E. P. Manche, *J. Appl. Polym. Sci.* **9**, 1895 (1965).
22. S. Straus and L. A. Wall, *J. Research Natl. Bur. Standards* **60**, 39 (1958).
23. T. G. Degteva, I. M. Sedova, and A. S. Kuzminiskii, *Vysok. Soed.* **5**, 378 (1963).
24. W. W. Wright, *Br. Polym. J.* **6**, 147 (1974).

SUPPLEMENTARY BIBLIOGRAPHY

High-temperature polymers containing cyclic functions, M. M. Koton, in *Advances in Macromolecular Chemistry,* Vol. 2, W. W. Pasika, (ed.), Academic Press, New York (1970).

Advances in polycondensation methods for the synthesis of thermostable polymers, S. V. Vinogradova and V. V. Korshak, *Russ. Chem. Rev.* **39**, 308 (1970).

Recent developments in thermally resistant polymers, R. L. Burns, *J. Oil Color Chem. Assn.* **53**, 52 (1970).

Contemporary state and prospects for the development of the production of heat-resistant polymers, M. M. Koton, *Plast. Massy* 59 (1970).

High-temperature resistant organic polymers, J. Idris Jones, *Chem. Brit.* **6**, 251 (1970).

Synthesis and properties of available heat-stable polymers, B. Sillion, *Collected Colloq. Semin. Inst. Fr. Petrole No.* **20**, 27 (1971).

New aspects of polycyclization, E. S. Krongauz, *Russ. Chem. Rev.* **42**, 857 (1973).

Thermally stable polymers, C. S. Marvel, *Appl. Polym. Symp.* **22**, 47 (1973).

Heat-resistant thermosetting polymers—a brief review, S. Oswitch, *Reinforced Plastics* **19** (6), 180 (1975); **19** (7), 215 (1975).

Structure and properties of heat-resistant polymers, A. A. Askadskii and G. L. Slonimskii, *Russ. Chem. Rev.* **44**, 767 (1975).

Problems of the chemistry of thermostable organic polymers, A. A. Berlin, *Russ. Chem. Rev.* **44,** 244 (1975).

High temperature resistant engineering plastics—properties, processing, and applications, H. Domininghaus, *Kunststoffe* **69,** 1 (1979).

2

THERMOSETTING POLYMERS

INTRODUCTION

As pointed out in the introductory chapter, one of the methods of increasing the softening point and glass transition temperatures of a polymer is by chemical cross-linking of the chains. The properties of such a system depend upon the structure of the main chains, the structure of the cross-links, and the number of the cross-links. Hence, many variants are possible. As the cross-link density increases, so normally does the heat resistance, but this is accompanied by decreases in impact strength, elongation at break, and reversible extensibility, and almost inevitably a compromise must be accepted. The cross-linking may be brought about either by addition or condensation reactions. The former have the advantage that no volatile by-products are evolved during cure, thus allowing simpler processing at lower pressures. Lower void contents are likely in the final material, and thick sections are much easier to fabricate. In theory all the functional groups should be used up in the cross-linking process, but in practice this is not so. Further reaction may therefore be possible if the material is later exposed to higher temperatures than those used in the cure cycle, with concomitant changes in properties.

The phenol–formaldehyde resins were the first of the thermosets to be developed (1909), and it is salutary to consider that, depending upon type and filler content, these may be rated for use at temperatures ranging from 120 to 260°C. The latter figure is not greatly surpassed by any of the more modern materials. This chapter gives a brief survey of the following types of thermosetting polymers in chronological order of development and exploitation: phenol-formaldehyde resins, melamine-formaldehyde resins, polyesters, epoxies, furans, bis-diene resins, vinyl esters, phenol-aralkyl (Friedel–Crafts) polymers, and the so-called PSP resin. All of them, at least in specific formulations, are capable of use at 150°C or higher temperatures, and this résumé of their properties will serve to put the perhaps more exotic developments into perspective.

These polymer systems will not be considered in great detail because whole books have been written on specific classes, such as the phenolics,[1-5] polyesters,[6-9] and epoxies.[10-15] Other cross-linked resin systems, such as the silicones and polyimides, are dealt with in separate chapters devoted to those classes of material.

PHENOL-FORMALDEHYDE RESINS

Preparation

Preparation involves the reaction of phenol, or a mixture of phenols, with formaldehyde.[1-5] Cresols, xylenols, and resorcinol have also been used, but to a much smaller extent than phenol itself. The low-molecular-weight prepolymers produced commercially can conveniently be divided into resols and novolacs. The former are produced when phenol and excess formaldehyde are reacted under alkaline conditions to give a complex mixture of mono- and poly-nuclear phenols with methylol substituent groups. An idealized structure is:

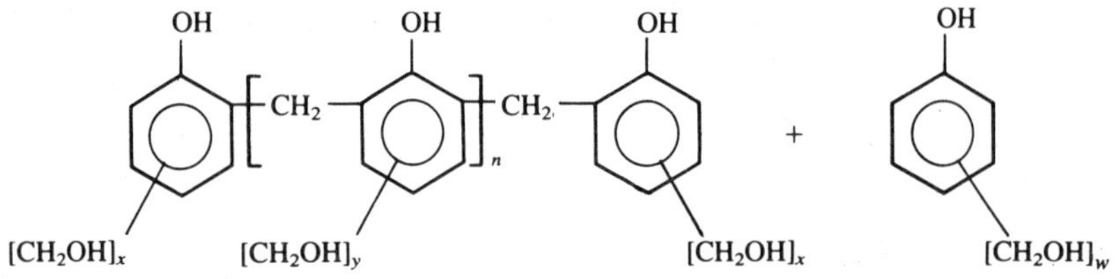

$$\underset{[CH_2OH]_x}{\overset{OH}{\bigcirc}}\!\!\!-\!\!\left[CH_2-\underset{[CH_2OH]_y}{\overset{OH}{\bigcirc}}\right]_n\!\!\!-CH_2-\underset{[CH_2OH]_x}{\overset{OH}{\bigcirc}} \quad + \quad \underset{[CH_2OH]_w}{\overset{OH}{\bigcirc}}$$

where $n = 0-2$, $w = 0-3$, $x = 0-2$, $y = 0-1$.

The resol is cured by heating, further condensation occurring by reaction of the methylol groups.

$$-CH_2OH + HOCH_2- \longrightarrow -CH_2-O-CH_2- + H_2O$$

$$-CH_2OH + H- \longrightarrow -CH_2- + H_2O$$

This is a disadvantage of the system in that it contains within itself the means of cross-linking, and precautions must be taken during resin synthesis to ensure that

gelation does not occur. The cross-linked network structure may be represented
as:

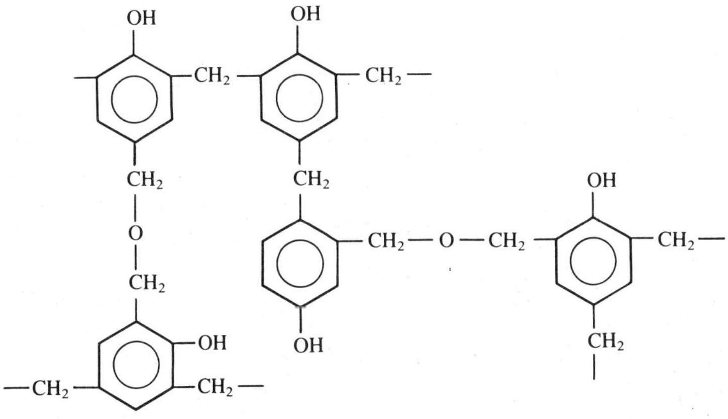

The novolacs are produced when formaldehyde and excess phenol are reacted under acid conditions. The resulting structure consists of polynuclear phenols linked by methylene groups, e.g.,

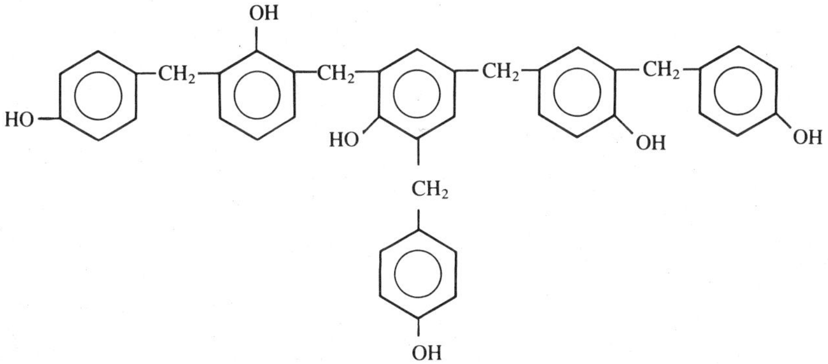

In contrast to the resols, reaction is complete and addition of a cross-linking
agent (normally hexamethylenetetramine, commonly known as hexamine) is
necessary for further reaction to occur. The mechanism of this cross-linking
process is still not fully understood, but the available evidence points to the network structure being of the form:

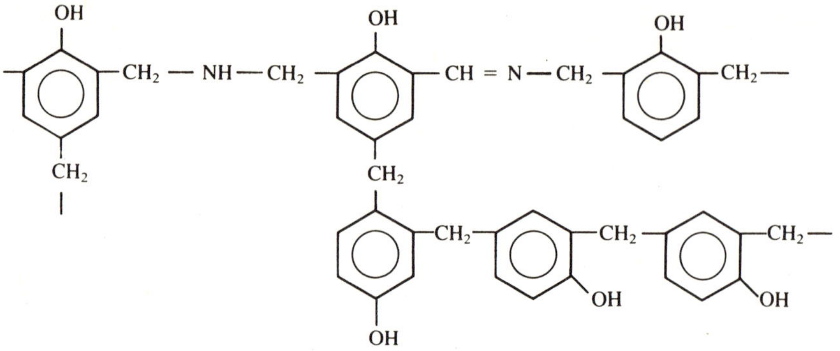

The thermo-oxidative stability of phenol-formaldehyde resins can be further improved by chemical modification. This involves etherification or esterification of the phenolic hydroxyl group, or complex formation with polyvalent elements. By suitable choice of reactants and reaction conditions, hydroxymethyl groups and phenolic hydroxyl groups can both be etherified. Inorganic polybasic acids, e.g., phosphoric and boric acid, are used to esterify phenol novolacs and this increases both heat and flame resistance. With boric acid and phosphoric acid the structures formed are of the types shown:

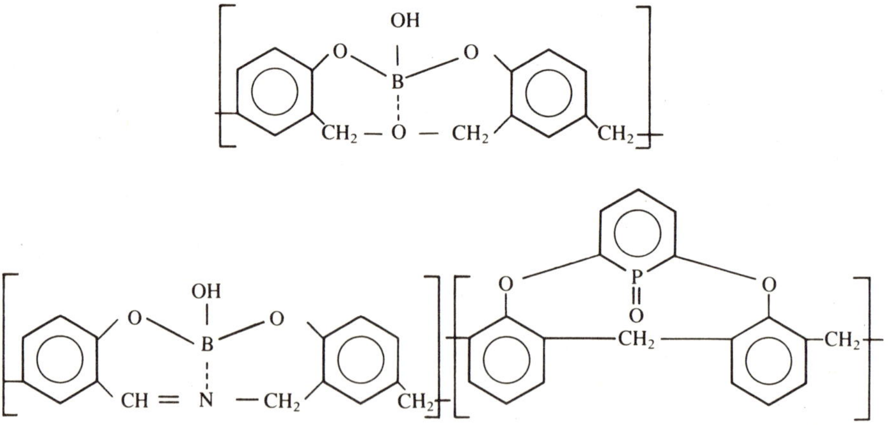

Reaction of phenolic resins with metal halides, metal alcoholates, or metal-organic compounds containing elements such as molybdenum, titanium, zirconium, tungsten, or aluminum, also reduces the rate of decomposition at elevated temperatures.

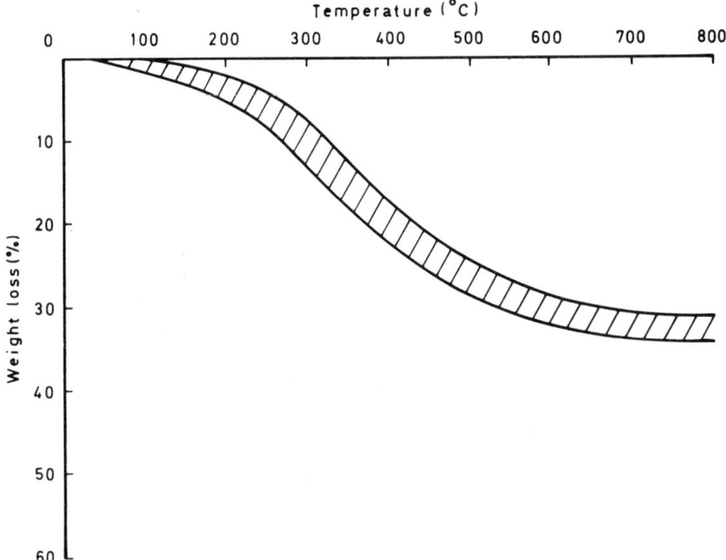

Figure 2.1.   *Thermogravimetry of post-cured phenol-formaldehyde resins of P/F ratios 2:1 to 10:1. (Heating rate 3.3°C/min in argon) Reference 16.*

Thermal Stability

Figure 2.1 shows the weight loss range for post-cured phenol-formaldehyde resins of widely differing phenol/formaldehyde ratios on heating in argon at a rate of temperature rise of 3.3°C per minute.[16] The maximum difference between formulations is never more than 5% at any temperature, and even at 800°C the total weight loss amounts to no more than 33%. Figure 2.2 compares the weight losses in air and in nitrogen using a heating rate of 2°C per minute. The much poorer thermo-oxidative stability of phenol-formaldehyde resins on a weight loss basis is very evident. The mechanism of thermo-oxidative degradation has been studied in considerable detail.[17] The first step is the oxidation of a methylene group to a hydroperoxide followed by decomposition of the latter to dihydroxy benzophenone and benzhydrol structures. The benzophenone structure then oxidizes further.

At temperatures between 250 and 400°C, water, carbon dioxide, and formaldehyde are the major volatile degradation products. At temperatures above 400°C, carbon dioxide, methane, hydrogen, toluene, phenols, cresols, and xylenols appear and become relatively more important as the temperature rises.

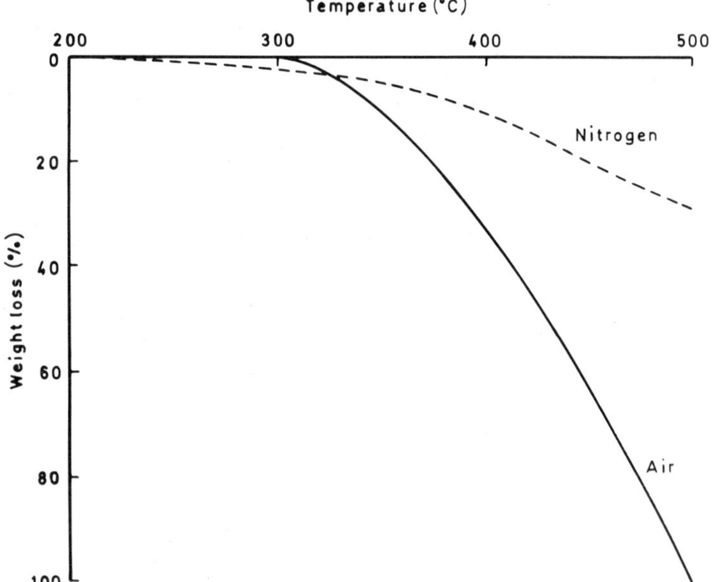

Figure 2.2. *Thermogravimetry of a phenol-formaldehyde resin in air and nitrogen. (Heating rate 2°C/min)*

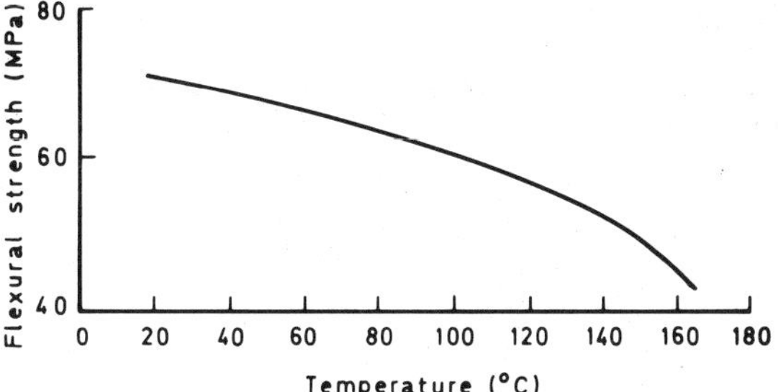

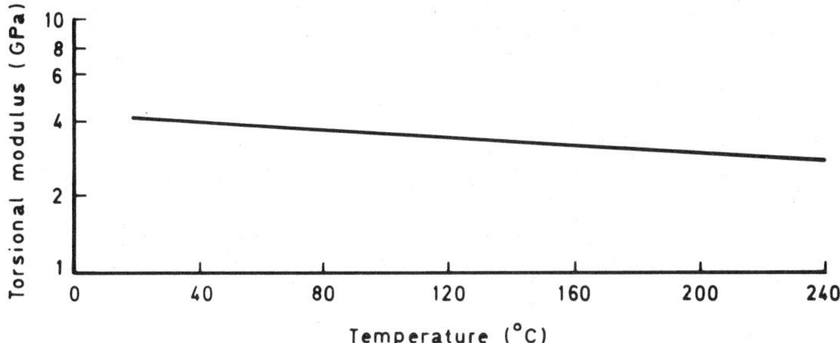

Figure 2.3. Mechanical properties as a function of temperature of carbon-filled phenolic moldings. Reference 18.

Elevated Temperature Properties

Figure 2.3 shows the change in mechanical properties as a function of temperature of carbon-filled phenolic moldings.[18] Although there would appear to be an accelerated decrease in flexural strength above about 120°C, the torsional modulus is relatively unchanged even at 240°C. The range of heat resistance attainable in phenolic moldings is illustrated in Figure 2.4, which compares lifetime plots for general purpose and high-temperature resistant grades.[18] The criterion adopted is the time at temperature at which 70% of the flexural strength is retained. It is noteworthy that the difference between the two grades lessens as the time at temperature is increased. For 100 hours use there is a difference of approximately 100°C between the temperature limits for the two grades, whereas for 10,000 hours use this has been reduced to about 30°C.

Various properties of a phenolic resin/glass fiber laminate before and after heating at 260°C for 30 minutes are listed in Table 2.1.

Except for flexural strength and modulus the retention of properties is at least 90%. Relatively little change in tensile and compressive creep rupture strengths is in fact seen over 1,000 hours at 260°C. This means that above about

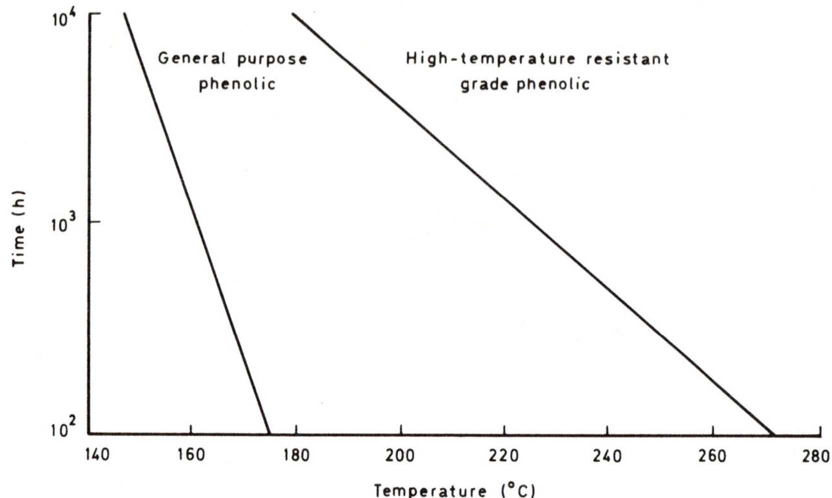

Figure 2.4. Lifetime at temperature plots for phenolic moldings based upon 70% retention of flexural strength. Reference 18.

125°C the strength retention of a phenolic resin/glass fiber laminate is superior to that of the aluminum alloys commonly used as aircraft structural materials. The changes in flexural, compressive, tensile strength, and modulus of a silane-modified phenolic resin/glass fiber laminate are shown[20] in Figure 2.5 for temperatures up to 535°C. On the basis of these data, phenolic resin composites would appear to have excellent retention of properties at elevated temperatures.

TABLE 2.1. Properties of a CTL 91-LD Phenolic Resin/181 Glass Fiber Laminate (Reference 19)

Property	Room temperature	Value at 260°C after 30 minutes at 260°C	Property retention (%)
Flexural strength (MPa)	600	449	75
Flexural modulus (GPa)	33	26	79
Tensile strength (MPa)	400	359	90
Tensile modulus (GPa)	23	21	91
Compressive strength (MPa)	345	319	92
Shear strength (MPa)	286	266	93
Izod impact strength (J/mm notch)	0.8	0.8	100
Dielectric constant at 1 MHz	4.0	3.98	—
at 10 MHz	3.57	4.16	—
Power factor at 1 MHz	0.0098	0.0055	—
at 10 MHz	0.010	0.0126	—

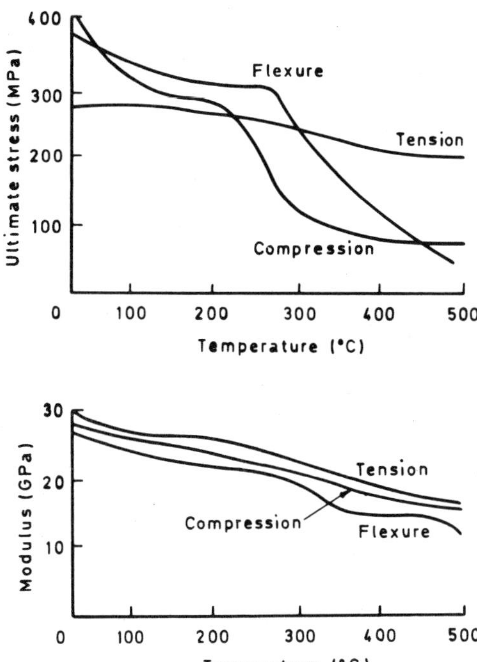

Figure 2.5. Change in mechanical properties of a silane-modified phenolic resin/glass fiber (181 Volan A) laminate with temperature. (Tested at temperature after 30 minutes at temperature) Reference 20.

A drawback to phenolic resin systems is, however, evident if heat aging is carried out for protracted time periods. Figure 2.6 contains flexural strength data for H-5834 phenolic resin/glass fiber laminates aged and tested at temperatures for periods up to nearly 10,000 hours.[21] Although retention of strength is very good initially, if the aging is carried on for a sufficiently long time, varying from 1,000 hours at 200°C to 10 hours at 371°C, there is then a catastrophic loss in properties.

MELAMINE-FORMALDEHYDE RESINS

Preparation

The production of melamine-formaldehyde resins began in the mid-1930s. Under slightly alkaline conditions melamine reacts with formaldehyde to give methylol derivatives with up to six methylol groups per molecule[22–24]:

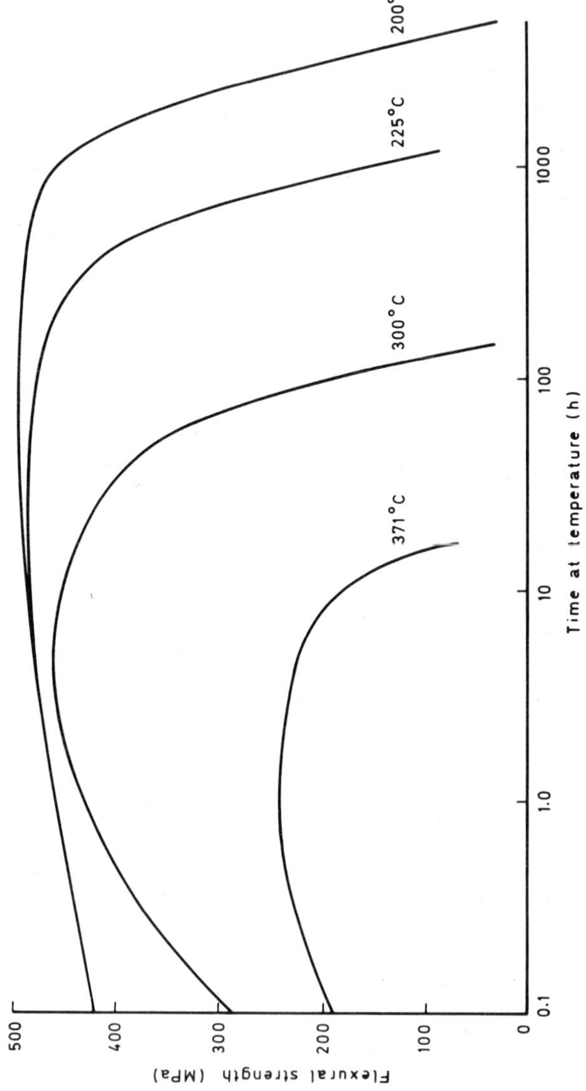

Figure 2.6. Effect of aging at temperature on flexural strength of a phenolic resin/glass fiber laminate. (H-5834 phenolic resin, 181/A1100 glass cloth) Reference 21.

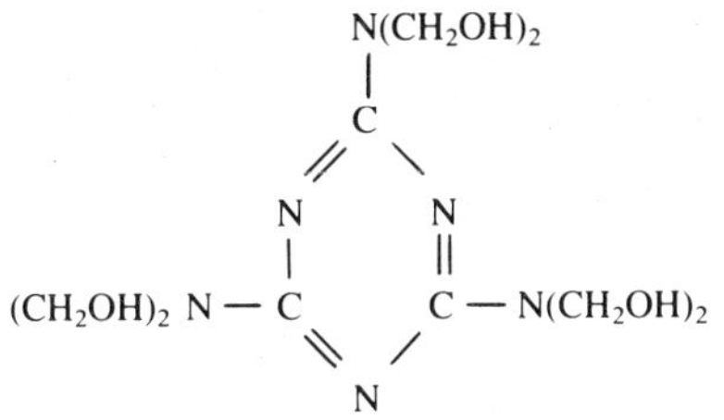

$$N(CH_2OH)_2 - C(=N)(N) \quad (CH_2OH)_2N-C \quad C-N(CH_2OH)_2 \quad N$$

If excess of formaldehyde (greater than $2:1$ molar) is used, then all the possible methylol melamines have been shown to be present in the reaction mixture. On heating, the methylol melamines undergo condensation reactions with the elimination of water and formaldehyde. The rate of reaction is strongly dependent upon pH. The mechanism of reaction is not fully understood, but it is thought that the methylol groups undergo self-condensation, or react with amino or imino groups. The final network structure may be represented as:

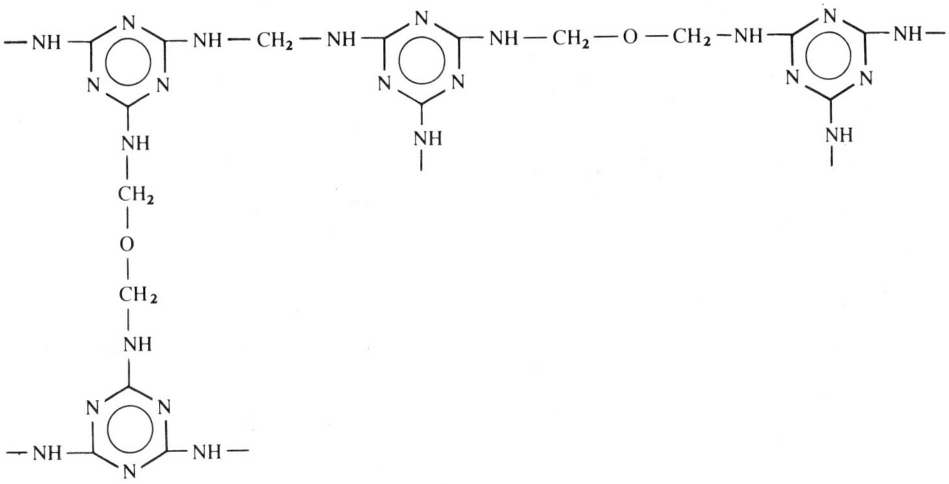

$$-NH-\underset{NH}{\overset{N}{\triangle}}-NH-CH_2-NH-\underset{NH}{\overset{N}{\triangle}}-NH-CH_2-O-CH_2-NH-\underset{NH}{\overset{N}{\triangle}}-NH-$$

$$CH_2-O-CH_2-NH-\underset{N}{\overset{N}{\triangle}}-NH-$$

Thermal Stability

Figure 2.7 shows thermogravimetric traces for a melamine-formaldehyde resin in air and nitrogen, both at a rate of temperature rise of 2°C per minute. The curves indicate that the mode of breakdown is very similar in air or nitrogen, and three distinct phases can be identified at 200–300°C, 380–400°C, and greater than 500°C, respectively. Isothermal weight loss experiments[25] (Figure 2.8) confirm that little degradation occurs below 200°C and that a major breakdown reaction sets in between 300 and 350°C. With respect to the mechanism of degradation and products formed, most of the available data relate to butylated materi-

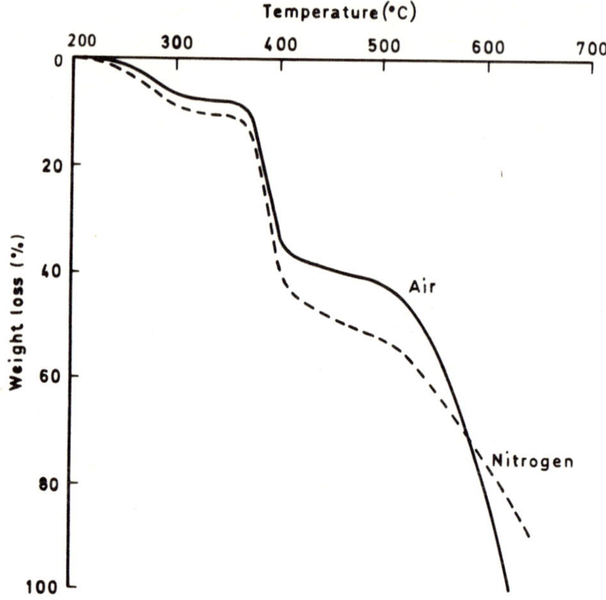

Figure 2.7.   Thermogravimetry of melamine-formaldehyde resin in air and nitrogen. (Heating rate 2°C/min)

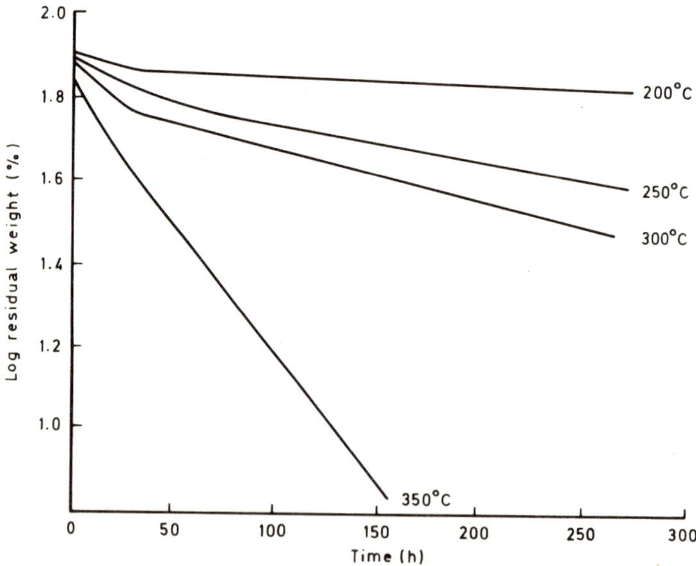

Figure 2.8. Effect of heat aging in air on the weight of a melamine-formaldehyde resin. Reference 25.

als, i.e., methylol melamines, which have subsequently been reacted with butanol under acid conditions to give ethers. In a typical process about half the available methylol groups are etherified. With such resins the major volatile products on thermo-oxidative degradation are carbon dioxide and butanol.[26] This indicates that the initial phase of breakdown involves the ether links. At higher temperatures, breakdown of methylene bridges and the triazine rings themselves commences.

$$\text{triazine} - NH - CH_2 - O - CH_2 - CH_2 - CH_2 - CH_3$$

$$\downarrow O_2$$

$$\text{triazine} - NH - \underset{\overset{|}{OOH}}{CH} \times O - CH_2 - CH_2 - CH_2 - CH_3$$

$$\downarrow$$

$$\text{triazine} - NH - COOH + HO - CH_2 - CH_2 - CH_2 - CH_3$$

$$\downarrow$$

$$\text{triazine} - NH_2 + CO_2$$

Elevated Temperature Properties

The only data available relate to the behavior of glass fiber/melamine resin laminates and are fairly old (30 years).[27,28] There is no reason to expect, however, that properties have greatly improved during this time interval. Table 2.2 gives flexural and compressive strength values of laminates at various elevated temperatures after different times of aging at temperature, and Figure 2.9 is a plot of the flexural strength results expressed as a percentage of the initial room temperature strength.

Figure 2.9 shows that the rate of change of properties measured at temperature is slow at least up to 200°C and for time periods of nearly 800 hours.[27]

TABLE 2.2. Mechanical Properties of Melamine Resin/Glass Fiber Laminates at Elevated Temperatures (Reference 27)

Aging time (hours) at temperature	Flexural strength (MPa) at a temperature (°C) of			
	100	150	200	250
1	215	193	181	135
2	226	195	164	128
4	224	197	152	116
24	224	192	140	113
48	254	175	131	108
72	244	169	130	108
192	255	184	133	71
384	253	181	124	38
768	243	168	127	21

Aging time (hours) at temperature	Compressive strength (MPa) at a temperature (°C) of			
	100	150	200	250
1	518	538	620	590
2	535	623	602	579
4	592	638	610	584
24	657	656	566	593
48	655	635	566	577
72	671	632	553	560
192	661	613	569	442
384	661	622	595	408
768	647	604	558	314

Figure 2.10 compares the behavior of melamine-, phenolic-, polyester-, and silicone-glass fiber laminates.[28] In contrast to the rapid fall in strength of the phenolic laminate at temperatures nearing 200°C, the melamine-formaldehyde and the other two laminate systems show a much more gradual loss in strength with rising temperature.

POLYESTER RESINS

Preparation

The production of these materials became established in the 1940s. Although many variants are now commercially available, the essential characteristic is a linear chain containing both ester groups and unsaturated groups that provide sites for subsequent cross-linking. The polyesters are normally prepared by reaction of saturated diols with a mixture of an unsaturated dibasic acid (or anhydride) and a saturated dibasic acid (or anhydride).[6–9] The function of the latter is

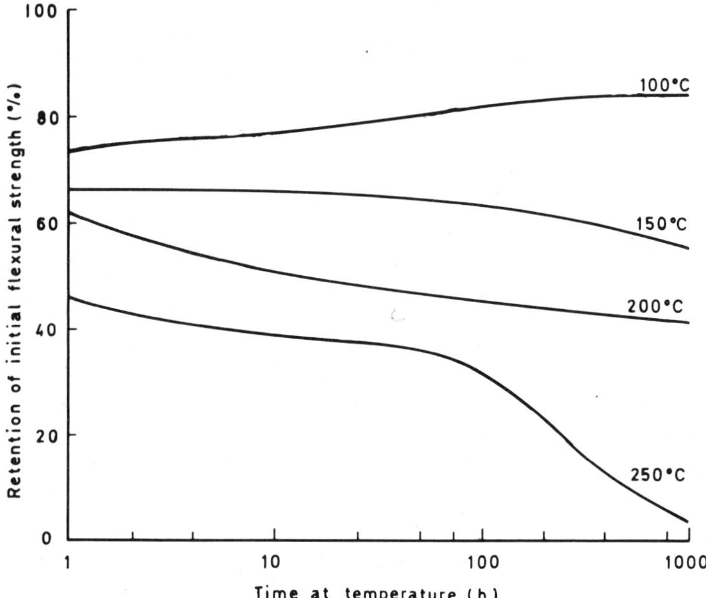

Figure 2.9. Flexural strength retention on heat aging of melamine-formaldehyde resin/ glass fiber laminates. Reference 27.

partly to decrease the cross-link density and hence the brittleness of the final product. The reactants most commonly used are propylene glycol, maleic anhydride, and phthalic anhydride, and these would give as an idealized structure:

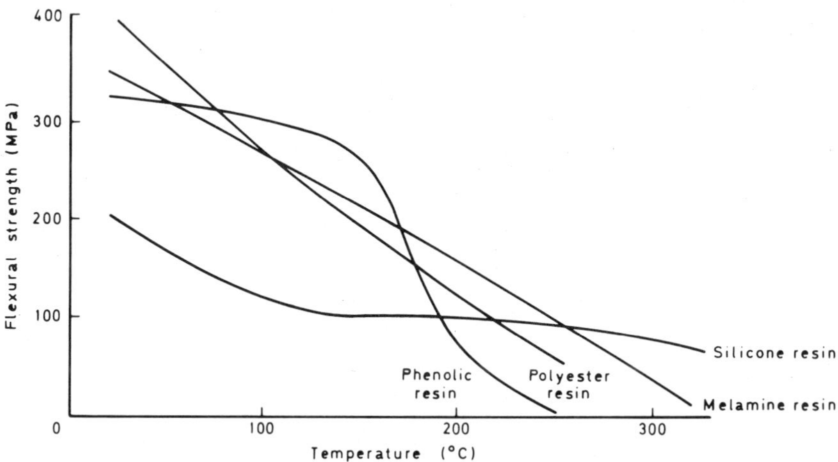

Figure 2.10. Flexural strength of glass fiber laminates at temperature after 200 hours at temperature. (Phenolic resin-Redux. Silicone resin-DC 804. Melamine resin-Melmac 405. Polyester resin-Plaskon 911-11) Reference 28.

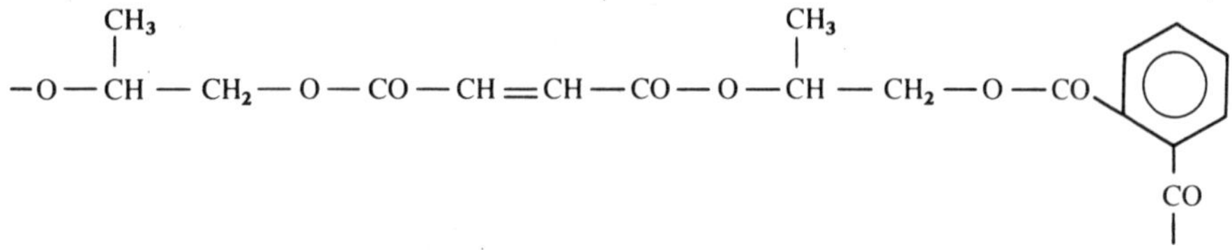

$$
\begin{array}{c}
\underset{|}{CH_3} \qquad\qquad\qquad\qquad\qquad\qquad\qquad \underset{|}{CH_3} \\
-O-CH-CH_2-O-CO-CH=CH-CO-O-CH-CH_2-O-CO-\text{(phenyl)}-CO-
\end{array}
$$

Improved heat resistance may be attained by the use of other components, e.g., neopentylene glycol (2,2-dimethylpropane-1,3-diol) instead of propylene glycol, and endomethylenetetrahydrophthalic anhydride (nadic anhydride) instead of phthalic anhydride.

The unsaturated chains may be linked directly, but the reaction is slow and only a low degree of cross-linking is attained. A cross-linking monomer is therefore added to overcome these limitations. Its addition also leads to a reduction in viscosity and hence to the easier impregnation of reinforcing fibers, besides affecting the properties of the cured resin. The most widely used monomer is styrene because of its compatibility with the polyester, its low viscosity, general ease of use, and relatively low price. The cross-linking reaction, which involves the formation of styrene bridge chains 1 to 3 units long between the polyester chains, is of the addition type, and no volatiles are evolved in the process. The reaction is brought about by the use of initiators (peroxides or azo compounds) that decompose to give free radicals. Two types of peroxide are commonly used, one decomposing under the influence of heat (e.g., benzoyl peroxide) and the second when mixed with an accelerator decomposing at room temperature (e.g., methylethylketone peroxide plus cobalt naphthenate). The latter system may, however, require a post-cure if optimum physical and mechanical properties are to be developed, especially at elevated temperatures. An idealized structure for the cross-linked polyester resin would therefore be:

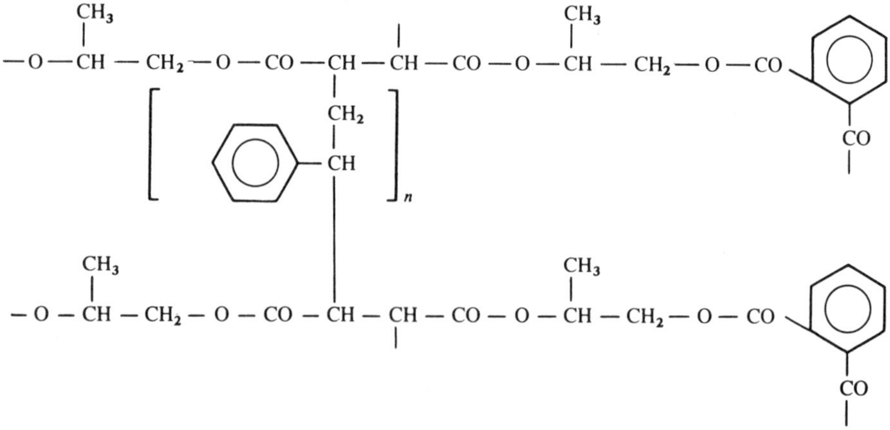

$$
\begin{array}{c}
\underset{|}{CH_3} \qquad\qquad\qquad | \qquad\qquad \underset{|}{CH_3} \\
-O-CH-CH_2-O-CO-CH-CH-CO-O-CH-CH_2-O-CO-\text{(phenyl)}-CO- \\
\left[\text{(phenyl)}-CH \overset{CH_2}{\underset{|}{|}} \right]_n \\
\underset{|}{CH_3} \qquad\qquad\qquad | \qquad\qquad \underset{|}{CH_3} \\
-O-CH-CH_2-O-CO-CH-CH-CO-O-CH-CH_2-O-CO-\text{(phenyl)}-CO-
\end{array}
$$

where n = 1 to 3.

Monomers other than styrene may be used to convey special properties; e.g., heat resistance is improved if multifunctional compounds such as diallylphthalate, divinyl benzene, or triallylcyanurate are substituted for styrene.

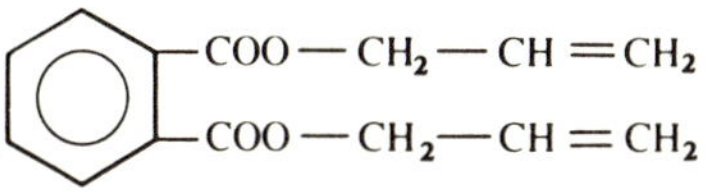

Diallylphthalate 　　　　　　　　　　　　　Divinylbenzene

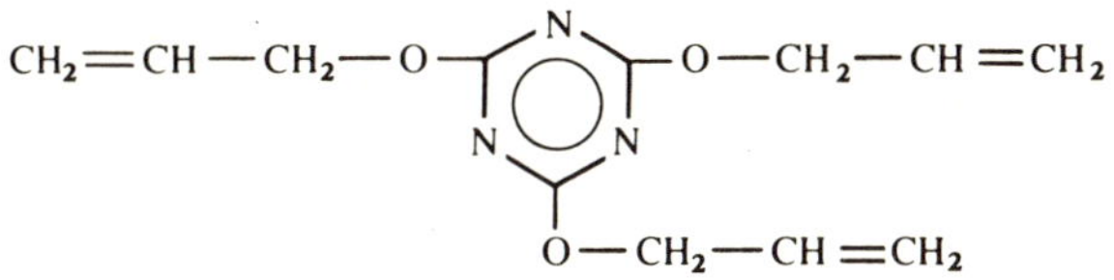

Triallylcyanurate

Thermal Stability

The kinetics of the thermal degradation of a polyester resin containing styrene as the cross-linking monomer have been studied by Anderson and Freeman in air and in argon[29] at a heating rate of 5°C per minute. The results were analyzed by the method of Freeman and Carroll and indicated a multistage reaction in air and a much simpler mechanism in argon. The kinetic parameters derived are summarized in Table 2.3 and the weight loss curves shown in Figure 2.11.

The principal volatile degradation products formed in air were carbon dioxide (67.0%), hydrogen (19.9%), methane (7.4%), and propylene (3.4%). The mechanism of thermo-oxidative degradation would appear to involve, as an initial step, hydroperoxidation of the alpha carbon of the styrenated branch of the polymer:

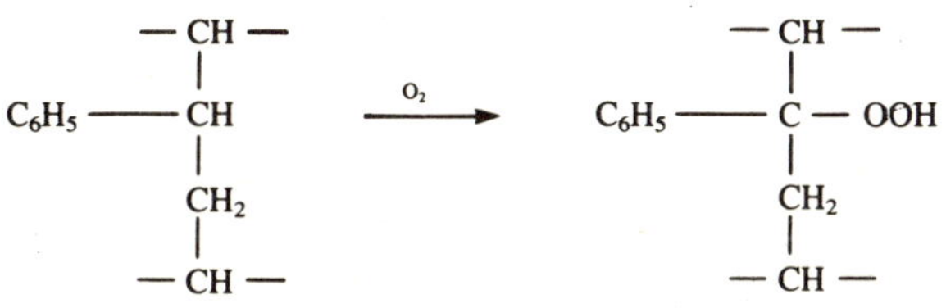

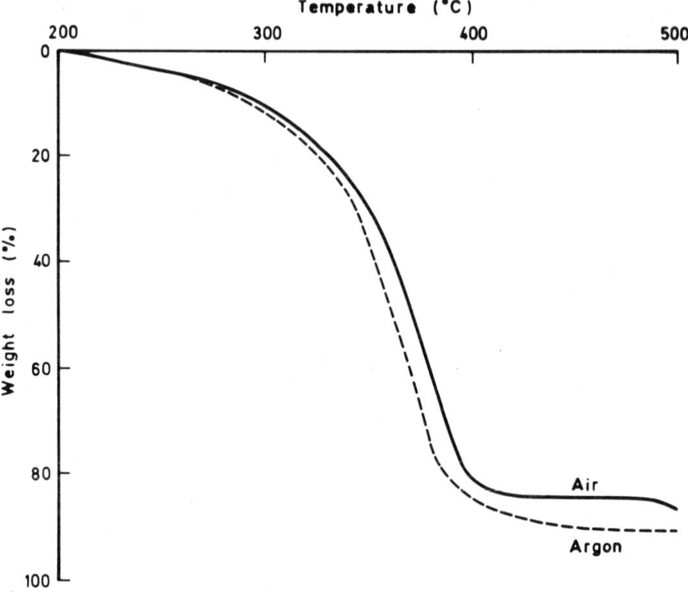

Figure 2.11. Thermogravimetry of a polyester resin (Laminac 4116) in air and argon. (Heating rate 5°C/min) Reference 29.

Scission of the hydroperoxide bond ensues, followed by rearrangement reactions:

TABLE 2.3. *Kinetic Parameters for Thermal Degradation of Laminac 4116 (Reference 29)*

Atmosphere	Temperature range ($°C$)	Order of reaction	Activation energy (kJ/mole)	Arrhenius factor (sec^{-1})
Air	200–260	0.4	80	1.3×10^5
Air	260–450	1.2	147	2.7×10^9
Air	450–550	1.0	332	4.2×10^{19}
Argon	200–450	1.0	84	4.8×10^3

At higher temperatures, scission of the chain occurs at the carboxyl oxygen yielding phthalic anhydride:

$$
\begin{array}{c}
\text{---} OOC\ C_6H_4\ CO|O\ CH_2\ CH\ OOC\ \overset{\displaystyle CH_3}{\underset{\displaystyle |}{CH}}\ \underset{\displaystyle |CH_3}{CH}\ COO\ \overset{\displaystyle CH_3}{\underset{\displaystyle |}{CH}}\ CH_2\ OOC\ \text{---} \\[1em]
C_6H_4(CO)_2O\ +\ {}^{\bullet}O\ CH_2\ \overset{\displaystyle CH_3}{\underset{\displaystyle |}{CH}}\ OOC\ \ldots\ldots\ldots
\end{array}
$$

Decarboxylation of the polyester also takes place and accounts for the formation of carbon dioxide and propylene.

Madorsky and Straus have examined the thermal degradation in a vacuum of a triallylcyanurate-based polyester at temperatures up to 1200°C and have compared the results with those for a phenolic resin and an epoxy novolac.[30] Figure 2.12 compares the stability of the three resins on the basis of the weight lost in a specific time at temperature. The similarity in behavior of the polyester and epoxy resins and the much smaller weight losses of the phenolic resin are evident. The overall activation energies for degradation were as follows: polyester resin, 151 kJ/mole; phenolic resin, 76 kJ/mole; and epoxy novolac, 214 kJ/mole. Table 2.4 gives the analysis of the volatile products produced when the three resins were pyrolyzed in a vacuum at 360, 500, 800, and 1200°C. The table serves to show the complexity of the products formed and the way in which the relative proportion of these varies with the temperature of pyrolysis.

Elevated Temperature Properties

The data included have been restricted to polyester systems cross-linked with triallylcyanurate, as these are the most thermally stable. Table 2.5 shows the effect of heat aging upon the flexural properties of glass cloth laminates made with Laminac 4233, one of the first TAC-polyesters to gain commercial accep-

TABLE 2.4. Analysis of Volatile Products from Pyrolysis of Three Resins in Vacuum at Different Temperatures (Reference 30)

| | Weight (%) of Total Volatiles | | | | | | | | | | | |
| | Polyester resin | | | | Phenolic resin | | | | Epoxy novolac resin | | | |
Product	360°C	500°C	800°C	1200°C	360°C	500°C	800°C	1200°C	360°C	500°C	800°C	1200°C
Hydrogen	—	—	0.7	1.3	—	—	5.2	2.8	—	—	0.8	2.1
Carbon monoxide	3.3	2.9	9.2	7.2	—	3.5	9.4	2.2	4.7	3.1	11.2	25.9
Carbon dioxide	62.4	41.4	29.3	29.4	0.5	5.5	2.4	1.2	16.2	6.0	3.7	1.8
Methane	—	0.7	1.8	3.4	—	4.3	11.8	4.9	1.0	0.8	1.6	4.3
Acetylene	—	—	—	0.1	—	—	—	1.9	—	—	—	2.5
Ethylene	—	0.6	6.5	14.8	—	—	1.8	1.6	—	—	3.0	3.0
Acetone	—	—	—	—	6.7	17.6	—	—	0.9	2.2	—	—
Propene	0.8	—	2.5	1.6	4.0	—	0.8	0.8	6.5	2.3	2.2	0.1
Propanol	—	—	—	—	10.9	11.1	—	—	—	0.3	—	—
Butane	—	—	—	—	—	—	6.8	—	—	—	1.6	—
Butene	—	—	1.5	—	—	—	—	—	—	0.2	—	—
Butadiene	—	—	0.3	1.4	—	—	—	0.1	—	—	—	0.2
Butanol	—	—	—	—	2.9	—	—	—	—	—	—	—
Pentadiene	—	—	2.6	—	—	—	—	—	—	—	—	—
Benzene	—	—	1.3	4.7	—	2.5	0.7	9.0	—	—	—	8.1
Toluene	—	—	—	0.1	—	4.7	1.2	1.5	—	—	—	0.1
Fraction volatile at pyrolysis temperature but not at room temperature	33.5	52.6	44.3	38.2	75.0	49.9	59.9	72.8	63.6	82.0	73.1	50.9

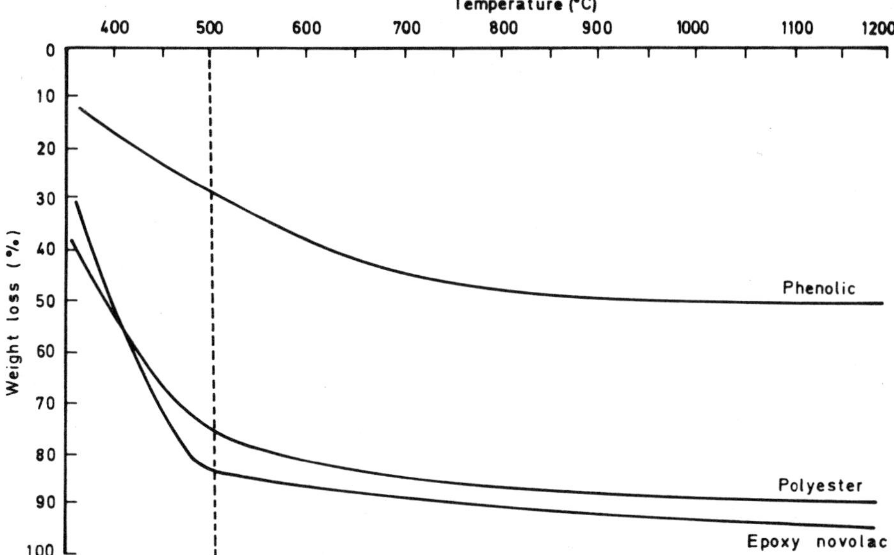

Figure 2.12. Comparison of thermal stabilities of polyester, phenolic, and epoxy novolac resins. (Based on weight loss after 30 minutes at temperature up to 500°C and 5 minutes at temperature thereafter) Reference 30.

tance. The glass cloth was ECC with finish 136. After 200 hours at 260°C, there was still greater than 50% retention of initial properties.

Table 2.6 gives properties after heat aging for different times at different temperatures for laminates made from Vibrin 136A resin and 181 cloth with a Garan finish. Here again there was greater than 50% retention of properties at least up to 500 hours at 260°C. Table 2.7 gives test data obtained at elevated temperatures for a PDL-7-669 resin and Y227 cloth in the heat-cleaned state. There was 96%, 77%, and 35% retention of the initial flexural strength after 200 hours at 150, 200, and 250°C, respectively, despite a 25% loss in resin content at the highest temperature used. It should be noted that the finish used on the glass cloth can have quite a substantial influence upon the retention of mechanical

TABLE 2.5. Effect of Heat Aging on Flexural Properties of Laminac 4233/Glass Cloth Laminates (Reference 31)

Exposure time at 260°C (hours)	Flexural strength (MPa)		Flexural modulus (GPa)	
	at RT	at 260°C	at RT	at 260°C
3	348	207	23	14
50	242	191	24	15
100	230	209	24	17
200	213	195	23	17

TABLE 2.6. Effect of Heat Aging on Flexural Properties of Vibrin 136A/Glass Cloth Laminates (Reference 32)

Exposure temperature (°C)	Exposure time (hours)	Flexural strength (MPa)	**Flexural modulus (GPa)**	Tensile strength (MPa)	Compressive strength (MPa)
260	0.5	276	18	276	207
260	192	276	17	—	—
260	500	297	—	—	—
260	1000	214	—	—	—
260	2000	83	—	—	—
316	0.5	152	—	—	—
370	0.5	76	—	—	—

properties at elevated temperatures. If a Garan-treated Y227 cloth was used instead of a heat-cleaned cloth with the PDL-7-669 resin, then a 64% retention of the initial flexural strength was observed after 200 hours at 250°C. (Compare a 35% retention with the heat-cleaned cloth.)

Figure 2.13 compares the flexural strength after 100 hours at different temperatures of laminates made from heat-cleaned cloth and a number of different resin systems. The good performance of the TAC-polyester is evident.

TABLE 2.7. Effect of Heat Aging on Flexural Properties of PDL-7-669/Glass Cloth Laminates

Test temperature (°C)	Flexural strength (MPa) after time at temperature of				
	1 hour	8 hours	50 hours	100 hours	200 hours
150	299	320	331	325	329
200	270	305	256	264	266
250	195	227	158	112	121
300	175	153	43	14	—

	Flexural modulus (GPa) after time at temperature of				
	1 hour	8 hours	50 hours	100 hours	200 hours
150	18	17	17	18	18
200	17	16	16	17	18
250	15	15	14	12	14
300	14	11	—	—	—

	Weight loss (%) after time at temperature of				
	1 hour	8 hours	50 hours	100 hours	200 hours
150	0.76	0.87	0.78	0.74	0.68
200	0.52	1.06	2.40	3.2	5.1
250	1.38	2.18	7.21	9.5	11.3
300	4.54	12.0	21.1	27.3	31.3

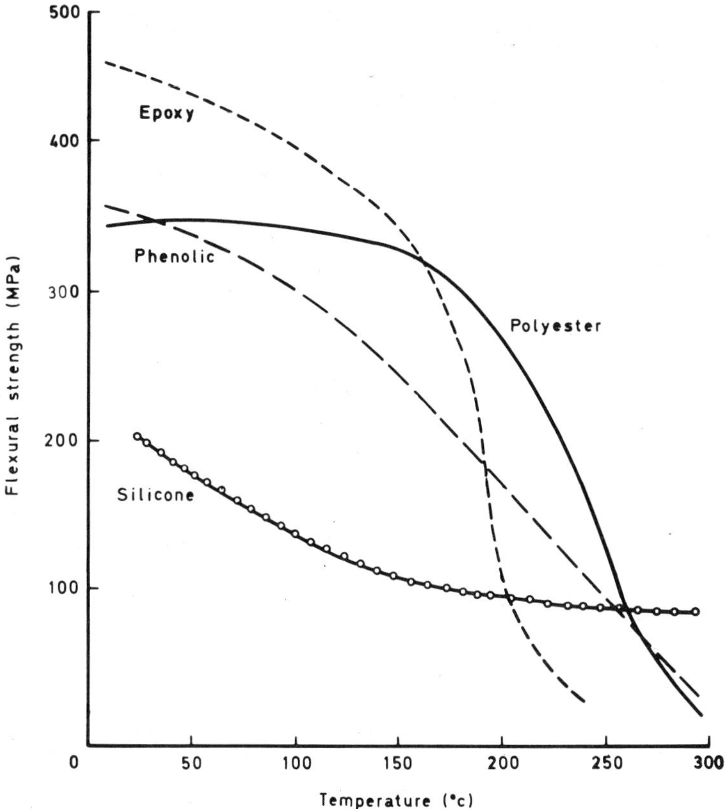

Figure 2.13. *Flexural strength of various glass-cloth laminates after 100 hours at temperature. (All heat-cleaned Y227 cloth. Polyester resin-PDL-7-669. Epoxy resin-Epikote 828. Phenolic resin-V17085. Silicone resin MS 2106.)*

EPOXY RESINS

Preparation

Epoxy resins[10-15] may be defined as those materials in which chain extension and cross-linking occurs through reactions of the epoxy group, i.e.,

$$-\overset{\displaystyle O}{\overset{\diagup\diagdown}{CH-CH_2}}$$

These resins were first developed in the 1940s and, although a number of different types are now commercially available, the major share of the market is still
held by systems based upon the reaction products of epichlorhydrin and
2,2-bis(4'-hydroxyphenyl) propane (Bisphenol-A):

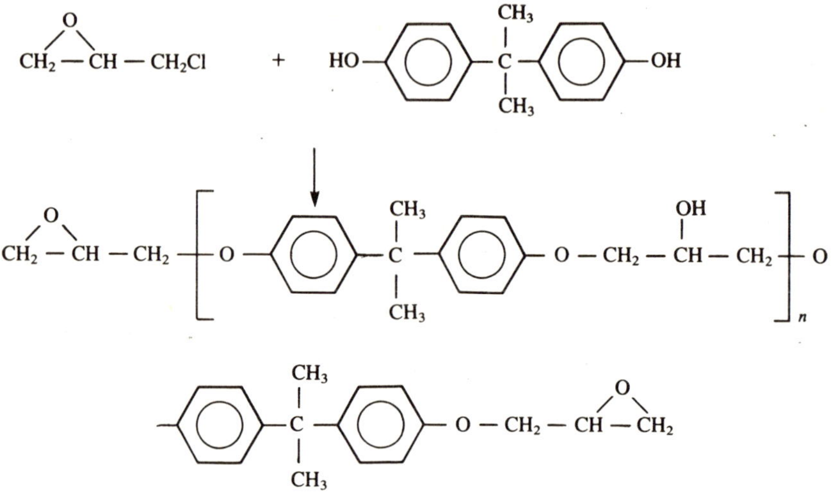

where $n = 0$ to 12.

Depending upon the value of n, the resins are either liquids or low-melting-
point solids. The products with low values of n are the most widely used because, as n increases, melt viscosity also increases and solubility decreases, thus
making processing more difficult. Cross-linking under the influence of heat alone
is not sufficiently rapid, and hence a curing agent must be added. The epoxy
system is very versatile; various types of curing agent may be used, creating the
possibility of tailoring to some extent the final network structure to meet specific
requirements.

Homopolymerization of the prepolymer is possible in the presence of a boron trifluoride complex.

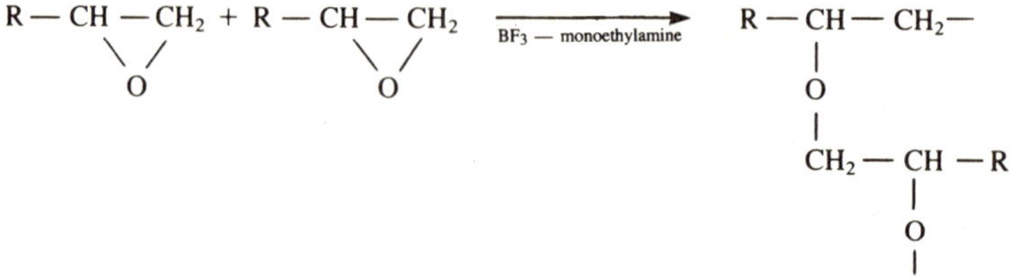

Most other curing agents actually copolymerize with the epoxy (or hydroxyl) groups. The compounds used to the greatest extent are acid anhydrides and polyfunctional amines, and their action may be represented as follows:

It is difficult to make generalizations with respect to structure and properties in epoxy resins because the type of curing agent used has such a marked effect upon behavior; e.g., the curing agent largely determines the chemical resistance. Anhydride cure appears to give good thermal stability, chemical resistance (except to alkalies), and electrical insulating properties. The aliphatic amines give fast cures and can be effective at room temperature. The aromatic amines are less reactive and require elevated temperature cure, but give products with higher heat resistance. Greater thermal stability is also attained by using base epoxies with multifunctionality or very compact structures in order to increase the crosslink density in the final cured product. Examples of such structures are:

Tetraglycidyl derivative
of diaminodiphenylmethane

Epoxy novolac

Cycloaliphatic diepoxides

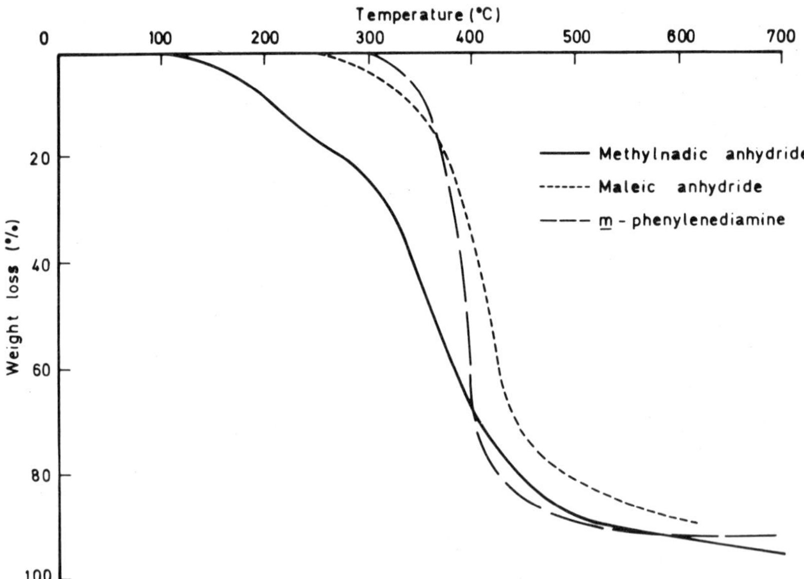

Figure 2.14. Thermogravimetry in vacuum of a bisphenol-A epoxy resin cured with different agents. (Heating rate 5°C/min) Reference 33.

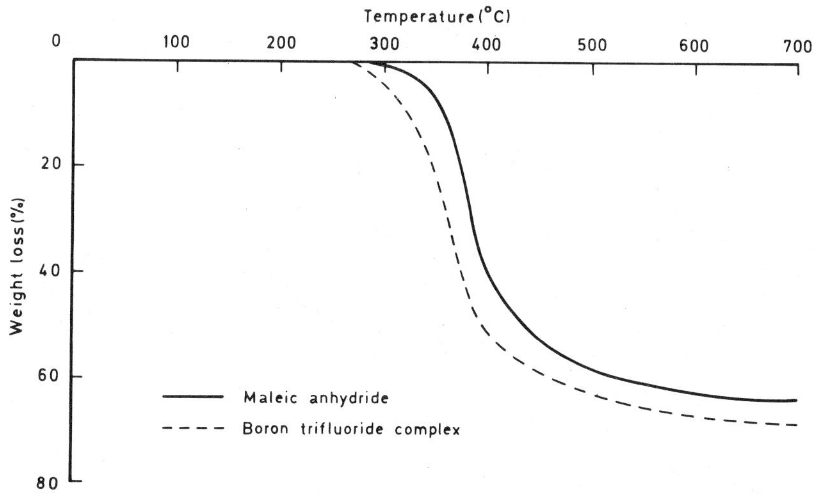

Figure 2.15. Thermogravimetry in air of an epoxy resin based upon the tetraglycidyl ether of tetraphenylethane cured with different agents. (Heating rate 2.5°C/min) Reference 34.

Thermal Stability

As stated earlier, the properties of epoxy resins are very much governed by the curing agent used. This is illustrated for thermal stability in Figure 2.14, which shows weight loss curves for a bisphenol-A epoxy resin cured with three different compounds.[33] There is a difference of over 100°C in the initial temperatures for breakdown. Figure 2.15 depicts similar curves, obtained in air rather than in vacuum, for an epoxy resin based upon the tetraglycidyl ether of tetraphenylethane and using two different curing agents.[34] In this case the difference in initial temperatures for breakdown is about 40°C.

The products and mechanism of thermal degradation of both uncured and cured epoxy resins have been studied in some detail in inert conditions, but much less work has been done on thermo-oxidative degradation. Table 2.8 compares the volatile products obtained on pyrolysis at 350°C in vacuum of a bisphenol-A epoxy and an epoxy novolac with diaminodiphenylmethane (DDM) and nadic methyl anhydride (NMA).[35] The distribution of products is quite different, with the two curing agents indicating different mechanisms of decomposition. Water is the main product from the DDM cured systems, indicating dehydration as a major reaction. With the NMA cured systems carbon dioxide and methylcyclopentadiene are produced in abundance. It is suggested that the diester groupings undergo beta elimination with release of carbon dioxide and an olefin:

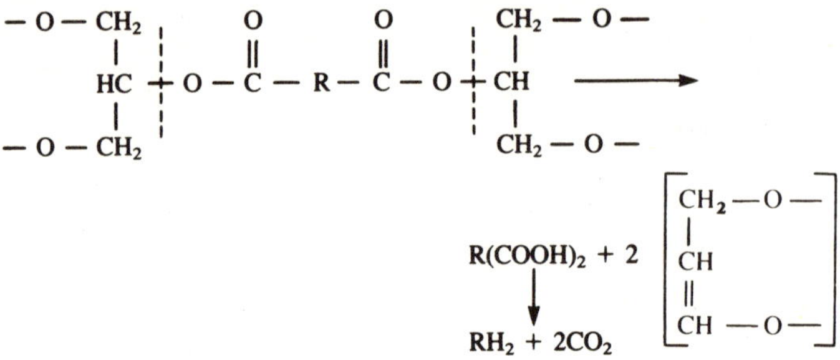

$$-O-CH_2 \atop HC \atop -O-CH_2} \Big| \; O \!=\! C-R-C \!=\! O \Big| \; CH_2-O- \atop CH \atop CH_2-O-} \longrightarrow$$

$$R(COOH)_2 + 2 \begin{bmatrix} CH_2-O- \\ CH \\ \| \\ CH-O- \end{bmatrix}$$

$$\downarrow$$

$$RH_2 + 2CO_2$$

TABLE 2.8. *Volatile Products from Pyrolysis at 350°C of Epoxy Resins (mole per unit) (Reference 35)*

	Bisphenol-A epoxy cured with		Epoxy novolac cured with	
Component	DDM	NMA	DDM	NMA
Hydrogen	—	4.21	—	6.06
Methane	0.14	1.12	—	0.59
Water	96.70	19.00	97.70	38.40
Carbon monoxide	—	5.16	—	8.89
Ethylene	0.13	1.11	—	0.85
Carbon dioxide	0.79	23.40	1.70	21.90
Methyl chloride	—	1.41	—	0.71
Cyclopentadiene	0.03	2.52	—	1.05
Methylcyclopentadiene	1.44	38.90	—	19.10
Toluene	0.58	0.76	0.56	1.03

DDM = 4,4′-diaminodiphenylmethane
NMA = nadic methylanhydride

In air the rate of degradation is proportional to the original epoxide functionality, and hence to the amount of curing agent or the linkages formed during the curing reaction. In the case of amine-cured resins, the oxidatively weak linkage is thought to be the amino-alcohol portion of the resin backbone:

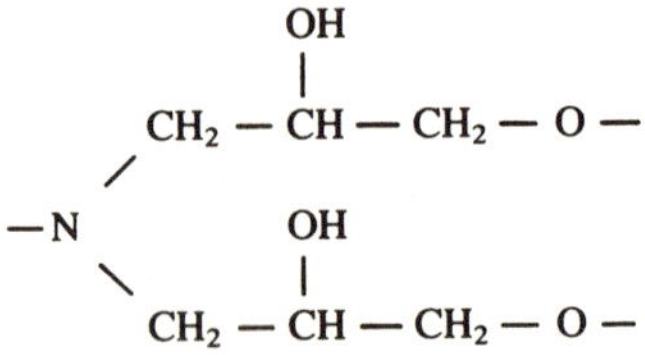

$$-N \Big\langle \; \begin{array}{c} OH \\ | \\ CH_2-CH-CH_2-O- \\[4pt] OH \\ | \\ CH_2-CH-CH_2-O- \end{array}$$

Greater oxidative stability is obtained by using phthalic anhydride as the curing agent.

Elevated Temperature Properties

The importance of using the correct curing agent to obtain the optimum high temperature properties with an epoxy resin is reemphasized by the results shown in Figures 2.16 and 2.17. The first figure gives the continuous operating temperatures based on the criterion of end of useful life after a weight loss of 16% for a bisphenol-A epoxy cured with five different compounds.[36] The superiority of an anhydride cure over an amine cure is again evident. The second figure plots flexural strength retention at temperature for bisphenol-A epoxy resin/glass cloth laminates made with four different curing agents.[37] The anhydride cures also result in superior strength retention at temperature. A comparison of the properties of cast samples of a number of different epoxy resin types all cured with aromatic diamines is made in Table 2.9. The best retention of properties at 150°C is observed with the tetrafunctional epoxy resin system. The cycloaliphatic resin system has high initial strength, but this falls off very markedly at temperature. A diaminodiphenyl sulphone cure of a bisphenol-A resin results in rather better

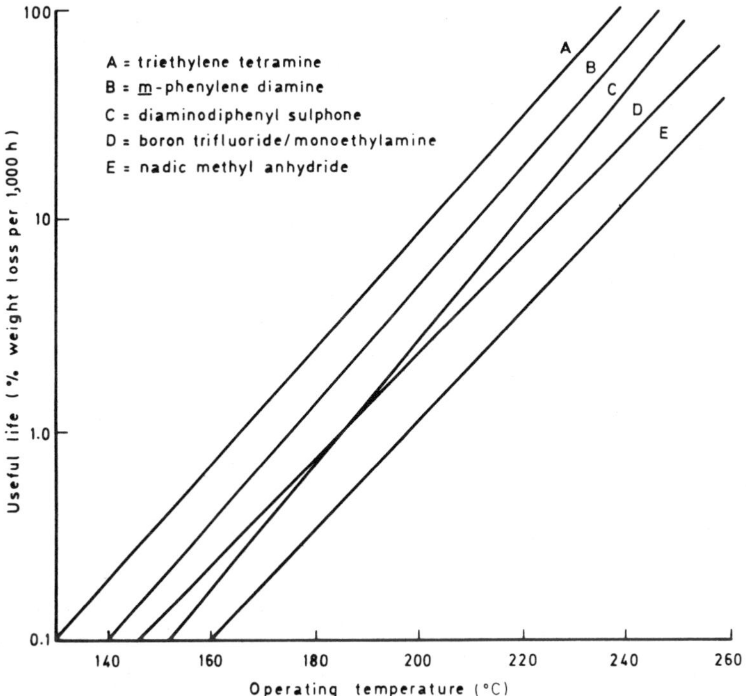

Figure 2.16. Continuous operating temperatures for a bisphenol-A epoxy resin cured with different agents. (Based on 16% weight loss as end point) Reference 36.

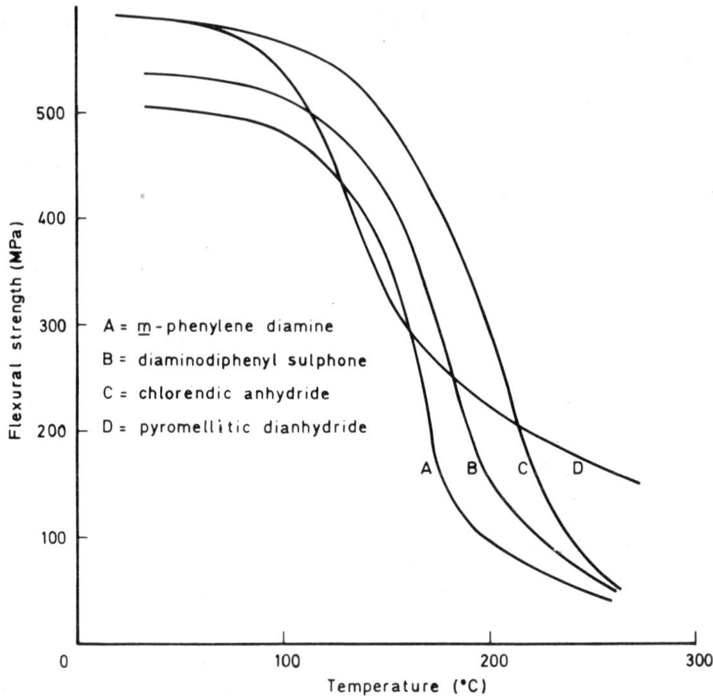

Figure 2.17. Flexural strength retention at temperature of bisphenol-A epoxy resin/glass cloth laminates cured with different agents. Reference 37.

high temperature properties than a diaminodiphenyl methane cure. The relatively poor performance at elevated temperatures of the cycloaliphatic resin is confirmed by long-term heat aging tests at 150°C on carbon fiber-reinforced samples (Figure 2.18). In contrast, the flexural strength of the bisphenol-A-based laminates is relatively unchanged between 5,000 and 30,000 hours aging and is appreciably higher than the initial room temperature strength value.

The behavior of an epoxy novolac system is illustrated in Table 2.10, which gives the hoop tensile strength and interlaminar shear strength of filament wound NOL-ring specimens made from a nadic methyl anhydride-cured epoxy novolac resin and glass fiber and aged at 260°C. After 192 hours at 260°C there is 76%, 66%, and 34% retention of tensile strength (a fiber-dominated property) when tested at 23°C, 260°C, and 399°C, respectively. The comparable figures for interlaminar shear strength retention (a resin-dominated property) are 21%, 19%, and 9%.

TABLE 2.9. Comparison of Properties of Cast Epoxy Resins (Reference 38)

Property	Bisphenol-A epoxy DDM cure	Bisphenol-A epoxy DDS cure	Cyclo aliphatic epoxy MPD cure	Tetrafunctional epoxy DDS cure
Tensile strength (MPa)				
at 20°C	53	59	89	41
150°C	19 (36)	37 (63)	78 (9)	33 (80)
Tensile modulus (MPa)				
at 20°C	2750	3070	6280	4100
150°C	1540 (56)	1470 (48)	432 (7)	3230 (79)
Compressive strength (MPa)				
at 20°C	> 111	107	227	178
150°C	> 29	63 (59)	65 (29)	115 (65)
Compressive modulus (MPa)				
at 20°C	2670	2000	4130	2930
150°C	721 (27)	1280 (64)	1290 (31)	2200 (75)
Flexural strength (MPa)				
at 20°C	116	102	159	86
150°C	41 (35)	49 (48)	84 (53)	86 (100)
Flexural modulus (MPa)				
at 20°C	2730	2790	6450	3920
150°C	1680 (62)	1160 (42)	2640 (41)	3200 (82)
Impact strength (MN/mm)				
at 20°C	0.21	0.17	0.21	0.083
150°C	0.19 (90)	0.21 (100)	0.15 (71)	0.073 (88)
Elongation at break (%)				
at 20°C	4.9	3.3	2.1	1.1
150°C	2.7	8.0	12.3	1.0

Figures in parentheses are percentage retention.

DDM = diaminodiphenylmethane   DDS = diaminodiphenylsulphone

MPD = m-phenylene diamine

TABLE 2.10. Mechanical Properties of NOL-Ring Specimens Made from an Epoxy Novolac Resin and Glass Fiber (Reference 39)

Test temperature (°C)	Hoop tensile strength (MN/m^2) after time (hours) at 260°C of						
	0	6	12	24	48	96	192
23	1083	973	1076	959	966	945	821
260	—	890	835	849	821	773	718
399	580	607	600	780	669	683	373

	Interlaminar shear strength (MN/m^2) after time (hours) at 260°C of						
	0	6	12	24	48	96	192
23	53	33	33	31	37	28	11
260	24	17	17	19	13	17	10
399	5	4	3	4	3	9	5

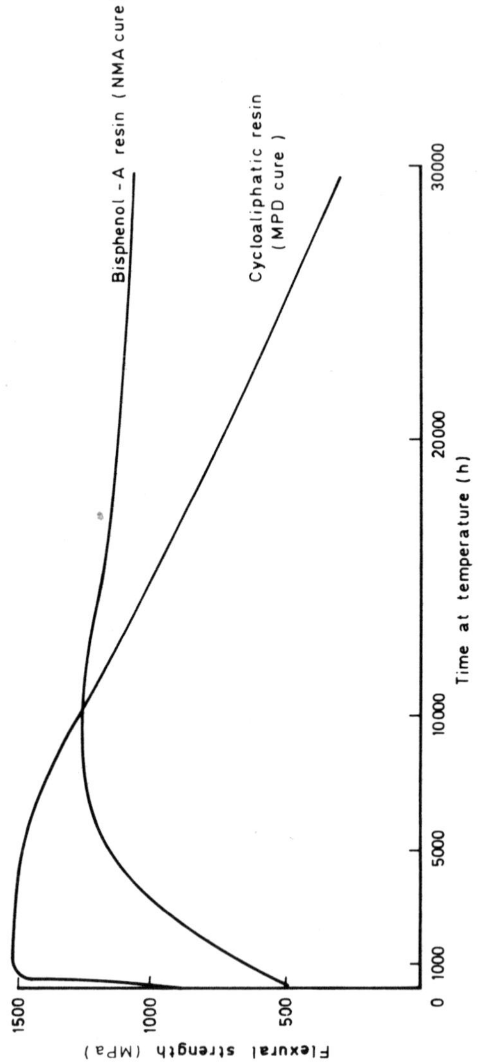

Figure 2.18. Flexural strength retention at 150°C of epoxy resin/carbon fiber laminates after aging at 150°C.

FURAN RESINS

Preparation

Although furan resins have been used for over 20 years, it is only comparatively recently that formulations have been developed which give composites
having acceptable curing characteristics and mechanical properties.[40–42] The
resins are produced by the acid catalyzed polymerization of furfuryl alcohol. The
polymerization is a condensation reaction involving the elimination of water.

$$n\left[\underset{O}{\bigcirc}-CH_2OH\right] \longrightarrow \underset{O}{\bigcirc}-CH_2\left[\underset{O}{\bigcirc}-CH_2\right]_{n-2}\underset{O}{\bigcirc}-CH_2OH + (n - 1)H_2O$$

The product of the required degree of condensation is neutralized and dehydrated
under reduced pressure. It may be diluted with further monomer. Cross-linking is
effected by the action of more acid catalysts; weak acids (e.g., phosphoric acid)
require elevated temperatures for cure to take place, but strong acids (e.g.,
sulphuric acid) cause reaction at room temperature. A post-cure is necessary for
the development of optimum properties. The mechanism of cross-linking is not
certain, but since unsaturation is lost in the process, the final network structure
may be of the form:

Thermal Stability

Figure 2.19 shows a thermogravimetric trace for a furan resin compared
with that for a polyester resin.[43] The furan resin is superior on a weight loss
basis because of the formation of a stable carbonaceous residue. This is facilitated if the temperature is raised slowly, or in a series of steps, and Figure 2.20
demonstrates that if the resin is held successively for two hours at 100, 200, 300,
400°C, etc., then the total weight lost even at 1000°C is only of the order of
40%.[44] Figure 2.21 is a graph of the yield of volatile degradation products dur-

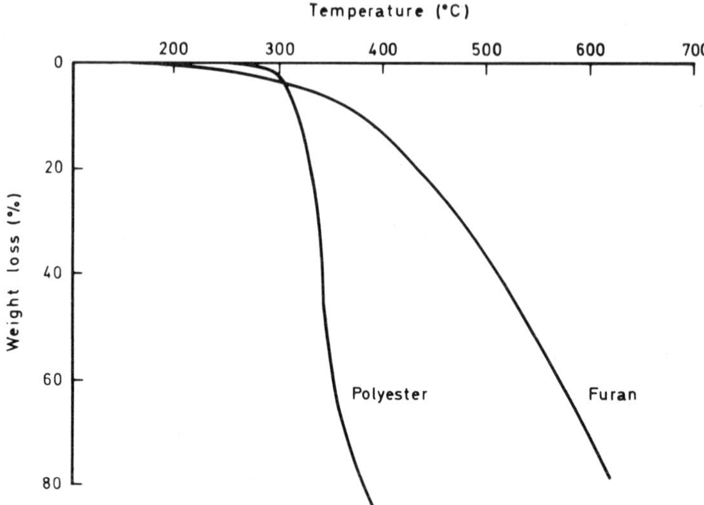

Figure 2.19. *Thermogravimetry of a furan resin compared with a polyester resin. (Heating rate not specified) Reference 43.*

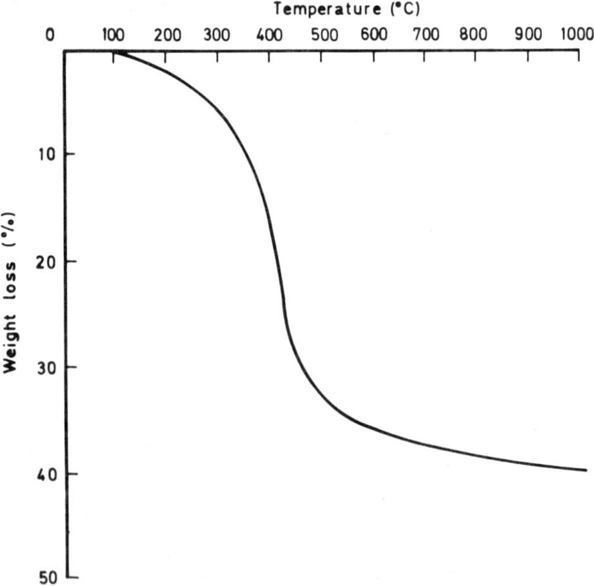

Figure 2.20. *Weight loss of a furan resin held successively for 2 hours at 100°C intervals. Reference 44.*

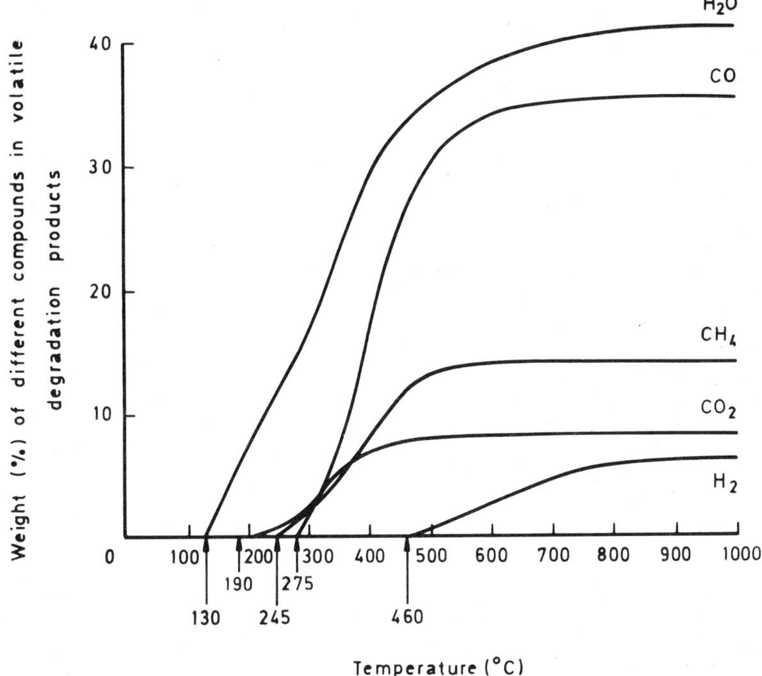

Figure 2.21. *Yield of volatile products from a furan resin held successively for 2 hours at 100°C intervals. Reference 44.*

ing this stepwise heating process. The major products in order of importance are water, carbon monoxide, methane, carbon dioxide, and hydrogen, and they commence to be evolved at 130, 275, 245, 190, and 460°C, respectively. The oxidative degradation of furan resins has been studied in some detail by Conley and Metil using infrared spectrometric techniques.[45] They concluded that the initial oxidation reaction is the formation of substituted bifuryl ketonic species, which subsequently undergo scission, producing substituted furoic acid. The mechanism is therefore very similar to that for phenolic resins.

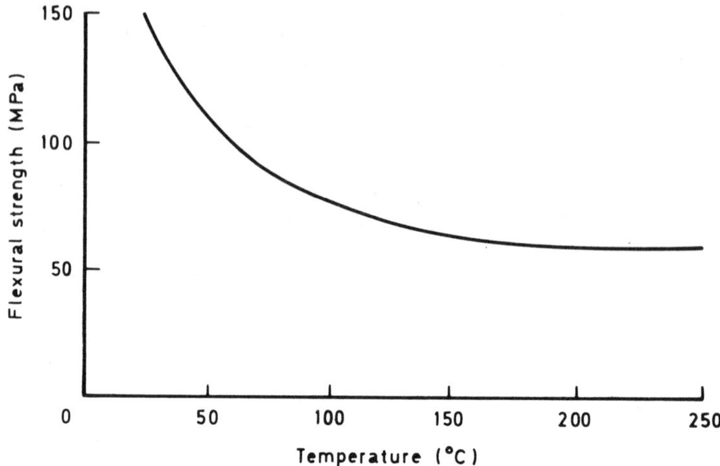

Figure 2.22. Flexural strength of furan resin/glass fiber laminates at elevated temperatures. Reference 42.

Elevated Temperature Properties

The available data relate to the properties of glass fiber laminates.[41,42] These consist of three layers of chopped strand mat and two layers of tissue mat with an overall glass content of 25%. The curing schedule was the very mild one of 1 hour at 40°C, 1 hour at 60°C, and 2 hours at 80°C. Figure 2.22 shows the variation in flexural strength with temperature, and Table 2.11 gives flexural strength and modulus figures and tensile strength and modulus figures for two different furan resin/glass fiber laminates at temperatures up to 260°C. The property measurements were made after 15 minutes at temperature.

TABLE 2.11. Mechanical Properties of Furan Resin/Glass Fiber Laminates at Elevated Temperatures (Reference 41)

Resin designation	Temperature (°C)	Flexural strength (MPa)	Flexural modulus (GPa)	Tensile strength (MPa)	Tensile modulus (GPa)
RP 100A	20	154	3.83	83	4.65
	150	75	1.65	55	3.32
	230	59	1.47	52	3.47
	260	57	1.70	52	3.38
QX 300	20	138	5.25	89	4.89
	150	79	2.72	75	4.00
	230	75	2.49	50	3.68
	260	63	2.55	63	2.76

In addition to good retention of mechanical properties at elevated temperatures, the furan resins have very good chemical resistance, being attacked by only concentrated sulphuric acid, hydrofluoric acid, strong caustic soda solutions, and oxidizing media. They also show very good fire resistance with very low emission of smoke.

BIS-DIENE RESINS

Preparation

An interesting class of cross-linkable nonpolar resins was described[46] in the late 1960s. The resins are based upon oligomeric bis-cyclopentadienyl compounds, which can react at elevated temperatures by Diels–Alder polyaddition, thus giving cross-linked products without evolution of volatiles. Two moles of cyclopentadienyl sodium are reacted with one mole of an organic dihalide:

where R may be aliphatic, aromatic, or SiR'_2.

The compound so formed reacts further, and the resins as supplied have the general formula:

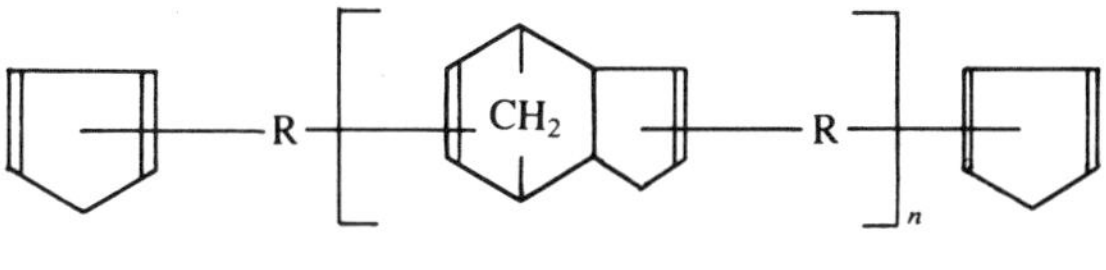

where n = 2 to 8.

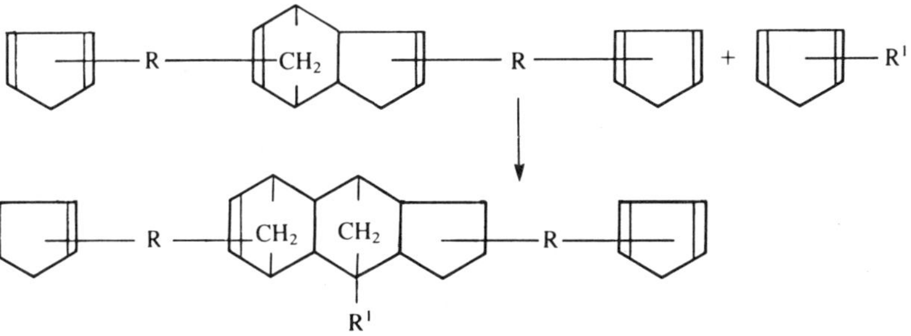

On heating at 180–200°C, monocyclopentadienyl radicals are split out and these react with unchanged oligomer to give tricyclopentadienyl nuclei, which are cross-linked through R bridges. At 200°C the process is complete in 3 to 4

minutes. At 180°C considerably longer times are required and an accelerator such as di-tert-butylperoxy-butane may be added to hasten gelling.

Thermal Stability

Results have not been published on the stability of the resin alone or on its mode of thermal degradation. Figure 2.23 shows the weight loss as a function of time and temperature of a glass fiber laminate made from D-556 glass with A-172 finish and a resin designated X355/2558 which contained a mixture of aliphatic and aromatic R groups. The figure also contains weight loss curves for a silicone resin/glass laminate and an epoxy resin/glass laminate. The data indicate good stability of the bis-diene, at least up to 150°C.

Elevated Temperature Properties

All the properties cited are for glass laminates of the composition detailed above. The laminates were made by dip impregnation of the glass cloth followed by drying at 130–150°C. The prepregs are flexible and dry and can be stored for several months. Cure was carried out at 180–200°C under a pressure of 2–5 MPa. Because of the nonpolar nature of the polymer, special attention has been devoted to the electrical characteristics. Table 2.12 lists some of the properties of the glass laminates before and after immersion in water at different temperatures.

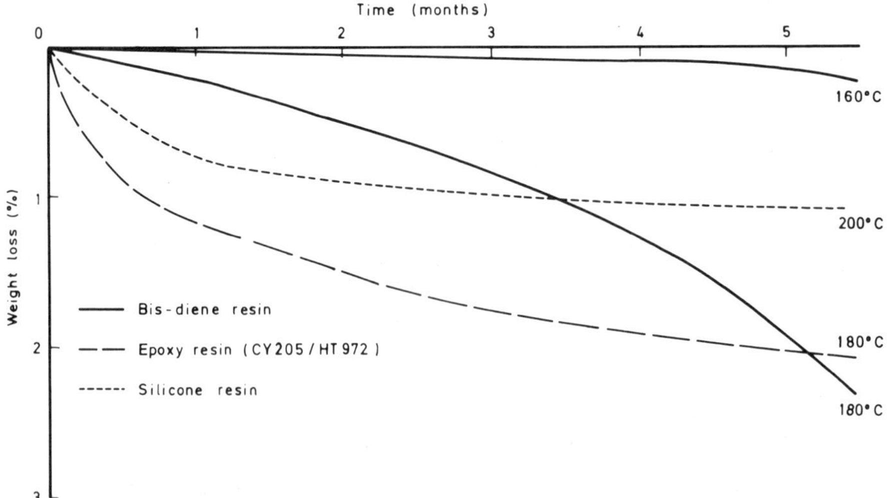

Figure 2.23. Isothermal weight loss of resin/glass fiber laminates. Reference 46.

TABLE 2.12. *Properties of X355/2558 Bis-Diene Resin D-556 A 172 Glass Laminates (Reference 46)*

Property	Value
Flexural strength	330–370MPa
Water absorption after 24 hours at 23°C	0.03–0.07%
Breakdown voltage parallel to laminate after 48 hours in water at 50°	45–60kV
Dielectric loss factor at 10^6 Hz	
as supplied	0.0035
after 24 hours in water at 23°C	0.0045
after 48 hours in water at 50°C	0.0050
Dielectric constant at 10^6 Hz	
as supplied, after 24 hours in water at 23°C and after 48 hours in water at 50°C	3.0–3.3
Surface resistivity after 90 hours at 35°C and 95% RH	10^{13} Ω
Volume resistivity after 90 hours at 35°C and 95% RH	10^6 Ω cm

Figures 2.24 and 2.25 show the changes in flexural strength and the changes of dielectric loss factor, respectively, after heat aging for different times at different temperatures. Figure 2.24 also includes the results for an epoxy resin/glass fiber laminate and a silicone resin/glass fiber laminate tested at the highest temperature used for the bis-diene system. Retention of mechanical properties of the bis-diene laminate is very good up to 200°C. Even at 225°C the flexural strength is comparable with that of the silicone laminate after 28 days at temperature, and is vastly superior to that of the particular epoxy laminate tested (type NEMA

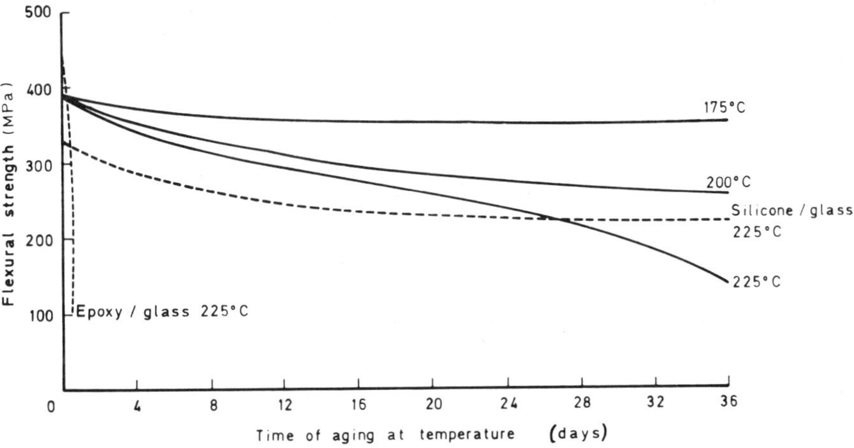

Figure 2.24. *Retention of flexural strength after heat aging of a bis-diene resin/glass fiber laminate. Reference 46.*

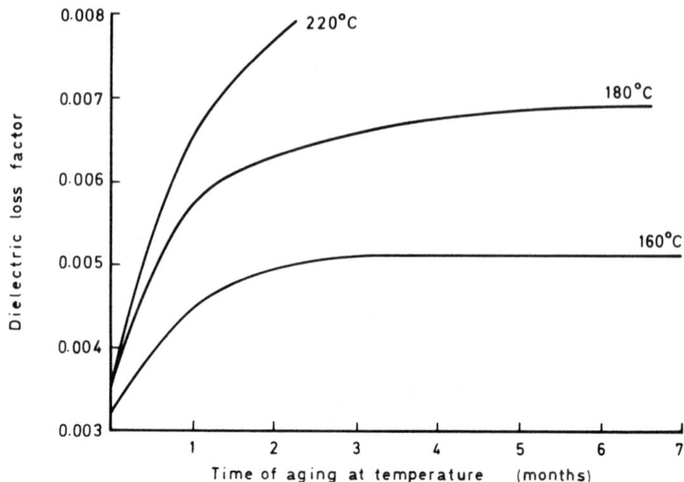

Figure. 2.25. Change in dielectric loss factor after heat aging of a bis-diene resin/glass fiber laminate. Reference 46.

FR-4). The dielectric loss factor results are in agreement with the mechanical data indicating that thermal degradation of the bis-diene resin is occurring at 220°C. If 2% weight loss is considered a limiting property and a lifetime of 25,000 hours is desired, then a temperature of approximately 150°C is the upper limit for permanent use.

A possible use of the material is as printed circuit boards. It is noteworthy that, after exposure to the following continuous cycle, there was no evidence of blistering or delamination:

(i) 1 hour immersion in boiling water
(ii) 2 min immersion in a cold mixture at −40°C
(iii) 6 min on the surface of a tin bath at 260°C
(iv) 2 min immersion in a cold mixture at −70°C
(v) 6 min immersion in a tin bath at 270°C
(vi) 2 min immersion in a cold mixture at −70°C.

VINYL ESTER RESINS

Preparation

These resins,[47–49] which may be regarded as hybrids of epoxy and polyester resin chemistry, were developed in the 1960s. They are made by reacting an epoxy resin with an ethylenically unsaturated carboxylic acid (usually acrylic or methacrylic acid) in order to give terminal unsaturation.

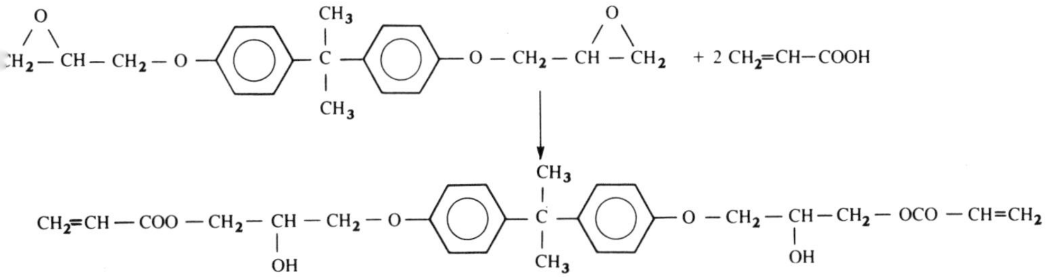

$$CH_2-CH-CH_2-O-\!\!\bigcirc\!\!-C(CH_3)_2-\!\!\bigcirc\!\!-O-CH_2-CH-CH_2 \quad + \ 2\ CH_2{=}CH{-}COOH$$

$$\downarrow$$

$$CH_2{=}CH-COO-CH_2-CH(OH)-CH_2-O-\!\!\bigcirc\!\!-C(CH_3)_2-\!\!\bigcirc\!\!-O-CH_2-CH(OH)-CH_2-OCO-CH{=}CH_2$$

While the vinyl-terminated resins are capable of homopolymerization, they are more commonly reacted with a monomer such as styrene using free radical initiation procedures exactly as for the unsaturated polyesters. The terminally located unsaturated groups are claimed to be particularly accessible for reaction, and the effect of steric hindrance on the cross-linking process is minimal. The idealized network structure using styrene as the added monomer is therefore:

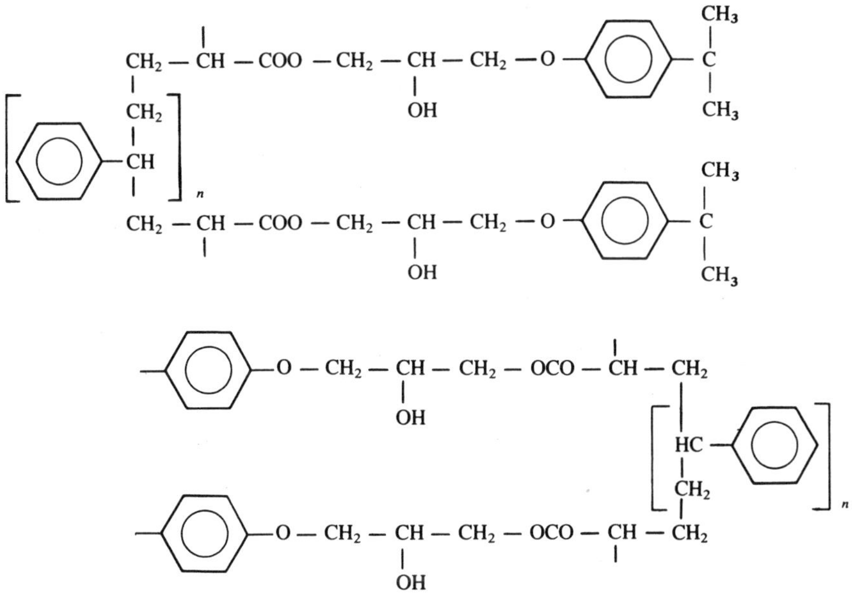

$$\left[\bigcirc{-}CH(CH_2{-})\right]_n \; CH_2-CH(CH_2)-COO-CH_2-CH(OH)-CH_2-O-\!\!\bigcirc\!\!-C(CH_3)_2$$

$$CH_2-CH-COO-CH_2-CH(OH)-CH_2-O-\!\!\bigcirc\!\!-C(CH_3)_2$$

$$-\!\!\bigcirc\!\!-O-CH_2-CH(OH)-CH_2-OCO-CH(-CH_2)\left[HC(\bigcirc){-}CH_2\right]_n$$

$$-\!\!\bigcirc\!\!-O-CH_2-CH(OH)-CH_2-OCO-CH-CH_2$$

Advantages claimed for this system are as follows:

(i) The epoxy resin backbone conveys high resilience and allows controlled molecular weight for low viscosity.

(ii) The terminal vinylic unsaturation is very reactive and its location ensures a more uniform final polymer structure.

(iii) The secondary hydroxyl groups promote wetting of and bonding to reinforcing fibers.

(iv) The smaller number of ester groups compared with a polyester resin (35–50% fewer per unit of molecular weight) results in lower water absorption and improved chemical resistance.

(v) If methacrylic acid is reacted with the epoxy, the methyl groups present shield the ester linkages still further, thus increasing the hydrolytic stability. In theory of course any epoxy resin could be used, but most work to date has been on bisphenol-A-based materials, epoxy novolacs, and cycloaliphatic diepoxides. The multifunctional epoxy novolacs once again give better high-temperature performance.

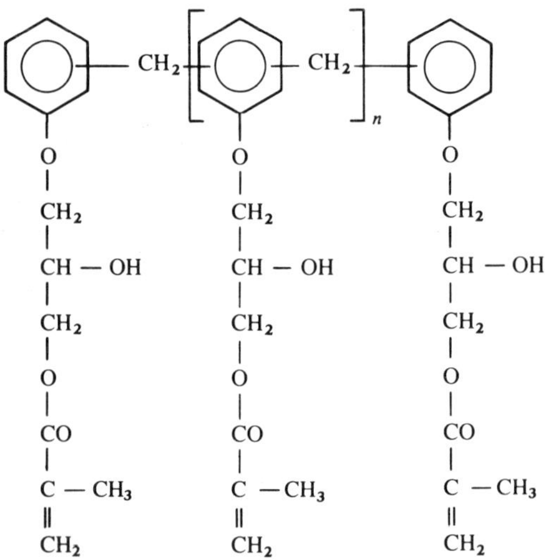

Although styrene is the cross-linking monomer most commonly used, data have also been reported for formulations containing diallylphthalate and divinylbenzene as reactive monomers.

Thermal Stability

No detailed information appears to have been published on the thermal stability of the vinyl ester resins or their mode of thermal, or thermo-oxidative, breakdown. One paper cites weight losses of 0.6% and 2.0% after 500 hours at 150°C, and 2.0% and 6.0% after 500 hours at 205°C for vinyl esters based upon epoxy novolacs and bisphenol-A, respectively.[50] In both cases styrene was used as the cross-linking monomer.

Elevated Temperature Properties

A bisphenol-A-based vinyl ester cross-linked with styrene (Epocryl DRH-480, Shell Chemical Co.), although having a heat distortion temperature of only 120°C, retained 93% of its initial flexural strength when tested at room temperature after 500 hours at 150°C, and 78% of the initial value after 500 hours at 205°C. Table 2.13 gives flexural strength and modulus and weight loss values after heat aging of a number of Epocryl resin formulations.[47]

Figure 2.26 shows the retention of flexural strength at temperature of glass fiber laminates made using Epocryl 12 cross-linked with either styrene or divinylbenzene and cured either thermally or by irradiation.[47] The system containing divinylbenzene and cured by irradiation is clearly superior at elevated temperatures.

Table 2.14 compares the behavior of glass fiber laminates made using vinyl-ester resins based upon bisphenol-A and an epoxy novolac[49] (Derakanes, Dow Chemical Co.).

The superiority of the epoxy novolac type for high temperature use is clearly evident.

Retention of electrical properties of the vinyl esters is also good and Figure 2.27 shows an electrical strength lifetime plot (based upon a 200 V/mil endpoint) of an experimental vinyl ester resin (designated XD 7156) compared with that of

TABLE 2.13. *Flexural Strength and Modulus after Heat Aging of Cast Epocryl Resins (Reference 47)*

Property	Resin designation			
	Epocryl 12	Epocryl 21	Epocryl DRH-321	Epocryl DRH-322
Heat distortion temperature (°C)	148	102	97	92
Weight loss (%)				
After 500 hours at 150°C	1.2	1.7	1.4	2.0
After 500 hours at 200°C	3.5	6.7	5.3	6.0
Flexural strength (MPa) at 23°C				
Initial	128	136	132	131
After 500 hours at 150°C	129	127	122	110
After 500 hours at 200°C	93	103	100	108
Flexural modulus (GPa) at 23°C				
Initial	4.05	3.37	3.21	3.10
After 500 hours at 150°C	3.70	3.35	2.95	2.95
After 500 hours at 200°C	3.28	3.28	3.17	3.12

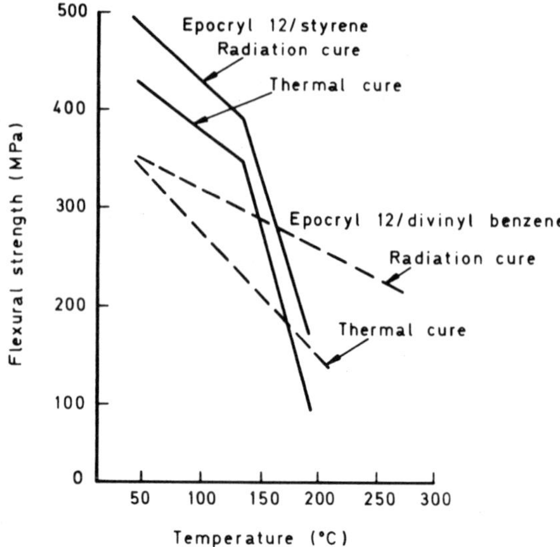

Figure 2.26. Elevated temperature properties of Epocryl resin/glass fiber laminates. Reference 47.

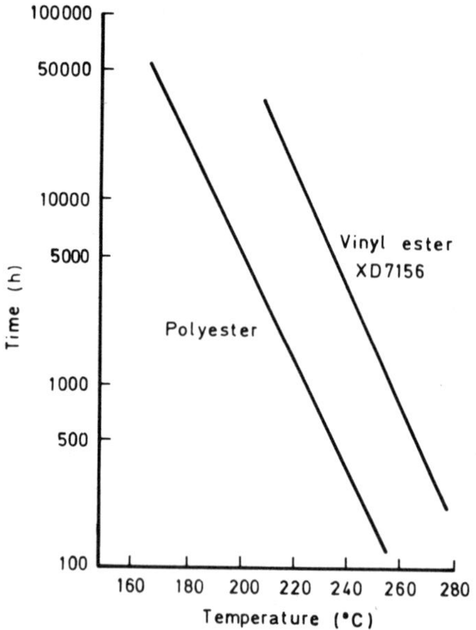

Figure 2.27. Electrical strength lifetime of a vinyl ester and a polyester resin (based on a 200 V/mil end point). Reference 48.

TABLE 2.14. Elevated Temperature Properties of Derakane Vinyl Ester/Glass Fiber Laminates (Reference 49)

Property	Bisphenol-A type	Epoxy novolac type
Heat distortion temperature (°C)	99–102	143–152
Flexural strength (MPa)		
Room temperature	204	165
66°C	196	—
93°C	189	169
107°C	101	—
121°C	34	166
149°C	22	145
163°C	—	83
177°C	—	55
Flexural modulus (GPa)		
Room temperature	7.1	8.6
66°C	7.0	—
93°C	5.9	8.1
107°C	3.4	—
121°C	1.6	7.3
149°C	1.6	5.7
163°C	—	4.2
177°C	—	3.5
Tensile strength (MPa)		
Room temperature	143	124
66°C	173	—
93°C	150	128
107°C	126	—
121°C	81	130
149°C	53	117
163°C	—	100
177°C	—	76
Tensile modulus (GPa)		
Room temperature	12.0	11.4
66°C	12.5	—
93°C	10.3	11.8
107°C	7.7	—
121°C	5.2	11.8
149°C	—	7.2
163°C	—	6.3
177°C	—	5.0

180°C rated polyester.[48] It has been claimed that in rigid insulation applications vinyl ester materials have maintained adequate electrical strength and dimensional stability for more than 10 years at 200°C. Vinyl ester resins made from a novolac have good corrosion resistance and have seen service for 10 years at 172°C in equipment handling chlorine and hydrogen chloride wet vapor.

PHENOL-ARALKYL RESINS

Preparation

In the early 1960s, it was discovered[51-52] that α,α'-dichloro-p-xylene in the presence of a mild Friedel–Crafts catalyst would react with a variety of compounds containing phenyl, phenylene, or phenoxy groups to yield cross-linked products which were termed Friedel-Crafts resins.[53] As an example of the process, the condensation of diphenyl with α,α'-dichloro-p-xylene is shown here:

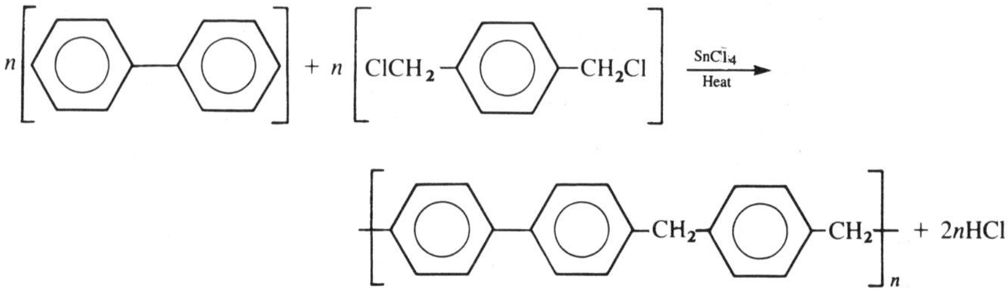

With a suitable choice of reactant concentrations, linear, soluble, fusible prepolymers are obtained, which readily cross-link on heating with further quantities of the p-xylene compound. The kinetics of the reaction and the structure of the condensation products have been studied in detail by Grassie and Meldrum.[54-57] They found that the reactivity of the second chloromethyl group was much greater than that of the first and that the reaction mixture rapidly became very complex with the formation of many isomers at each molecular weight level.

A disadvantage of this system is the evolution of hydrochloric acid during cure, but the properties of the resins were sufficiently promising to stimulate a search for alternative reactants. These were found in the aralkyl ethers and specifically in α,α'-dimethoxy-p-xylene[58-59]:

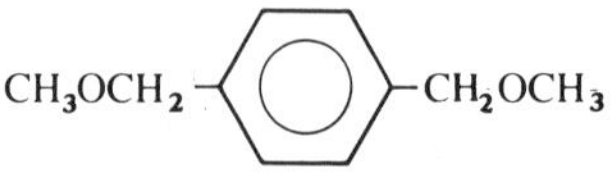

This reacts in a similar manner to the dichloro compound, but methanol is eliminated instead of HCl. The reactivity of the dimethoxy derivative with a wide range of aromatic, heterocyclic, and organometallic compounds has been studied.[52] For example, a wire coating enamel with the trade name Caldura was developed from the reaction product of α,α'-dimethoxy-p-xylene with diphenyl ether.[60] Undoubtedly the greatest success in this area, however, has been achieved by the condensation of the dimethoxy compound with phenols.[61,62] As an example of this the reaction with phenol itself is depicted below:

$$OH-C_6H_5 + CH_3-O-CH_2-C_6H_4-CH_2-O-CH_3 \xrightarrow[\text{Heat}]{\text{SnCl}_4}$$

$$HO-C_6H_4\left[-CH_2-C_6H_4-CH_2-C_6H_3(OH)-\right]_n-CH_2-C_6H_4-CH_2-C_6H_4-OH$$

The prepolymer can then be cross-linked by heating with hexamine, or a di- or poly-epoxide. This class of thermosets, termed phenol-aralkyl resins, is commercially available under the trade name of Xylok. The structure of the prepolymer derived from phenol is very similar to that of a phenolic novolac, the essential difference being that the phenolic nuclei are linked by xylylene instead of methylene bridges. This effectively results in a lower concentration of hydroxyl groups, and this is reflected in differences in properties such as chemical reactivity, chemical resistance, heat resistance, water absorption, and electrical properties. The rate of cure with hexamine is slower than that of the novolacs, and the reduced polar nature of the resin results in improved electrical properties, improved chemical resistance, and lower water uptake. More importantly in the context of this book, the thermal stability is also considerably increased.

A drawback of the hexamine cure system is that ammonia is released as a by-product of the reaction and hence it is difficult to fabricate thick sections. This has been overcome by heating the phenol-aralkyl prepolymers in the presence of

$$\cdots CH_2-C_6H_2(CH_2\cdots)(CH_2\cdots)-OH + CH_2\underset{\diagdown O \diagup}{-}CH-R-CH\underset{\diagdown O \diagup}{-}CH_2 \quad + \quad HO-C_6H_2(\cdots CH_2)(\cdots CH_2)-CH_2\cdots$$

$$\xrightarrow[\text{Imidazole}]{\text{Heat}}$$

$$\cdots CH_2-C_6H_2(CH_2\cdots)(CH_2\cdots)-O-CH_2-\underset{OH}{CH}-R-\underset{OH}{CH}-CH_2-O-C_6H_2(\cdots CH_2)(\cdots CH_2)$$

certain cycloaliphatic diepoxides and imidazole accelerators.[63] Cure then occurs with negligible evolution of volatiles. The heat resistance is, however, reduced in comparison with the hexamine-cured materials.

It is also possible to epoxidize the phenolic hydroxyl groups by reaction with epichlorhydrin, thus forming the equivalent of an epoxy novolac, which can then be cured by any of the conventional epoxy resin cure mechanisms.

Thermal Stability

Very little has been published on the thermal stability and mechanism of degradation of the commercially available polymers. The literature data relate to the polymers produced by reaction of α,α'-dichloro-p-xylene with aromatic compounds. Figure 2.28 shows the weight loss curve for an α,α'-dichloro-p-xylene/benzene polymer in nitrogen at a heating rate of 10°C/min.[64] The weight loss in the 100–500°C region is attributed to loss of residual solvent and monomer, true thermal breakdown only occurring at > 500°C. This has been confirmed by a detailed study[65] of the thermal degradation of polymers formed

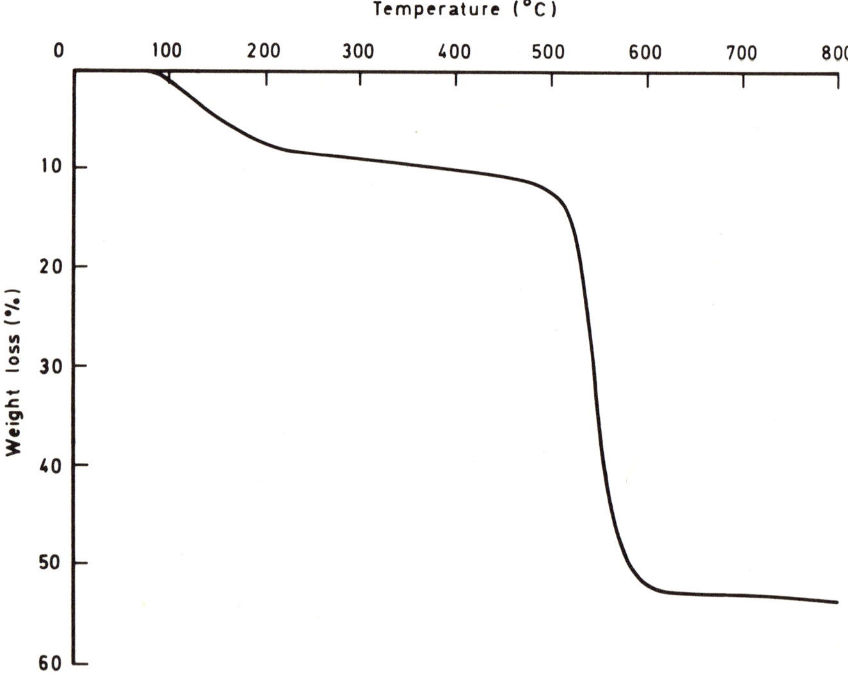

Figure 2.28. Thermogravimetry of an α,α'-dichloro-p-xylene/benzene polymer in nitrogen. (Heating rate 10°C/min) Reference 64.

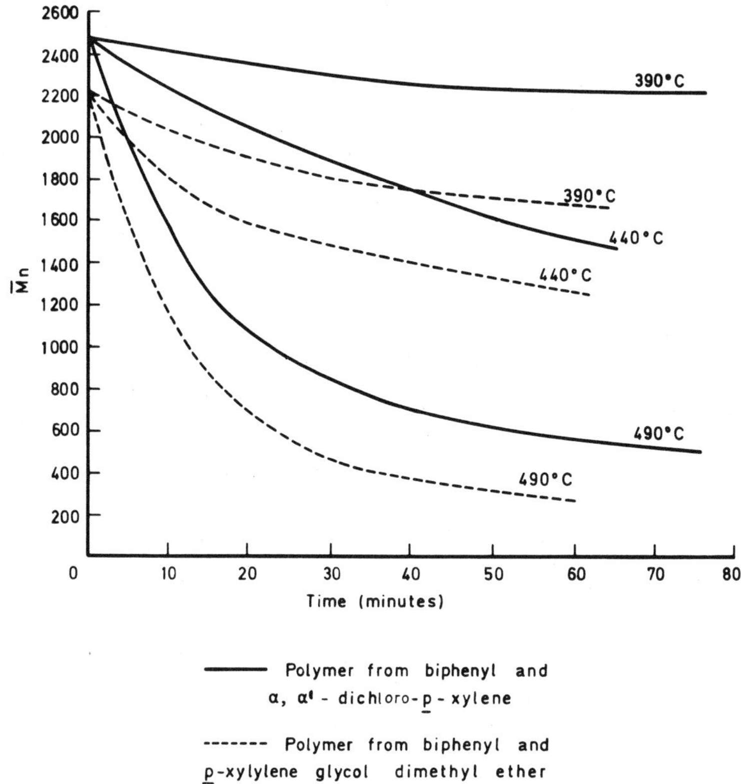

Figure 2.29. Change in number average molecular weight with time at temperature. Reference 65.

by reaction of biphenyl with either α,α'-dichloro-p-xylene or p-xylylene glycol dimethyl ether, degradation processes being very slow below 450°C. The thermal stability of the polymer formed using the dichloro compound (A) is slightly greater than that obtained using the dimethoxy derivative (B), both on the basis of weight loss and change in molecular weight (Figure 2.29). The overall activation energy for thermal degradation of polymer (A) is 240 kJ/mole and for polymer (B) 269 kJ/mole. Analysis of the volatile degradation products formed by holding the polymers successively for 1 hour at 400, 450, and 500°C shows that the major products are benzene, toluene, and xylene (Table 2.15), together with minor amounts of methane and, in the case of (B), methanol. These compounds arise from chain scission reactions. Secondary products include substituted 9-phenyl fluorenes.

The rather sparse evidence for oxidizing conditions indicates that the primary step is the conversion of the methylene groups between the rings to car-

TABLE 2.15. Volatile Products from Thermal Degradation of Polymers Based on Biphenyl (Reference 65)

Polymer	Degradation temperature (°C)	Volatiles (wt %)	Composition of Volatiles (mole %)		
			Benzene	Toluene	Xylene
Biphenyl +	400	<0.1	68.6	14.5	16.9
α,α′-dichloro-*p*-	450	0.5	78.2	13.2	8.6
xylene	500	2.8	22.7	36.4	40.9
Biphenyl +	400	0.2	28.8	21.9	49.3
p-xylylene glycol	450	0.7	49.2	35.5	15.3
dimethyl ether	500	3.2	18.2	39.0	42.1
(Polymer held successively at each temperature for 1 hour.)					

bonyl. The rate of this process is very similar for phenolic and phenol-aralkyl resins.

Elevated Temperature Properties

Glass cloth laminates are made by impregnating the cloth with a 60% solids solution of the phenol-aralkyl resin in methylethylketone. This is precured for 10 minutes at 125–140°C for subsequent press cure. In the press, temperatures of 150–175°C, pressures of 0.7–6.9 MPa, and times of 1–2 hours are required, depending upon the precure given, the thickness of the laminate, and the precise reinforcement used. For optimum strength retention at elevated temperatures, the glass cloth used should have been treated with an aminosilane finish, and a post-cure is essential (Table 2.16).

Figure 2.30 compares the flexural strength retention at room temperature and at 250°C of various glass cloth laminates that have been aged for periods of up to 1,000 hours at 250°C.[52] From this it can be seen that the phenolic laminate has high initial strength, but poor thermal stability. The epoxy and silicone laminates have relatively poor strengths at elevated temperature, but the retention of properties with time at temperature is good. Only the Xylok and polyimide laminates show both high mechanical strength and good retention of this property.

TABLE 2.16. Effect of Post-Cure on Elevated Temperature Properties of Xylok 210 Glass Cloth Laminates (Reference 66)

Time of post-cure at 175° (hours)	Flexural strength (MPa) 175°C	Measured at 250°C
0	217	—
24	616	234
48	650	297
72	675	419

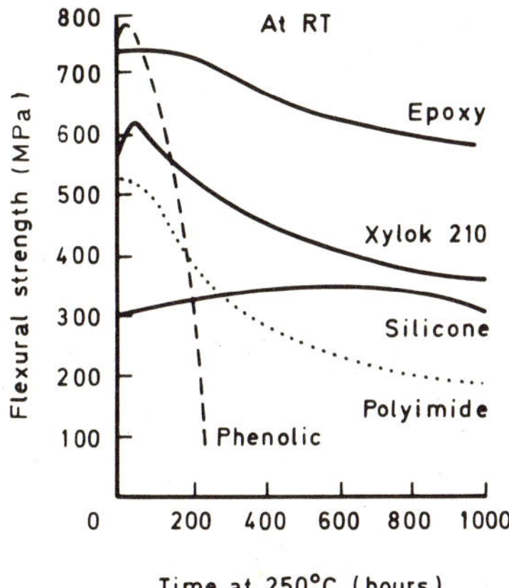

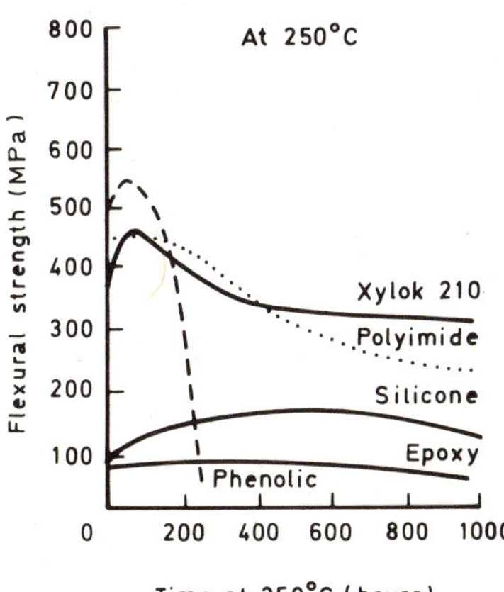

Figure 2.30. Comparison of flexural strength at RT and 250°C of various glass cloth laminates aged at 250°C. Reference 52.

The Xylok material retains > 80% of the initial strength after 1,000 hours at 250°C and > 50% after 2,000 hours at 250°C. Fifty per cent of strength is still retained after 800 hours at 275°C and 300 hours at 300°C. On the basis of a 50% retention of strength, a lifetime of 20,000 hours would be predicted at approximately 180°C. Good retention of electrical properties is also demonstrated at temperatures at least up to 200°C (Table 2.17).[66]

Table 2.18 compares the electric strength life at 250°C of Xylok 210 glass cloth laminates with that of other systems.[66]

The epoxide-cured Xylok resins, although not as thermally stable as the hexamine-cured variety, nevertheless show very little change in electrical properties up to 180°C (Figure 2.31).[52] The superiority over an epoxy novolac composite is evident. After 2,500 hours at 180, 200, 220, or 240°C, the Xylok 237

TABLE 2.17. Retention of Electrical Properties of Xylok 210 Glass Cloth Laminates at Elevated Temperatures (Reference 66)

Property		Temperature of test (°C)			
	25	50	100	150	200
Permittivity at 60 Hz	5.24	5.23	5.24	5.22	—
10^3 Hz	4.92	4.93	4.93	4.87	4.83
10^6 Hz	4.77	4.83	4.89	4.89	4.87
Loss tangent at 60 Hz	0.0037	0.0036	0.0036	0.0064	—
10^3 Hz	0.0063	0.0043	0.0037	0.0039	0.0063
10^6 Hz	0.0107	0.0109	0.0086	0.0061	0.0047
Loss factor at 60 Hz	0.0194	0.0189	0.0189	0.0335	—
10^3 Hz	0.0310	0.0211	0.0182	0.0190	0.0304
10^6 Hz	0.0510	0.0560	0.0420	0.0298	0.0229

TABLE 2.18. Comparison of Electrical Strength of Xylok 210 Glass Cloth Laminates with Other Systems (Reference 66)

Property	Phenolic	Special phenolic	Epoxy	Epoxy novolac	Acrylic	Silicone	Polyimide	Xylok 210
Electric strength at 20°C (MV/m)	20	30	26	28	4	26	22	28–33
Electric strength life at 250°C (hour)	144	216	700	430	0	750	144	1000–1400

glass cloth laminates retain 95, 92, 70, and 47%, respectively, of their initial strength.

A molding formulation (Xylok 225) is also available that can be used with a wide range of fillers—asbestos, glass, silica, calcium silicate, mica, graphite, synthetic fibers, aluminum, bronze, and ferric oxide. Since the chemical structure of Xylok 225 is very similar to that of a phenolic-novolac, and the same curing mechanism can be used for both, they may be blended without compatibility or processing problems arising. Figure 2.32 compares the flexural strength retention on aging at 250°C of asbestos-filled moldings based upon Xylok 225, a phenolic-novolac, and blends of the two in varying proportions.[52] The Xylok 225 molding has a strength retention as a function of time approximately eight times greater than that of the epoxy novolac material. Blending the two, Xylok:phenolic, in ratios of 1:3, 1:1, and 3:1 results in improvements in time of strength retention by factors of 2.4, 4.0, and 6.7 over the unmodified phenolic, respectively.

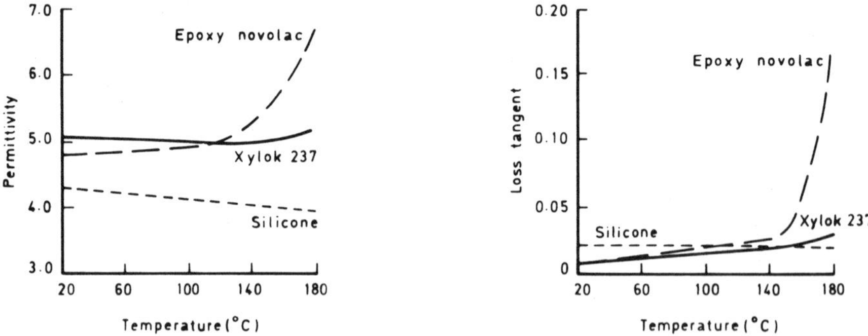

Figure 2.31. Effect of temperature on permittivity and loss tangent measured at 50 Hz of various glass cloth laminates. Reference 52.

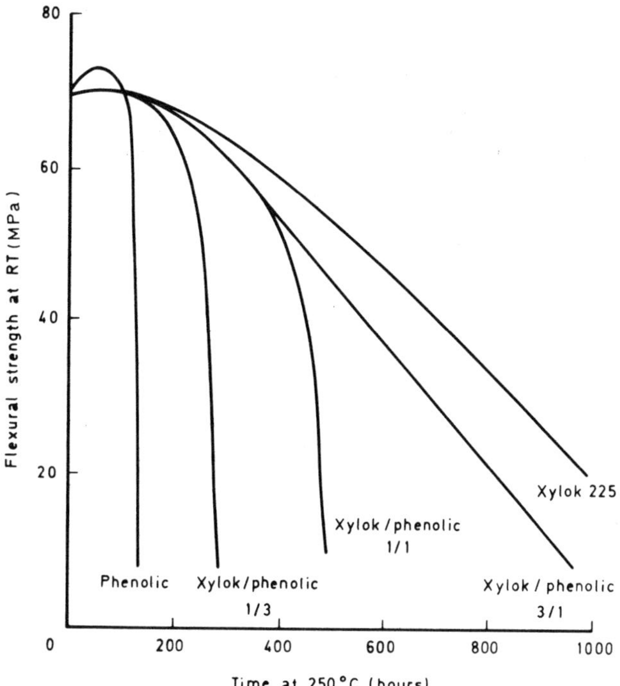

Figure 2.32. Effect of aging at 250°C on the flexural strength of asbestos-filled phenolic and Xylok 225/phenolic resin blends. Reference 52.

In addition to their good heat resistance, phenol-aralkyl materials also possess very good flame resistance and low smoke-generating characteristics. At room temperature they are little affected by organic solvents, alkalies, inorganic, or organic acids, but at 90°C there is some attack by mineral acids, strong caustic alkalies, and organic bases.

PSP RESINS

Preparation

During the late 1970s a new class of heterocyclic aromatic polymer was synthesized at ONERA and has been called PSP [poly(styryl pyridine)] resin.[67–69] The system is based upon the condensation of aromatic aldehydes with methylated derivatives of pyridine, 2,4,6-trimethyl pyridine (collidine) being particularly favored.

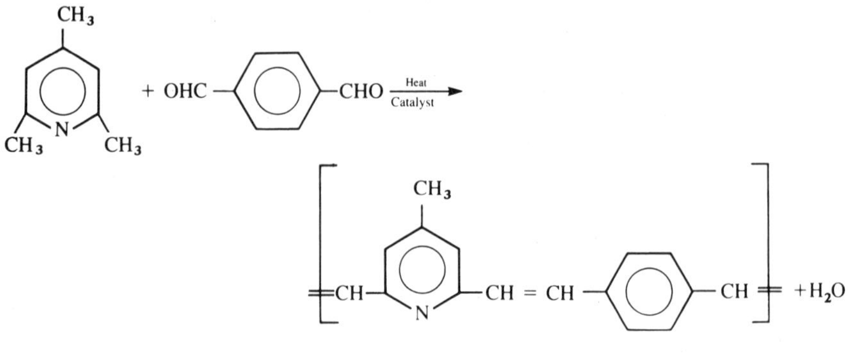

By restricting the reaction time, a liquid resin is obtained that is suitable for impregnation of various fibrous reinforcements. Extension of reaction time leads progressively to solids soluble in ethanol or acetone/ethanol mixtures, solids only soluble in polar solvents such as N,N-dimethylformamide or N-methylpyrrolidone, or solids more suitable for compression or injection molding. Cross-linking of the prepolymer is effected by heating above 150°C and preferably above 200°C, optimum mechanical properties only being developed by a post-cure of at least two hours at 250°C. Presumably the mechanism of cross-linking is a straightforward addition reaction between the unsaturated groups in the polymer chain yielding a network structure of the form:

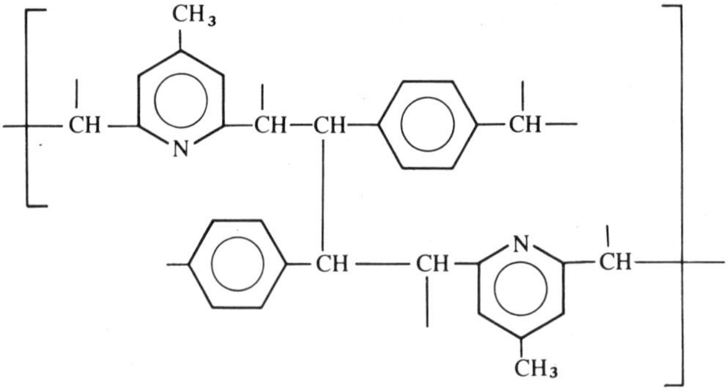

Thermal Stability

Figure 2.33 shows the dynamic weight loss curves (rate of temperature rise 5°C/min) for a PSP resin in argon and in air.[67] Thermal degradation commences at about 300°C in both atmospheres and 10% weight loss is attained at 415°C in air and at 435°C in argon. Rapid breakdown continues in air and by 600°C over 90% of the initial weight has been lost. In argon, however, stabilization occurs

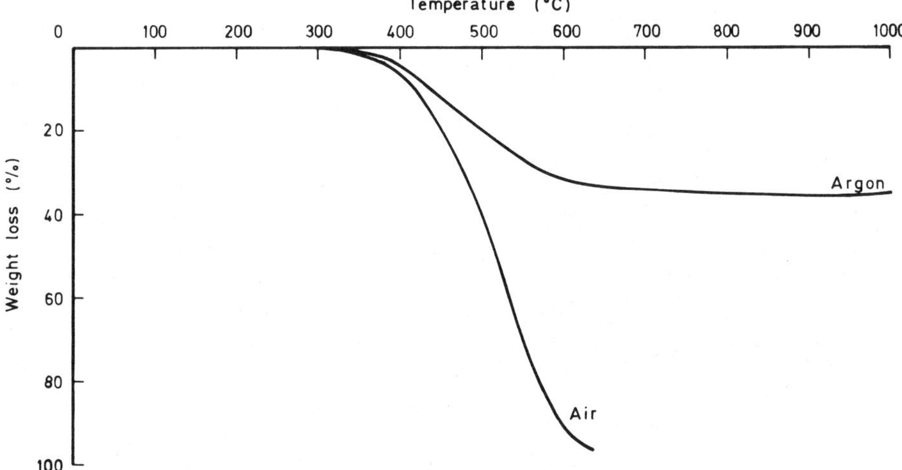

Figure 2.33. Thermogravimetry of PSP resin in argon and in air. (Heating rate 5°C/min) Reference 67.

and even at 1000°C the total loss in weight is no more than 35%. The mechanism of thermal degradation does not appear to have been studied as such, but experiments on glass fiber-reinforced laminates indicate that in air, carbon dioxide is a major decomposition product, carbon monoxide a minor decomposition product, and hydrogen cyanide is only formed in trace amounts.[67]

Elevated Temperature Properties

A resin PSP 6022 is now supplied commercially by SNPE (Société Nationale des Poudres et Explosifs) in grades for fiber impregnation from the melt or from solution and for injection or compression molding.[70] For solution impregnation, a 75% solids content in methylethylketone is used. The fibers are dried at 100°C to a 10–12% volatiles content. At this stage the life of the prepreg at room temperature is about 3 months. In a press cure the prepreg is heated to 200°C over 20 minutes and then held at this temperature, and a pressure of 0.5–1 MPa is applied after 45 minutes. The temperature is held at 200°C for about 3 hours. For elevated temperature use, a post-cure is necessary: a minimum of 2 hours at 250°C for use up to 150°C and of 16 hours at 250°C, or 3 hours at 300°C, for higher temperatures. Table 2.19 shows the influence of post-cure conditions upon the flexural strength of PSP/HT-S carbon fiber laminates at room temperature and at 250°C. An intermediate treatment at 225°C is advantageous in maximizing room temperature strength.

TABLE 2.19. Influence of Post-Cure Conditions on Flexural Strength of PSP/HT-S Laminates (Reference 69)

	Flexural strength (MPa) after post-cure of			
Temperature (°C)	2 hours at 250°C	16 hours at 250°C	16 hours at 225°C	16 hours at 225°C + 16 hours at 250°C
23	1690	1260	1550	1640
250	830	1330	1150	1340

Figure 2.34 shows the effect of heat aging at different temperatures on the flexural strength of the PSP/HT-S laminates.[69] These data indicate that continuous use at 200°C for periods of at least 10,000 hours may be considered. Above this temperature the strength retention is much more limited; e.g., a loss of 50% of initial strength occurs in 5,000 hours at 225°C, 2,000 hours at 250°C, and in less than 16 hours at 400°C. Table 2.20 gives figures for flexural strength, flexural modulus, and interlaminar shear, measured at room temperature and at 250°C after different times of aging at 250°C, of a T300/PSP 6022N laminate that had been post-cured for 16 hours at 250°C. Performance was not affected by up to 2,000 thermal cycles between ambient temperature and 225°C (5-minute hold at each extreme).

Most data with this resin system have been obtained using carbon fiber as reinforcement, but Table 2.21 shows that equally good results at high temperature may be obtained with glass, Kevlar, or boron fibers. All the figures are for unidirectional laminates.

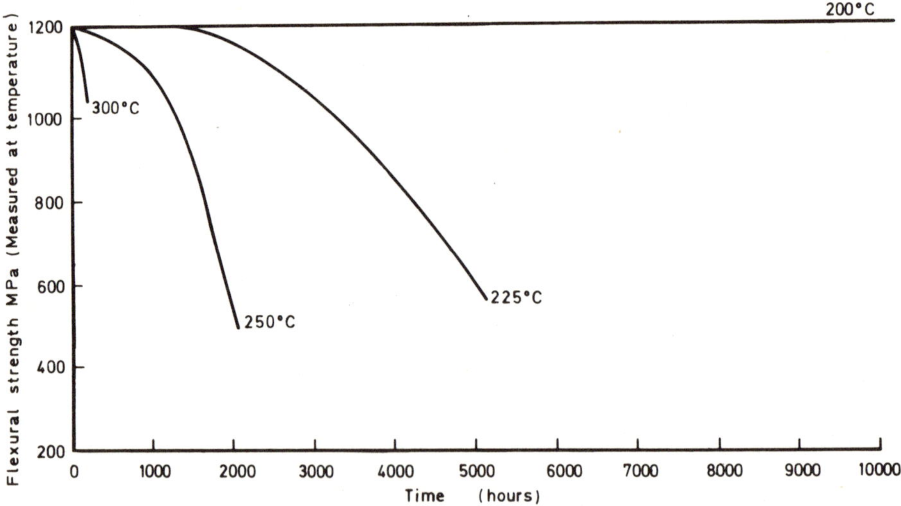

Figure 2.34. Effect of heat aging on the flexural strength of a PSP/HT-S carbon fiber laminate. Reference 69.

TABLE 2.20. Effect of Thermal Aging on Properties of PSP/T300 Carbon Fiber Laminates (Reference 70)

Time at 250°C	Flexural strength (MPa)		Flexural modulus (MPa)		Interlaminar shear strength (MPa)	
(hours)	RT	250°C	RT	250°C	RT	250°C
0	1400	1300	107,000	105,000	85	64
250	1100	1200	93,000	95,000	84	65
500	—	—	—	—	79	63
1000	1000	1000	89,000	103,000	62	56
1500	—	—	—	—	60	40

TABLE 2.21. Mechanical Properties at 20°C and 250°C of PSP 6030 Resin Laminates Made with Different Reinforcements (Reference 69)

Property (MPa)	Temperature (°C)	Carbon fiber				Glass fiber A1100	Kevlar Boron 49	fiber
		HT-S	A-S	HM-S	T300			
Flexural	320	1700	1400	1100	1800	1000	600	1900
strength	250	1250	1000	—	1480	1000	—	1900
Flexural	20	110,000	90,000	150,000	134,750	40,000	80,000	—
modulus	250	110,000	90,000	—	121,700	40,000	—	—
Shear strength	20	90	100	50	90	50	40	50
	250	70	70	40	62	40	—	50

It has since been realized that water pickup can have quite drastic effects upon the elevated temperature properties of polymers. Even if no chemical action is involved, absorbed water acts as a plasticizing agent, thus lowering the glass transition temperature of the polymer and hence its upper temperature limit of use. PSP/HT-S laminates have been tested by boiling in water, as well as by an exposure cycle that included a "thermal spike," i.e., exposure to 150°C for 10 minutes once during the course of a cycle.[69] The effect of boiling water immersion on flexural and shear strength at room temperature and at 250°C is shown in Table 2.22. An equilibrium water content of 0.7% was attained within 24 hours.

TABLE 2.22. Effect of Immersion in Boiling Water on the Mechanical Properties of PSP/HT-S Laminates (Reference 69)

Property (MPa)	Temperature (°C)	Time of immersion in boiling water (hours)			
		0	350	750	1000
Flexural strength	20	1650	1650	1600	1650
	250	1200	1200	—	1250
Shear strength	20	90	95	86	88
	250	70	55	—	55

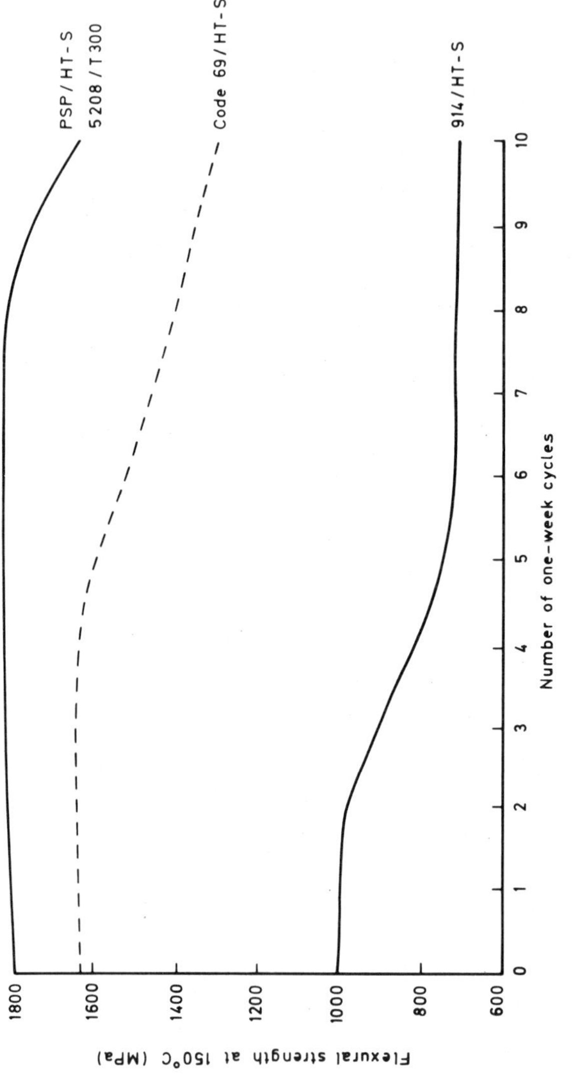

Figure 2.35. Flexural strength at 150°C of PSP/HT-S carbon fiber laminates after humidity cycling. Reference 69.

Flexural strength was not affected even by 1,000 hours immersion, whereas shear strength fell by about 20% in the first 100 hours and thereafter remained unchanged.

The cyclic exposure involved the following one-week cycle:

> 48 hour immersion in water at room temperature,
> 48 hour air drying at room temperature (10 min at
>   150°C),
> 72 hours in air at room temperature.

Figure 2.35 shows data for the flexural strength measured at 150°C of laminates that have received up to 10 weekly cycles, including a thermal spike.[69] There is, in fact, relatively little loss in strength for cycles with and without the spike. Figure 2.35 also includes curves for some high performance epoxy resin/carbon fiber composites for comparison purposes. The PSP/HT-S laminates show equally good behavior.

PHENOLIC FIBERS

Before leaving the area of thermosetting polymers, brief mention should be made of the fact that phenolic fibers are commercially available under the trade name Kynol. These are nonmelting and nonburning, and in a flame they give a 60% carbon residue and carbon dioxide and water as the major volatile products.[71] They can be used in blends with other fibers to impart flame resistance, their limiting oxygen index being 35–36. Figure 2.36 shows the weight loss of Kynol fibers in air at 150, 200, and 250°C. These curves indicate that oxidative degradation commences above 150°C and that this figure should be regarded as the upper temperature limit for long-term use. Some improvement can be obtained by the addition of compounds preventing peroxide formation. Table 2.23 gives fiber properties after aging at elevated temperatures in nitrogen and in air. Even after 1,000 hours at 250°C, or 100 hours at 350°C in inert atmosphere, the tensile strength and modulus are relatively unchanged, but there is a substantial reduction in the elongation at break.

SUMMARY

1. Various thermosetting resin systems are available, giving among them a whole range of properties, e.g., chemical resistance, good electrical characteristics, low fire hazards, low smoke emission, low pressures and temperatures for cure, coupled with good thermal stability. Each system has its own sphere of applications.

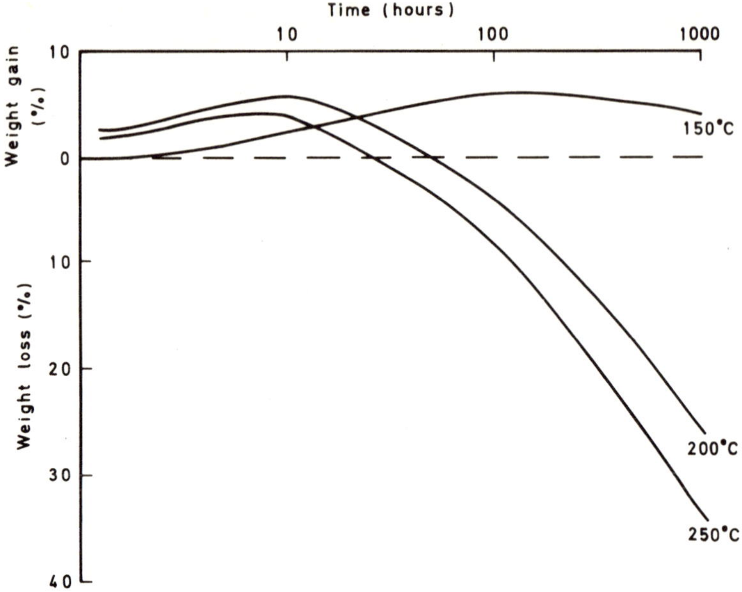

Figure 2.36. Isothermal weight loss of Kynol fibers in air at different temperatures. Reference 71.

2. The thermal stability can be increased by increasing the cross-link density, but this is normally accompanied by increased brittleness and hence a falloff in impact properties.

3. A disadvantage of thermosetting resins in general is that chemical reactions must take place during fabrication. Because of this, however, the starting

TABLE 2.23. Properties of Kynol Fibers after Aging at Elevated Temperatures (Reference 71)

Aging condition	Atmosphere	Tenacity (g/den)	Modulus (g/den)	Elongation at break (%)
Unaged		1.9	48	33.6
350 hours at 250°C	Nitrogen	1.9	44	7.3
500 hours at 250°C	Nitrogen	1.9	42	9.2
1000 hours at 250°C	Nitrogen	1.9	46	8.3
100 hours at 350°C	Nitrogen	1.8	43	6.8
500 hours at 150°C	Air	0.9	47	4.2
1200 hours at 150°C	Air	0.6	37	1.6
350 hours at 250°C	Air	0.7	35	2.9
500 hours at 250°C	Air	0.7	49	2.6
1000 hours at 250°C	Air	0.4	30	1.2

materials can be of low molecular weight and hence of low viscosity, giving good flow at low pressures.

4. Some of the thermosets cure by condensation reactions and this may create problems during fabrication because of the volatiles evolved. It may be difficult to produce thick sections unless high pressures or protracted cure cycles are used.

5. The heat resistance of systems such as the epoxies and polyesters, where substantial amounts of curing agent or cross-linking monomers are used, may be dominated by the structure of the hardener or comonomer used.

6. Advantages may accrue through the coreaction of two different thermosetting resins, or through the blending of a thermoplastic with a thermoset, and the development of further combinations of these types is likely.

REFERENCES

1. R. W. Martin, *The Chemistry of Phenolic Resins*, John Wiley & Sons, Inc., New York (1956).
2. N. J. L. Megson, *Phenolic Resin Chemistry*, Butterworths, London (1958).
3. D. F. Gould, *Phenolic Resins*, Reinhold Publishing Corp., New York (1959).
4. A. A. K. Whitehouse, E. G. K. Pritchett, and G. Barnet, *Phenolic Resins*, Iliffe Books Ltd., London (1967).
5. A. Knop and W. Scheib, *Chemistry and Application of Phenolic Resins*, Springer-Verlag, Berlin (1979).
6. J. R. Lawrence, *Polyester Resins*, Reinhold Publishing Corp., New York (1962).
7. H. V. Boenig, *Unsaturated Polyesters: Structure and Properties*, Elsevier, Amsterdam (1964).
8. B. Parkyn, F. Lamb, and B. V. Clinton, *Polyesters*, Vol. 2, *Unsaturated Polyesters and Polyester Plasticisers*, Iliffe Books Ltd., London (1967).
9. P. F. Bruins (ed.), *Unsaturated Polyester Technology*, Gordon and Breach, New York (1976).
10. I. Skeist, *Epoxy Resins*, Reinhold Publishing Corp., New York (1958).
11. H. Lee and K. Neville, *Handbook of Epoxy Resins*, McGraw-Hill, New York (1967).
12. P. F. Bruins (ed.), *Epoxy Resin Technology*, Interscience Publishers, New York (1968).
13. W. G. Potter, *Epoxide Resins*, Iliffe Books, London (1970).
14. R. F. Gould (ed.), *Epoxy Resins* (Advances in Chemistry Series No. 92), American Chemical Society, Washington (1970).
15. C. A. May and Y. Tanaka (eds.), *Epoxy Resins: Chemistry and Technology*, Marcel Dekker, New York (1973).
16. H. W. Lochte, E. L. Straus, and R. T. Conley, *J. Appl. Polym. Sci.* **9**, 2799 (1965).
17. R. T. Conley, Thermosetting resins, in *Thermal Stability of Polymers*, R. T. Conley, (ed.), Marcel Dekker, New York (1970), Vol. 1, Ch. 11, p. 457.
18. Reference 5, p. 149.
19. W. Brenner, D. Lum, and M. W. Riley, in *High-Temperature Plastics*, Reinhold Publishing Corp., New York (1962), p. 59.
20. H. J. Doyle and S. C. Harrier, Phenolics and Silicones, in *Handbook of Fibreglass and Advanced Plastics Composites*, G. Lubin, (ed.), Van Nostrand Reinhold Co., New York (1969), Ch. 4, p. 98.
21. Reference 19, p. 69.
22. J. F. Blais, *Amino Resins*, Reinhold Publishing Corp., New York (1959).

23. C. P. Vale and W. G. K. Taylor, *Aminoplastics,* Iliffe Books Ltd., London (1964).
24. C. P. Vale, Amino resins, in *Developments with Thermosetting Plastics,* A. Whelan and J. A. Brydson, (eds.), Applied Science Publishers, London (1975), Ch. 2, p. 13.
25. W. R. Moore and E. Donnelly, *J. Appl. Chem.* **13,** 537 (1963).
26. Reference 17, p. 505.
27. N. Fried, R. R. Winans, and L. E. Sieffert, *ASTM Proceedings* **50,** 1383 (1950).
28. B. M. Axilrod and M. A. Sherman, *J. Research Nat. Bur. St.* **44,** 65 (1950).
29. D. A. Anderson and E. S. Freeman, *J. Appl. Polym. Sci.* **1,** 192 (1959).
30. S. L. Madorsky and S. Straus, *Mod. Plast.* **38** (6), 134 (1961).
31. Reference 19, p. 85.
32. Reference 19, p. 86.
33. H. C. Anderson, *J. Appl. Polym. Sci.* **6,** 484 (1962).
34. Reference 11, pp. 6–36.
35. Reference 11, pp. 6–38 and 6–39.
36. Reference 11, pp. 17–22.
37. Reference 19, p. 47.
38. D. B. S. Berry, B. I. Buck, A. Cornwell, and L. N. Phillips, *Handbook of Resin Properties, Part A, Cast Resins,* Yarsley Testing Laboratories, Ashstead (1975).
39. H. Lee and K. Neville, Epoxy resins, in *Encyclopedia of Polymer Science and Technology,* H. F. Mark, N. G. Gaylord, and N. M. Bikales, (eds.), Interscience Publishers, New York (1967),Vol. 6, p. 259.
40. A. T. Radcliffe, Furane resins, in *Developments with Thermosetting Plastics,* A. Whelan and J. A. Brydson, (eds.), Applied Science Publishers, London (1975), Ch. 5, p. 58.
41. A. T. Radcliffe and T. J. Lens, *Reinforced Plastics Group (Plastics Institute) Conference,* New and Improved Resin Systems, London (1973), Paper 4.
42. P. A. Downing, *Chem. Eng. (London)* **331,** 272 (1978).
43. J. E. Selley, *Proc. 29th SPI Reinforced Plastics/Composites Institute Conference* (1974), Paper 23-A.
44. E. Fitzer, W. Schaefer, and S. Yamada, *Carbon* **7,** 643 (1969).
45. R. T. Conley and I. Metil, *J. Appl. Polym. Sci.* **7,** 1083 (1963).
46. H. Rembold, *Kunststoffe* **60,** 879 (1970).
47. R. E. Young, Vinyl ester resins, in *Unsaturated Polyester Technology,* P. F. Bruins, (ed.), Gordon and Breach, New York (1976), p. 315.
48. M. E. Kelley, Vinyl ester resin applications, in *Unsaturated Polyester Technology,* P. F. Bruins, (ed.), Gordon and Breach, New York (1976), p. 343.
49. T. F. Anderson and V. B. Messick, Vinyl ester resins, in *Developments in Reinforced Plastics—1,* G. Pritchard, (ed.), Applied Science Publishers, London (1980), Ch. 2, p. 29.
50. J. E. Carey and M. B. Launikitis, *Proc. 28th SPI Reinforced Plastics/Composites Institute Conference* (1973), Paper 8-A.
51. A. G. Edwards, Friedel-Crafts resins, in *Developments with Thermosetting Plastics,* A. Whelan and J. A. Brydson, (eds.), Applied Science Publishers, London (1975), Ch. 4, p. 41.
52. G. I. Harris, Phenol-aralkyl and related polymers, in *Developments in Reinforced Plastics—1,* G. Pritchard, (ed.), Applied Science Publishers, London (1980), Ch. 4, p. 87.
53. L. N. Phillips, RAE Technical Report CPM 3 (1963).
54. N. Grassie and I. G. Meldrum, *Eur. Polym. J.* **5,** 195 (1969).
55. N. Grassie and I. G. Meldrum, *Eur. Polym. J.* **6,** 499 (1970).
56. N. Grassie and I. G. Meldrum, *Eur. Polym. J.* **6,** 513 (1970).
57. N. Grassie and I. G. Meldrum, *Eur. Polym. J.* **7,** 17 (1971).
58. L. N. Phillips, U.K. Patent No. 1,094,181 (1967).

59. G. I. Harris and H. S. B. Marshall, U.K. Patent No. 1,099,123 (1968).
60. J. C. Paxton, Ministry of Defense (Procurement Executive), D. Mat. Report 190 (1973).
61. G. I. Harris and F. Coxon, U.K. Patent No. 1,150,203 (1969).
62. G. I. Harris, *Br. Polym. J.* **2**, 270 (1970).
63. G. I. Harris and A. G. Edwards, U.K. Patent No. 1,305,551 (1973).
64. N. Grassie and I. G. Meldrum, *Eur. Polym. J.* **7**, 1253 (1971).
65. B. Ellis and P. G. White, *Br. Polym. J.* **9**, 15 (1977).
66. Trade Brochure on Xylok 210, Albright and Wilson Ltd. (1971).
67. M. Ropars and B. Bloch, *La Recherche Aérospatiale* **2**, 103 (1977).
68. M. Ropars, B. Bloch, and B. Malassine, Paper presented at the Fifth European Conference on Plastics and Rubbers, Paris (1978).
69. B. Bloch and M. Ropars, *23rd Natl. SAMPE Symposium* **23**, 836 (1978).
70. PSP 6022 Resin and Preimpregnated High Performance Fibers, Provisional Technical Data Sheet, Société Nationale des Poudres et Explosifs.
71. J. Economy and L. Wohrer, Phenolic fibers, in *Encyclopedia of Polymer Science and Technology,* H. F. Mark, N. G. Gaylord, and N. M. Bikales, (eds.), Interscience Publishers, New York (1971), Vol. 15. p. 365.

SUPPLEMENTARY BIBLIOGRAPHY

General

Heat resistant thermosetting polymers—a brief review, S. Oswitch, *Rein. Plast.* **19**, 180 (1975) and **19**, 215 (1975).
Thermally stable organic matrices for use in composites, N. C. W. Judd and W. W. Wright, *Rep. Prog. Appl. Chem.* **59**, 87 (1975).
High temperature resistant engineering plastics—properties, processing, and applications, H. Domininghaus, *Kunststoffe* **69**, 1 (1979).
Recent advances in the properties and applications of thermosetting materials, International Conference, Plastics & Rubber Institute, Coventry (1979).
High-temperature properties of thermally stable resins, G. J. Knight, *Developments in Reinforced Plastics—1,* G. Pritchard, (ed.), Applied Science Publishers, London (1980), Ch. 6, p. 145.

Phenolic Resins

Les résines phénoliques modifiés par des additions minérales, S. Kohn, ONERA Technical Publication No. 324 (1966).
Phenolic resins, W. A. Keutgen, in *Encyclopedia of Polymer Science and Technology,* H. F. Mark, N. G. Gaylord, and N. M. Bikales, (eds.), Interscience Publishers, New York (1969), Vol. 10, p. 73.
Ways of increasing the heat resistance of phenol-formaldehyde polymers and materials based on them, D. V. Gvozdev, A. B. Blyumenfeld, B. M. Kovarskaya, M. S. Akutin, and Y. M. Budnitskii, *Plast. Massy* (1) 28 (1980), translated in *International Polymer Science and Technology* **7**, T/37 (1980).

Melamine-Formaldehyde Resins

Amino resins, G. Widmer, in *Encyclopedia of Polymer Science and Technology*, H. F. Mark, N. G. Gaylord, and N. M. Bikales, (eds.), Interscience Publishers, New York (1965), Vol. 2, p. 1.

Polyester Resins

Polyesters unsaturated, H. V. Boenig, in *Encyclopedia of Polymer Science and Technology*, H. F. Mark, N. G. Gaylord, and N. M. Bikales, (eds.), Interscience Publishers, New York (1969), Vol. 11, p. 129.

Epoxy Resins

Epoxy resins, H. Lee and K. Neville, in *Encyclopedia of Polymer Science and Technology*, H. F. Mark, N. G. Gaylord, and N. M. Bikales, (eds.), Interscience Publishers, New York (1967), Vol. 6, p. 209.

The thermal degradation of epoxide resins, D. P. Bishop and D. A. Smith, *Ind. Eng. Chem.* **59,** 32 (1967).

Further aspects of the thermal degradation of epoxide resins, M. A. Keenan and D. A. Smith, *J. Appl. Polym. Sci.* **11,** 1009 (1967).

Furan Resins

Furan polymers, K. J. Siegfried, in *Encyclopedia of Polymer Science and Technology*, H. F. Mark, N. G. Gaylord, and N. M. Bikales, (eds.), Interscience Publishers, New York (1967), Vol. 7, p. 432.

Fiberglass reinforced furan composites—a unique combination of properties, K. B. Bozer, L. H. Brown, and D. D. Watson, *Proc. 26th SPI Reinforced Plastics/Composite Institute Conference* (1971), Paper 2-C.

High temperature and combustion properties of furan composites, K. B. Bozer and L. H. Brown, *Proc. 27th SPI Reinforced Plastics/Composite Institute Conference* (1972), Paper 3-C.

Vinyl Ester Resins

DERAKANE 470-45, a new high temperature corrosion resistant resin, T. E. Cravens, *Proc. 27th SPI Reinforced Plastics/Composites Institute Conference* (1972), Paper 3-B.

Vinyl ester resins, P. Varco and M. J. Seamark, *Reinforced Plastics Group (Plastics Institute) Conference,* New and Improved Resin Systems, London (1973), Paper 3.

New high performance corrosion resistant resin, R. J. Lewandowski, E. C. Ford, D. M. Longenecker, A. J. Restaino, and J. P. Burns, *Proc. 30th SPI Reinforced Plastics/Composites Institute Conference* (1975), Paper 6-B.

A new heat resistant vinyl ester resin, M. B. Launkitis, *Proc. 31st SPI Reinforced Plastics/ Composites Institute Conference* (1976), Paper 15-C.

Phenol-Aralkyl Resins

Friedel-Crafts Resin/Carbon Fiber Composites.
    Part 1. A preliminary assessment, B. M. Parker, RAE Tech. Report 70200 (1970).
    Part 2. Toluene and terphenyl resins, B. M. Parker, RAE Tech. Report 72029 (1972).
    Part 3. Xylene resins, B. M. Parker, RAE Tech. Report 72220 (1972).
    Part 4. Diphenyl oxide resins, B. M. Parker, RAE Tech. Memo Mat. 217 (1975).
    Part 5. Chemical resistance, B. M. Parker, RAE Tech. Report 75115 (1975).
    Part 6. Mechanical properties, B. M. Parker, RAE Tech. Report 76051 (1976).
Friedel-Crafts thermosetting resins, B. M. Parker and L. N. Phillips, *Reinforced Plastics Group (Plastics Institute) Conference,* New and Improved Resin Systems, London (1973), Paper 8.
Friedel-Crafts resin composites for hostile environments, G. I. Harris, *Reinforced Plastics Group (Plastics Institute) Conference,* New and Improved Resin Systems, London (1973), Paper 12. Also published by G. I. Harris, A. G. Edwards, and B. G. Huckstepp, in *Plastics and Polymers,* (December 1974), p. 239.
Xylok resins—their properties and applications, G. I. Harris, A. G. Edwards, and F. Coxon, *Chimica Petrolchimica* **6,** 403 (1976).
Xylok resins—a high performance family of resins, G. Buchi and R. Kultzow, Paper presented at SPE Antech Conference on High Performance Plastics, Cleveland, 1976.
Reinforced phenol-aralkyl resin composites for demanding applications, G. I. Harris and A. G. Edwards, Paper presented at The Reinforced Plastics Congress, Brighton, 1978.

PSP Resins

Flammability, smoke, and smoke gas properties of materials made with PSP 6030 type resins, B. Malassine, *23rd Natl. SAMPE Symposium* **23,** 929 (1978).
PSP 6022 resin, a solution for the electrical problems posed by potential release of free carbon/graphite fibers into the environment, B. Malassine, *24th Nat. SAMPE Symposium* **24,** 1 (1979).
High temperature reinforced plastic radome manufacturing by an injection technique using PSP resin, B. Bloch, ONERA TP No. 1980-68, Paper presented at the 15th Symposium on Electromagnetic Windows, Atlanta, June 18–20, 1980.
Fabrication of heat-resistant composite materials by injection of PSP resin into a glass cloth reinforcement, B. Bloch, ONERA TP No. 1980-88, Paper presented at the 3rd International Conference on Composite Materials, Paris, August 26–29, 1980.
SEM contribution to the study of the fracture behavior of composites with a PSP thermostable resin, J. P. Favre and M. Ropars, ONERA TP. No. 1980-89, Paper presented at the 3rd International Conference on Composite Materials, Paris, August 26–29, 1980.

3

FLUORINE-CONTAINING POLYMERS

INTRODUCTION

The synthesis of fluorine-containing polymers and their development as possible heat-resistant materials has been largely prompted by the very high thermal stability of polytetrafluoroethylene (PTFE), whose preparation was first reported in a patent dated 1941.[1] The thermal stability of PTFE has been attributed to the high carbon-fluorine bond strength (487 kJ/mole in CF_4 compared with 418 kJ/mole for C-H bonds in CH_4)[2] and to the shielding effect that the highly electronegative fluorine atoms have on the carbon backbone. The commercial introduction of PTFE was followed in the early 1950s by that of polychlorotrifluoroethylene, and in the 1960s by that of polyvinylidene fluoride and polyvinyl fluoride. A thermoplastic copolymer is also available based upon tetrafluoroethylene and hexafluoropropene. Copolymer formulations have been especially productive in giving elastomers, and the following combinations are the basis of commercial products: vinylidene fluoride/chlorotrifluoroethylene, vinylidene fluoride/hexafluoropropene, vinylidene fluoride/hexafluoropropene/ tetrafluoroethylene, propene/tetrafluoroethylene, pentafluoropropene/tetrafluoroethylene, perfluoromethylvinylether/tetrafluoroethylene. An hepta-fluorobutylacrylate-based elastomer has also been made. It will be noted that all the polymers so far mentioned are produced by vinylic addition reactions. Condensation polymers—polyhexafluoropentylene adipate and polyhexafluoropentylene adipate/isophthalate—were marketed by the Hooker Electrochemical Company, but only for a short time and only in development quantities. It will also be noted that all the polymers cited contain fluoro-alkyl groups; there are no fluoroaromatic or fluoro-heterocyclic moieties present, despite considerable research work on such materials. The main reason for this is that such data as are available

TABLE 3.1. Thermal Stabilities of Some Fluorine and Nonfluorine-Containing Model Compounds (References 3, 4)

Compound	$T_D{}^a$ (°C)	Compound	$T_D{}^a$ (°C)
$C_6F_5C_6F_5$	538	$C_6H_5C_6H_5$	543
$p\text{-}C_6F_5C_6F_4C_6F_4C_6F_5$	392	$p\text{-}C_6H_5C_6H_4C_6H_4C_6H_5$	426
$m\text{-}C_6F_5C_6F_4C_6F_4C_6F_5$	425	$m\text{-}C_6H_5C_6H_4C_6H_4C_6H_5$	458
$p\text{-}C_6F_5OC_6F_4C_6F_4OC_6F_5$	360	$p\text{-}C_6H_5OC_6H_4C_6H_4OC_6H_5$	439–453
$p\text{-}C_6F_5SC_6F_4SC_6F_5$	328	$p\text{-}C_6H_5SC_6H_4SC_6H_5$	365
$m\text{-}C_6F_5SC_6F_4SC_6F_5$	324	$m\text{-}C_6H_5SC_6H_4SC_6H_5$	375
$p\text{-}C_6F_5NHC_6F_4C_6F_4NHC_6F_5$	325	$p\text{-}C_6H_5NHC_6H_4C_6H_4NHC_6H_5$	411
$p\text{-}C_6F_5OOCC_6F_4COOC_6F_5$	318–324	$p\text{-}C_6H_5OOCC_6H_4COOC_6H_5$	353
$m\text{-}C_6F_5OOCC_6F_4COOC_6F_5$	300	$m\text{-}C_6H_5OOCC_6H_4COOC_6H_5$	362
$C_6F_5OCH_2C_6F_4CH_2OC_6F_5$	227	$C_6H_5OCH_2C_6H_4CH_2OC_6H_5$	221

^aT_D is the temperature at which the rate of pressure increase in an isoteniscope is 0.84 mm Hg/minute.

on perfluoroaromatic compounds and polymers indicate that there is little difference between their thermal stabilities and those of unfluorinated analogs; in fact, the latter frequently appear to be more stable. Table 3.1 compares the thermal stabilities of some perfluoroaromatic compounds and their unfluorinated analogs derived from isoteniscope measurements of the temperatures (T_D) at which the rate of pressure increase due to decomposition is 0.84 mm Hg/minute.

Accordingly, the remainder of this chapter is devoted to the aliphatic fluorine-containing polymers and copolymers. It is divided into two sections; the first deals with thermoplastic, and the second with elastomeric products.

It has proved more convenient to deal with those fluorine-containing polymers in which fluorine substitution is merely a variant on a particular structure elsewhere in this review. Mention of fluorine-containing materials will, therefore, also be found in the sections covering polyheteroaromatics, polysiloxanes, poly(carborane-siloxanes), and polyphosphazenes.

FLUORINE-CONTAINING THERMOPLASTICS

Preparation

The polymers to be considered are polytetrafluoroethylene, the copolymer of tetrafluoroethylene and hexafluoropropene, polychlorotrifluoroethylene, and polyvinylidene fluoride. Brief mention will also be made of polyvinylfluoride for comparison purposes and because it is commercially available, although it should not be classed as a heat-resistant polymer, as its upper-use temperature is only about 110°C.

The preferred methods for polymerization of tetrafluoroethylene involve aqueous media and pressures in the range 0.7 to 6.9 MPa. Suitable initiators include sodium, potassium, and ammonium persulphates, oxygen, hydrogen peroxide, and some organic peroxy compounds.[5-7] Oxidation-reduction systems based on persulphates with either bisulphites or ferrous ions have also been used. The product is a linear polymer with a high degree of crystallinity (93 to 98%). Number average molecular weight estimates derived from end-group determinations place the polymers of commercial interest in the range 400,000 to 10,000,000. Polytetrafluoroethylene (PTFE) is made in France, Germany, Italy, Japan, the U.K., the U.S.A., and the U.S.S.R. under a variety of trade names, e.g., Algoflon, Fluon, Ftorlon 4, Halon, Hostaflon, Polyflon, Soreflon, Teflon, and Tetran, and is supplied in granular, fine powder, or aqueous dispersion form.

A copolymer of tetrafluoroethylene and hexafluoropropene resulted from the search for a material possessing the desirable properties of PTFE, but which was easier to process. It can be made under similar conditions to those used for the production of PTFE.[6] Heavy metal fluorides, such as AgF_2, AsF_3, CoF_3, and PbF_4 have also proved effective initiators of copolymerization over the temperature range 0 to 150°C. Careful control of copolymer composition is necessary to maintain a balance between mechanical properties and satisfactory melt viscosities for processing. As polymerized, the material has a degree of crystallinity of about 70%, but this is reduced to 40–60% after processing. Injection molding, extrusion, superior stress-crack resistant and aqueous dispersion grades have been marketed. The main difference between these is in molecular weight and hence melt viscosity. The base copolymer is available from Dupont under the trade name Teflon FEP.

Chlorotrifluoroethylene is polymerized commercially using bulk, suspension, or emulsion techniques.[7,8] Bulk polymerization generally takes place in small static reactors with halogenated acyl peroxides as initiators. Suspension polymerization is carried out in water with either inorganic or organic peroxides as initiators. Bisulphites, together with copper, iron, or silver compounds as promoters may also be used. Pressures of 0.4 to 1.7 MPa and temperatures of 20–80°C are normal for both suspension and emulsion polymerizations. The latter employs inorganic peroxides as initiators and highly halogenated acid salts as emulsifying agents. The polymers have degrees of crystallinity of 40 to 70%. They are produced in France, Germany, Japan, the U.S.A., and the U.S.S.R. under the trade names Aclar, Daiflon, Ftorlon 3, Hostaflon, Kel-F, Plaskon, and Voltalef.

Polymerization of vinylidene fluoride can be carried out in suspension, emulsion, or solution at pressures between 1.2 to 31 MPa and temperatures from 10 to 150°C.[7,9] In suspension, aqueous recipes may be used with or without colloidal dispersants and with organic percarbonate or peroxy compounds as initiators. Emulsion polymerization requires the use of a chemically stable fluorinated surfactant and persulphates or peroxides as initiators. Solution polymerization is possible in a variety of solvents and with a number of free radical initiators. The product has a crystalline content of about 68% and is marketed by Pennwalt Corporation under the trade name Kynar.

In contrast to some of the monomers already cited, vinyl fluoride is quite difficult to polymerize and normally requires pressures of the order of 5 to 104 MPa.[7,10] Suspension and emulsion techniques have been favored, although Ziegler–Natta catalysis has also been investigated. The latter technique gives polymers with degrees of crystallinity between 40 and 50%. Dupont sells polyvinylfluoride as Tedlar.

Thermal Stability

A comparison is made in Figures 3.1 and 3.2 of the thermal stabilities of these fluorine-containing polymers in vacuum and oxygen, respectively.[11,12] The comparison is based upon the weight lost in 2 hours at constant temperature; the exception is for polyvinylfluoride where the data of Madorsky *et al.*[13] have been used; this is for heating to pyrolysis temperature over 45 minutes followed by 30 minutes at temperature. In vacuum the order of thermal stability is:

$$[CF_2CF_2]_n \approx [CF_2CF_2/CF_3CFCF_2]_n > [CF_2CH_2]_n > [CF_2CFCl]_n$$

The curve for $[CHFCH_2]_n$ is intermediate between those of $[CF_2CH_2]_n$ and $[CF_2CFCl]_n$, but if the results had been obtained under the same experimental conditions, the curve for $[CHFCH_2]_n$ would be displaced towards lower temperatures. In oxygen the polymers behave quite differently. Whereas polytetrafluoroethylene and polychlorotrifluoroethylene show little change in stability in oxidizing conditions, the copolymer of tetrafluoroethylene and hexafluoropropene and polyvinylidene fluoride are very adversely affected. This is perhaps more clearly demonstrated in Table 3.2, which quotes the temperatures at which 25% weight loss occurs either in vacuum or under 300 mm oxygen pressure.

The poorer stability of the copolymer can be attributed to the chain branching and the presence of tertiary fluorine atoms—polyhexafluoropropene itself decomposes below 300°C.[14] The lower thermo-oxidative stability of polyvinylidene fluoride can be ascribed to the presence of methylene groups. Polyvinylfluoride would be expected to have even lower stability. Table 3.3 gives the available kinetic data for these polymers in inert and oxidizing atmospheres.

Because of its commercial importance, unique properties, and very high stability, the thermal breakdown of polytetrafluoroethylene has been widely investi-

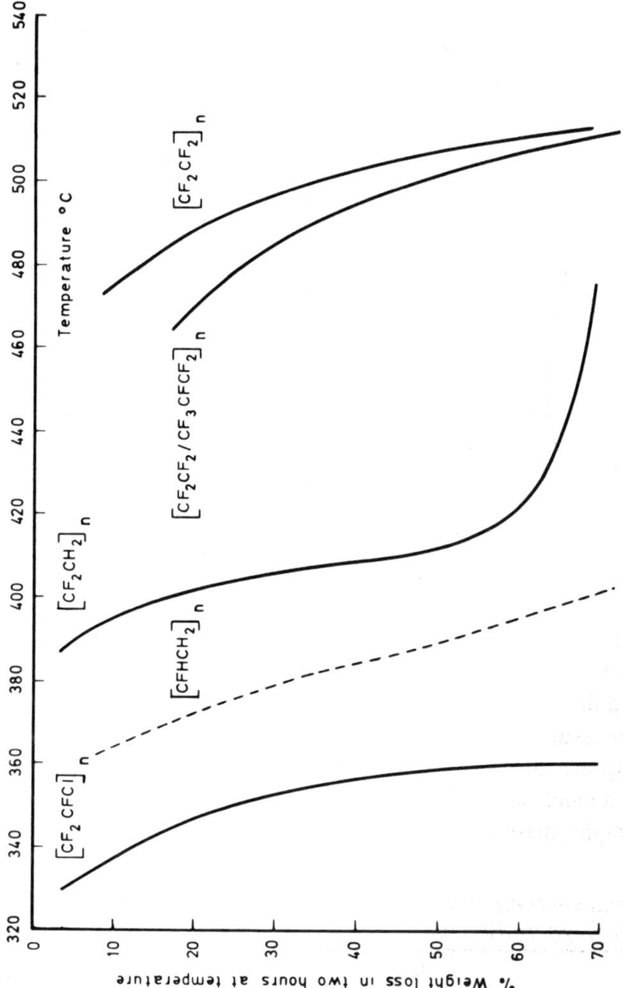

Figure 3.1. Comparison of thermal stabilities of fluorine-containing polymers in vacuum. (Data for [CFHCH₂]ₙ is for 45 minutes of heating to temperature followed by 30 minutes at temperature.)

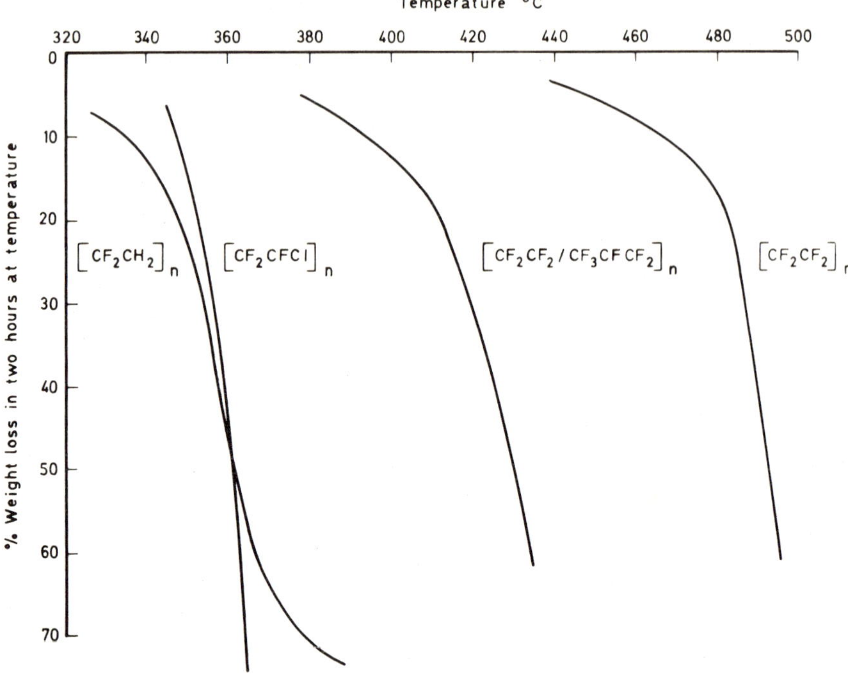

Figure 3.2. Comparison of thermal stabilities of fluorine-containing polymers in oxygen.

gated and discussed. The position up to 1971 is summarized in Wall's book on fluoropolymers.[15] Most studies in inert atmospheres have concluded that decomposition is a first-order reaction. The products of decomposition, though dependent upon pressure, are relatively independent of temperature. In vacuum, over the temperature range 500–800°C, the major product is tetrafluoroethylene (90–97%) with a much smaller amount (3–6%) of hexafluoropropene. Pyrolysis-gas chromatography-mass spectrometry shows that there are also small amounts

TABLE 3.2. Relative Thermal Stabilities of Fluorine-Containing Polymers in Vacuum and Oxygen (References 11, 12)

Polymer	Temperature (°C) for 25% weight loss in two hours		Temperature difference (°C)
	In vacuum	In oxygen	
Polytetrafluoroethylene	494	482	−12
Copolymer of tetrafluoroethylene and hexafluoropropene	481	417	−64
Polyvinylidenefluoride	403	354	−49
Polychlorotrifluoroethylene	349	355	+6

TABLE 3.3. Kinetic Data for Degradation of Fluorine-Containing Polymers in Vacuum and Oxygen (References 11, 12)

Polymer	Atmosphere	Activation energy (kJ/mole)	Arrhenius factor (sec^{-1})	Rate of weight loss at 350°C (% per min)
Polytetrafluoroethylene	Vacuum	319	10^{18}	2×10^{-6}
Polytetrafluoroethylene	Oxygen	332	10^{20}	3×10^{-6}
Copolymer of tetrafluoroethylene and hexafluoropropene	Vacuum	307	10^{18}	2×10^{-6}
Copolymer of tetrafluoroethylene and hexafluoropropene	Oxygen	277	10^{18}	1.2×10^{-3}
Polyvinylidene fluoride	Vacuum	298	10^{20}	5×10^{-3}
Polyvinylidene fluoride	Oxygen	122	10^{10}	3.4×10^{-1}
Polychlorotrifluoroethylene	Vacuum	210	10^{15}	3.0×10^{-1}
Polychlorotrifluoroethylene	Oxygen	277	10^{21}	2.2×10^{-1}
Polyvinylfluoride	Vacuum	223	10^{12}	2×10^{-2}

of up to 34 perfluoroalkanes, perfluoroalkenes, and perfluorocyclic compounds.[16] As the pressure is increased, the yield of monomer falls and that of dimer increases. With the proper choice of pressure, temperature, and residence time in the furnace, high yields of hexafluoropropene and perfluoro-isobutene (toxic) can be obtained. The course of the reaction is changed if the reaction products are allowed to accumulate. The nature of the surface in contact with the polymer can also have an influence. The molecular weight of the polymer, as measured by changes in specific gravity, decreases during pyrolysis.

Mass spectrometric studies have also been made of the volatile products of degradation in air or oxygen.[16–19] In complete contrast to the results in inert atmospheres is the virtual absence of monomer. The main oxidation products are carbonyl fluoride (highly toxic), carbon dioxide, tetrafluoromethane, and silicon tetrafluoride, the last named arising from reaction of fluorine-containing compounds with the glass of the apparatus. Carbonyl fluoride (COF_2) is the major product below about 650°C; at higher temperatures it disproportionates to yield CO_2 and CF_4. The activation energy for evolution of COF_2 in air is 183–205 kJ/mole. Oxidative degradation has also been studied under ignition, or burning conditions.[18,19] The auto-ignition temperature of PTFE under the particular flow conditions used was 575°C in air and 512°C in oxygen. In air, prior to combustion, both saturated and unsaturated fluorocarbons were formed, together with COF_2 and CO_2. Once combustion was established, COF_2, CO, and saturated fluorocarbons were the most abundant species.[19] When PTFE was burnt in a flow of oxygen, it was suggested that elemental fluorine was present in the gas close to the decomposing surface.[18]

Various attempts have been made to improve the thermal stability of PTFE. These include (i) polymerization in the presence of fluorocarbon catalysts or polymerization by photochemical means in order to eliminate possible labile end-groups, (ii) the inclusion of a number of different structural units in the polymer chain to promote chain transfer of the free radicals active in depolymerization, (iii) the inclusion of molecules capable of promoting chain transfer.[20-22] Only approach (iii) showed any measure of success, and this involved pyrolysis in the presence of gases such as ClF_3 and IF_5, which dissociated at the pyrolysis temperatures to give fluorine atoms.

Decomposition of the copolymer of tetrafluoroethylene (TFE) and hexafluoropropene (HFP) shows that it has a molar composition of approximately 80% TFE and 20% HFP. The main decomposition products in an inert atmosphere are the same as observed with PTFE, i.e., tetrafluoroethylene, hexafluoropropene, and perfluorocyclobutane.[16] Figure 3.3 compares the yield of these three compounds as a function of temperature for both polymers. The main differences between the two are that the copolymer begins to decompose about 100°C lower than PTFE and that HFP is produced in much larger quantities at all degradation temperatures. Above about 650°C, the fluorocarbons are reacting with the glass of the reaction vessel to produce SiF_4, CO, and CO_2, and this accounts for the falloff in product yield shown in the figure. A simultaneous thermogravimetry-mass spectrometry experiment indicates that the copolymer decomposes in two stages. At temperatures between 450 and 525°C, hexafluoropropene is by far the major component in the volatile products. Above 525°C, TFE, HFP, and cyclo-C_4F_8 are all being formed simultaneously. The activation energies for evolution of HFP in the first and second stages are 296 kJ/mole and 360 kJ/mole, respectively, and the order of reaction changes from 1.0 to 0.5. The order of reaction for production of TFE in the second stage is also 0.5, and the activation energy (354 kJ/mole) is very similar to that for HFP.

In air or oxygen the major degradation products are COF_2 and CO_2, with relatively small amounts of CF_4, TFE, and HFP. A two-stage reaction was again observed in a simultaneous thermogravimetry-mass spectrometry experiment, but the activation energy for formation of COF_2 was essentially the same in both stages—202 kJ/mole.

In contrast to the fully fluorinated polymers, the thermal degradation in vacuum of polychlorotrifluoroethylene yields only 28% of breakdown products that are volatile at room temperature.[23] These comprise largely monomer, together with small amounts of C_3F_5Cl and $C_3F_4Cl_2$. The remainder of the degradation products, of average molecular weight 900, are volatile at pyrolysis temperatures, but not at room temperature. It is probable with this polymer that a chlorine atom is first eliminated, followed by chain scission and the production of some monomer by a depolymerization reaction. Transfer reactions involving the liberated chlorine atoms are also liable to occur both intermolecularly and

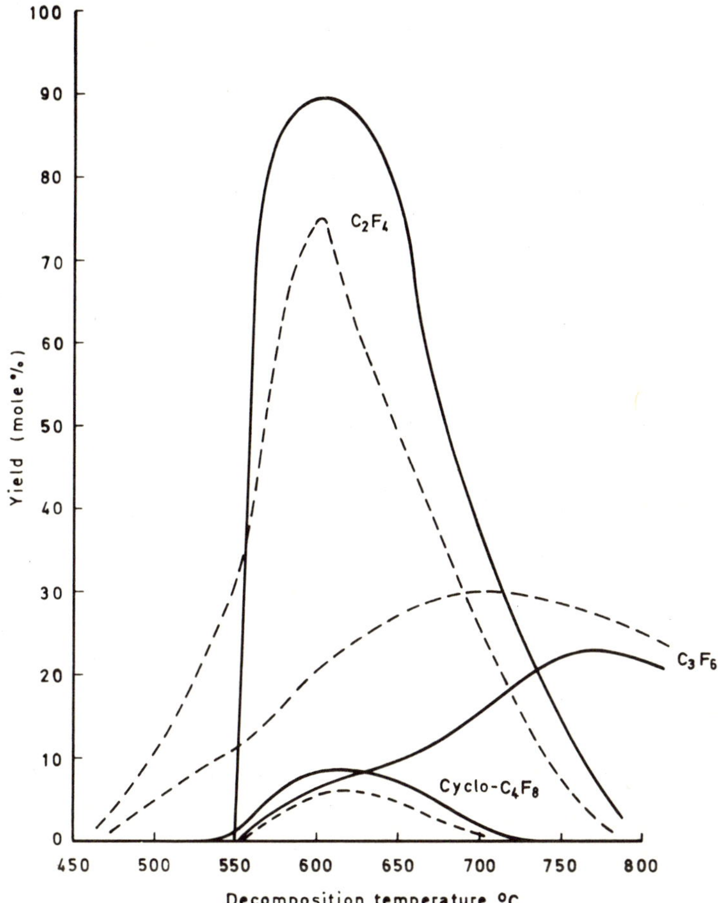

Figure 3.3.   Comparison of product yields for decomposition of PTFE (——) and a copolymer of TFE and HFP (---) in helium. Reference 16.

intramolecularly. Infrared studies have shown that in vacuum $-CF = CF_2$ groups are formed, which confirms the elimination of chlorine atoms.[24] The products probably contain compounds of the type $Cl[C_2F_3Cl]_n$ $CF = CF_2$. In air, the initially formed $-CF = O$ and $-CCl = O$ groups subsequently hydrolyze to carbonyl groups.

The hydrofluoro polymers degrade by two simultaneous processes: chain scission and elimination of hydrogen fluoride (HF) to give an unsaturated chain. Whereas polyvinylfluoride breaks down completely to volatiles on pyrolysis in vacuum, polyvinylidene fluoride stabilizes at about 65% weight loss.[13] This is related to the more rapid formation of conjugated double bonds in the latter case.

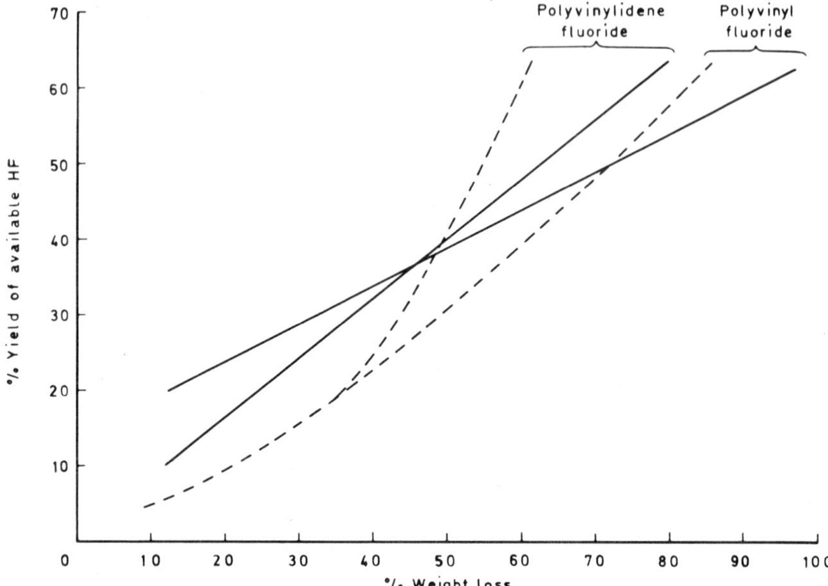

Figure 3.4. Comparison of HF yields from hydrofluoroethylene polymers obtained by two experimental techniques. Mass spectrometer vacuum ——, fluoride ion specific electrode nitrogen ----. Reference 25.

The production of HF has been monitored indirectly by mass spectrometric analysis of the SiF_4 formed by reaction between HF and the SiO_2 of the glass apparatus.[13] It has also been measured directly using a fluoride ion selective electrode.[25] Figure 3.4 shows the yields of HF as a function of total weight loss obtained by the two different experimental techniques. Considering the tremendous difference in experimental procedure, the agreement is very good. A comparison of the HF yields for polyvinylfluoride and polyvinylidene fluoride in air and nitrogen determined with the fluoride ion electrode is given in Table 3.4.

TABLE 3.4. HF Yields from Hydrofluoroethylene Polymers (Reference 25)

| Polymer | Atmosphere | Temperature (°C) | | Final total yield of HF(%) |
		Initial HF yield	1% HF yield	
Polyvinylfluoride	Air	153	276	93.7
	Nitrogen	177	337	70.8
Polyvinylidene fluoride	Air	214	386	75.4
	Nitrogen	284	396	70.5

Fabrication

Difficulties in processing are encountered with many fluoropolymers. Polytetrafluoroethylene has a very high melt viscosity even at 380°C and cannot be fabricated by conventional injection or extrusion methods. Instead, techniques must be employed that are similar to those used for ceramics or powdered metals. Compression molding involves compression of the dry powder at ambient temperatures using pressures of 10–40 MPa, followed by sintering at 365–385°C and controlled cooling from these temperatures to impart the desired degree of crystallinity to the final product. In contrast, the copolymer of tetrafluoroethylene and hexafluoropropene can be processed by normal injection molding and melt-extrusion procedures.

Compression molding at 250–275°C and pressures up to 21 MPa are preferred for polychlorotrifluoroethylene if the best physical properties are to be obtained. Care must also be taken in the cooling of moldings, as this governs the degree of crystallinity and hence the physical properties. Slow cooling results in a material with higher specific gravity, greater hardness, and greater strength at elevated temperatures. Polychlorotrifluoroethylene can also be processed by injection molding or melt extrusion, but these may necessitate a prior reduction in the molecular weight of the polymer.

All the normal fabrication processes are possible with polyvinylidene fluoride. Temperatures and pressures used are in the ranges 175–215°C and 7–70 MPa, respectively. Polyvinylfluoride, however, presents problems because of the high degree of order and the large number of hydrogen bonds present. This results in a melting point that is close to the decomposition temperature. Successful processing requires the use of so-called "latent solvents," which disrupt the molecular order and hence lower the temperature necessary for fabrication. There are no practical room temperature solvents for polyvinylfluoride, and the "latent solvents" solvate the polymer only at elevated temperatures.

Elevated Temperature Properties

The maximum recommended service temperatures for continuous use of these fluoropolymers are presented in Table 3.5.

Intermittent use is possible some 25°C above these values. A comparison of some of the mechanical properties of the three most stable of the polymers at room and elevated temperatures is made in Tables 3.6–3.8.

Polytetrafluoroethylene (Table 3.6) possesses useful properties over one of the widest temperature ranges (over 500°C) known for any polymer. It can be used up to 260°C and yet remains tough down to −273°C. It is insoluble in all

TABLE 3.5. Maximum Service Temperatures for Continuous Use of Fluoropolymers

Polymer	Maximum temperature (°C) for continuous use
Polytetrafluoroethylene	260
Tetrafluoroethylene/hexafluoropropene copolymer	205
Polychlorotrifluoroethylene	175
Polyvinylidene fluoride	150
Polyvinylfluoride	110

TABLE 3.6. Some Mechanical Properties of Polytetrafluoroethylene (Reference 6)

Property	Temperature (°C)	Value (MPa)
Tensile yield strength	23	9.0
Tensile yield strength	70	5.5
Tensile yield strength	120	3.5
Tensile modulus	23	345
Tensile modulus	100	69
Flexural modulus	23	621
Flexural modulus	55	400
Flexural modulus	100	193
Compressive yield strength	23	11.7
Compressive yield strength	55	9.0
Compressive yield strength	100	4.8

TABLE 3.7. Some Mechanical Properties of Tetrafluoroethylene/Hexafluoropropene Copolymer (Reference 6)

Property	Temperature (°C)	Value (MPa)
Tensile yield strength	23	12.4
Tensile yield strength	70	6.9
Tensile yield strength	120	3.5
Tensile modulus	23	414
Tensile modulus	100	69
Flexural modulus	23	655
Flexural modulus	55	345
Flexural modulus	100	110
Flexural modulus	200	41
Compressive yield strength	23	15.2
Compressive yield strength	55	11.0
Compressive yield strength	100	3.5

TABLE 3.8. Some Mechanical Properties of Polychlorotrifluoroethylene (Reference 8)

		Value for sample cooled	
Property	Temperature (°C)	Quickly	Slowly
Tensile strength (MPa)	23	38.6	37.3
	70	20.7	23.8
	125	3.6	3.7
Tensile modulus (MPa)	23	1056	1277
	70	400	572
	125	37.3	104
Flexural modulus (MPa)	23	1311	1753
	70	373	1028
	125	89.7	221
Flexural strength (MPa)	23	53.8	73.8
	70	16.2	34.2
	125	4.8	11.4
Elongation at break (%)	23	190	125
	70	375	390
	125	>450	>450

solvents and hence highly resistant to chemical attack. It has high dielectric strength, low dielectric constant and loss factor, good antistick and frictional properties. One drawback is cold flow—the polymer yields under stress (9–14 MPa) at room temperature. This can be alleviated to some extent by the use of fillers. The copolymer (Table 3.7) can also be used over a wide temperature range (about 450°C), the upper limit being determined by lack of stiffness and not thermal degradation. Like polytetrafluoroethylene, it is chemically inert and has a low coefficient of friction. It has very good weathering properties.

The presence of the chlorine atom in the polymer (Table 3.8) reduces its upper temperature limit of use compared with the fully fluorinated polymers. Polychlorotrifluoroethylene has the lowest known water permeability of all thermoplastics.

FLUORINE-CONTAINING ELASTOMERS

Preparation

The various types of fluorine-containing elastomers that have attained some commercial promotion are listed in Table 3.9.

The polyheptafluorobutylacrylate (1F4) and the two Hooker condensation polyesters containing relatively long aliphatic hydrocarbon sequences do not

TABLE 3.9. Fluorine-Containing Elastomers

Polymer	Principal repeating units	Trade names
Polyheptafluorobutylacrylate	$-CH_2CH(COOCH_2C_3F_7)-$	1F4
Vinylidene fluoride/chloro-trifluoroethylene copolymer	$-CH_2CF_2/CF_2CFCl-$	KEL—F, VOLTALEF SKF
Vinylidene fluoride/hexafluoro-propene copolymer	$-CH_2CF_2/CF_3CFCF_2-$	VITON, FLUOREL TECNOFLON, SKF
Vinylidene fluoride/pentafluoro-propene copolymer	$-CH_2CF_2/CF_3CFCFH-$	TECNOFLON
Tetrafluoroethylene/propene copolymer	$-CF_2CF_2/CH_3CHCH_2$	AFLAS
Vinylidene fluoride/hexafluoro-propene/tetrafluoroethylene terpolymer	$-CH_2CF_2/CF_3CFCF_2/CF_2CF_2-$	VITON: DAI—EL
Vinylidene fluoride/pentafluoro-propene/tetrafluoroethylene terpolymer	$-CH_2CF_2/CF_3CFCFH/CF_2CF_2-$	TECNOFLON
Tetrafluoroethylene/perfluoro-methylvinylether/3rd monomer terpolymer	$-CF_2CF_2/CF_3OCFCF_2/-$	KALREZ
Polyhexafluoropentylene adipate	$-[(CH_2)_4COOCH_2(CF_2)_3CH_2OCO]-$	
Polyhexafluoropentylene adipate/isophthalate	$-[(CH_2)_4COOCH_2(CF_2)_3CH_2OCO$ $C_6H_4COOCH_2(CF_2)_3CH_2OCO]-$	HOOKER POLYESTERS

have the same high temperature stability as do the more highly fluorinated polymers and were only produced commercially for a short time. They will therefore not be considered further.

The remaining polymers, together with the polysiloxanes, comprise the most thermally stable elastomeric systems currently available. They are all prepared by free-radical emulsion polymerization using peroxy initiators, temperatures of 80–125°C, and pressures of 2–10 MPa. Persulphate/bisulphite recipes and fluorinated emulsifiers such as ammonium perfluorooctanoate are commonly used. Molecular weight is controlled by variation of the monomers to initiator ratio or by the use of chain transfer agents. Careful control of the composition of the co- or terpolymers is necessary if optimum elastomeric properties are to be attained. With too great a proportion of any of the monomers, the products are plastics rather than rubbers.

Thermal Stability

A comparison is made in Figures 3.5 and 3.6 of the thermal stabilities of the fluorine-containing elastomers in nitrogen and air, respectively, based upon the weight lost in two hours at constant temperature. The comparison is not strictly

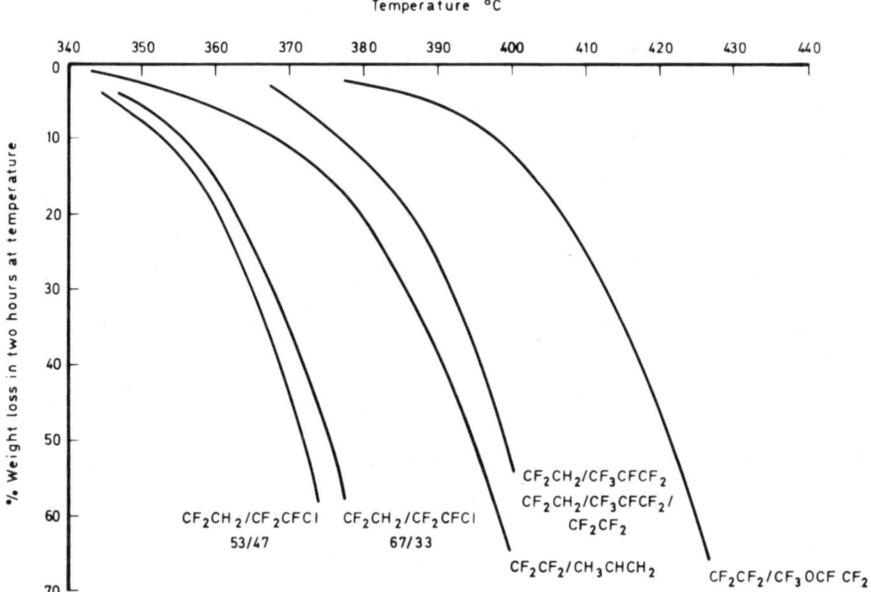

Figure 3.5.   Comparison of thermal stabilities of fluoroelastomers in nitrogen.

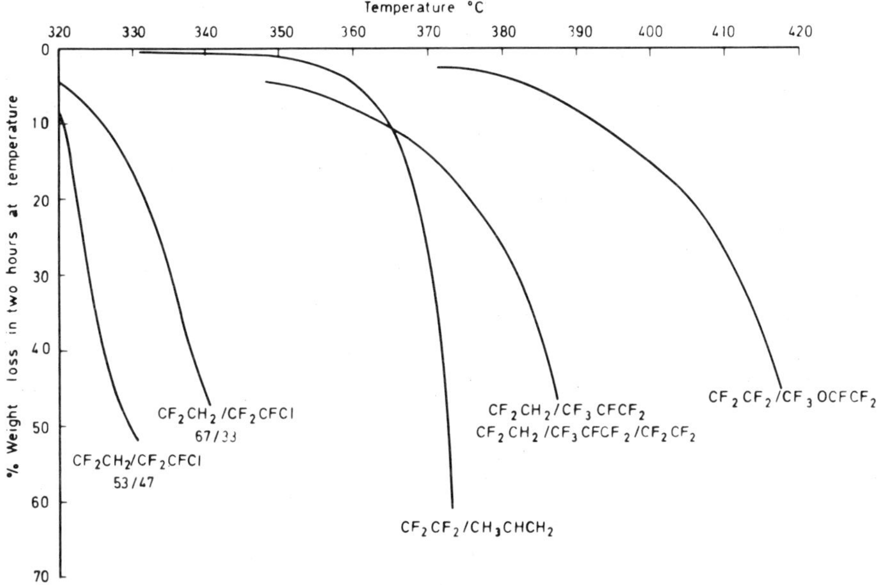

Figure 3.6.   Comparison of thermal stabilities of fluoroelastomers in air.

TABLE 3.10. Relative Thermal Stabilities of Fluorine-Containing Elastomers in Nitrogen and Air

Copolymer	Temperature (°C) for 25% weight loss in 2 hours		Temperature difference (°C)
	In nitrogen	In air	
CF_2CF_2/CF_3OCFCF_2	410	409	−1
CF_2CH_2/CF_3CFCF_2	389	379	−10
$CF_2CH_2/CF_3CFCF_2/CF_2CH_2$	389	379	−10
CF_2CF_2/CH_3CHCH_2	383	370	−13
$CF_2CH_2/CF_2CFCl[67/33]$	365	334	−31
$CF_2CH_2/CF_2CFCl[53/47]$	363	323	−40

valid for the CF_2CF_2/CF_3OCFCF_2 material, as this is only supplied in compounded form. The order of thermal stability is the same in both inert and oxidizing atmosphere and is $CF_2CF_2/CF_3OCFCF_2 > CF_2CH_2/CF_3CFCF_2 = CF_2CH_2/CF_3CFCF_2/CF_2CF_2 > CF_2CF_2/CH_3CHCH_2 > CF_2CH_2/CF_2CFCl$. The chlorine-containing copolymers are more adversely affected by the presence of oxygen (Table 3.10) than the nonchlorine-containing materials. This is in contrast to the behavior of polychlorotrifluoroethylene itself (see Table 3.2).

Table 3.11 gives the available kinetic data for these polymers derived from weight loss curves.

With respect to products of decomposition, only the copolymers CF_2CH_2/CF_3CFCF_2 and CF_2CH_2/CF_2CFCl have been studied in detail. Wall and Straus[26] showed that the CF_2CH_2/CF_3CFCF_2 copolymer broke down to 95% volatiles at 500°C. Of these decomposition products, about 30–40% were volatile at room temperature and comprised largely hydrogen fluoride (HF). Degteva et al.,[27] working at lower temperatures, found only a slight weight loss (~1%)

TABLE 3.11. Kinetic Data for Degradation of Fluorine-Containing Elastomers

Copolymer	Atmosphere	Activation energy (kJ/mole)	Arrhenius factor (sec^{-1})	Rate of weight loss at 350°C (% per min)
CF_2CF_2/CF_3OCFCF_2	Nitrogen	261	10^{21}	2×10^{-3}
CF_2CF_2/CF_3OCFCF_2	Air	281	10^{20}	4×10^{-3}
CF_2CH_2/CF_3CFCF_2	Vacuum	193	10^{13}	4×10^{-2}
CF_2CH_2/CF_3CFCF_2	Oxygen	(176)	(10^{11})	3×10^{-2}
$CF_2CH_2/CF_3CFCF_2/CF_2CF_2$	Vacuum	189	10^{13}	5×10^{-2}
$CF_2CH_2/CF_3CFCF_2/CF_2CF_2$	Oxygen	—	—	$\sim 7 \times 10^{-2}$
CF_2CF_2/CH_3CHCH_2	Nitrogen	198	10^{15}	3×10^{-2}
CF_2CF_2/CH_3CHCH_2	Air	177	10^{15}	2.6×10^{-1}
$CF_2CH_2/CF_2CFCl[67/33]$	Vacuum	227	10^{16}	6×10^{-2}
$CF_2CH_2/CF_2CFCl[67/33]$	Oxygen	155	10^{11}	6.9×10^{-1}
$CF_2CH_2/CF_2CFCl[53/47]$	Vacuum	210	10^{15}	1.2×10^{-1}
$CF_2CH_2/CF_2CFCl[53/47]$	Oxygen	151	10^{11}	1.3

after 40 hours at 320°C, with 0.1% HF being evolved. Extensive cross-linking of the gumstock had, however, occurred. The copolymer is completely destroyed between 360 and 400°C, yielding 20% of products volatile at room temperature (SiF_4, CF_3H, CF_2CH_2, fluorocarbons of unknown structure, CO, CO_2) and 70% of products of molecular weight 500–600 containing isolated and conjugated $—CF = CH —$ bonds. When the copolymer is cross-linked by reaction with amines or peroxides to form an elastomer, the thermal stability is decreased, the effect being most pronounced with an amine cure.[28] The effect of curing conditions upon the yield of HF from the copolymer will be discussed in more detail later.

The copolymers of CF_2CH_2 and CF_2CFCl also yield HF as the major light volatile product, hydrogen chloride (HCl) being present in much smaller quantities; at 400°C the ratio is about 7 to 1.[26] Degteva *et al.* have made an intensive investigation of the degradation of a CF_2CH_2/CF_2CFCl copolymer both in vacuum and oxygen.[29–33] In vacuum there is little change until 250°C, when evolution of the hydrogen halides commences together with cross-linking of the chains. Above 300°C chain scission occurs and the polymer is converted to a black, viscous mass. Between 250 and 360°C the relative abundance of HCl and HF in the products changes—at 250–300°C the ratio is 2 HCl/1 HF, whereas at 360°C the ratio is 3 HF/1 HCl. The hydrogen halide evolution can be profoundly affected by the presence of the oxides of aluminum, iron, zinc, cobalt, or titanium, all of which accelerate the rate. In the presence of oxygen, the principal degradation products are the hydrogen halides and gaseous and liquid low (500) molecular weight compounds.

Elastomer Formulation

The curing systems of greatest importance for the copolymers containing vinylidene fluoride as one of the components are either aliphatic diamine derivatives in combination with basic metal oxides, or aromatic dihydroxy compounds in combination with basic metal oxides and hydroxides and strong alkyl, or aryl bases, or their derivatives.[34] The most commonly used amine derivatives are hexamethylenediamine carbamate:

$$\left[{}^{+}H_3N - (CH_2)_6 - N {\overset{\displaystyle COO^-}{\underset{\displaystyle H}{\Big\langle}}} \right]$$

and N,N′-dicinnamylidene-1,6-hexanediamine:

$$\left[\begin{array}{c} \text{C}_6\text{H}_4 \end{array}-\text{CH}=\text{CH}-\text{CH}=\text{N(CH}_2)_6\text{N}=\text{CH}-\text{CH}=\text{CH}-\text{C}_6\text{H}_4\right]$$

Detailed studies have been made of the mechanism of cure of fluoroelastomers with a diamine in conjunction with a metallic oxide such as MgO.[35-36] It is considered to occur in three stages:

1. Bases, e.g., metallic oxide, react with the polymer chains and create unsaturation by the elimination of HF:

$$\begin{array}{ccc} | & & | \\ \text{CH}_2 & -\text{HF} & \text{CH} \\ | & & \| \\ \text{CF}_2 & \longrightarrow & \text{CF} \\ | & & | \\ \text{CH}_2 & & \text{CH}_2 \\ | & & | \end{array}$$

2. The diamines react with the unsaturated groups forming cross-links as shown:

$$\begin{array}{ccc} | & & | \qquad\qquad | \\ \text{CH} & & \text{CH}_2 \qquad\quad \text{CH}_2 \\ | & & | \qquad\qquad | \\ 2\;\;\text{CF} & +\;\text{H}_2\text{N(CH}_2)_6\text{NH}_2 \longrightarrow & \text{CF}-\text{NH(CH}_2)_6\text{NH}-\text{CF} \\ | & & | \qquad\qquad | \\ \text{CH}_2 & & \text{CH}_2 \qquad\quad \text{CH}_2 \\ | & & | \qquad\qquad | \end{array}$$

3. During post-cure further dehydrofluorination occurs at these cross-link sites, producing unsaturation, which may further react:

$$\begin{array}{ccc} | \qquad\qquad | & & | \qquad\qquad | \\ \text{CH}_2 \qquad\quad \text{CH}_2 & & \text{CH}_2 \qquad\quad \text{CH}_2 \\ | \qquad\qquad | & & | \qquad\qquad | \\ \text{CF}-\text{HN(CH}_2)_6\text{NH}-\text{CF} & \xrightarrow{-2\,\text{HF}} & \text{C}=\text{N(CH}_2)_6\text{N}=\text{C} \\ | \qquad\qquad | & & | \qquad\qquad | \\ \text{CH}_2 \qquad\quad \text{CH}_2 & & \text{CH}_2 \qquad\quad \text{CH}_2 \\ | \qquad\qquad | & & | \qquad\qquad | \end{array}$$

Water in the system, produced by the reaction of HF with the metallic oxide, is simultaneously removed.

$$2\,\text{HF} + \text{MgO} \longrightarrow \text{MgF}_2 + \text{H}_2\text{O}$$

In contrast to the above, the tetrafluoroethylene-propene copolymer is cured using organic peroxides. Other peroxide curing fluoroelastomers have recently been described.[37]

The elastomer based upon tetrafluoroethylene and perfluoro-(methylvinylether) contains a third monomer to facilitate cross-linking,[38] e.g., $CF_2{=}CF{-}O{-}(CF_2)_4CN$, $CF_2{=}CF{-}O{-}(CF_2)_4{-}COOCH_3$, $C_6F_5{-}O{-}CF(CF_3){-}CF_2{-}O{-}CF{=}CF_2$, or $C_6F_5{-}O{-}(CF_2)_3{-}O{-}CF{=}CF_2$. Diamines may be used to cross-link the polymers containing perfluoro-phenoxy- and carbomethoxy- groups, but for those containing cyano groups, tetraphenyl tin or silver oxide catalysts are used.

In addition to the curing agent and metallic oxide, a typical formulation contains a filler and a processing aid. A representative formulation follows[39]:

Elastomer	100 parts by weight
Metal oxide	3–20 parts by weight
Filler	5–50 parts by weight
Curing agent	1–3 parts by weight
Cure promoter	1.5–6 parts by weight
Processing aid	0.5–3 parts by weight

The function of the metallic oxide during cure has already been outlined. It also serves to reduce the amount of HF evolved during subsequent service at elevated temperatures. A number of oxides have been used, e.g., ZnO, PbO, MgO, and CaO, each having advantages and disadvantages, but MgO is generally reckoned to give the best heat resistance. The most common filler is a medium thermal carbon black, as this gives the best balance of physical properties, compression set, and processability.

Cure promoters such as calcium hydroxide are used with bisphenol curing agents. Processing aids, when used, may be very low molecular weight fluoroelastomers or polyethylenes.

At high temperatures fluoroelastomers eliminate HF as a degradation product, and this can create problems such as stress-corrosion if the fluorocarbon is in contact with a metal such as titanium. The effect of variation in the formulation of a vinylidene fluoride/hexafluoropropene/tetrafluoroethylene terpolymer upon HF evolution has been studied in some detail over the temperature range 200–275°C in air.[40] The rate of HF evolution is affected by the diamine curing agent, the carbon black filler, and the acid acceptor used, and their effects are interrelated. Cross-linking the gumstock with a diamine promotes HF evolution. The initial rate of HF production from the cross-linked material is greater than that from the base polymer, and the rate remains essentially unchanged over the time scale of the experiments. In contrast, the rate of evolution from the gumstock decreases steadily with time (Figure 3.7). The addition of 20% of a medium thermal black to the cross-linked polymer produces a further slight reduction in stability. Dissimilar blacks behave differently, and Figure 3.8 shows the effect of various blacks upon the HF yield from formulations that also contain 15% MgO. It can be seen that the channel and furnace blacks have relatively

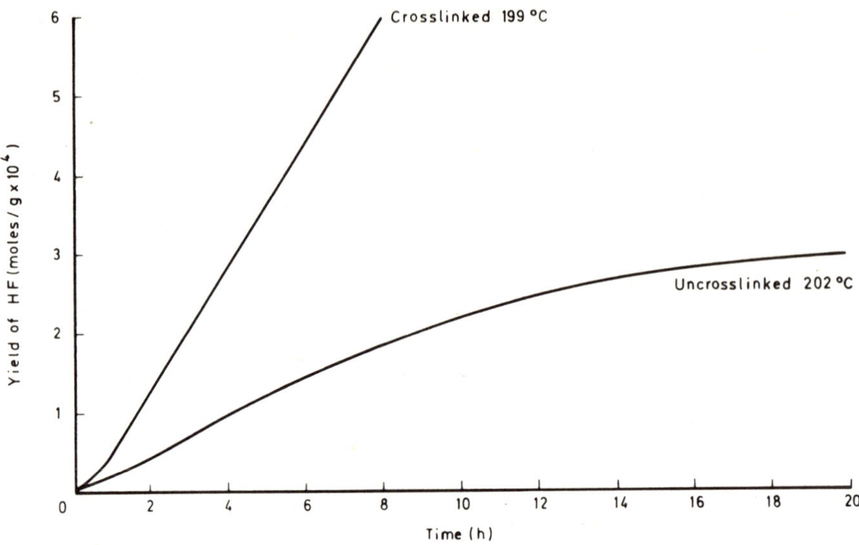

Figure 3.7. Comparison of HF yield from a vinylidene fluoride/hexafluoropropene/tetra-fluoroethylene terpolymer before and after cross-linking with a diamine. Reference 40.

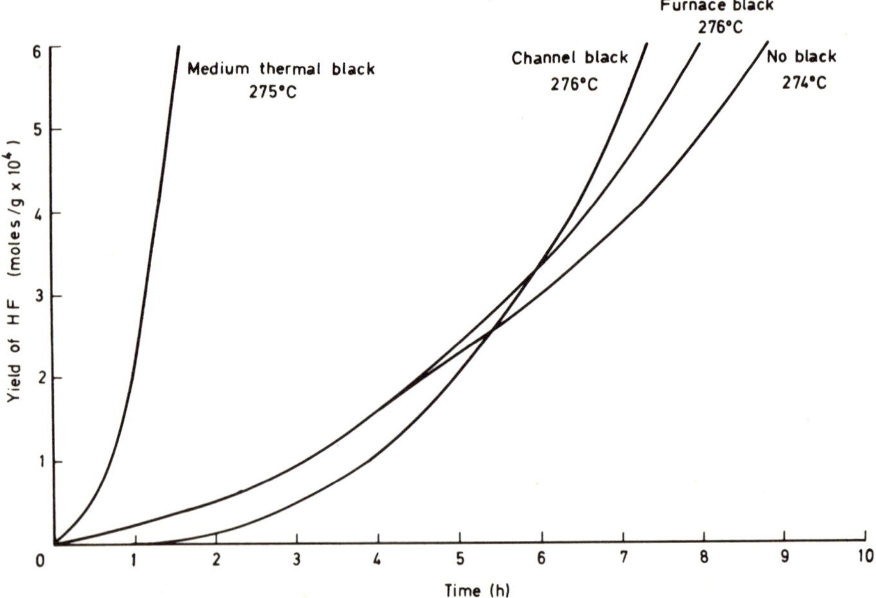

Figure 3.8. Comparison of HF yield from a vinylidene fluoride/hexafluoropropene/tetra-fluoroethylene terpolymer containing different carbon blacks. Reference 40.

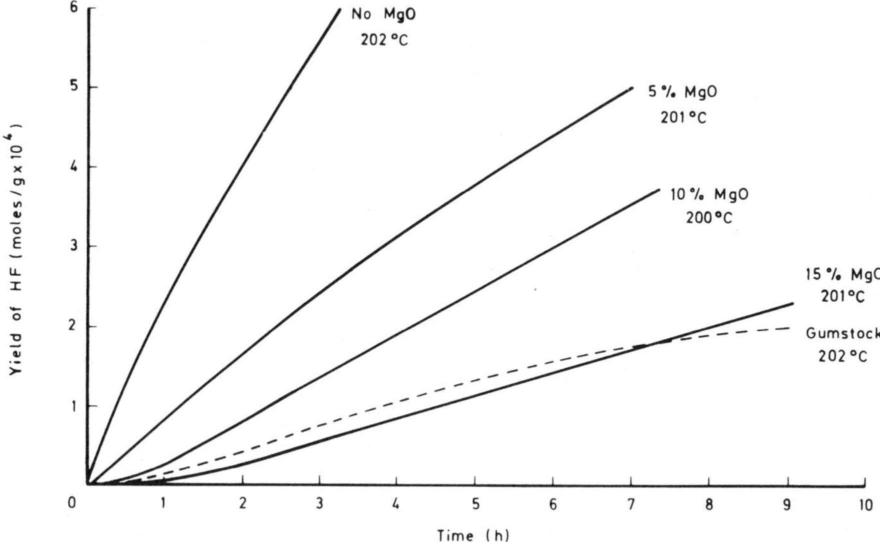

Figure 3.9. *Comparison of HF yield from a vinylidene fluoride/hexafluoropropene/tetra-fluoroethylene terpolymer containing different amounts of MgO. Reference 40.*

little effect on the efficiency of MgO as an acid acceptor, the results being very little different from those for a formulation containing no black. This is certainly not the case for the medium thermal black, where the stability is quite severely reduced. These differences must be related to the differing chemical nature of the surface of the blacks. The presence of MgO as an acid acceptor does not prevent HF elimination; it merely reduces its rate of evolution, and a loading of 15% is necessary to give a result comparable with that of the gumstock alone (Figure 3.9).

Basic oxides, hydroxides, and carbonates other than those of magnesium may be used as HF acceptors, and experiments show that a number of these, when used in conjunction with a medium thermal black, give far better stability. The relative efficiencies of these alternatives at three different temperatures (taking MgO as 1.0) are compared in Table 3.12. These compounds do not function solely as HF acceptors, but also play a vital part in the curing reaction, and it must be ascertained that in their presence an adequate cross-link density is attained. When this parameter is measured, it transpires that the use of CaO, $Ca(OH)_2$, $Mg(OH)_2$, $CaCO_3$, Li_2CO_3, $MgCO_3$, Na_2CO_3, and K_2CO_3 results in a molecular weight between cross-links very similar to that realized with MgO. This finding in conjunction with the data of Table 3.12 establishes that $Mg(OH)_2$, $MgCO_3$, CaO, $Ca(OH)_2$, $CaCO_3$, and Li_2CO_3 are the preferred additives when HF evolution is to be minimized. In the temperature range 250–275°C, $Ca(OH)_2$ gives the optimum results of the compounds studied,

TABLE 3.12. Comparison of Efficiencies of Compounds as HF Acceptors (Reference 40)

Additive (15% by weight)	Relative efficiency compared with MgO at		
	275°C	250°C	225°C
MgO	1.0	1.0	1.0
TiO_2	0.2	0.2	0.2
Al_2O_3	1.5	1.7	1.8
BeO	1.5	2.2	1.6
B_2O_3	2.1	0.8	2.7
Li_2O	1.5	3.4	3.4
CaO	25.6	>48	38.8
$Al(OH)_3$	2.2	2.7	2.2
LiOH	6.9	10.6	32.2
$Mg(OH)_2$	37.0	44.3	31.6
$Ca(OH)_2$	42.5	>64	54.7
K_2CO_3	0	0.1	0.7
Na_2CO_3	2.6	3.4	17.6
$MgCO_3$	12.2	25.5	115.3
Li_2CO_3	21.4	32.4	137.6
$CaCO_3$	41.3	56.6	87.4

reducing HF evolution by a factor of ~50 compared with MgO. At lower temperatures, $MgCO_3$ or Li_2CO_3 would be preferred, as they show more than a hundredfold improvement over MgO.

As might be surmised from the above results, when a number of different fluoroelastomers are examined the evolution of HF from them is more affected by the compounding of the gumstock than by the chemical nature of the gumstock itself.[41] One fact which has emerged from as yet unpublished data is that low yields of HF appear to result if a polymer is cured with a peroxide or a bisphenol rather than with a diamine (Figure 3.10). This may be related to the mechanism of diamine cure already outlined.

Elevated Temperature Properties

A very good summary of the properties of the elastomers based upon vinylidene fluoride/chlorotrifluoroethylene copolymers and vinylidene fluoride/hexafluoropropene copolymers has been made by Stivers.[42] A selection of data has been made from his review. Table 3.13 compares properties of two vinylidene fluoride/chlorotrifluoroethylene elastomers before and after aging at 205°C.

Different curing agents were used for the two elastomers—benzoyl peroxide for the Kel-F 5500 and hexamethylenediamine-carbamate for the Kel-F

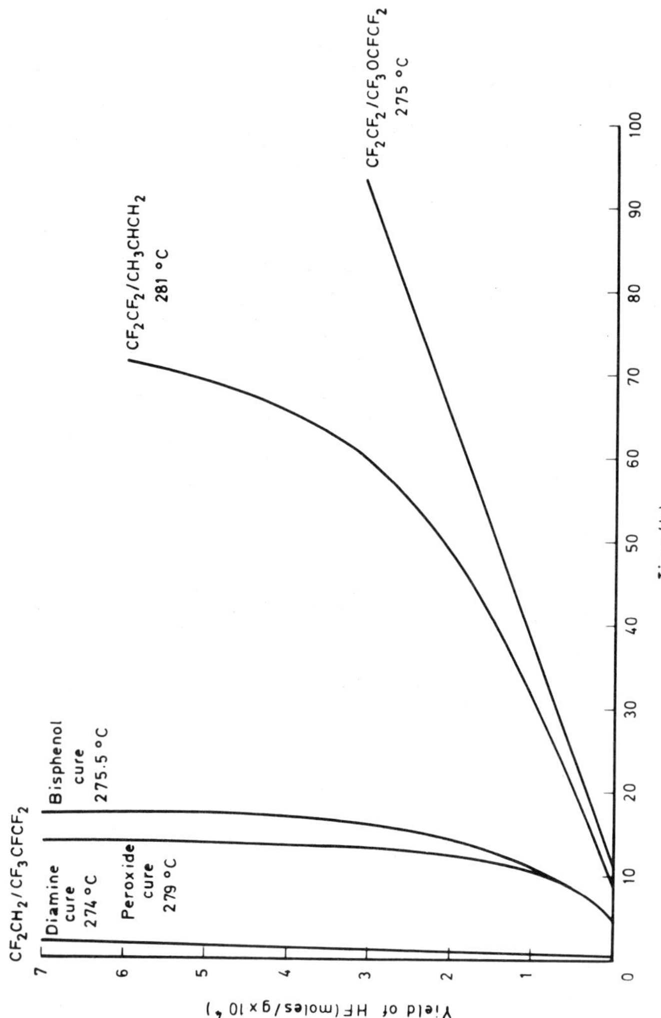

Figure 3.10. Comparison of HF yield from different compounded fluoroelastomers.

TABLE 3.13. Effect of Heat Aging on Properties of Vinylidene Fluoride Chlorotrifluoroethylene Elastomers (Reference 42)

Property	KEL-F 5500		KEL-F 3700	
	Original value	After heating at 205°C for 14 days	Original value	After heating at 205°C for 14 days
Tensile strength (MPa)	16.8	9.5	15.9	17.5
Elongation (%)	410	630	300	490
Shore A hardness	60	69	60	70

3700—otherwise the formulations and press and oven cure were identical. The results indicate that the Kel-F 3700 has superior strength retention. Its compression set resistance is also superior; figures of 25 and 40% are obtained after 70 hours at 120 and 150°C, respectively, whereas the comparable figures for the Kel-F 5500 are 50 and 60%. The Kel-F type elastomers have very good resistance to strong oxidizing acids but are not as easy to process as those based on vinylidene fluoride/hexafluoropropene copolymers. The latter also have superior physical properties. Some typical values are given in Table 3.14.

All the figures in Tables 3.13 and 3.14 are for tests made at room temperature after aging at the temperatures stated. One is more often interested in the properties at temperature, and Table 3.15 gives some data for Viton B.

There is a rapid loss in tensile strength as the temperature is increased from 25 to 260°C, and this is accompanied by a loss of 12 points in Shore A hardness.

If elastomers are to be used in seal applications, an important feature is resistance to compression set. This is greatly affected in some of the fluoroelastomers by the time and temperature of oven post-cure. Table 3.16

TABLE 3.14. Effect of Heat Aging on Properties of Vinylidene Fluoride Hexafluoropropene Elastomers (Reference 42)

Elastomer	Aging condition	Tensile strength (MPa)	Elongation (%)	Shore A hardness
Fluorel 2140	—	16.6	310	68
	15 days at 260°C	10.0	150	82
	16 hours at 315°C	9.2	265	70
Viton A	—	15.0	470	68
	100 days at 230°C	6.9	160	87
	20 days at 260°C	8.6	100	94
	2 days at 315°C	7.2	60	91
Viton B	—	15.5	410	74
	100 days at 230°C	4.3	480	75
	20 days at 260°C	3.8	400	83
	2 days at 315°C	3.5	240	83
	1 day at 345°C	4.0	15	91

TABLE 3.15. Properties of Viton B at Elevated Temperature
(Reference 42)

Temperature°C	Tensile strength (MPa)	Elongation (%)	Shore A hardness
25	16.9	330	75
150	3.5	120	65
260	2.1	80	63

TABLE 3.16. Effect of Post-Cure Conditions on the Compression Set Resistance
of a Viton A Elastomer (Reference 42)

Post-cure 24 hours at	Compression set % after 24 hours at						
	−30°C	0°C	70°C	120°C	177°C	205°C	230°C
177°C	100	43	17	15	38	—	—
205°C	100	33	17	13	17	41	95
260°C	100	53	15	10	9	20	60

shows the variation in compression set of a Viton A formulation at different temperatures as a function of post-cure conditions.

Compression set also depends upon the duration of a test, and the effect of compression time is given in Table 3.17.

Because of the importance of this property, considerable effort has been devoted to the production of fluoroelastomers with lower compression set. This has been achieved by the use of curing systems based on calcium hydroxide and specially prepared hydroquinone. The improvement that has resulted from the use of these systems is illustrated in Table 3.18.

The compression set resistance of the best of these systems as a function of temperature is depicted in Figure 3.11. Some typical figures for another low compression set elastomer, also cured with a bisphenol, are given in Table 3.19.

Recently a family of peroxide-curable fluoroelastomers has been developed[37] that, though capable of vulcanization at atmospheric pressure, remains essentially void-free. These do not have the good compression set resistance of the LCS type, but still possess excellent thermo-oxidative stability (Table 3.20).

TABLE 3.17. Effect of Compression Time on the Compression Set Resistance
of a Viton A Elastomer (Reference 42)

Compression time	Compression set % at a temperature of							
	−30°C	−18°C	24°C	70°C	120°C	177°C	205°C	230°C
1 Day	100	54	18	17	13	17	41	85
3 Days	100	62	20	—	15	30	80	95
7 Days	100	59	22	20	20	50	—	95

TABLE 3.18. Improvement in Compression Set Resistance with Cure System (Reference 43)

Elastomer type	Cure system	Compression set (%) after 70 hours at 200°C
Viton A	Hexamethylenediamine-carbamate	47
Viton A-HV	Hexamethylenediamine-carbamate	35
Viton A	Hydroquinone/Super 6	35
Viton A-HV	Hydroquinone/Super 6	24
Viton E-60	Hydroquinone/Super 6	24
Viton E-60C	New cure system	13

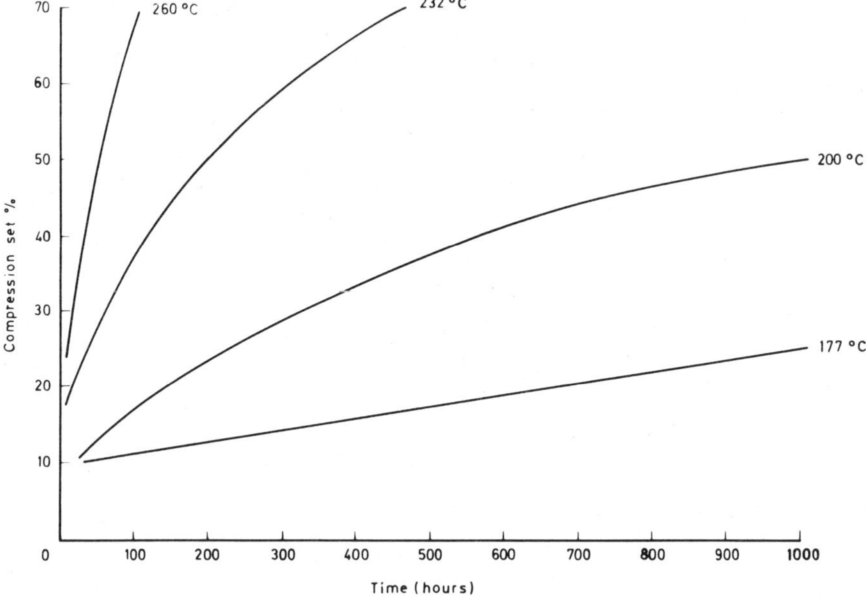

Figure 3.11. Compression set resistance of Viton E-60C at elevated temperatures. Reference 43.

TABLE 3.19. Effect of Heat Aging on the Properties of an LCS Elastomer Fluorel 2160 (Reference 42)

Aging condition	Tensile strength (MPa)	Elongation (%)	Shore A hardness
—	12.8	185	67
70 hours at 250°C	11.7	185	68
	Compression set (%)		
70 hours at 25°C	4		
168 hours at 175°C	7		
70 hours at 200°C	12		
168 hours at 200°C	19		

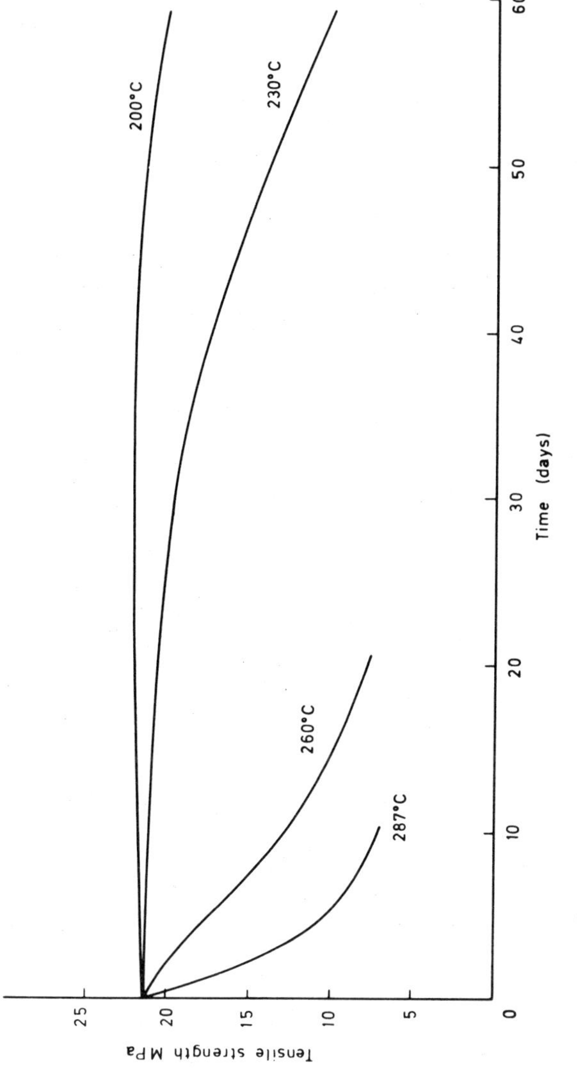

Figure 3.12.   Effect of heat aging on the tensile strength of Aflas 100 vulcanizates.

TABLE 3.20. Effect of Heat Aging on the Properties of a Peroxide-Cured Fluoroelastomer FKM B-910 (Reference 37)

Aging condition	Tensile strength (MPa)	Elongation (%)	Durometer A hardness
—	13.8	175	78
70 hours at 275°C	11.4	220	74
	Compression set (%)		
70 hours at 200°C	30		
70 hours at 232°C	50		

All the data given so far have been for elastomers containing vinylidene fluoride as one of the comonomers. Another peroxide vulcanizable system is that based on copolymers of tetrafluoroethylene and propene. Figures 3.12 and 3.13 illustrate the effect of heat aging at different temperatures upon the tensile strength and elongation at break of Aflas 100 vulcanizates, and Figure 3.14 compares the compression set of two formulations as a function of time at 200°C. (These data were taken from a commercial brochure on Aflas materials.) The greatest thermal stability, however, is attained in the elastomer systems based upon the copolymer of tetrafluoroethylene and perfluoro (methylvinylether) because of the lack of hydrogen in the structure. The properties of this elastomer (Kalrez) have been described in a number of papers.[38,44-46] The thermal stability of the copolymer decreases as the proportion of perfluoro (methylvinylether) in it increases. For optimum heat resistance in the vulcanizate, a press cure of 30–60 minutes at 170–200°C must be followed by a stepwise post-cure in an oven over a five-day period up to a final temperature of 285°C. The effect of temperature on tensile strength and elongation at break is given in Table 3.21, and Figures 3.15 and 3.16 show the change in these properties after aging for different times at temperature.

Further data are listed in Table 3.22, including the changes in modulus at 100% extension and in Durometer A hardness. Table 3.23 gives compression set figures at elevated temperatures for vulcanizate in the form of pellets or O-rings.

A brief mention should be made here of the possibility of stabilization of fluorine-containing elastomers. Conventional low-molecular-weight antioxidants can be lost from rubber vulcanizates by volatilization out at high temperatures, or by leaching out if the material is in contact with hot fluids over long periods of time. One method of preventing this is to link the antioxidant directly to the polymer chain. This requires a compound that can react chemically with the rubber during vulcanization while retaining its antioxidant function intact. In the case of the partially hydrogenated fluoroelastomers, this can be achieved using a molecule of the type:

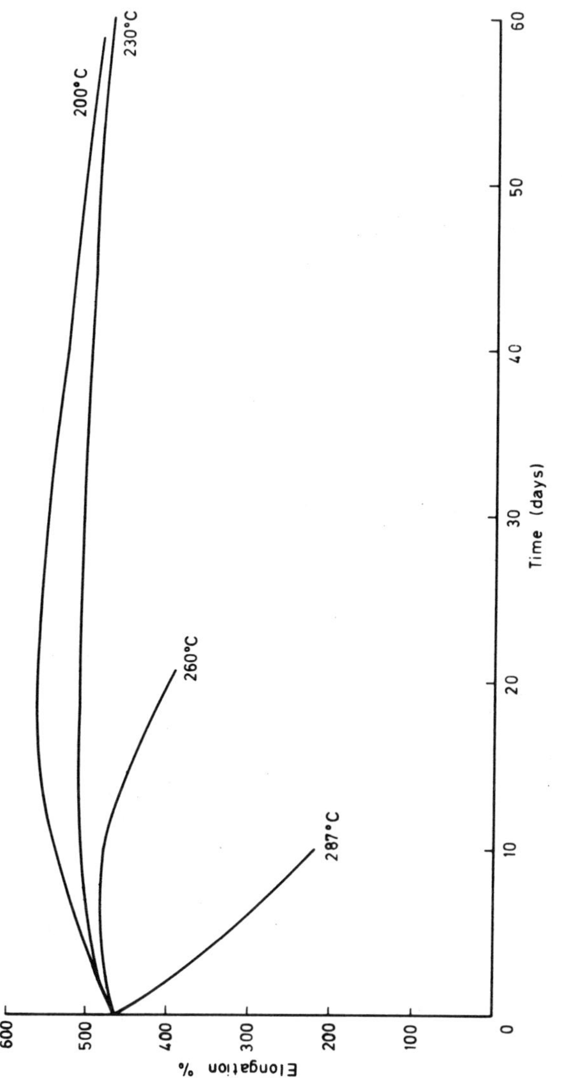

Figure 3.13. Effect of heat aging on the elongation to break of Aflas 100 vulcanizates.

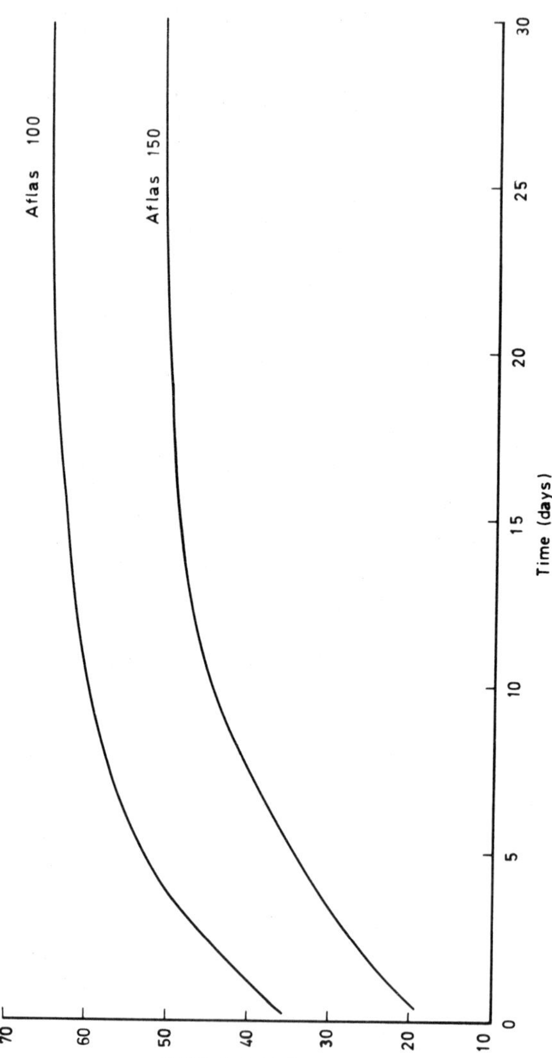

Figure 3.14. Comparison of compression set resistance of two Aflas formulations at 200°C.

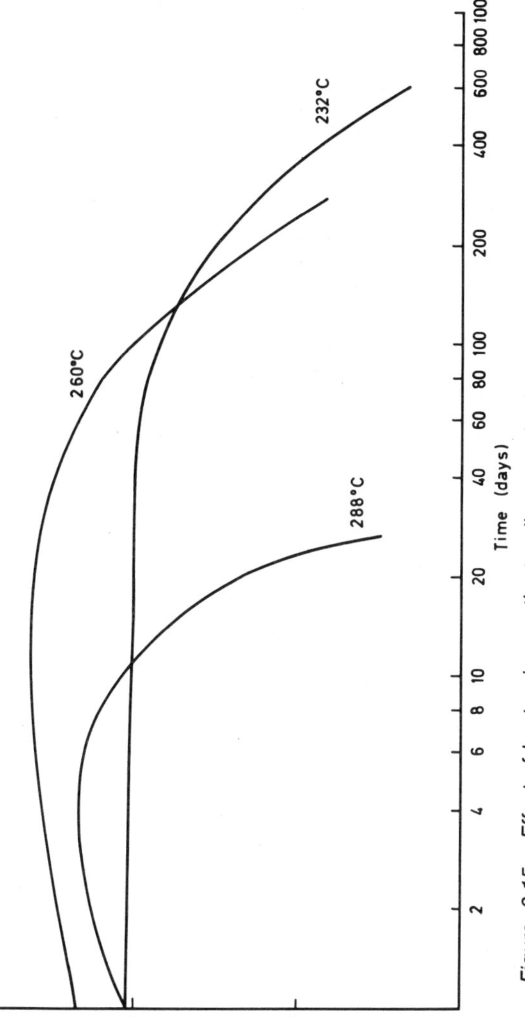

Figure 3.15. Effect of heat aging on the tensile strength of a Kalrez elastomer. Reference 45.

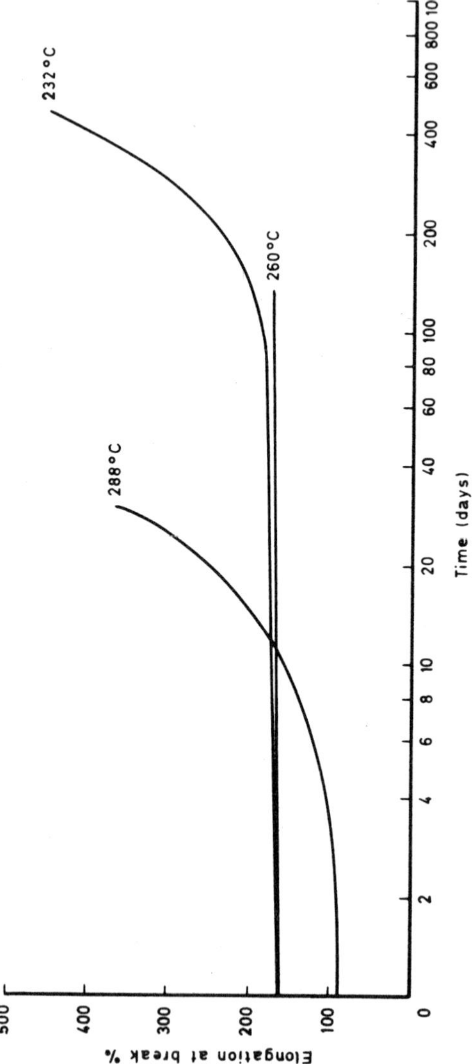

Figure 3.16.   Effect of heat aging on the elongation to break of a Kalrez elastomer. Reference 45.

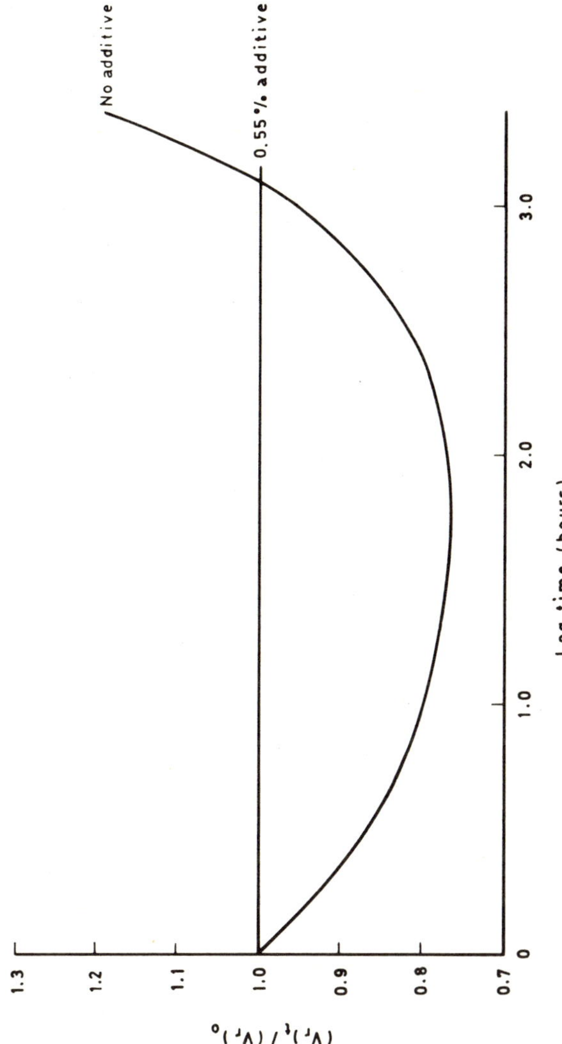

Figure 3.17. Effect of aging at 250°C on the network chain density of a fluoroelastomer. Reference 47.

TABLE 3.21. Effect of Temperature on Properties of Kalrez Elastomer (Reference 45)

Temperature (°C)	Tensile strength (MPa)	Elongation at break (%)
−65	58.1	13
−45	50.2	42
25	23.8	190
120	5.1	90
205	2.6	90

TABLE 3.22. Effect of Time at Temperature on Properties of Kalrez Elastomer (Reference 46)

Aging condition	Tensile strength (MPa)	Elongation at break (%)	100% Modulus (MPa)	Durometer A hardness
—	18.3	120	13.1	85
84 days at 232°C	21.4	175	10.3	88
112 days at 260°C	16.9	235	6.7	87
28 days at 288°C	11.2	320	6.2	86
7 days at 315°C	13.8	230	6.9	86

TABLE 3.23. Effect of Time at Temperature on Compression Set Resistance of Kalrez Elastomer (Reference 45)

	Compression set % of	
Aging condition	Pellets	O-rings
70 hours at room temperature	34	—
70 hours at 121°C	32	25
70 hours at 205°C	34	37
70 hours at 232°C	34	43
70 hours at 260°C	41	40
70 hours at 288°C	49	62

TABLE 3.24. Effect of Network-Bound Antioxidant on the Compression Set Resistance of a Fluoroelastomer (Reference 47)

	Compression set (%)			
Aging conditions	No additive	0.5% additive	1.0% additive	20% additive
70 hours at 100°C	18	9	9	9
70 hours at 150°C	21	12	9	8
70 hours at 200°C	47	26	24	21

With this compound, attachment to the network occurs via addition of the primary amine end-groups to unsaturated bonds in the chain (see earlier for the cure of a fluoroelastomer with a diamine), and the antioxidant activity arises from the aryl-NH groups. The effectiveness of this additive as an antioxidant has been demonstrated in fluoroelastomers at temperatures up to 250°C in air, and it has been shown that once the compound is incorporated into the vulcanizate during cure, it cannot be subsequently removed by solvent extraction.[47] Figure 3.17 shows the change in the network density (degree of cross-linking) of a fluoroelastomer vulcanizate with and without 0.55% of additive, and Table 3.24 shows the influence of varying amounts of additive upon the compression set resistance of O-rings. Quite substantial improvement in property retention is evident.

SUMMARY

1. Polytetrafluoroethylene is the most thermally stable linear addition polymer known. Its very stability means that special methods have to be used in processing it.

2. All structural modifications to polytetrafluoroethylene involving the introduction of perfluoroalkyl branches, or the substitution of fluorine atoms by chlorine or hydrogen, cause a diminution in thermal stability.

3. Similarly, in the elastomer systems the fully fluorinated copolymer of tetrafluoroethylene and perfluoro(methylvinylether) is the most stable.

4. The formulation used is vitally important in these systems. It has a profound effect on such properties as compression set resistance and HF elimination.

5. The fluoroelastomers, together with the polysiloxanes, comprise the most generally useful heat-resistant rubbers currently available.

REFERENCES

1. R. J. Plunkett, U.S. Patent No. 2,230,654.
2. T. L. Cottrell, *The Strengths of Chemical Bonds,* Butterworths, London (1958).
3. J. Thrower and M. A. White, *Polym. Preprints* **7,** 1077 (1966).
4. W. Cummings, E. R. Lynch, and E. B. McCall, Paper presented at the 3rd International Symposium on Fluorine Chemistry, Munich, 1965.
5. C. A. Sperati and H. W. Starkweather, *Fortschr. Hochpolym.–Forsch.* **2,** 465 (1961).
6. D. I. McCane, Tetrafluoroethylene polymers, in *Encyclopedia of Polymer Science and Technology,* H. F. Mark, N. G. Gaylord, and N. M. Bikales, (eds.), Interscience (1970), Vol. 13, p. 623.
7. J. E. Fearn, Polymerization of fluoro-olefins, in *Fluoropolymers,* L. A. Wall, (ed.), Wiley–Interscience (1972), Ch. 1, p. 1.

8. R. P. Bringer, Chlorotrifluoroethylene polymers, in *Encyclopedia of Polymer Science and Technology,* H. F. Mark, N. G. Gaylord, and N. M. Bikales, (eds.), Interscience (1967), Vol. 7, p. 204.

9. J. E. Dohany, A. A. Dukert, and S. S. Preston, Vinylidene fluoride polymers, in *Encyclopedia of Polymer Science and Technology,* H. F. Mark, N. G. Gaylord, and N. M. Bikales, (eds.), Interscience (1971), Vol. 14, p. 600.

10. F. S. Cohen and P. Kraft, Vinyl fluoride polymers, in *Encyclopedia of Polymer Science and Technology,* H. F. Mark, N. G. Gaylord, and N. M. Bikales, (eds.), Interscience (1971), Vol. 14, p. 522.

11. J. M. Cox, B. A. Wright, and W. W. Wright, *J. Appl. Polym. Sci.* **8,** 2935 (1964).

12. J. M. Cox, B. A. Wright, and W. W. Wright, *J. Appl. Polym. Sci.* **8,** 2951 (1964).

13. S. L. Madorsky, V. E. Hart, S. Straus, and V. A. Sedlak, *J. Research Natl. Bur. Std.* **51,** 327 (1953).

14. S. Straus and L. A. Wall, *SPE Trans.* **4,** 56 (1964).

15. L. A. Wall, Thermal decomposition of fluoropolymers, in *Fluoropolymers,* L. A. Wall, (ed.), Wiley–Interscience (1972), p. 381.

16. S. Morisaki, *Thermochimica Acta* **25,** 171 (1978).

17. R. E. Kupel, M. Nolan, R. G. Keenan, M. Hite, and L. D. Scheel, *Anal. Chem.* **36,** 386 (1964).

18. C. P. Fenimore and G. W. Jones, *J. Appl. Polym. Sci.* **13,** 285 (1969).

19. K. L. Paciorek, R. H. Kratzer, and J. Kaufman, *J. Polym. Sci., Polym. Chem. Ed.* **11,** 1465 (1973).

20. R. E. Florin, L. A. Wall, D. W. Brown, L. A. Hymo, and J. D. Michaelsen, *J. Res. Natl. Bur. Std.* **53,** 121 (1954).

21. L. A. Wall and J. D. Michaelsen, *J. Res. Natl. Bur. Std.* **56,** 27 (1956).

22. J. D. Michaelsen and L. A. Wall, *J. Res. Natl. Bur. Std.* **58,** 327 (1957).

23. S. L. Madorsky and S. Straus, *J. Res. Natl. Bur. Std.* **55,** 223 (1955).

24. L. I. Tarutina and T. S. Dunaevskaya, *Vysok. Soed.* **4,** 276 (1962).

25. T. R. F. W. Fennell, G. J. Knight, and W. W. Wright, *Proceedings 3rd ICTA* **3,** 245 (1971).

26. L. A. Wall and S. Straus, *J. Res. Natl. Bur. Std.* **65A,** 227 (1961).

27. T. G. Degteva, I. M. Sedova, K. A. Khamidov, and A. S. Kuzminskii, *Vysok. Soed.* **7,** 1198 (1965).

28. K. L. Paciorek, W. G. Lajiness, and C. T. Lenk, *J. Polym. Sci.* **60,** 141 (1962).

29. T. G. Degteva, *Vysok. Soed.* **3,** 671 (1961).

30. T. G. Degteva, I. M. Sedova, and A. S. Kuzminskii, *Vysok. Soed.* **5,** 378 (1963).

31. T. G. Degteva, I. M. Sedova, and A. S. Kuzminskii, *Vysok. Soed.* **5,** 1485 (1963).

32. T. G. Degteva and A. S. Kuzminskii, *Vysok. Soed.* **5,** 1417 (1963).

33. T. G. Degteva and A. S. Kuzminskii, *Sov. Rubber Technol.* **23,** (2) 13 (1964).

34. R. G. Arnold, A. L. Barney, and D. C. Thompson, *Rubber Chem. Technol.* **46,** (3) 619 (1973).

35. J. F. Smith, *Rubber World* **142,** 102 (1960).

36. J. F. Smith and G. T. Perkins, *Rubber and Plast. Age* **42,** 59 (1961); *J. Appl. Polym. Sci.* **16,** 460 (1961).

37. J. E. Alexander and H. Omura, *Elastomerics* **110,** 19 (1978).

38. G. H. Kalb, R. W. Quarles, and R. S. Graff, *Appl. Polym. Symp.* **22,** 127 (1973).

39. J. H. Brown, *Progress in Rubber Technology* **39,** 31 (1976).

40. W. W. Wright, *Br. Polym. J.* **6,** 147 (1974).

41. G. J. Knight and W. W. Wright, *Br. Polym. J.* **5,** 395 (1973).

42. D. A. Stivers, Fluorocarbon rubbers, in *Rubber Technology* (2nd ed.), M. Morton, (ed.), Van Nostrand Reinhold, N.Y. (1973), Ch. 16, pp. 407–439.

43. A. L. Moran and D. B. Pattison, *Rubber Age* **107,** (7) 37 (1971).

44. A. L. Barney, W. J. Keller, and N. M. Gulick, *J. Polym. Sci. A 1* **8,** 1091 (1970).

45. A. L. Barney, G. H. Kalb, and A. A. Khan, *Rubber Chem. Technol.* **44,** (3) 660 (1971).
46. A. M. Houston, *Mat. Eng.* **80,** (7) 54 (1974).
47. E. Kay, D. K. Thomas, and W. W. Wright, Paper presented at the International Rubber Conference, Brighton, U.K., 1972.

SUPPLEMENTARY BIBLIOGRAPHY

Vinylidene fluoride-hexafluoropropylene copolymer. A thermally stable elastomer, S. Dixon, D. R. Rexford, and J. S. Rugg, *Ind. Eng. Chem.* **49,** 1687 (1957).

Fluorine-containing polymers—I. Fluorinated vinyl polymers with functional groups, condensation polymers, and styrene polymers, W. Postelnek, L. E. Coleman, and A. M. Lovelace, *Fortschr. Hochpolym.-Forsch.* **1,** 75 (1958).

The molecular structure of perfluorocarbon polymers. II. Pyrolysis of polytetrafluoroethylene, J. C. Siegle, L. T. Muus, T–P Lin, and H. A. Larsen, *J. Polym. Sci. A 2,* 391 (1964).

Thermal degradation of a Viton A type elastomer in the temperature range 250–400°C, T. G. Degteva, I. M. Sedova, K. A. Khamidov, and A. S. Kuzminskii, *Vysok. Soed.* **7,** 1198 (1965).

Thermal degradation of polyvinylidene fluoride and a copolymer of vinylidene fluoride with hexafluoropropylene, L. A. Oksentevich and A. N. Pravednikov, *Vysok. Soed. Krat. Soobsch.* **10B,** 49 (1968).

Fluorocarbon polymers, W. W. Wright, in *Thermal Stability of Polymers,* R. T. Conley, (ed.), Marcel Dekker (1970), Vol. 1, Ch. 9, p. 287.

Fluorocarbon resins (patent review), M. W. Raney, Chemical Process Review No. 51, Noyes Data Corp. (1971).

Fluorine-containing polyethers, polyesters, and polythio-compounds, B. F. Malichenko, *Russian Chem. Rev.* **40,** 301 (1971).

Polycondensation reactions of fluorinated organic compounds, R. M. Gitina, E. L. Zaitseva, and A. Y. Yakubovich, *Russian Chem. Rev.* **40,** 679 (1971).

Fluoropolymers, L. A. Wall, (ed.), Wiley–Interscience, New York (1972).

Recent developments in fluoroelastomers, J. M. Blackman, *Polym. J. (Singapore)* **3,** 23 (1972).

Long-term aging of elastomers. Chemorheology of Viton B fluorocarbon elastomer, S. H. Kalfayan, R. H. Silver, A. A. Mazzeo, and S. T. Liu, *JPL Quarterly Tech. Rev.* **2,** (3) 32 (1972).

High-temperature aging of fluoropolymers, D. W. Brown and L. A. Wall, *J. Polym. Sci., Polym. Chem. Ed.* **10,** 2967 (1972).

The thermal degradation of hydrofluoro polymers, G. J. Knight and W. W. Wright, *J. Appl. Polym. Sci.* **16,** 683 (1972).

The thermal degradation of perfluoro polymers, G. J. Knight and W. W. Wright, *J. Appl. Polym. Sci.* **16,** 739 (1972).

Study of dehydrofluorination of fluorine-containing rubbers at 200–300°C, P. B. Kiroshko and M. B. Grimblat, *Kauch. i Rezina* **3,** 6 (1974).

Hydrofluorocarbon sealants with improved low temperature and stress corrosion properties, W. F. Anspach, *J. Elastomers Plast.* **7,** 3 (1975).

Vacuum evaluation of a new perfluoroelastomer, M. F. Zabielski and P. R. Blaszuk, *J. Vac. Sci. Technol.* **13,** 644 (1976).

POLYMERS WITH AROMATIC RINGS IN THE CHAIN

INTRODUCTION

As stated earlier, poly-*p*-phenylene would appear to be a good standard of what can be achieved with respect to a thermally stable organic polymer. As it is arguably the simplest of the aromatic ring systems, its investigation has played an important part in the development of heat-resistant polymers. Commercial products based on polyphenylenes have appeared from time to time but have met with only limited success, due in the main to difficulties in fabrication. Improvement has not been achieved solely by substitution in the ring, and hence recourse has had to be made to the use of flexibilizing, solubilizing linking groups. Many such groups have been studied (Table 4.1). Of these, $-CH_2CH_2-$, $-O-$, $-S-$, $-SO_2-$, $-CO-$, $-CO-O-$, $-CO-O-CO$, and $-CO-NH-$ have proved the most useful, and they are present, either alone or in combination, in a variety of com-

TABLE 4.1. Linking Groups Used between Aromatic Rings

$-CH_2-$	$-CO-$	$-CO-NH-$	
$-CH_2CH_2-$	$-CO-O-$	$-CO-NH-NH-CO-$	
$-C(CH_3)_2-$	$-CO-O-CO-$	$-NH-CO-CO-NH-$	
$-CH=CH-$	$-O-CO-O-$	$-NH-CS-NH-$	
$-C\equiv C-$			
$-C(CN)=CH-$			
$-[CF_2]_n-$			
$-C(CF_3)_2-$			
$-O-$	$-NH-$	$-S-$	$-SO-$
	$-N=N-$	$-S-S-$	$-SO_2-$
			$-SO_2-O-$
			$-SO_2-NH-$

mercially important materials. Some of these, however, cannot be classed as heat-resistant.

POLYPHENYLENES

Preparation

The synthesis of polyphenylenes is possible by a number of routes. The first experiments to achieve molecular weights in excess of 1000 employed a Wurtz–Fittig reaction between 1,4-dichlorobenzene and metallic sodium, or a liquid potassium–sodium alloy in dioxane solution[1-3]:

A benzene soluble fraction of molecular weight 2200–2800 was obtained, which did not melt at least up to 550°C.

Another method involved the polymerization of 1,3-cyclohexadiene and dehydrogenation of the polymer so formed:

A high molecular weight (5,000–10,000) poly-1,3-cyclohexadiene was formed by the reaction of 1,3-cyclohexadiene with a Ziegler catalyst derived from tri-isobutylaluminum and titanium tetrachloride at ambient temperatures.[4,5] Dehydrogenation of this was best carried out using chloranil in refluxing xylene.

It is also possible to polymerize benzene under mild reaction conditions, temperatures of 35–50°C for 15 minutes, using an aluminum chloride-cupric chloride catalytic system.[6-8] The mechanism is believed to be an oxidative, cationic polymerization:

Products from all these reactions have been of relatively low molecular weight and difficult to fabricate. Hence the synthetic routes to them have not been used commercially. Marketable materials have utilized entirely different concepts. During the 1960s, Monsanto Chemicals Ltd. manufactured a polyphenylene resin laminating system. This was based upon the reaction of aromatic disulphonyl chlorides with bi- or terphenyls (Monsanto Polyphenylene Resin brochure). At high temperatures both HCl and SO_2 are split out during polymer formation:

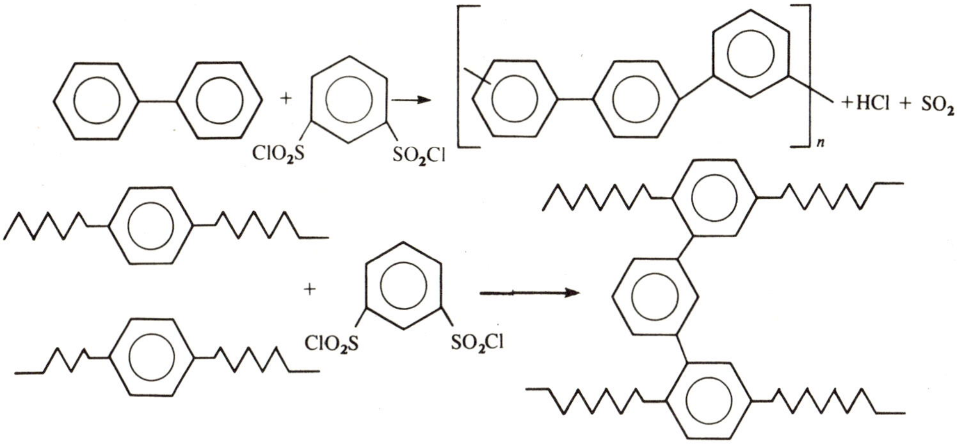

The process depends upon interrupting the reaction at such a stage that a soluble prepolymer is formed. This is used for impregnating fibrous reinforcements, and the cross-linking is then achieved by heating to above 275°C.

More recently, Hercules, Inc. has developed[9,10] a polyphenylene resin system designated H-resin, which is soluble and which exhibits melt flow at 70–150°C. This consists of an acetylenic terminated oligophenylene with the type of structure shown:

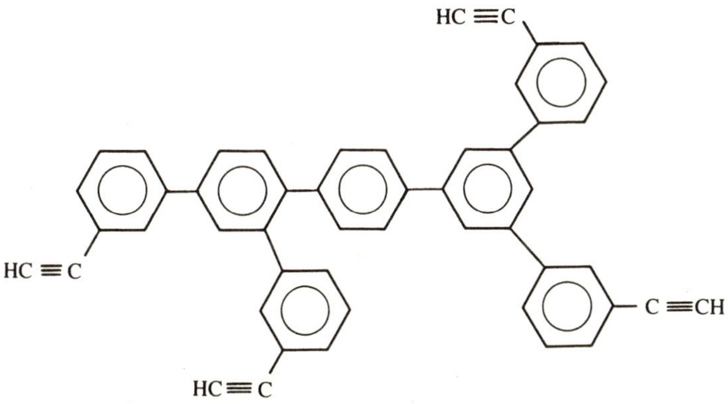

It was thought that cure of this compound took place at elevated temperatures by trimerization of the acetylenic groups to form phenylene rings. Later work on acetylene-terminated quinoxaline compounds[11] and polyimide resins,[12] however, demonstrated that in both those cases cross-linking proceeded by a number of different routes simultaneously, trimerization playing only a relatively minor role.

Thermal Stability

Thermogravimetry in air and in nitrogen of the poly-*p*-phenylenes derived from cyclohexadiene,[13] benzene,[13] and an acetylene-terminated oligophenylene[9] (Figure 4.1) shows that there is little difference in the stability of the polymers in nitrogen, but that in air the polymer from the acetylenic precursor is markedly less stable. This must reflect the presence of residual unsaturation. These weight loss curves indicate that little degradation is occurring in nitrogen at least up to 500°C. Data from isothermal experiments,[14] however, prove that appreciable weight loss takes place in two hours at temperatures below 380°C (Figure 4.2). (Wherever possible, these data for poly-*p*-phenylene will be used

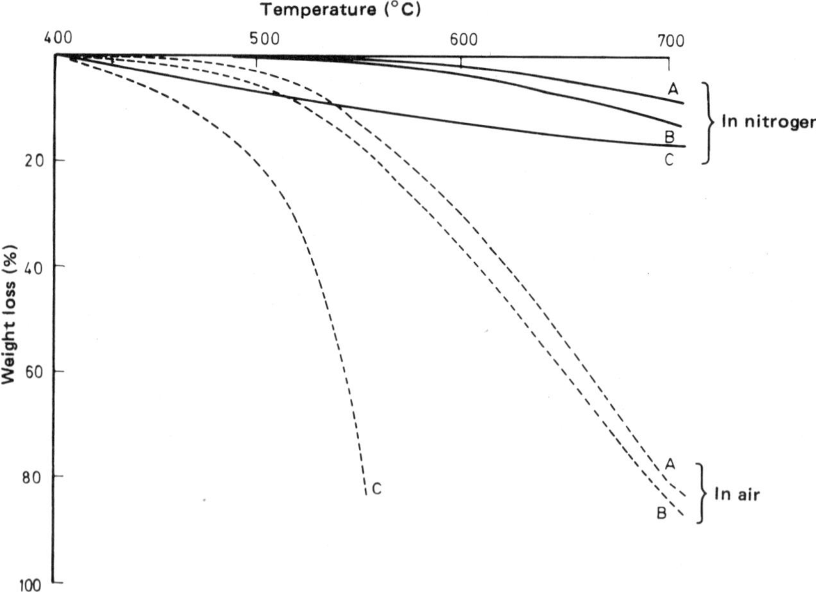

Figure 4.1.   Thermogravimetry of poly-p-phenylene. A, from cyclohexadiene. B, from benzene. C, from acetylene-terminated oligophenylene. (Heating rate 10°C/min)

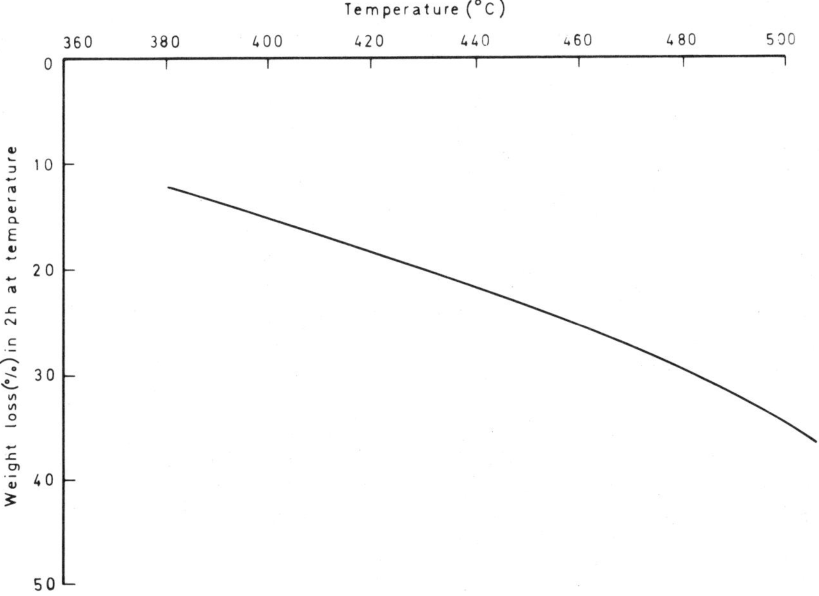

Figure 4.2. Thermal stability of poly-p-phenylene in an inert atmosphere. Reference 14.

for comparison purposes with those from the other aromatic ring-containing polymers.)

Volatile products from the thermal degradation at temperatures up to 620°C of the poly-*p*-phenylene prepared from benzene using ferric chloride as catalyst have been analyzed by mass spectrometry.[15] The results are given in Table 4.2.

The polymer obviously contained chlorine arising from the ferric chloride catalyst and oxygen from some unknown source. The formation of methane shows that the phenylene rings must be breaking down below 450°C. The production of hydrogen implies cross-linking of rings via dehydrogenation. Other workers[6,16] have noted the formation of biphenyl, terphenyl, quaterphenyl, and quinquephenyl by inter-ring bond scission.

TABLE 4.2. Volatile Products from the Thermal Degradation of Poly-1,4-Phenylene (Reference 15)

Temperature range (°C)	Total wt loss (%)	Wt volatiles (%)	Volatile products (mole %)						
			H_2	CH_4	H_2O	CO	HCl	CO_2	C_6H_6
20–450	8.0	4.6	36.9	6.5	16.6	4.5	23.5	8.3	3.7
450–550	10.5	4.3	88.5	5.4	2.2	1.4	—	—	2.5
550–620	2.2	~2.2	85.6	8.9	1.9	3.1	—	—	0.5

TABLE 4.3. *Typical Properties of Polyphenylene Resin/Chrysotile Asbestos Laminate*

Resin content (weight %)	50 ±2
Specific gravity	1.4–1.7
Flexural strength at 25°C	207–255 MPa
Flexural strength at 300°C	173–207 MPa
Flexural modulus	12.4–17.3 GPa
Tensile strength at 25°C	138–186 MPa
Tensile strength at 200°C	104–138 MPa
Tensile strength at 300°C	69–97 MPa

Elevated Temperature Properties

The polyphenylene resin produced by Monsanto was used to fabricate laminates from asbestos felt and from high modulus and high strength carbon fibers. To make the asbestos laminates, preimpregnated felts were placed between the platens of a press held at 310°C. After 6 minutes, during which HCl and SO_2 were evolved, a pressure of 3.5 to 7.0 MPa was applied and held for 4 hours, bumping the press three times in the first 9 minutes. Pressure was not released until the platen temperature had fallen below 100°C. A post-cure of 24 hours at 275°C, 300°C, and 325°C was then carried out. Typical properties are quoted in Table 4.3. (These and the data in Tables 4.4 and 4.5 are taken from a Monsanto Technical Bulletin on Polyphenylene Resin.)

Using as criterion the time taken for the flexural strength to fall below a value of 34.5 MPa, lifetimes of the laminates can be calculated for various temperatures. The figures obtained are listed in Table 4.4.

The data for the carbon fiber composites are given in Table 4.5. Good resistance is shown to acids, alkalies, hydraulic and lubricating fluids, and to boiling water or steam.

The polyphenylene system designated as "H-resin" exhibits melt flow above 70°C, rapidly sets at temperatures above 140°C, but needs a short exposure to temperatures in the range 220 to 250°C to be considered as fully cured. Typical properties of a cured unfilled resin sample are cited in Table 4.6.

TABLE 4.4. *Calculated Lifetimes of Polyphenylene Resin/ Asbestos Laminates*

Temperature (°C)	Lifetime (Time for flexural strength to fall below 34.5 MPa)
200	6–12 years
240	7000–15000 hours
300	1000–1500 hours
350	200–400 hours
400	30–150 hours

TABLE 4.5. Properties of Polyphenylene Resin/Carbon Fiber Composites

Property	Type 1 (high modulus) carbon fiber		Type 2 (high strength) carbon fiber	
	Untreated	Surface treated	Untreated	Surface treated
Resin content (weight %)	30–50	30–50	30–50	30–50
Fiber volume (%)	45–55	45–55	40–50	40–60
Specific gravity	1.4–1.5	1.4–1.53	1.4–1.52	1.4–1.55
Flexural strength (MPa)				
at 25°C	621	800	1104	1587
150°C	—	—	725	—
240°C	380	—	—	—
300°C	—	—	—	345
at 25°C after 2000 hours at 300°C	345	—	—	—
300°C after 2000 hours at 300°C	186	—	—	—
Flexural modulus (GPa)	83–104	83–159	62–83	62–76
Interlaminar shear strength (MPa)	21–31	21–36	28–43	35–58

As with the Monsanto material, mechanical properties are relatively unaffected by prolonged immersion in various organic liquids, acid, or alkaline solutions at temperatures at least up to 90°C.

TABLE 4.6. Typical Properties of Cured H-Resin (Reference 9)

Density (g/ml)	1.145
Flexural strength (MPa)	
at 23°C	48–138
at 360°C	41–55
Flexural modulus (GPa)	
at 23°C	4.8–8.3
at 360°C	3.5–4.8
Volume resistivity (ohm-cm)	
at 23°C	5×10^{17}
at 100°C	5×10^{16}
Dielectric constant	
60 Hz at 23°C	3.0
60 Hz at 100°C	3.2
1 MHz at 23°C	3.2
1 MHz at 100°C	3.3
Dissipation factor	
60 Hz at 23°C	0.0016
60 Hz at 100°C	0.0008
1 MHz at 23°C	0.0025
1 MHz at 100°C	0.0020
Limiting oxygen index	55
Continuous use temperature in air (°C)	215
Short term use temperature in air (°C)	350

POLYXYLYLENES

Preparation

Poly-*p*-xylylene was first prepared in sufficiently high molecular weight for film formation by the pyrolysis of *p*-xylene.[17] High temperatures (700–1,000°C) and reduced pressures (1–5 mm Hg) were required.

The product is extremely insoluble and only melts with degradation at 400–410°C. This intractability was attributed to a high degree of crystallinity coupled with some cross-linking.[18] Alternative synthetic routes were investigated in the hope of obtaining a more tractable material. These included a Wurtz reaction with bis(halogenomethyl)benzene[19,20] and the Hoffman degradation of *p*-methylbenzyl-trimethylammonium halides[21]:

The resultant polymers were indeed more soluble. However, in practice, a pyrolysis route was adopted as the basis for an industrial process, in which *p*-xylene is pyrolyzed in the presence of steam at atmospheric pressure, followed by the quenching of the pyrolysis gases in an aromatic hydrocarbon, for example, xylene or toluene.[22,23] This yields a dimer—di-*p*-xylylene—via a quinone dimethide intermediate.

The crystalline dimer is purified and converted to polymer by vaporization at 250°C and 0.1 mm Hg, followed by pyrolysis at 650–680°C and 0.5 mm Hg; finally it is condensed on a cold surface at 0.2 mm Hg, when spontaneous polymerization occurs. The polymerization is in some ways unique, as it takes place in two completely separate steps: the cleavage of CH_2–CH_2 bonds to give an intermediate (a quinone dimethide), which is stable in the vapor phase, followed

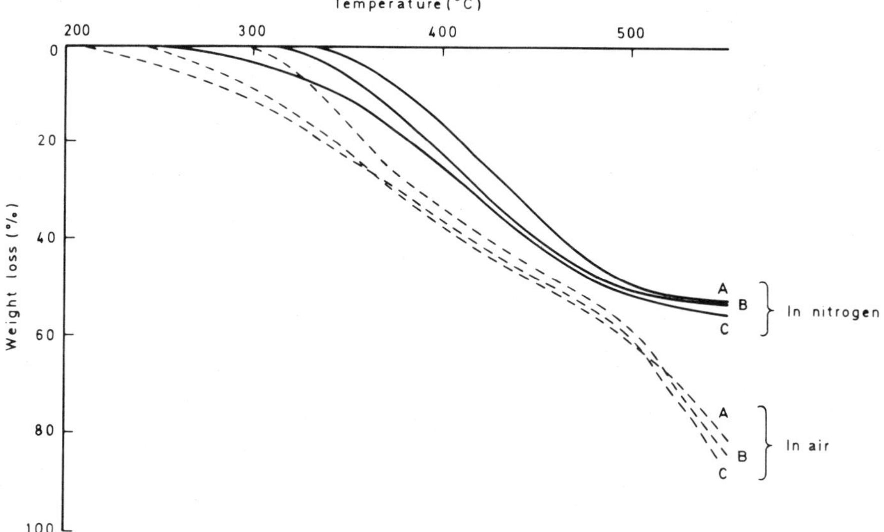

by spontaneous polymerization on condensation to yield a high-molecular-weight (500,000) polymer. The mechanism of the reaction has been studied in some detail.[24] It is also possible to carry through the whole process starting with mono- or dichloro-p-xylene without loss of the chlorine atoms. Union Carbide Corporation markets three products: poly-p-xylylene (Parylene N), poly-monochloro-p-xylylene (Parylene C), and polydichloro-p-xylylene (Parylene D).

Thermal Stability

Thermogravimetric curves for poly-p-xylylenes prepared by the three different routes[25] are shown in Figure 4.3. There is apparently relatively little difference in thermal stability of the samples in either nitrogen or air, although weight loss for the polymer prepared by the Wurtz reaction commences at lower temperatures in both cases. The curves occur in a temperature range approximately 200°C lower than that observed for poly-p-phenylenes (see Figure 4.1). Isother-

Figure 4.3. Thermogravimetry of poly-p-xylylenes. A, from Hoffman reaction. B, from pyrolysis of p-xylene. C, from Wurtz reaction. (Heating rate 10°C/min) Reference 25.

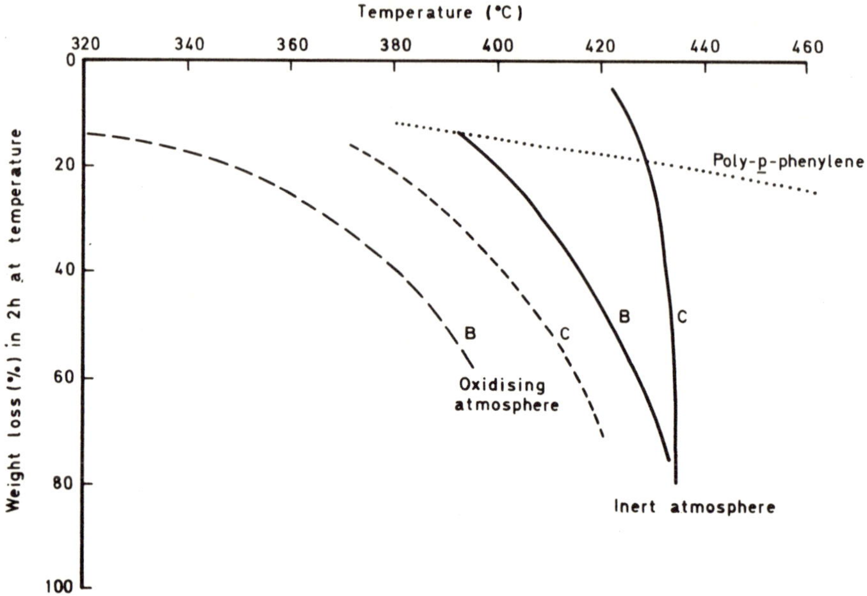

Figure 4.4. Comparison of thermal stabilities of poly-p-xylylenes. B, from pyrolysis of p-xylene. C, from Wurtz reaction.

mal results, however, indicate that the material prepared via the Wurtz reaction route is somewhat more stable than that synthesized by pyrolysis,[14, 26] and that the difference in thermal stability of poly-p-xylylene and poly-p-phenylene is not as great as would have been predicted from the dynamic weight loss curves (Figure 4.4). It should be noted that Schaefgen,[27] who studied the degradation in solution of poly-p-xylylene prepared by the Hoffman method, claimed that it was an order of magnitude more stable than the pyrolysis product. In solution the decomposition occurred by a first-order random scission process with an activation energy of 244 kJ/mole. Madorsky and Straus[28] also found that isothermal weight loss data for degradation in vacuum of a poly-p-xylylene (made by pyrolysis of p-xylene) obeyed first-order kinetics up to 60–80% conversion. For the bulk sample, the overall activation energy for breakdown was 319 kJ/mole. The volatile degradation products over the temperature range 419–464°C comprised p-xylene (78.6 mole %), toluene (8.0 mole %), methylethylbenzene (6.8 mole %), p-methylstyrene (3.9 mole %), and benzene (2.7 mole %).

Elevated Temperature Properties

Poly-p-xylylene coatings are claimed by the manufacturer to be usable at temperatures up to 150°C for extended periods in air and at temperatures up to 220°C in an inert atmosphere. Table 4.7 compares the ambient temperature prop-

TABLE 4.7. *Typical Properties of Polyxylylene Polymers*

Property	Poly-*p*-xylylene Parylene N	Poly-monochloro-*p*-xylylene Parylene C	Polydichloro-*p*-xylylene Parylene D	Epoxy	Silicone
Density (g/ml)	1.11	1.289	1.418	1.11–1.40	1.05–1.23
Tensile strength (MPa)	45	69	76	28–90	6–7
Elongation at break (%)	30	200	10	3–6	100
24 hour water absorption (%)	0.06	0.01	—	0.08–0.15	0.12 (7 days)
Melting or heat distortion temperature (°C)	405	280	>350	up to 220	up to 300
Linear coefficient of expansion (10^{-5}/°C)	3.5	6.9	—	4.5–6.5	25–30
Dielectric strength (V/mil)	7000	5600	5500	2300	2000
Volume resistivity 23°C 50% RH (ohm cm)	1×10^{17}	6×10^{16}	2×10^{16}	1×10^{14}	1×10^{15}
Surface resistivity 23°C 50% RH (ohm cm)	10^{13}	10^{14}	5×10^{16}	5×10^{13}	3×10^{13}
Dielectric constant					
60 Hz	2.65	3.15	2.84	4.2	2.6
10^3 Hz	2.65	3.10	2.82	3.9	2.6
10^6 Hz	2.65	2.95	2.80	3.4	2.6
Dissipation factor					
60 Hz	0.0002	0.020	0.004	0.03	0.0005
10^3 Hz	0.0002	0.019	0.003	0.03	0.0004
10^6 Hz	0.0006	0.013	0.002	0.04	0.0008

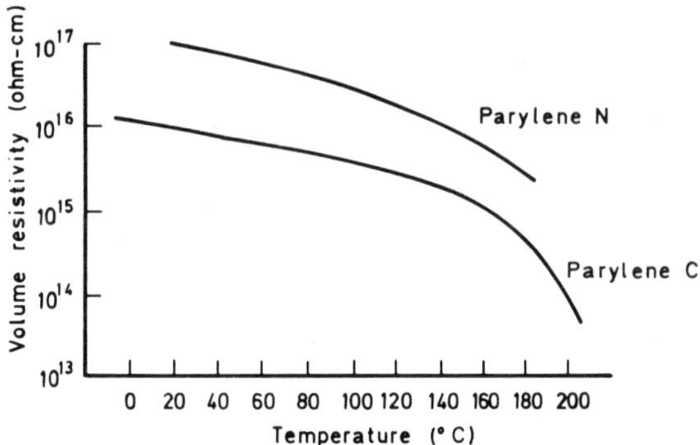

Figure 4.5. Volume resistivity of polyxylylenes as a function of temperature.

erties of the three Parylene materials with those of an epoxy and a silicone coating formulation. The data are taken from a technical brochure on Parylene conformal coatings. The good electrical properties of the polyxylylenes are immediately evident, and they find application in the electrical and electronics fields. Figures 4.5–4.7 show the changes in volume resistivity, dielectric con-

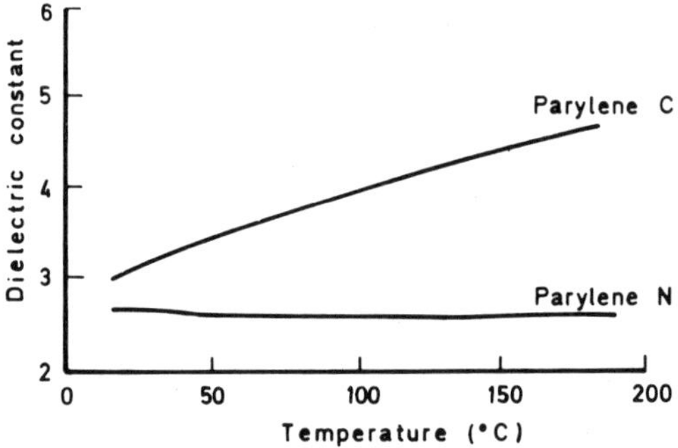

Figure 4.6. Dielectric constant of polyxylylenes as a function of temperature.

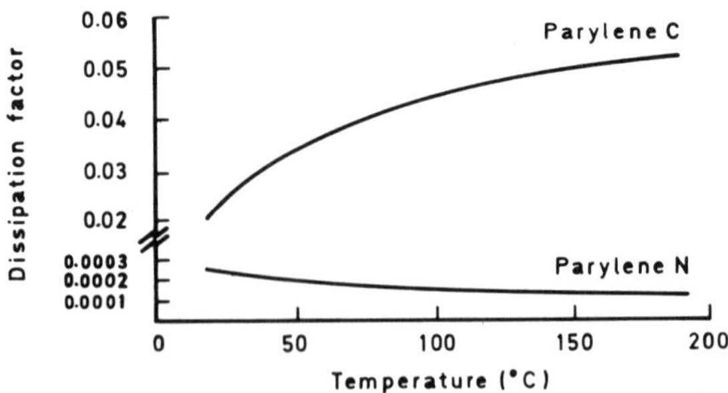

Figure 4.7. Dissipation factor of polyxylylenes as a function of temperature.

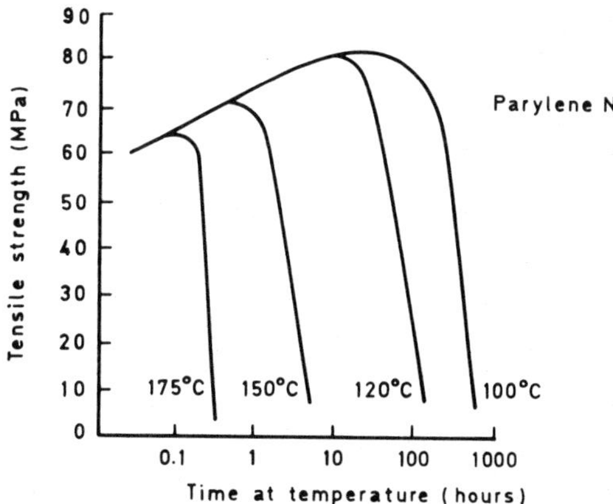

Figure 4.8.   Tensile strength of polyxylylene as a function of time at temperature.

stant, and dissipation factor as a function of temperature for Parylene N and C, and Figure 4.8 shows the change in tensile strength with time of aging at different temperatures for Parylene N. Electrical properties are preserved long after mechanical strength is lost.

AROMATIC POLYESTERS

Preparation

The synthesis of an aromatic polyester involves the reaction of an aromatic diol (or its derivative) with an aromatic diacid (or its derivative), and the techniques most widely used have been melt polycondensation, high-temperature solution polymerization, and low-temperature interfacial polymerization. Russian workers have been especially active in this field, terming the materials "polyarylates." Korshak's book,[29] published in Russian in 1969 and in English in 1971, lists no less than 140 aromatic polyester structures that had been synthesized and investigated up to that date. A number of these have been produced on a "commercial scale" in Russia under the names Aman, Esteran, and Tesan. The Aman series of polyarylates has been used as high-temperature, self-lubricating, antifriction materials. Western activity on aromatic polyesters has been at a much lower level and there has been relatively little commercial exploitation. The development of a polymer based upon *p*-hydroxybenzoic acid was first reported[30,31] in 1970 and its synthesis and structure were disclosed[32] in detail in 1976:

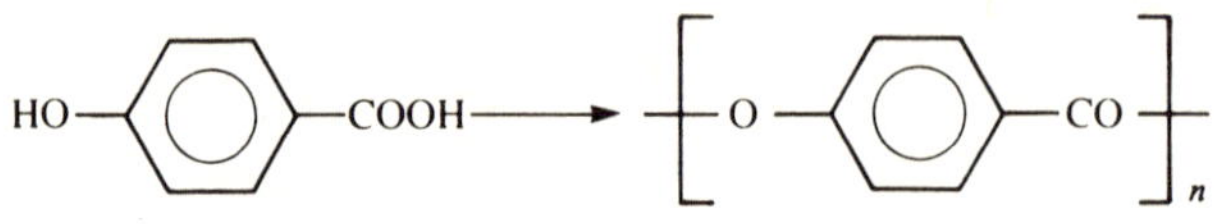

$$HO-\underset{}{\bigcirc}-COOH \longrightarrow \left[O-\underset{}{\bigcirc}-CO\right]_n$$

Early work to prepare this polymer was complicated by the fact that three competing reactions occur at elevated temperature: decarboxylation at about 200°C, esterification (i.e., polymerization) at > 250°C, and etherification and decarboxylation at > 250–300°C. Successful synthesis depended upon the use of the phenyl ester, or the *p*-acetoxy derivative of *p*-hydroxybenzoic acid and a heat exchange medium, which permitted careful control of the polymerization reaction. Blocking of the carboxyl group eliminates acid-catalyzed etherification reactions and facilitates polymerization by ester interchange:

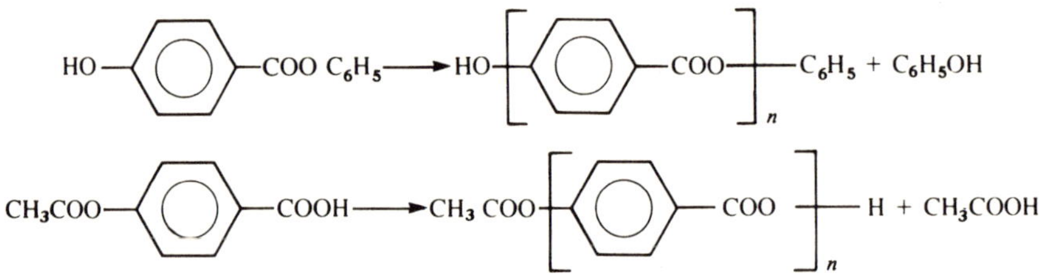

$$HO-\underset{}{\bigcirc}-COO\ C_6H_5 \longrightarrow HO\left[\underset{}{\bigcirc}-COO\right]_n C_6H_5 + C_6H_5OH$$

$$CH_3COO-\underset{}{\bigcirc}-COOH \longrightarrow CH_3\ COO\left[\underset{}{\bigcirc}-COO\right]_n H + CH_3COOH$$

Using highly purified acetoxybenzoic acid, single crystals of the polymer could be prepared of a double helix structure.

The homopolymer is sold under the trade name of Ekonol by the Carborundum Company. In addition to this, two copolyesters are made under the designations Ekkcel C-1000 (for compression molding) and Ekkcel I-2000 (for injection molding). The former is stated to be the condensation product of *p*-hydroxybenzoic acid, terephthalic acid, and a bisphenol, whereas the latter contains isophthalic acid instead of terephthalic acid and the ratio of the components is different[33,34]:

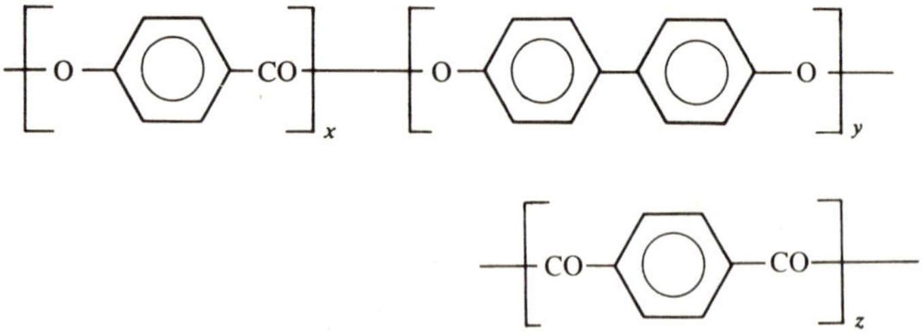

$$\left[O-\underset{}{\bigcirc}-CO\right]_x\left[O-\underset{}{\bigcirc}-\underset{}{\bigcirc}-O\right]_y$$

$$\left[CO-\underset{}{\bigcirc}-CO\right]_z \qquad \text{EKKCEL C-1000}$$

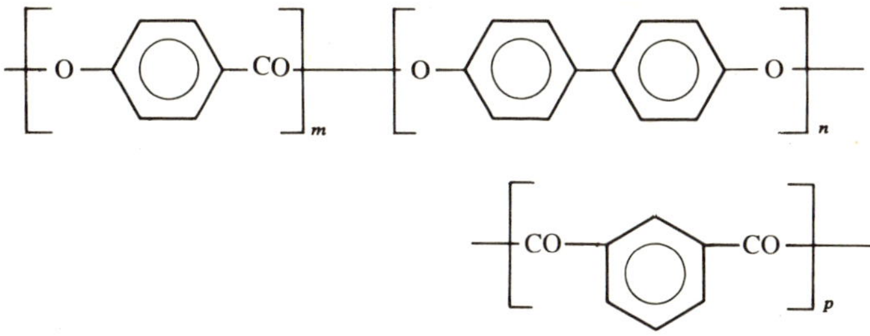

The Unitika Company of Japan in 1973 introduced a polyarylate called U-Polymer.[35] With the structure given below, it is presumably based upon the reaction between bisphenol-A and terephthalic acid. A number of grades are available.

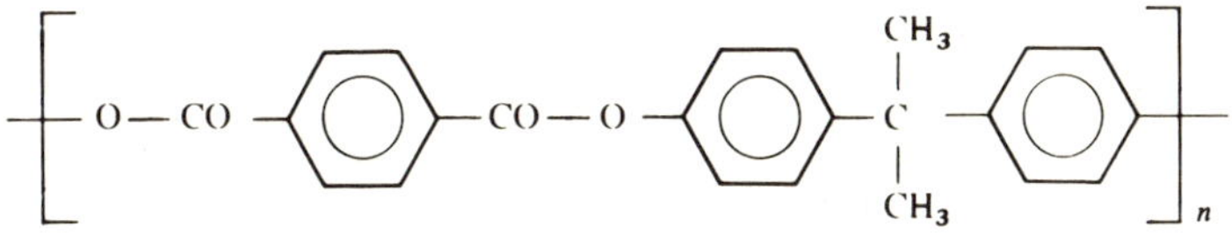

More recently (1979) the Solvay Company has marketed an aromatic polyester under the trade name Arylef U 100. No details of its chemical composition are yet available.

Thermal Stability

Poly(p-hydroxybenzoic acid) has a highly crystalline structure, the X-ray diffraction pattern of which remains unchanged up to 330°C. Above this temperature, loss of some of the crystal structure occurs, but this is not a crystalline melting point since the major peak in the X-ray scan is retained.[36] In fact, the polymer does not melt below 550°C, where it decomposes rapidly. A comparative study has been made of the thermal stability of Ekonol and the two Ekkcel polymers using dynamic and isothermal weight loss measurements.[37] The isothermal data are summarized in Figures 4.9 and 4.10, which show the weight loss in two hours at temperature against temperature for an inert and oxidizing atmosphere, respectively. The good stability of the polymers in both conditions is evident. The copolyesters are more stable than the homopolymer, especially in nitrogen. In air the discrepancy is less marked, largely because of a decrease in stability of the copolymers, that of the homopolymer being relatively unaffected by the presence of oxygen up to 30–40% weight loss. The overall activation energies for degradation are listed in Table 4.8.

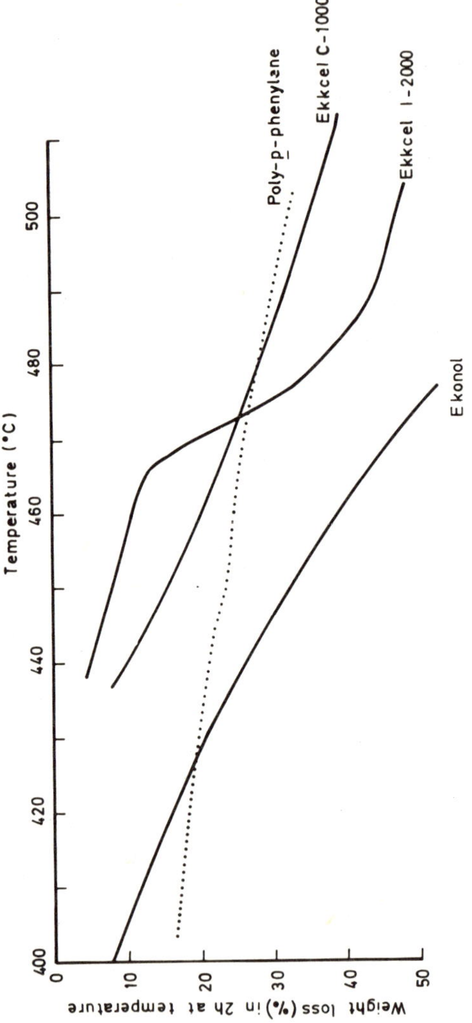

Figure 4.9.   Comparison of thermal stability of polymers based on p-hydroxybenzoic acid in nitrogen. Reference 37.

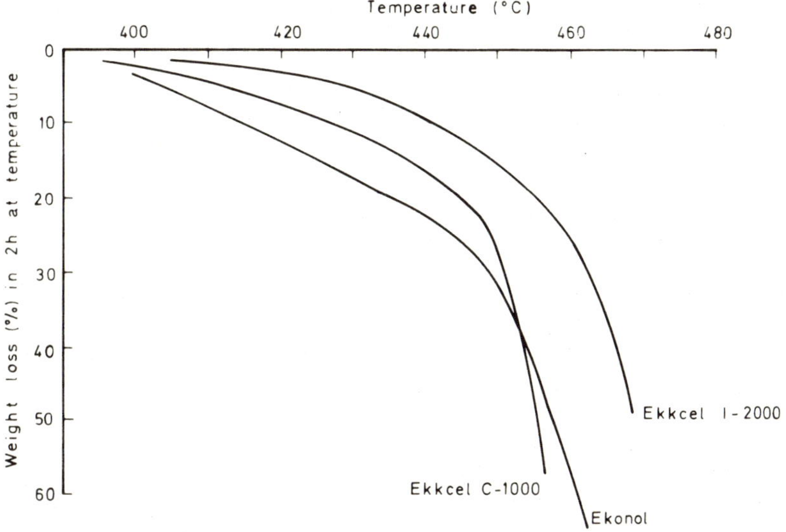

Figure 4.10.   Comparison of thermal stability of polymers based on p-hydroxybenzoic acid in air. Reference 37.

TABLE 4.8. Activation Energies for Degradation of p-Hydroxybenzoic Acid Polymers (Reference 37)

Polymer	Activation energy (kJ/mole) for 5–10% weight loss interval	
	In nitrogen	In air
Ekonol	172	123
Ekkcel C-1000	235	226
Ekkcel I- 2000	279	215

The mechanism and products of degradation do not appear to have been studied in any detail. Changes in composition of Ekonol after aging in air for various times at 425°C have been followed by elemental analysis of the residues.[36] The carbon, hydrogen, and oxygen contents remained virtually unchanged up to at least 65% weight loss, as did the infrared spectra. Ehlers *et al.*[38] have, however, analyzed the products of degradation in vacuum of polymers (**A**) and (**B**), which are structurally isomeric with Ekonol and U-Polymer, i.e.,

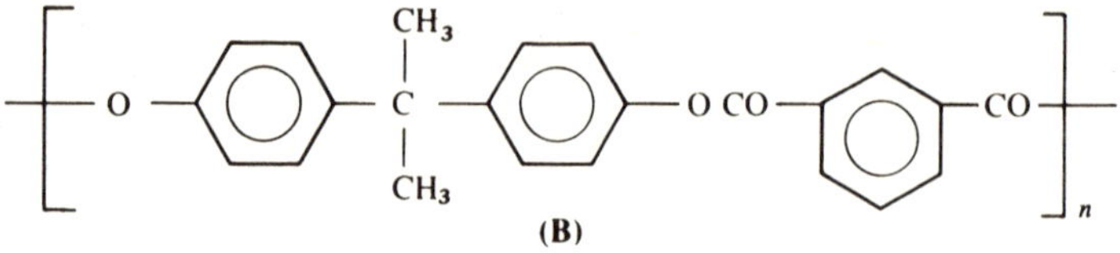

$$\left[\!-\!O\!-\!\bigcirc\!-\!\underset{\underset{CH_3}{|}}{\overset{\overset{CH_3}{|}}{C}}\!-\!\bigcirc\!-\!O\,CO\!-\!\bigcirc\!-\!CO\!-\!\right]_n$$

(B)

The volatile degradation products from these two polymers are given in Table 4.9.

As can be seen from these figures, polymer B, as a consequence of the presence of the isopropylidene group, is appreciably less thermally stable than polymer A. The major part of the weight loss can be attributed to the sublimation of hydroquinone in the case of polymer A and bisphenol-A in the case of polymer B. In both cases the major volatile products are CO and CO_2, arising from rupture of the ester linkage.

Elevated Temperature Properties

Rather unusual fabrication methods (for a polymeric material) are necessary for Ekonol. Compressive sintering has been extensively used—this involves temperatures of the order of 370°C, pressures of 70 MPa, and times ranging from minutes to an hour, depending on the thickness of sample. Moldings may be ground and resintered. High-energy-rate forging cuts fabrication time to 6–10 seconds. Here a preform is heated to 155–260°C and then forged using an energy of 14–136 kJ. Other advantages of forging are that microporosity is reduced, thicker specimens can be fabricated, and higher filler loadings are possible. Coatings are produced by plasma spraying in which the powdered polymer is fed into a plasma arc and sprayed onto metal substrates.

The Ekkcel copolymers, as mentioned earlier, can be compression or injection molded. The conditions for compression molding are the same as for compression sintering. Injection molding requires temperatures of 385–400°C and pressures of 34–69 MPa. A comparison of the properties of compression sintered Ekonol with those of some other plastics with high-temperature capability is made in Table 4.10. The modulus of elasticity of the polyester is appreciably higher than that of the other materials cited. This advantage is maintained at higher temperatures (Figure 4.11). The high thermal conductivity is important for bearing applications. In these the polymer is normally filled with polytetrafluoroethylene, molybdenum disulphide, or graphite.

The properties of compression (Ekkcel C-1000) and injection (Ekkcel I-2000) molded samples are summarized in Table 4.11.

Data on the elevated temperature properties of U-Polymer and Arylef are very sparse, but heat distortion temperatures in both cases of 175°C under 1.8 MPa load indicate that the upper temperature limits for use must be well below those for the Ekonol and Ekkcel materials.

TABLE 4.9. Volatile Products from the Thermal Degradation of (A) Poly (p-Phenyleneisophthalate-co-Terephthalate) and (B) the Polymer from Isophthalic Acid and 4,4'-Dihydroxydiphenyldimethylmethane (Reference 38)

	Temperature range	Total wt loss	Wt volatiles	Volatile products (Mole %)							
	(°C)	(%)	(%)	H_2	CH_4	H_2O	CO	C_2H_6	CO_2	C_6H_6	$C_6H_5CH_3$
(A)	20–450	52.3	29	2.0	0.4	0.4	61.7	—	34.8	0.7	—
	450–550	8.0	8	10.9	1.8	0.7	57.9	—	27.2	1.5	—
	550–620	3.2	3	22.7	3.6	1.8	50.3	—	21.0	0.6	—
(B)	20–375	63.4	14	0.8	5.9	1.6	35.0	—	55.4	1.2	0.1
	375–450	5.1	5	7.6	11.7	0.8	41.4	—	33.3	4.2	1.0
	450–550	5.0	5	48.2	15.4	0.6	24.2	1.3	8.7	1.5	0.1

TABLE 4.10. Properties of Ekonol and Other High-Temperature Capability Plastics (Reference 36)

Property	Polyester (Ekonol)	Polysulphone (Astrel 360)	Polyimide (Vespel)	Polytetra-fluoroethylene
Density (g/ml)	1.44	1.36	1.40	2.13
Flexural strength (MPa)	74	119	81–97	—
Flexural modulus (GPa)	7.1	2.7	3.2	0.6
Compressive strength (MPa)	226	124	166	7
Dielectric strength (V/mil)	660	300	430	620
Dielectric constant	3.8	3.9	3.6	2.1
Dissipation factor ($\times 10^4$)	2	30	34	3
Volume resistivity (ohm cm)	$> 10^{15}$	10^{13}	10^{16}–10^{17}	$> 10^{18}$
Water absorption 24 hours at RT (%)	0.02	0.22	0.30	0.01
Coefficient of static friction	0.10–0.16	—	0.25–1.2	0.05–0.08
Thermal conductivity [10^{-4} cal/(sec)(cm)^2(°C/cm)]	18.0	6.0	6.0	6.0

AROMATIC POLYAMIDES

Preparation

Synthesis of aromatic polyamides is normally achieved by reaction of aromatic diamines with aromatic diacids (or their derivatives), and the techniques used are the same as for the aromatic polyesters, i.e., melt polycondensation,

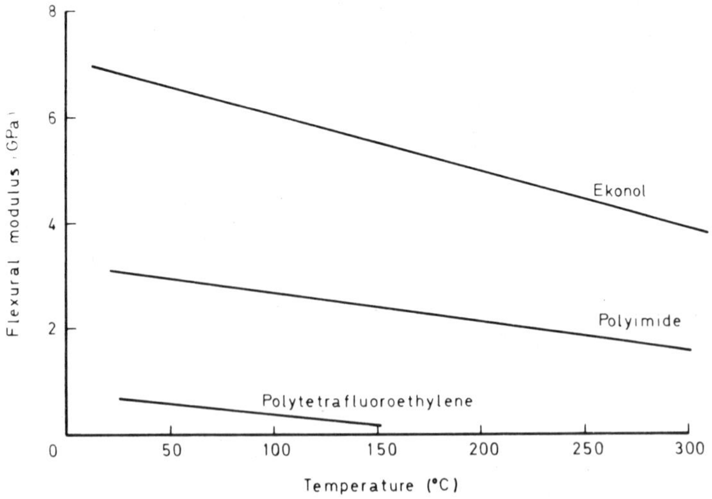

Figure 4.11. Flexural modulus vs. temperature for Ekonol, polyimide, and polytetrafluoroethylene. Reference 36.

TABLE 4.11. Properties of Compression and Injection Molded Poly(Hydroxybenzoic Acid) Copolymers (Reference 33)

Property		Ekkcel C-1000	Ekkcel I-2000
Tensile strength	at 23°C (MPa)	69	97
	at 260°C (MPa)	21	21
Tensile modulus	at 23°C (GPa)	1.3	2.4
Flexural strength	at 23°C (MPa)	104	117
	at 260°C (MPa)	35	28
Flexural modulus	at 23°C (MPa)	3.1	4.8
	at 260°C (MPa)	0.9	1.4
Compressive strength at 23°C (MPa)		—	124
Compressive modulus at 23°C (GPa)		—	3.5
Heat distortion temperature (°C), 1.8 MPa		300°C	293°C
Water absorption 24 h at RT (%)		0.040	0.025
Dielectric constant at 10^3 Hz and 23°C		3.68	3.16
Dissipation factor at 10^3 Hz and 23°C		0.0085	0.010
Dielectric strength (V/mil)		450	350

high-temperature solution polymerization, and low-temperature interfacial polymerization. There has been intense activity in the area mainly aimed at producing fibers, and Korshak's book[29] lists over 70 polyamide structures. Despite this activity only two polymer systems have achieved really large-scale production. The first of these is the condensation product of *m*-phenylene diamine and isophthaloyl chloride, first marketed by Dupont Ltd. in 1962 under the trade name of Nomex.[39,40] A similar material named Fenilon is made in the U.S.S.R.

The polymer is relatively insoluble and is spun, for example, from hot dimethylacetamide containing 3% of calcium chloride. The second polymer system is either poly(*p*-benzamide) or poly(1,4-phenylene terephthalamide), or a mixture of both structural units. Gan *et al.*[41] suggest that the material originally introduced as Fibre B, or PRD-49, by Dupont in 1970 was poly(*p*-benzamide), but that the higher tenacity material available from 1971 onwards is poly(1,4-phenylene terephthalamide). This conclusion is based upon infrared and NMR spectroscopic data.

poly(*p*-benzamide)

poly(1, 4-phenylene terephthalamide)

Brown and Ennis,[42] however, using pyrolysis/gas chromatography/mass spectrometry, believe that both structural units are present in Kevlar 49. These para-linked polymers can be prepared in a mixture of dimethylacetamide and hexamethylphosphoramide saturated with dry lithium chloride.[41] The maximum polymer concentration in this solution is only 2–3%. Of considerable importance is the fact that the polymer forms lyotropic liquid crystalline solutions and that the spinning of these gives fibers of high stiffness and strength.[42–44] The U.S.S.R. produces similar fibers under the designations Vniivlon-N and SVM.

Thermal Stability

Kevlar 49 and Nomex fibers have been studied using differential thermal analysis, thermogravimetry, and thermomechanical analysis.[42] Weight loss curves in nitrogen and air are compared in Figure 4.12. The initial weight losses

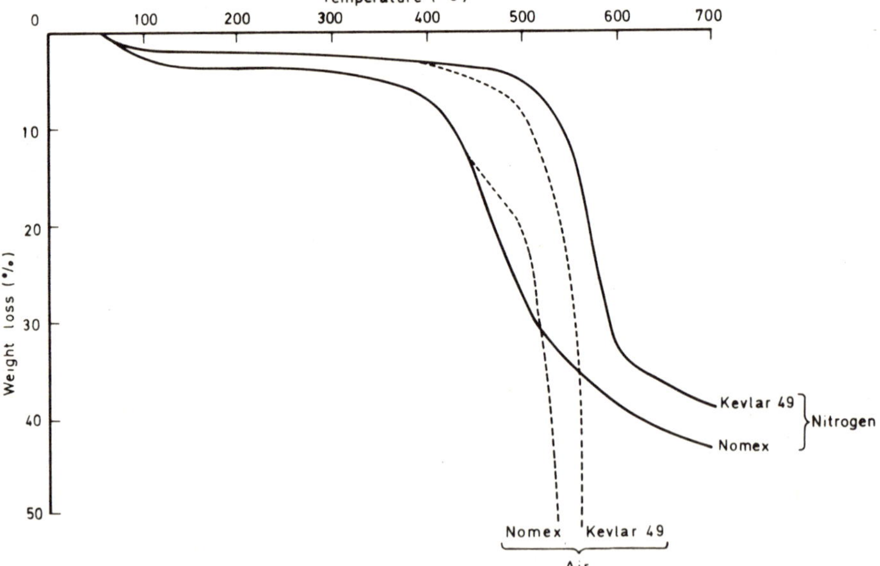

Figure 4.12. Thermogravimetry of Kevlar 49 and Nomex in air and nitrogen. (Heating rate unspecified) Reference 42.

TABLE 4.12. Volatile Decomposition Products
of Poly(p-Phenylene Terephthalamide) at 435°C
(Reference 48)

Product	Relative amount (arbitrary units)
Carbon dioxide	100
Carbon monoxide	60
Water	59
Benzene	3.3
Aniline	2.9
Benzonitrile	2.6
Hydrogen cyanide	2.5
Formamide	1.6
Cyanhydric acid	1.6
Benzoic acid	1.1

below 100°C are ascribed to water. Thereafter there is little change until rapid degradation occurs above 400°C for Nomex and above 500°C for Kevlar 49. The superior thermal stability of a para–para-linked compared with a meta–meta-linked polymer has been reported previously.[45–47] As can be seen in Figure 4.12, the initial stages of degradation are not greatly affected by the presence of oxygen. The volatile decomposition products of a poly(p-phenylene-terephthalamide) at 435°C have been analyzed by mass spectrometry[48] and the results are summarized in Table 4.12.

Solid phase degradation products include p-phenylene diamine and N-amino phenylbenzamide. The mechanisms that have been proposed to account for some of these products are as follows[49]:

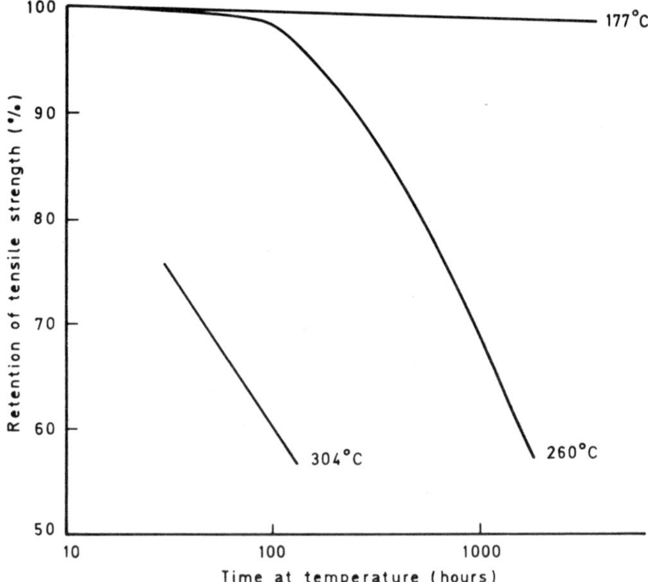

They all involve initial scission of the ester linkage.

Elevated Temperature Properties

The two books by Lee, Stoffey, and Neville[50] and by Frazer[13] between them give a substantial amount of information on the elevated temperature properties of Nomex fibers and paper, and only a representative selection of these data will be given here. Figure 4.13 shows the strength retention of fiber with

Figure 4.13. Strength retention of Nomex fiber after aging at elevated temperatures. (Tested at RT) Reference 13.

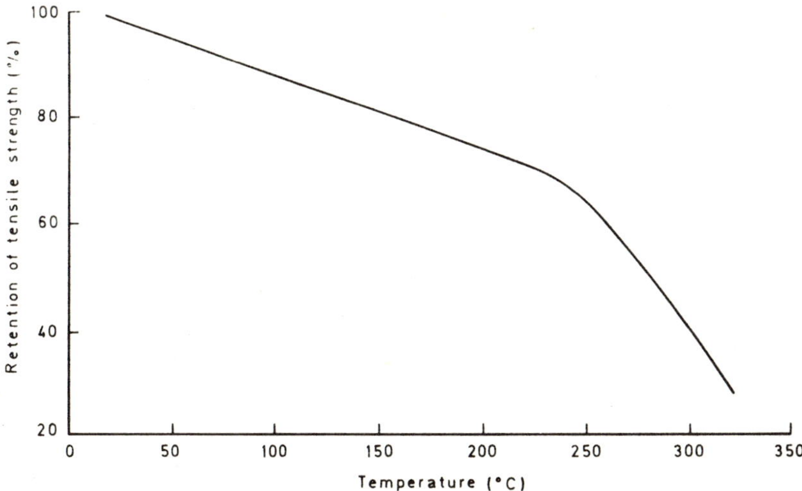

Figure 4.14.   Strength retention of Nomex paper as a function of temperature. Reference 13.

time of aging at different temperatures. There is very good retention of strength at room temperature after several thousand hours at 177°C. When tested at temperature after equilibrating there, approximately 50% of the initial strength and modulus of filament yarns are retained, even at 285°C. At the same time, the elongation at break increases by about 30%. (Initial values are as follows: tenacity, 5 g/den; modulus, 145 g/den; and elongation at break, 17%.) Nomex paper behaves in very similar fashion; Figure 4.14 shows retention of tensile strength as a function of temperature. There is once again 50% retention up to 280°C. Based on this concept of 50% strength retention, lifetimes of 100 hours would be expected at 315°C, 1,000 hours at 285°C, and 10,000 hours at 260°C. Approximately the same values are obtained using the criteria of 50% retention of the initial elongation at break and the time required for the dielectric strength to fall to a value of 300 V per mil from an initial value of 800 V per mil.

The Kevlar fibers have been used mainly as reinforcement in various resin matrices, and the properties of the resultant composites depend as much upon the resin used as upon the fiber. (See other chapters for the properties of the resins.) Figure 4.15 illustrates the effect of temperature upon the properties of a PRD-49-III yarn (the early designation of Kevlar), the measurements being made in air after 30 minutes at temperature.[51] The initial property levels were tensile strength, 3.3 GPa; tensile modulus, 132 GPa; and elongation at break, 2.2%. The long-term heat resistance limit is claimed to be about 240°C; at this temperature there is a 30% loss in tensile strength after 450 hours at temperature. The Limiting Oxygen Index for both Nomex and Kevlar is about 0.29.

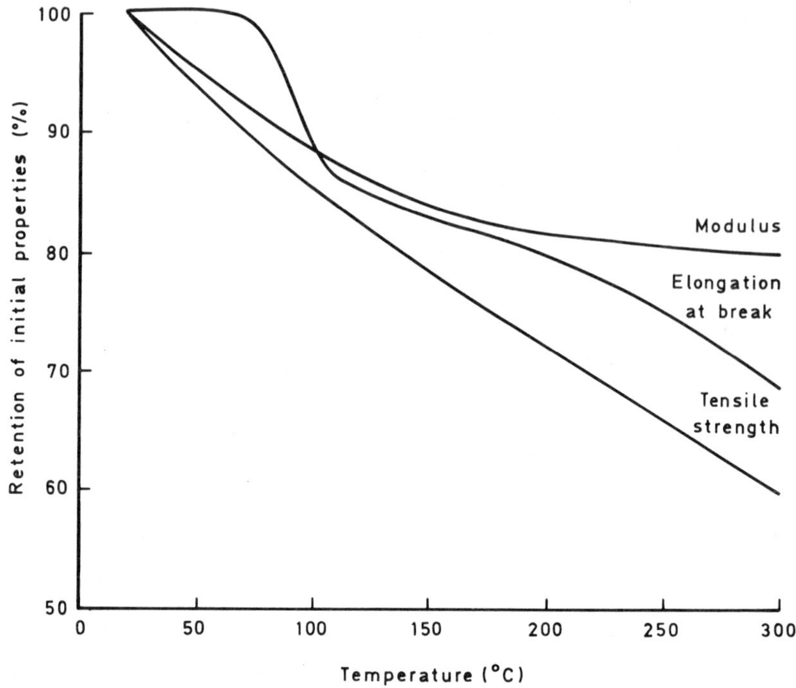

Figure 4.15. Property retention of PRD-49-III yarn as a function of temperature. Reference 51.

The properties of laminates made using the same epoxy resin and either glass cloth or Kevlar cloth as reinforcement are summarized in Table 4.13. As can be seen from the figures, the Kevlar laminate has superior tensile strength and modulus at room temperature when dry and wet, and at 150°C when dry. Its compressive strength is, however, inferior to that of the glass laminate, especially at elevated temperature.

TABLE 4.13. Properties of Laminates Made from an Epoxy Resin and Glass or Kevlar Cloth (Reference 51)

Property		Glass cloth laminate	Kevlar cloth laminate
Density g/ml		1.80	1.33
Tensile strength	RT dry (MPa)	435	518
	RT wet (MPa)	380	504
	at 150°C (MPa)	345	352
Tensile modulus	RT dry (GPa)	24.8	31.1
	RT wet (GPa)	22.1	27.6
	at 150°C (GPa)	23.5	31.7
Compressive strength	RT dry (MPa)	428	255
	at 150°C (MPa)	421	97

POLY(PHENYLENE SULPHIDE)

Preparation

The synthesis, structure, and properties of oligo- and poly-(arylene sulphides) have recently been reviewed.[52] Interest in these materials was stimulated by the production by Macallum[53] of phenylene sulphide polymers by the reaction of sulphur, sodium carbonate, and dichlorobenzene in a sealed vessel at 275–360°C:

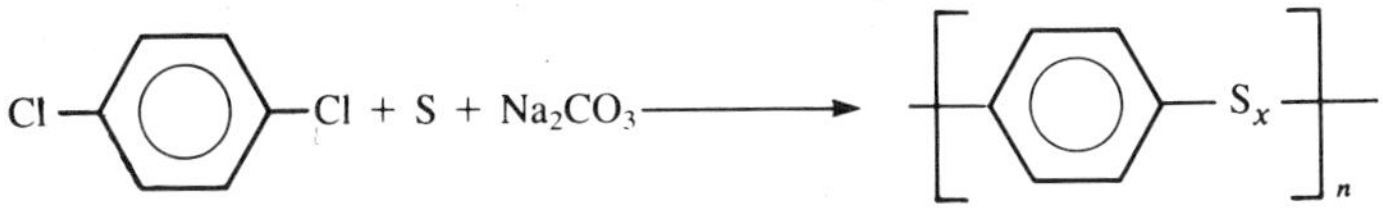

The composition and physical properties of the polymers obtained depend very much upon the ratio of the reactants. A molar ratio of sulphur and sodium carbonate to p-dichlorobenzene of about 1.5:1 gives hard, high-softening point materials with a value of x in the empirical formula of approximately one. As the proportion of sulphur in the reactant mix is increased, so is the value of x, and finally rubbery polymers are produced in which x is 2 to 5. The mechanism of this polymerization has been studied in detail by Lenz and Carrington.[54,55] They concluded that it involved a complex series of reactions, the main features of which are as follows:

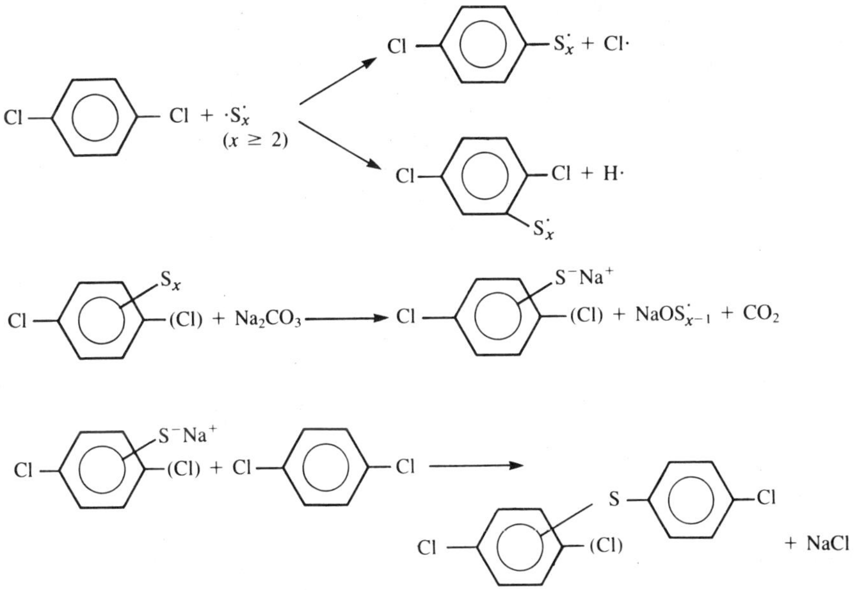

$$Cl\left[\!\!\left[\!\!\bigcirc\!\!\right]\!\!-S\right]_n + Na_2S \longrightarrow Na^+S^-\left[\!\!\left[\!\!\bigcirc\!\!\right]\!\!-S\right]_n + NaCl$$

$$Na^+S^-\left[\!\!\left[\!\!\bigcirc\!\!\right]\!\!-S\right]_n + Cl-\!\bigcirc\!-Cl \longrightarrow Cl\left[\!\!\left[\!\!\bigcirc\!\!\right]\!\!-S\right]_{n+1} + NaCl$$

In addition to this complexity, the total reaction is highly exothermic and hence difficult to control. As a consequence, an alternative route to poly(phenylene sulphide) was developed via the self-condensation of metal halothiophen-oxides[56]:

$$X-\!\bigcirc\!-S^-M^+ \longrightarrow \left[\!\!\left[\!\!\bigcirc\!\!\right]\!\!-S\right]_n + MX$$

The reaction is carried out at 200–250°C under nitrogen in the solid state, or in pyridine. The order of reactivity of the halogens (group X)[57] is I > Br > F ~ Cl, and of the metal cations (group M^+), $Cu^+ > Li^+ > Na^+ > K^+$. It proved difficult to remove from the polymers the salt MX, e.g., copper bromide, produced as a by-product of the reaction, and real commercialization had to await the development of another route. In this, *p*-dichlorobenzene is reacted with sodium sulphide in a polar solvent.[58] This process is the basis for the production of the poly(phenylene sulphides) made by Phillips Petroleum Co. under the trade name of Ryton. The polymer

$$Cl-\!\bigcirc\!-Cl + Na_2S \xrightarrow[\text{Solvent}]{\text{Heat}} \left[\!\!\left[\!\!\bigcirc\!\!\right]\!\!-S\right]_n + 2NaCl$$

is highly crystalline and soluble only to a limited extent at elevated temperature in some aromatic, chlorinated aromatic, or heterocyclic compounds, e.g., 1-chloronaphthalene. While this linear polymer possesses moderate mechanical strength, it can be converted into a tougher product by heat treatment in air. This process results in chain extension and cross-linking; the "cured" polymer is tough, ductile, and extremely insoluble.[59] The latter factor and the limited solubility of the initial polymer make precise determination of the mechanism of reaction almost impossible.

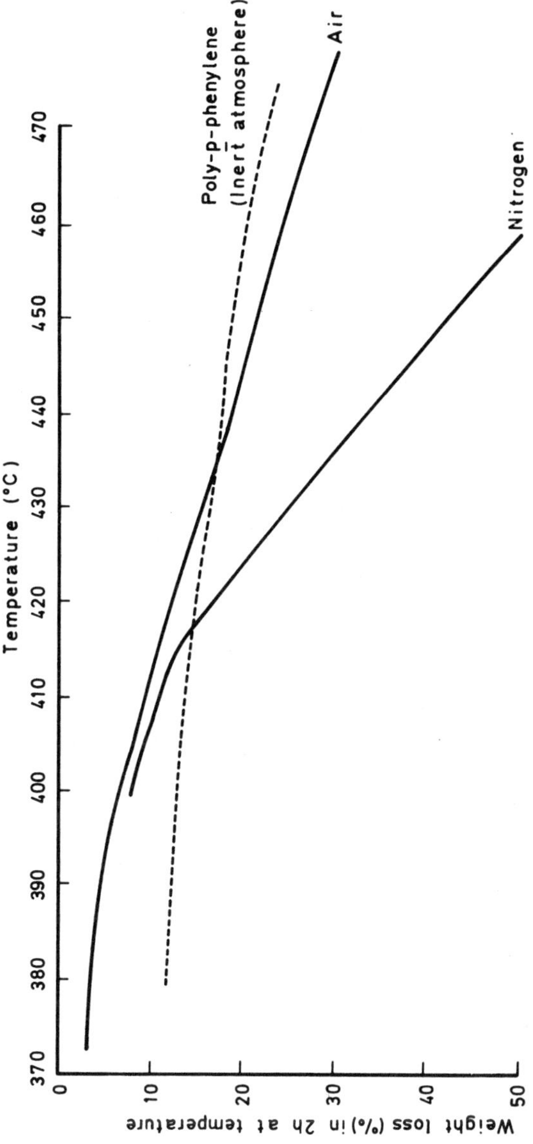

Figure 4.16. Thermal stability of poly(phenylene sulphide) in air and nitrogen.

Thermal Stability

Figure 4.16 compares the thermal stability of poly(phenylene sulphide) in air and nitrogen on a weight loss basis. Whereas the behavior is very similar up to 410°C with about 10% weight loss, above that temperature the weight loss in air becomes progressively less than in nitrogen. This is brought about by the cross-linking process mentioned above. The stability in air is approximately the same as that of poly-p-phenylene in an inert atmosphere. Pyrolysis in an evacuated ampoule of poly(phenylene sulphide) at 350°C for 1 hour gave relatively little volatile material.[60] Traces of SO_2, CO_2, and COS were detected using a mass spectrometer, and by heating the residual material to 275°C, low intensity peaks were observed corresponding to C_6H_5Br, $C_6H_4Br_2$, $C_6H_5SC_6H_5$, $BrC_6H_4SC_6H_5$, $BrC_6H_4SC_6H_4Br$, and dibenzthiophene. (The polymer sample used had obviously been prepared by the self-condensation of a metal bromothiophenoxide.) As the pyrolysis temperature was raised, the proportion of dibenzthiophene increased and hydrogen was also detected. Another study[61] using stepwise degradation gave the results cited in Table 4.14.

From these figures, the majority of the weight loss can be attributed to compounds volatile at the temperature of pyrolysis, but not sufficiently volatile to be detected in the mass spectrometer. Hydrogen sulphide is the predominant volatile product at the lower temperatures and hydrogen at the higher temperatures studied. Dibenzthiophene and compounds such as C_6H_5SH, $C_6H_5SC_6H_5$, $C_6H_5SC_6H_4SH$, and $C_6H_5SC_6H_4SC_6H_5$ were detected in the liquid-to-waxy condensate products. No bromine-substituted chain fragments were found. More recent investigations[62,63] have studied the direct pyrolysis of poly(phenylene sulphide) and oligomeric model compounds in the ion source of a mass spectrometer. In the case of the polymer, the major products observed were chain scission fragments of molecular weight up to 758. The variations in concentration of these products with temperature of pyrolysis (rate of temperature rise 50°C/minute) are shown in Figure 4.17. Quite different reaction products are obviously detected using the different experimental techniques. With sealed ampoules, secondary reactions are possible. The formation of the various products

TABLE 4.14. Volatile Products from the Thermal Degradation of Poly(Phenylene Sulphide) (Reference 61)

Temperature range (°C)	Total wt loss (%)	Wt volatiles (%)	Volatile products (mole %)							
			H_2	CH_4	H_2O	CO	H_2S	CO_2	SO_2	C_6H_6
20–450	56.5	9.3	8.5	0.4	1.4	1.0	84.5	1.1	1.8	1.3
450–550	2.5	~2.5	59.7	3.1	1.2	2.9	29.8	0.2	—	3.1
550–620	3.0	2.8	61.4	2.8	1.4	2.3	28.7	0.2	—	3.2

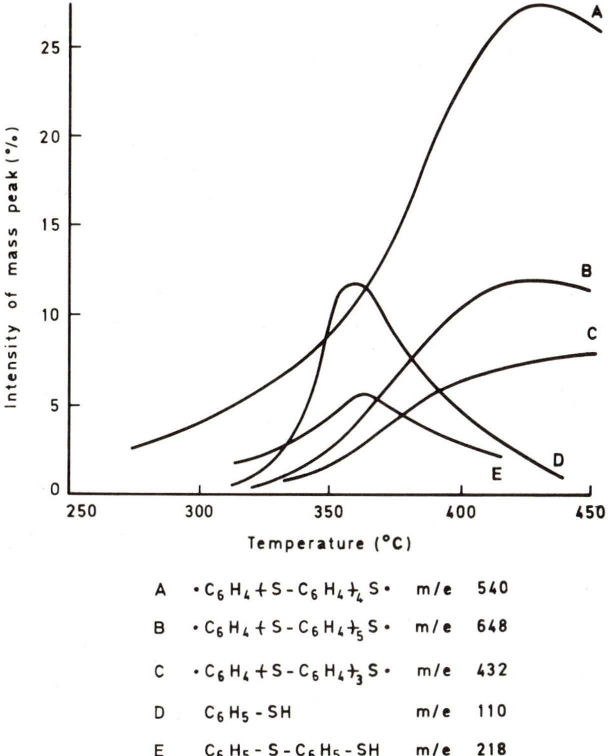

A $\cdot C_6H_4{+}S{-}C_6H_4{\displaystyle +}_4 S\cdot$ m/e   540

B $\cdot C_6H_4{+}S{-}C_6H_4{\displaystyle +}_5 S\cdot$ m/e   648

C $\cdot C_6H_4{+}S{-}C_6H_4{\displaystyle +}_3 S\cdot$ m/e   432

D $C_6H_5{-}SH$ m/e   110

E $C_6H_5{-}S{-}C_6H_5{-}SH$ m/e   218

Figure 4.17. Mass spectral thermal analysis of poly(phenylene sulphide). (Heating rate 50°C/min) Reference 62.

can readily be explained by a combination of chain-scission and transfer processes:

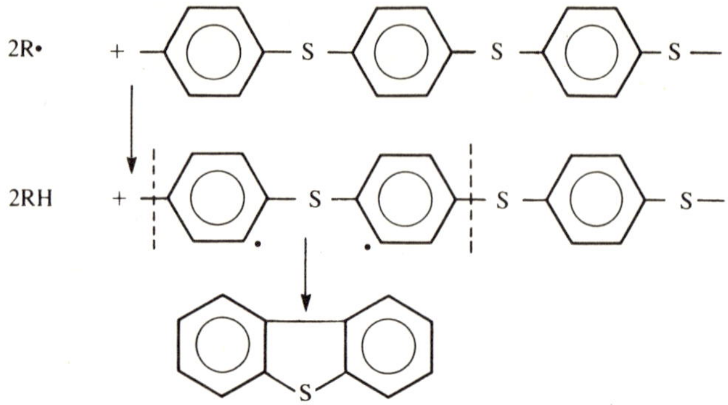

The chain transfer process can also lead to chain branching and cross-linking:

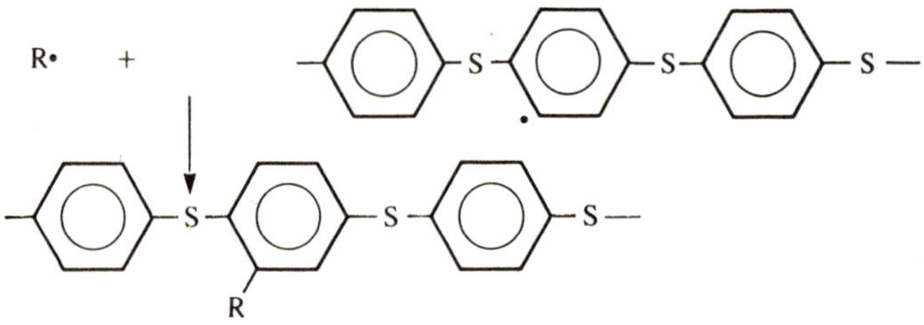

Elevated Temperature Properties

Poly(phenylene sulphide) is available in injection molding, compression molding, and surface coating grades. It has also been used in blends with other polymers as a bearing material and as the resin matrix in reinforced plastics. Table 4.15 compares the properties of unfilled and glass-filled injection molded specimens, and Table 4.16 gives figures for strength retention of glass-filled test pieces with time of aging at 232°C. Processing involves melt temperatures of 315–360°C, mold temperatures of 20–200°C, and injection pressures of 75–150 MPa. (The data of Tables 4.15, 4.16, and 4.17 and Figures 4.18 and 4.19 are taken from a technical brochure on Ryton PPS.)

There is still considerable strength retention after over one year's continuous exposure to 232°C.

Figure 4.18 shows the retention of tensile strength up to 250°C for both unfilled and glass-filled moldings. The effect of annealing (heating for 1 hour at 200°C) on property retention is also included. This improves tensile strength retention of the glass-filled material at elevated temperatures and also results in a significantly higher flexural modulus over the entire temperature range (Figure 4.19). The long-term thermal stability of poly(phenylene sulphide)-based coatings is illustrated by the weight loss results at 260°C in air given in Table 4.17.

TABLE 4.15. Properties of Poly(Phenylene Sulphide) Moldings

Property	Unfilled material (Ryton R–6)	40% Glass fiber-filled (Ryton R–4)
Density (g/ml)	1.3	1.6
Tensile strength (MPa)	66	134
Elongation at break (%)	1.6	1.3
Flexural strength (MPa)	96	200
Flexural modulus (GPa)	3.79	11.71
Compressive strength (MPa)	110	145
Heat distortion temperature (°C)	135	>260
Water absorption (24 h at 25°C) (%)	0.01	< 0.01
Dielectric strength (KV/mm)	15	14
Volume resistivity (ohm cm)	4.5×10^{16}	4.5×10^{16}
Dielectric constant at 25°C 1 kHz	3.1	3.9
1 MHz	3.1	3.8
Dissipation factor at 25°C 1 kHz	0.0005	0.0010
1 MHz	0.0009	0.0013

TABLE 4.16. Strength Retention after Aging Ryton R–4 at 232°C

Time at temperature (hours)	Flexural strength (MPa)	% Strength retention	Tensile strength (MPa)	% Strength retention
0	200	—	131	—
700	166	84	82	63
1500	131	66	82	63
3000	124	63	82	63
5000	118	59	82	63
7200	89	45	75	57
9400	62	31	48	37

TABLE 4.17. Thermal Stability of Poly(Phenylene Sulphide)-Based Coatings at 260°C in Air

Time at temperature (hours)	Weight loss%	
	Coating A	Coating B
24	0.003	0.02
100	0.06	0.07
500	0.18	0.21
1000	0.50	0.34
1182	0.47 (cracked)	0.31
1686	—	0.95 (cracked)

Coating A = 3 parts poly(phenylene sulphide) (RYTON)
             1 part titanium dioxide

Coating B = 3 parts poly(phenylene sulphide) (RYTON)
             1 part titanium dioxide
             0.3 parts polytetrafluoroethylene

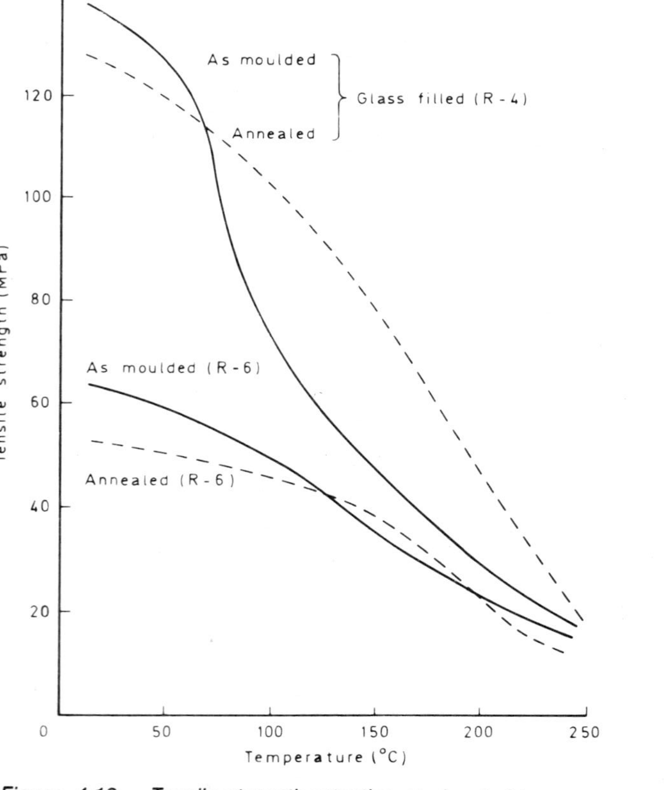

Figure 4.18. Tensile strength retention at elevated temperatures of poly(phenylene sulphide) moldings.

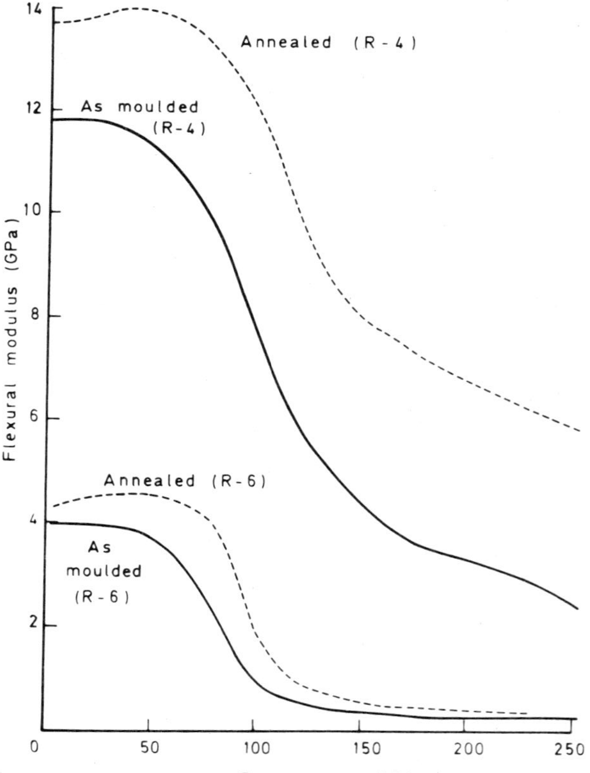

Figure 4.19. Flexural modulus retention at elevated temperatures of poly(phenylene sulphide) moldings.

The coatings require an oven bake at 370°C for periods on the order of 45 minutes.

Poly(phenylene sulphide) has been evaluated as a bearing material.[64] The best performance was attained with a composition of 55% poly(phenylene sulphide), 25% polytetrafluoroethylene, 10% graphite, and 10% lead oxide. The wear behavior of this formulation compared with that of an epoxy resin-based formulation is shown in Figure 4.20. The superiority of the poly(phenylene sulphide) composition at elevated temperatures is evident. Combinations of poly(phenylene sulphide) with other resins have also been examined as candidate-bearing materials.[65,66] Formulations tested include 34% poly-

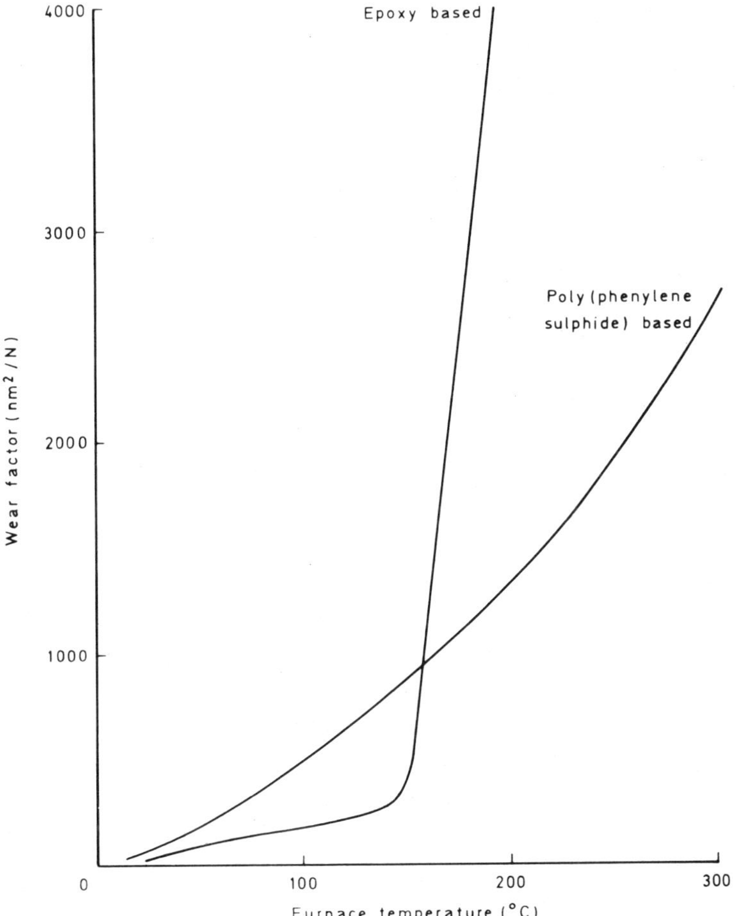

Figure 4.20. Wear rate of poly(phenylene sulphide) and epoxy resin-based bearings as a function of temperature. Reference 64.

TABLE 4.18. *Mechanical Properties of Ryton/T-300 Unidirectional Laminates (Reference 67)*

Property	Condition	Room temperature	121°C	177°C
Flexural strength (0°) MPa	Dry	1162	687	449
MPa	Wet	755	517	339
Flexural strength (90°) MPa	Dry	195	126	103
MPa	Wet	212	174	143
Flexural modulus (0°) GPa	Dry	84	86	73
GPa	Wet	86	75	66
Flexural modulus (90°) GPa	Dry	11	9.0	8.4
GPa	Wet	14	8.6	5.2
Tensile strength (0°) MPa	Dry	818	642	564
Transverse tensile strength MPa	Dry	103	78	67
Tensile modulus (0°) GPa	Dry	108	84	90
Transverse tensile modulus GPa	Dry	14	17	3.6

(phenylene sulphide), 51% polyimide resin, and 15% graphite,[65] and 65/35 to 35/65 mixes of poly(phenylene sulphide) and phenolic resin.[66] In the latter case, using a fluorosilicone lubricating oil, a bearing ran for 500 hours without failure at a speed of 0.5 m/s, a load of 570 KPa, and a temperature of 192°C.

Carbon fiber-reinforced composites have been made by a film stacking technique using poly(phenylene sulphide) as the matrix.[67] Mechanical properties of the laminates at temperatures up to 177°C are summarized in Table 4.18.

It is claimed that these figures indicate use temperatures up to at least 150°C and possibly higher for low load structures.

POLY(PHENYLENE ETHER SULPHONES)

Preparation

A polymer chain consisting only of phenylene rings linked by sulphone groups appears to be too rigid to show thermoplasticity.[68] Accordingly, other more flexible links must also be incorporated into the chain and attention has concentrated upon the ether unit. Currently four polymers of this poly(phenylene ether sulphone) type are commercially available:

1. $\left[-\!\!\left\langle\bigcirc\right\rangle\!-\!\underset{\underset{CH_3}{|}}{\overset{\overset{CH_3}{|}}{C}}\!-\!\left\langle\bigcirc\right\rangle\!-\!O\!-\!\left\langle\bigcirc\right\rangle\!-\!SO_2\!-\!\left\langle\bigcirc\right\rangle\!- \right]_n$ Udel *P*1700 (Union Carbide)

2. 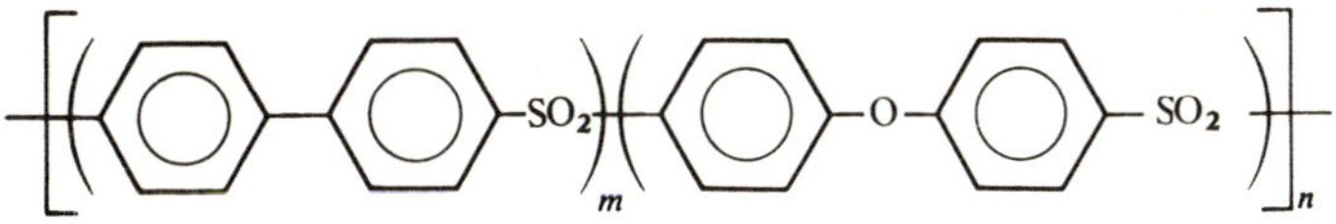 Astrel 360 (3M's)

$x > y$

3. 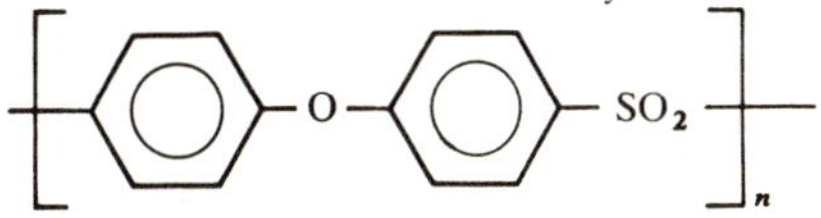 Polyethersulphone 200P and 300P (Victrex) (ICI)

4. Structure not disclosed, but infrared spectrum closest to that of polymer 3 — Radel (Union Carbide)

These polymers can be made either by polyetherification or polysulphonylation. The Udel material was the first to be produced commercially (1965) by the reaction of the disodium salt of bisphenol-A and 4,4′-dichlorodiphenylsulphone in a solvent such as dimethyl sulphoxide or sulpholane.[69] In this polyetherification route, the sulphone group is an essential part of the dihalide as it activates the halogens to attack by the phenoxide groups. Aromatic halides that do not contain powerful electron-withdrawing groups are unreactive and high polymers are not formed. The dipolar aprotic solvent also plays an important part in the reaction, as it dissolves both reactants and the polymer and increases the overall rate of reaction.

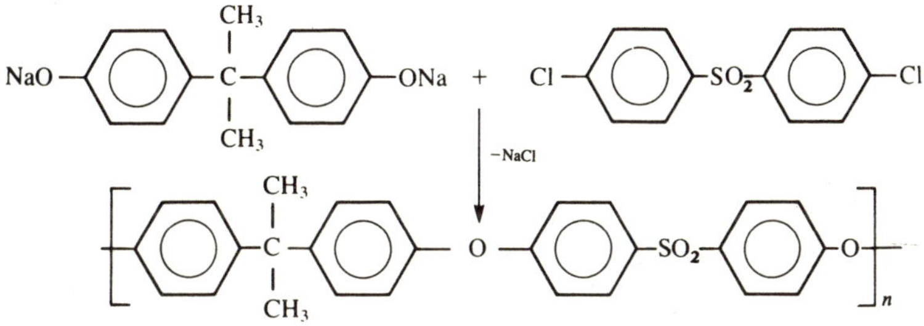

The Astrel 360 polymer (introduced in 1967) was synthesized by a polysulphonylation process involving the Friedel–Crafts polycondensation of aromatic sulphonyl chlorides with aromatic hydrocarbons[70]:

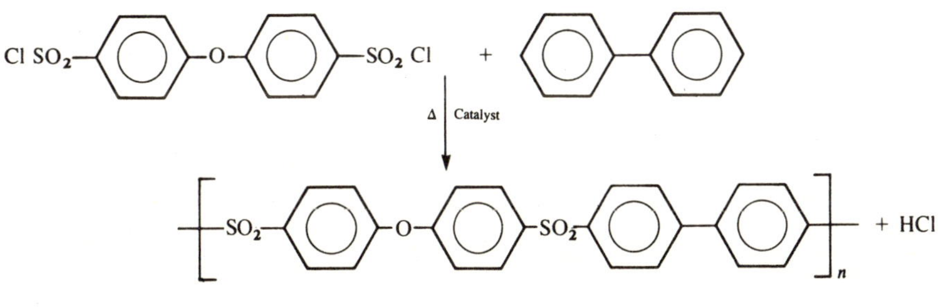

Much of the success of this synthesis is due to the unique activity of certain Friedel–Crafts catalysts. In contrast to acylation with carboxylic acid chlorides, which requires the use of molar quantities of Friedel–Crafts catalysts, sulphonylation with sulphonyl chlorides requires only small, catalytic amounts of Lewis acids such as ferric chloride, antimony pentachloride, molybdenum pentachloride, indium trichloride, or trifluoromethanesulphonic acid. The choice of catalyst is important, as it minimizes side reactions and the problem of catalyst removal from the polymers. The polymerization can be carried out in the melt, but it is better to use a diluent as reaction medium. The preferred solvents are nitrobenzene and dimethylsulphone.

Although ICI's polyethersulphone was not marketed until 1972, papers relating to the preparation of the polymer appeared several years before this.[68, 71–73] The polyetherification route is used because it leads to linear, controlled structures, whereas in polysulphonylation branching occurs. Polymerizations in the melt and in solution have been described.

Melt polycondensation:

$$F{-}\langle\bigcirc\rangle{-}SO_2{-}\langle\bigcirc\rangle{-}OK \xrightarrow[\text{1 hour}]{280^\circ} {-}\!\left[\langle\bigcirc\rangle{-}SO_2{-}\langle\bigcirc\rangle{-}O\right]_n$$

Solution polymerization:

$$KO{-}\langle\bigcirc\rangle{-}SO_2{-}\langle\bigcirc\rangle{-}OK \;+\; Cl{-}\langle\bigcirc\rangle{-}SO_2{-}\langle\bigcirc\rangle{-}Cl \xrightarrow[\text{260}^\circ\text{C}]{\text{Sulpholane}}$$

$$\left[O{-}\langle\bigcirc\rangle{-}SO_2{-}\langle\bigcirc\rangle\right]_n$$

$$\text{or} \quad Cl{-}\langle\bigcirc\rangle{-}SO_2{-}\langle\bigcirc\rangle{-}OK$$

The chemistry of these reactions has been extensively documented.[74–80] The fourth polysulphone, Radel, entered the market in 1976 and, as stated earlier, its structure and method of preparation have not been disclosed.

Two other routes to polysulphones that deserve mention, although they are not at present used commercially, are the oxidation of poly(p-phenylene sulphide) to yield poly(p-phenylene sulphone)[81] and the use of acetylene-terminated sulphone oligomers.[82] The latter have been evaluated as the matrix for carbon fiber-reinforced composites (see later under elevated temperature properties).

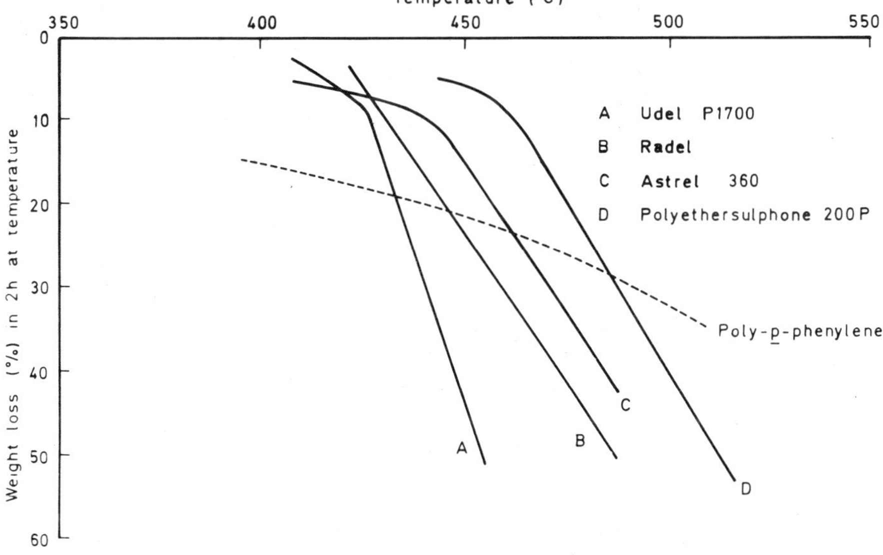

Thermal Stability

A comparison of the thermal stability of the commercially available polyethersulphones in nitrogen and in air, based upon the weight lost in two hours at elevated temperatures, is shown in Figures 4.21 and 4.22, respectively. Overall activation energies for degradation in both inert and oxidizing atmospheres, derived from the initial parts of the weight loss curves, are listed in Table 4.19. The order of thermal stability in nitrogen is 200P > Astrel 360 > Radel > Udel P1700. In air, however, the difference between the materials is much less

Figure 4.21. Thermal stability of polyethersulphones in nitrogen.

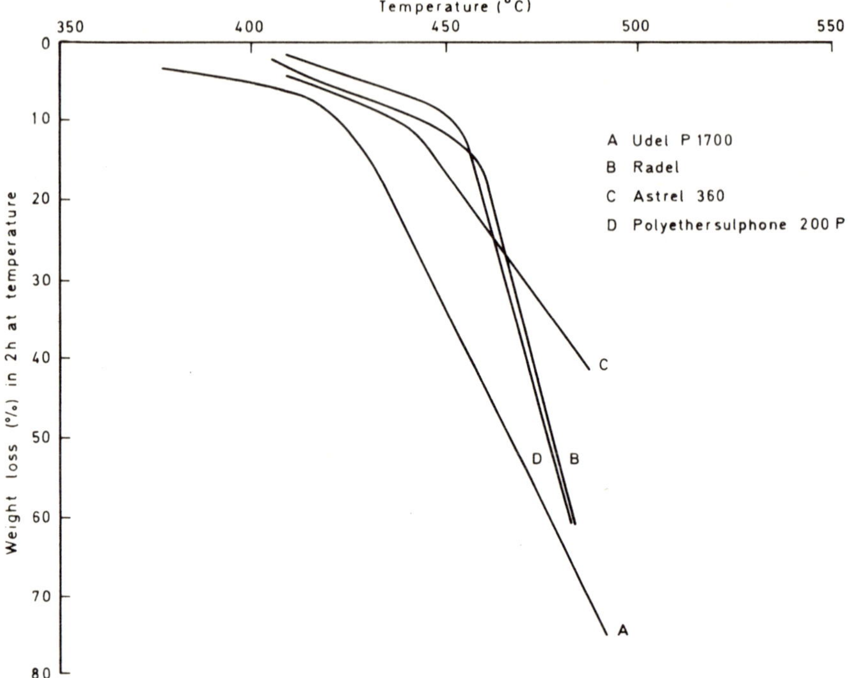

Figure 4.22. *Thermal stability of polyethersulphones in air.*

marked and 200P $\approx$ Astrel 360 $\approx$ Radel > Udel P1700. The presence of the isopropylidene group in Udel P1700 accounts for its somewhat lower stability.

A reason for the much lower activation energy for degradation in air of Radel cannot be advanced without knowing its structure and method of preparation.

The thermal degradation of the Udel P1700 polymer has been the subject of several studies.[61,83,84] It is generally agreed that the major volatile decomposition products at temperatures up to 450°C are sulphur dioxide and methane. As

TABLE 4.19. Overall Activation Energies
for Thermal Degradation of Polyethersulphones
(Reference 37)

Polymer	Activation energy (kJ/mole)	
	In nitrogen	In air
Udel P1700	270	137
Radel	270	78
Astrel 360	310	172
Polyethersulphone 200P	311	183

TABLE 4.20. Volatile Products from the Thermal Degradation of Polysulphone Udel P1700 (Reference 61)

Temperature range (°C)	Total wt loss (%)	Wt volatiles (%)	H_2	CH_4	H_2O	CO	H_2S	CO_2	SO_2	C_6H_6	$C_6H_5CH_3$
20–350	3.7	1.8	3.0	12.7	11.2	9.6	—	6.0	48.5	—	—
20–400	45.3	16.8	4.9	19.9	1.5	1.3	1.0	0.4	70.3	0.5	0.2
20–450	61.0	23.2	9.4	18.3	1.0	6.9	4.9	0.8	57.3	1.0	0.4
450–550	6.5	3.9	61.7	11.4	1.9	22.0	0.8	1.3	—	0.8	0.2
550–620	0.5	1.6	80.0	7.5	0.6	11.6	0.3	0.4	—	0.5	0.1

the temperature is raised still further, hydrogen, carbon dioxide, hydrogen sulphide, benzene, and toluene are also produced. Other products include diphenylethers and phenols. The results of mass spectrometric analysis of the volatile degradation products are given in Table 4.20.

In addition to scission processes giving rise to these compounds, cross-linking also takes place. If a polymer sample is heated in vacuum for periods greater than three hours, gel formation occurs, the quantity increasing with the time of heating and reaching a constant value of about 75% after twenty hours at temperature.[84] A mechanism as shown below accounts for the major features of the breakdown.

Russian workers have studied the thermal degradation of polysulphones of the same structure as Astrel 360 and 200P. The products of decomposition found and the reaction mechanisms proposed are very similar to those for Udel P1700, with the obvious exception of reactions involving the isopropylidene group.[85,86]

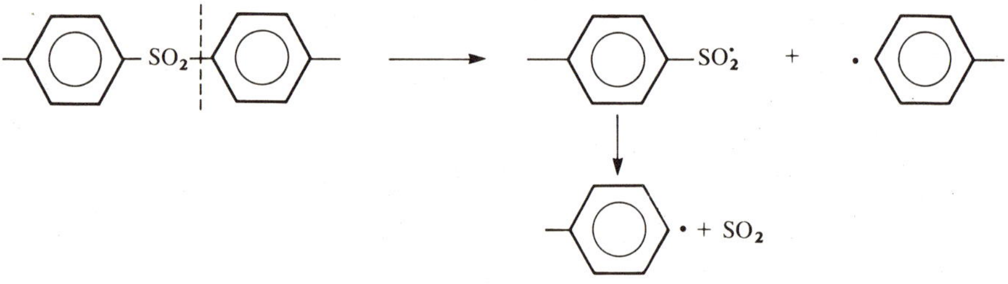

$$SO_2 + 3H_2 \longrightarrow H_2S + 2H_2O$$

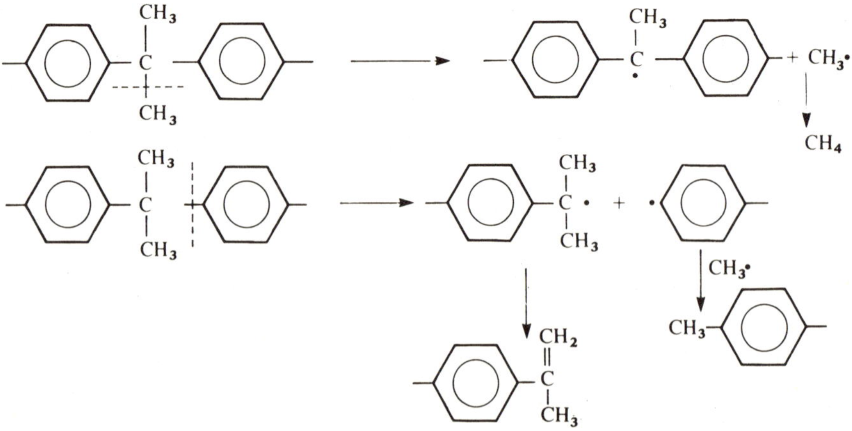

3.

It should be noted that, although the polysulphones have good thermal stability and good resistance to high energy radiation,[87] they are adversely affected by exposure to ultraviolet radiation.[88,89] This results in yellowing, decrease in molecular weight, elongation at break, and tensile strength. The change in properties has in fact been used as the basis of a UV dosimeter.[90]

Elevated Temperature Properties

Before citing properties at elevated temperature, it is relevant to compare the initial room temperature properties of the four polysulphones, and this is done in Table 4.21. (The data are taken from the manufacturer's technical brochures for the materials, as are the figures in Table 4.22 and Figures 4.23 and 4.24.) Differences that stand out are the high heat distortion temperature of

TABLE 4.21. Comparison of Properties of Polysulphone Polymers

Property	Units	Udel P1700	Astrel 360	200P	Radel
Specific gravity	—	1.24	1.36	1.37	1.29
Tensile strength	MPa	70	90	84	72
Tensile modulus	GPa	2.48	2.55	2.44	2.14
Elongation at yield	%	5–6	13	—	7
Flexural strength	MPa	106	119	129	86
Flexural modulus	GPa	2.69	2.73	2.57	2.28
Compressive strength	MPa	96	124	—	—
Compressive modulus	GPa	2.55	2.35	—	—
Izod impact (notched)	J/mm	0.7	2.7	0.85	6.4
Rockwell hardness	—	M69	M110	M88	—
Heat distortion temperature (1.82 MPa)	°C	174	275	203	204
Coefficient of thermal expansion	cm/cm/°C	5.5×10^{-5}	4.6×10^{-5}	5.5×10^{-5}	5.5×10^{-5}
Dielectric constant at 60 Hz	—	3.14	3.94	3.5	3.44
Dissipation factor at 60 Hz	—	0.008	0.003	0.001	0.0006
Volume resistivity	ohm-cm	5×10^{16}	3×10^{16}	10^{17}–10^{18}	9×10^{16}
Water absorption in 24 hours	%	0.22	1.8	0.43	—

TABLE 4.22. *Effect of Temperature on Mechanical Properties of Radel and Astrel 360*

Polymer	Property	Room temperature	Temperature 205°C		Room temperature after 21 days at 205°C
Radel	Tensile strength at yield (MPa)	72	28	(39)[a]	83
	Tensile modulus (GPa)	2.14	1.17	(55)	2.22
	Elongation at yield (%)	7.0	11.0		7.4
		Room temperature	260°C		
Astrel 360	Tensile strength (MPa)	90	28	(31)	—
	Tensile modulus (GPa)	2.55	—		—
	Elongation at yield (%)	13	7		—
	Flexural strength (MPa)	119	61	(51)	—
	Compressive strength (MPa)	124	52	(42)	—

[a]Figures in parentheses are percentage retention.

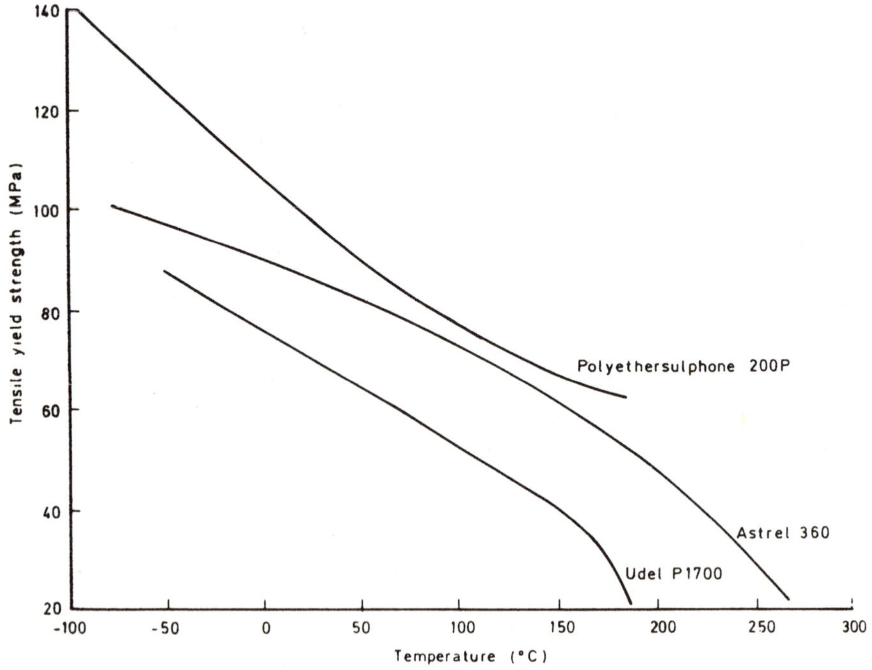

Figure 4.23. *Tensile strength of polyethersulphones as a function of temperature.*

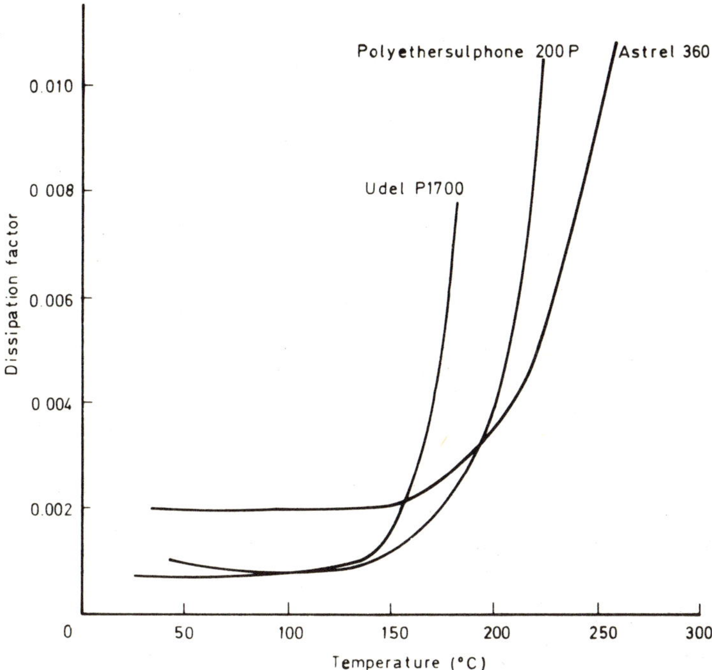

Figure 4.24. Dissipation factor at 60 Hz of polyethersulphones as a function of temperature.

Astrel 360 and the high impact strength (i.e., toughness) of Radel. Figures 4.23 and 4.24 compare the tensile strength at yield and the dissipation factor at 60 Hz as a function of temperature for three of the polymers. As can be seen, there is a slow but steady loss in tensile strength with rising temperature, the retention at 150°C for the two more heat-resistant polymers being about 70%. The dissipation factor is relatively unchanged for all three materials until the temperature is raised above 150°C. For Astrel 360 the curves for flexural and compressive strength lie parallel to the curve for tensile strength in Figure 4.23, but at a level about 35 MPa higher. Over the temperature range 20–260°C, the flexural modulus decreases by approximately 40% and the elongation at yield by 50%. In contrast, the flexural modulus of the less stable Udel P1700 has fallen by 55% at 200°C. Table 4.22 cites some properties of Radel at elevated temperature and includes some data for Astrel 360 by way of comparison.

Of more direct interest is the effect of aging at temperature upon properties. Figure 4.25 shows the effect of heating at 210°C for extended periods upon the properties of Udel P1700 measured at room temperature.[91] This gives half-lives (i.e., time for properties to fall to 50% of their initial value) at 210°C of > 187 days for tensile strength, 134 days for notched impact strength, and 44 days for

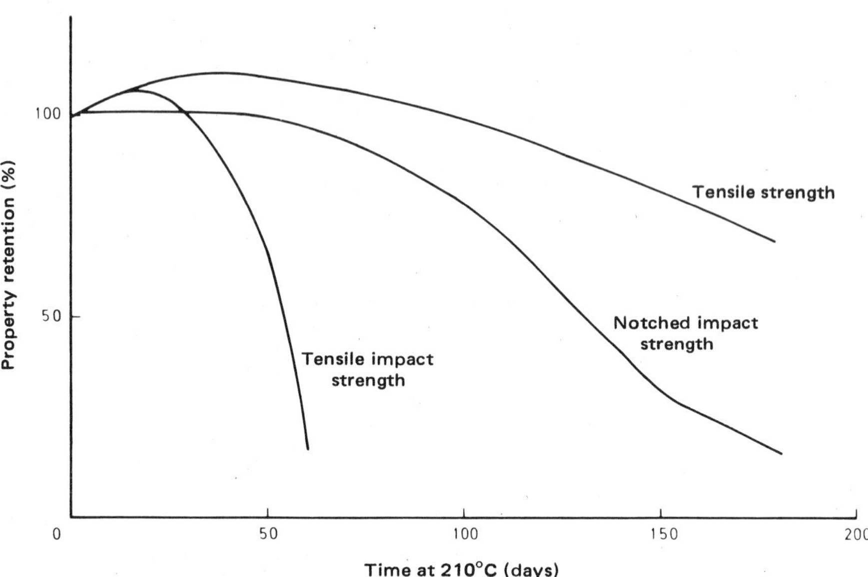

Figure 4.25. Effect of aging at 210°C on the mechanical properties of Udel P1700 polysulphone. Reference 91.

tensile impact strength. The 200P material effectively shows no loss in room temperature tensile strength after aging for a year at 150°C, and Astrel 360 behaves similarly up to 2,000 hours at 260°C. If the measurement is made at 260°C, there is a 66% loss in strength immediately, but relatively little change thereafter up to 2,000 hours. Data such as these indicate the possibility of continuous use at temperatures up to 150°C for Udel P1700, 175°C for polyethersulphone 200P, and 195–210°C for Astrel 360.

All the figures quoted so far relate to molded unreinforced material. More recently, however, the polysulphones have been investigated as thermoplastic matrices in fiber-reinforced composites using glass, carbon, or Kevlar fibers as reinforcement.[92–97] The advantages of using thermoplastic polymers in this way are lower fabrication costs, shorter processing times, longer shelf-lives, less waste, no chemical reactions involved in the processing operation, and tougher composites. Disadvantages are high fabrication temperatures (of necessity above the glass transition temperature of the polymer), creep at elevated temperatures, and lower chemical resistance. The effect of process variables upon the dry and wet shear strength of composites made from Udel P1700 and S-glass, AS carbon fibers, and Kevlar 49 has been dealt with in considerable detail by Scola and Roylance.[95] The aim was to improve fiber wetting and fiber distribution. Hoggatt[92] has also reported results on Udel P1700/Kevlar 49 and Polyethersulphone 200P/AS carbon fiber composites. Tests on the former were

TABLE 4.23. *Retention at Elevated Temperatures of Initial Mechanical Properties of Polyethersulphone 200P/AS Carbon Fiber Composites (Reference 92)*

Property	Temperature (°C)	Retention of initial room temperature value (%)
Tensile strength	175	90
Tensile modulus	175	100
Compressive strength	175	73
Compressive modulus	175	100
Flexural strength	160	58
Flexural strength	145	73
Flexural modulus	160	100
Interlaminar shear strength	160	51
Interlaminar shear strength	145	61

restricted to a maximum temperature of 120°C because of the nature of the reinforcement. Retentions of mechanical properties at temperatures up to 175° for the 200P/AS composites are listed in Table 4.23. Particularly noteworthy are the unchanged moduli. Table 4.24 gives values of mechanical properties of Radel/8H Satin T-300 carbon cloth composites, both dry and wet and after aging for 1,000 hours at 175°C. Good retention of properties at elevated temperature is shown by composites saturated with water. Similar behavior has been observed for acetylene-terminated sulphone/AS carbon fiber specimens,[82] which gave interlaminar shear strengths dry and wet (1.2% water content) of 62 and 53 MPa at room temperatures, and 43 and 33 MPa at 175°C (i.e., retentions of 85 and 77%, respectively). The most comprehensive comparison of the polysulphones, however, together with other thermoplastic resins as the matrix in glass-reinforced composites, has been made by Theberge and Cloud.[97] Their compilation details thermal and mechanical properties, tensile strength versus tempera-

TABLE 4.24. *Mechanical Properties of Radel/8H Satin T-300 Carbon Cloth Composites (Reference 96)*

Property	Test Condition	Test Temperature (°C) Room temperature	175
Flexural strength (MPa)	Dry	546	321 (59)[a]
	Wet	423	314 (74)
	Dry after 1000 hours at 175°C	575	453 (79)
Flexural modulus (GPa)	Dry	29	21 (72)
	Wet	24	25 (100)
	Dry after 1000 hours at 175°C	33	33 (100)
Interlaminar shear strength (MPa)	Dry	62	46 (74)
	Wet	59	37 (63)

[a]Figures in parentheses are percentage retention of equivalent room temperature values. Wet means a water content of 0.46–0.58%.

TABLE 4.25. Tensile Strength at Elevated Temperatures of Some Glass-Reinforced Thermoplastics (Reference 97)

Polymer type	Glass content (Wt %)	Tensile strength (MPa) at (°C)					
		23	95	150	175	205	230
Polysulphone (P1700)	40	119	103	16	8	—	—
Polyethersulphone (200P)	40	157	134	90	34	21	—
Polyarylsulphone (Astrel 360)	0	90	72	60	51	39	22
Poly(phenylene sulphide)	40	160	77	56	33	8	—
Poly-*p*-oxybenzoate	0	96	77	64	54	44	27
Polyimide	30	90	43	33	21	16	12
Polyamideimide	0	189	137	112	79	57	48
TFE/HFP copolymer	20	35	29	16	8	—	—
TFE/E copolymer	20	78	47	43	14	—	—

ture, tensile stress relaxation at elevated temperature, tensile strength after aging at 205 and 260°C, and chemical resistance to a variety of liquids for 7 days at room temperature, 3 days at 80°C, and 1 day at 150°C for glass fiber-reinforced polysulphone, polyethersulphone, polyarylsulphone, poly(phenylene sulphide), poly-*p*-oxybenzoate, polyimide, polyamideimide, polytetrafluoroethylene, tetrafluoroethylene/hexafluoropropene copolymer, tetrafluoroethylene/ethylene copolymer, a polyester, and nylon 6/6. We have extracted from their data tensile

TABLE 4.26. Effect of Aging at Elevated Temperature on the Tensile Strength of Some Glass-Reinforced Thermoplastics (Reference 97)

Polymer type	Glass content (Wt %)	Tensile strength (MPa) after aging at 205°C for time (hours) of						
		0	100	250	500	750	1000	1500
Polysulphone	40	140	104	69	54	30	—	—
Polyethersulphone	40	157	159	134	105	88	104	78
Polyarylsulphone	0	90	91	92	90	89	89	88
Poly(phenylene sulphide)	40	160	164	159	158	141	128	114
Poly-*p*-oxybenzoate	0	96	98	97	96	96	96	95
Polyimide	30	90	95	95	95	95	92	90
Polyamideimide	0	189	188	186	183	169	164	155
TFE/HFP copolymer	20	35	35	32	32	32	32	32
TFE/E copolymer	20	78	81	79	79	78	78	77

Polymer type	Glass content (Wt %)	Tensile strength (MPa) after aging at 260°C for time (hours) of						
		0	100	250	500	750	1000	1500
Polysulphone	40	14.0	melted	—	—	—	—	—
Polyethersulphone	40	157	108	102	99	95	98	72
Poly(phenylene sulphide)	40	160	113	110	107	104	100	95
Polyimide	30	90	104	99	92	88	83	77
TFE/E Copolymer	20	78	79	69	48	35	26	16

strength at elevated temperature and after aging at elevated temperature for those polymers with an Underwriters Laboratory rating of continuous use of at least 150°C. These figures are given in Tables 4.25 and 4.26. As can be seen, the performance of the completely aromatic sulphone polymers is very good.

POLY(PHENYLENE ETHER KETONES)

Preparation

Poly(phenylene ether ether ketone)—PEEK—has only recently become available commercially, being first marketed by ICI in development quantities in 1978.

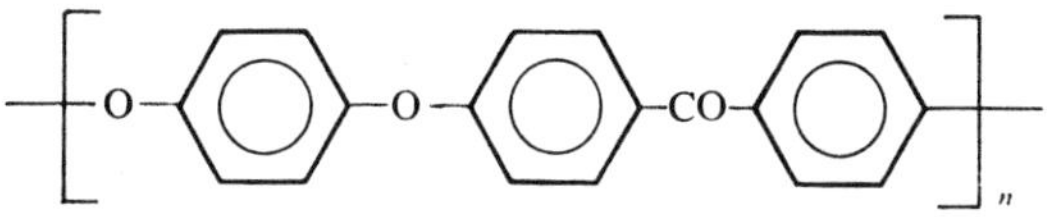

PEEK (Victrex)

It has proved more difficult to develop procedures yielding high-molecular-weight polymer than was the case with the poly(phenylene ether sulphones). This is due to different solid state morphology; whereas poly(phenylene ether sulphone) is amorphous, poly(phenylene ether ketone) is crystalline[98] and separates out during polymerization. Poly(phenylene ether ketones) may be prepared by acylation reactions using acid chlorides in the presence of Lewis acid catalysts.[99-103] The products, as isolated, tend to be of low molecular weight (intrinsic viscosities 0.2 to 0.5).

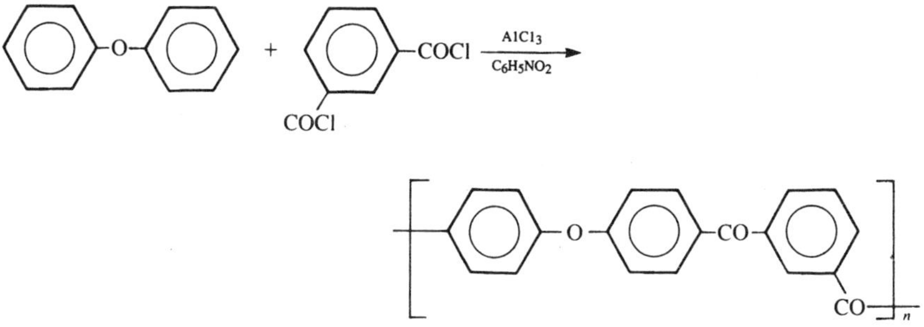

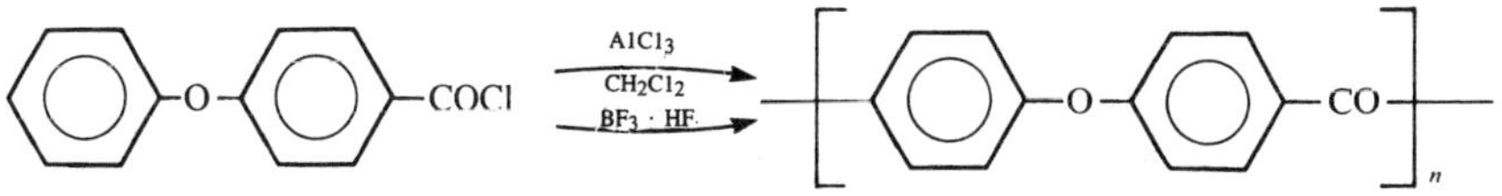

The BF$_3$HF system is exceptional in giving polymers with intrinsic viscosities in the range 1.0 to 2.8. ICI has employed a polyetherification route first described by Clendinning *et al.*;[104] by using certain diaryl sulphones as the polycondensation solvents at temperatures close to the polymers' melting points, ICI has prepared poly(arylether ketones) of high molecular weights[105] (reduced viscosities up to 6.0). The mechanism of the reaction is discussed in detail.

$$\text{Halide}-\!\!\bigcirc\!\!-\text{CO}-\!\!\bigcirc\!\!-\text{Halide} \quad + \quad \text{KO}-\!\!\bigcirc\!\!-\text{CO}-\!\!\bigcirc\!\!-\text{OK}$$

$$\Big\downarrow \ \text{C}_6\text{H}_5\text{SO}_2\text{C}_6\text{H}_5$$

$$-\!\!\Big[\!\!-\text{O}-\!\!\bigcirc\!\!-\text{CO}-\!\!\bigcirc\!\!-\text{O}-\!\!\bigcirc\!\!-\text{CO}-\!\!\bigcirc\!\!-\Big]_n\!\!-$$

Thermal Stability

Figure 4.26 illustrates the thermal stability of poly(phenylene ether ketone) PEEK in nitrogen and air on the basis of the weight lost in two hours at elevated temperature. If this graph is compared with the others of the same kind in this chapter, it can be seen that the stability of PEEK is excellent; in fact, the curves lie in the highest temperature range of any of the polymers studied. The thermal degradation of the polymer has not been studied in detail, but the combustion of PEEK-coated wire is reported[106] to give only carbon monoxide and carbon dioxide as combustion products, the temperature for maximum rate of carbon monoxide evolution being 550°C.

Elevated Temperature Properties

Little has been published to date on the elevated temperature properties of PEEK. The following information has been extracted from ICI provisional data sheets[106] and merely serves to give an indication of high-temperature behavior.

The estimated life of the PEEK wire insulation is at least 50,000 hours at 200°C. The coating is also extremely resistant to embrittlement by gamma radiation, its tolerance far exceeding that of polystyrene.

Details have been given of the conditions required for injection molding, rotational molding, and internal coating using PEEK. The elevated temperature

TABLE 4.27. *Selected Properties of Poly(Phenylene Ether Ketone) PEEK (Reference 106)*

Property	Units	Value
Glass transition temperature	°C	143
Melting point	°C	334
Processing temperature range	°C	370–400
Melt thermal stability at 400°C	hours	1
Compression molded unfilled specimens		
Tensile modulus at 150°C	GPa	1.1
Tensile modulus at 180°C	GPa	0.4
Tensile strength of notched samples at 23°C		
0.25 – 2.0 mm notch	MPa	112–132
Tensile strength of notched samples at 200°C		
0.25 – 2.0 mm notch	MPa	34
Creep in tension after 1 week at		
150°C and 10 MPa stress	%	1.73
180°C and 5 MPa stress	%	1.70
Limiting oxygen index	%	35
Glass-reinforced specimens		
(25% glass) Creep in tension after 1 week at		
180°C and 3 MPa stress	%	0.55
180°C and 6 MPa stress	%	0.82
180°C and 10 MPa stress	%	1.69
Heat distortion temperature at 1.81 MPa		
10% glass content	°C	209
20% glass content	°C	286
Film material		
Power factor 20 – 160°C at 50 Hz	—	0.001–0.004
Power factor 20 – 160°C at 10 Hz		0.002–0.003
Volume resistivity 20 – 140°C	ohm cm	$> 10^{15}$
Wire insulation		
Scrape abrasion at 150°C	Strokes	70–112
Dynamic cut through at 150°C	kg	33.0–39.4
Heat embrittlement at 200°C	—	No cracks
Air aging at 200°C followed by voltage proof test	—	Passed
Coating cohesion after heating twisted wires at 200°C for 6 hours	—	No sticking
Shrinkage of 305 mm length after heating at 200°C for 6 hours	mm	3
Current overload to give 300°C for 15 minutes	—	No smoke produced
Limiting oxygen index	%	40

properties of injection moldings containing 20 and 30% chopped strand carbon fiber are given in Table 4.28.

Laminates have been made by hot pressing layups consisting of alternate layers of PEEK film and glass cloth, or carbon fiber in continuous filament, mat,

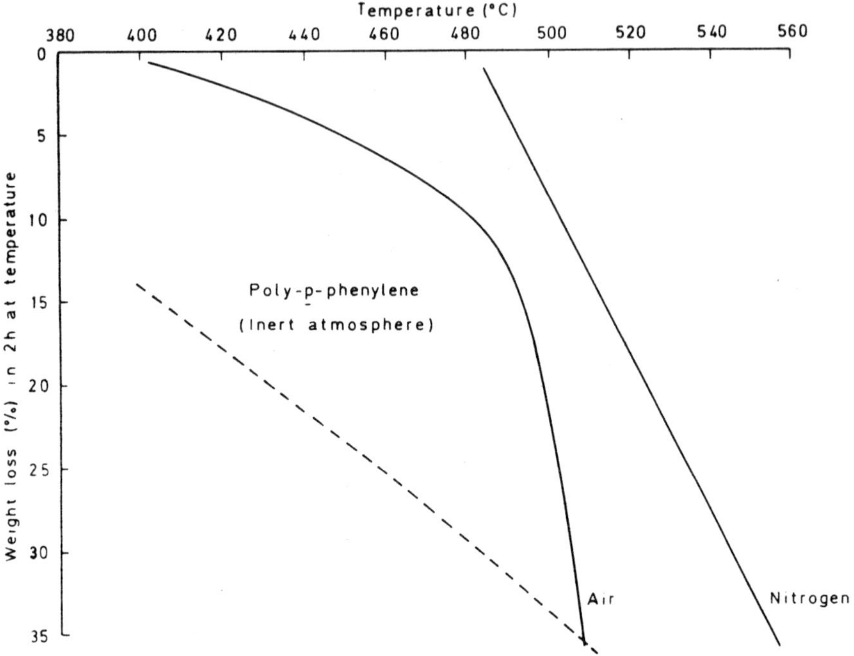

Figure 4.26. Thermal stability of poly(phenylene ether ketone) PEEK in nitrogen and air.

or woven cloth form. Properties of glass cloth laminates are listed in Table 4.29 and of carbon fiber laminates in Table 4.30.

It should be emphasized that these are preliminary results and processing has not yet been optimized.

TABLE 4.28. *Mechanical Properties at Elevated Temperatures of PEEK Reinforced with Chopped Strand Carbon Fibers (Reference 106)*

	20% Carbon fiber		30% Carbon fiber	
Temperature	Tensile strength (MPa)	Flexural modulus (GPa)	Tensile strength (MPa)	Flexural modulus (GPa)
23	165	12.5	215	15.5
100	127	9.5	185	12.2
150	82	4.0	107	10.0
200	33	2.5	67	3.5
250	30	—	49	—
300	29	—	38	2.1

TABLE 4.29. Mechanical Properties at Elevated Temperatures of PEEK/Glass Cloth Laminates (Reference 106)

Temperature (°C)	Flexural modulus (GPa)
23	14.7
100	14.2
150	13.0
200	11.2
300	9.0

OTHER POLYMERS WITH AROMATIC RINGS IN THE CHAIN

Although many poly(phenylene ethers)—also called poly(phenylene oxides)—have been synthesized, and the thermal stability of polymers such as poly(1,4-phenylene ether) containing no substituents in the aromatic ring has been shown to be very good,[107] the only polymer to receive extensive commercial development has been poly(2,6-dimethyl phenylene ether). This is prepared by oxidation of 2,6-xylenol, the reaction being catalyzed by a cuprous salt and a tertiary amine. The heat distortion temperatures of the various commercial grades of this polymer (Noryl, General Electric Company) lie between 100 and 150°C, and they are rated for continuous use at 110–120°C. Because of this low limit they will not be considered further here. For detailed reviews, readers are referred to the articles by Lee *et al.* [50] and A. S. Hay *et al.*[108]

$$n \left[\underset{CH_3}{\overset{CH_3}{\bigcirc}} OH \right] + \frac{n}{2} O_2 \longrightarrow \left[\underset{CH_3}{\overset{CH_3}{\bigcirc}} O \right]_n + nH_2O$$

Another aromatic polymer with a recommended maximum service temperature of only 135°C is the polycarbonate derived from the reaction of bisphenol-A

TABLE 4.30. Mechanical Properties at Elevated Temperatures of PEEK/Carbon Fiber Laminates (Reference 106)

Carbon fiber construction	Fiber content (w/w%)	Tensile strength (MPa)	Elongation at break (%)	Tensile modulus(GPa) at 23°C	200°C	300°C
Woven mat	50	360	1.6	28	14	8
Unidirectional tow	60	1280	1.3	106	53	32
Unidirectional tow	70	1600	1.3	128	64	38

and phosgene. This material is especially noteworthy for its good impact properties and is marketed by Bayer Polymeric Materials as Makrolon and by General Electric Company as Lexan. For reviews of the polycarbonates, see the books by Christopher and Fox,[109] by Schnell,[110] and the article by Bottenbruch.[111]

SUMMARY

1. Despite its central role in this area, attempts to produce useful forms of polyphenylene have not met with marked success. This relates to the difficulties of fabricating a material composed only of phenylene nuclei. As a consequence, all the successful developments have comprised phenylene rings linked by other groups.

2. The structures that have been commercialized are all relatively simple, and only in the case of poly(phenylene ether) and two of the poly(xylylenes) are there substituents on the rings.

3. Many of the polymers are highly crystalline, e.g., the polyesters, polyamides, poly(phenylene sulphide), and poly(phenylene ether ketone).

4. Unusual processing methods are involved in a number of cases. The polyxylylenes are formed by vapor phase deposition; the polyester homopolymer is fabricated by techniques more often used for ceramics; the properties of some of the polyamides are dependent upon liquid crystal formation.

5. Each of the polymer systems fulfills a particular role, and taken together they provide a remarkable spectrum of properties and high temperature applications. For example, they are used as coatings, films, fibers, molding materials, bearing materials, matrices for composites, and insulating coverings for wire.

REFERENCES

1. G. Goldfinger, *J. Polym. Sci.* **4**, 93 (1949).
2. G. A. Edwards and G. Goldfinger, *J. Polym. Sci.* **6**, 125 (1951).
3. G. A. Edwards and G. Goldfinger, *J. Polym. Sci.* **16**, 589 (1955).
4. C. S. Marvel and G. E. Hartzell, *J. Am. Chem. Soc.* **81**, 448 (1959).
5. C. S. Marvel, P. E. Cassidy, and S. Ray, *J. Polym. Sci. A* **3**, 1553 (1965).
6. P. Kovacic and A. Kyriakis, *Tetrahedron Lett.* 467 (1962).
7. P. Kovacic and R. M. Lange, *J. Org. Chem.* **28**, 968 (1963).
8. P. Kovacic, V. J. Marchionna, and J. P. Kovacic, *J. Polym. Sci. A* **3**, 4297 (1965).
9. L. C. Cessna and H. Jabloner, *J. Elastomers Plast.* **6**, 103 (1974).
10. T. M. Bednarski, J. Del Nero, R. H. Mayer, and J. A. Hagan, *Soc. Plast. Eng., Tech. Pap.* **21**, 90 (1975).

11. R. F. Kovar, G. F. L. Ehlers, and F. E. Arnold, *J. Polym. Sci., Polym. Chem. Ed.* **15,** 1081 (1977).

12. M. D. Sefcik, E. O. Stejskal, R. A. McKay, and J. Schaefer, *Macromolecules* **12,** 423 (1979).

13. A. H. Frazer, *High Temperature Resistant Polymers,* Wiley Interscience Publishers, New York (1968), pp. 46, 47.

14. J. M. Lancaster, B. A. Wright, and W. W. Wright, *J. Appl. Polym. Sci.* **9,** 1955 (1965).

15. G. F. L. Ehlers, K. R. Fisch, and W. R. Powell, *J. Polym. Sci. A 1* **7,** 2931 (1969).

16. D. N. Vincent, *Polym. Preprints* **28,** 3819 (1968).

17. M. Szwarc, *Discuss. Faraday Soc.* **2,** 46 (1947).

18. L. A. Errede and R. S. Gregorian, *J. Polym. Sci.* **60,** 21 (1962).

19. A. C. Farthing and C. J. Brown, *J. Chem. Soc.* 3270 (1953).

20. J. H. Golden, *J. Chem. Soc.* 1604 (1961).

21. T. E. Young, Brit. Pat. 807,196 (1959).

22. W. D. Niegisch, in *Encyclopedia of Polymer Science and Technology,* H. F. Mark, N. G. Gaylord, and N. M. Bikales, (eds.), Interscience Publishers, New York (1971), Vol. 15, p. 98.

23. S. Oldham, *SAMPE Q.* **10,** 1 (1979).

24. W. F. Gorham, *J. Polym. Sci. A 1* **4,** 3027 (1966).

25. Reference 13, pp. 63–65.

26. J. M. Lancaster and W. W. Wright, *J. Appl. Polym. Sci.* **11,** 1641 (1967).

27. J. R. Schaefgen, *J. Polym. Sci.* **41,** 133 (1959).

28. S. L. Madorsky and S. Straus, *J. Res. Nat. Bur. Std.* **55,** 223 (1955).

29. V. V. Korshak, *Heat-Resistant Polymers,* Israel Program for Scientific Translations, Jerusalem (1971).

30. J. Economy, B. E. Nowak, and S. G. Cottis, *Polym. Preprints* **11,** 332 (1970).

31. J. Economy, B. E. Nowak, and S. G. Cottis, *SAMPE J.* **6,** 21 (1970).

32. J. Economy, R. S. Storm, V. I. Matkovich, S. G. Cottis, and B. E. Nowak, *J. Polym. Sci., Polym. Chem. Ed.* **14,** 2207 (1976).

33. R. S. Storm and S. G. Cottis, *New Industrial Polymers,* R. D. Deanin, (ed.), ACS Symposium Series 4, American Chemical Society, Washington (1974), Chapter 12, p. 156.

34. Anon., *Eur. Plast. News,* **3,** (10) 31 (1976).

35. K. Hazama, *Jpn. Plast. Age,* 39 (July–August 1976).

36. J. Economy and S. G. Cottis, in *Encyclopedia of Polymer Science and Technology,* H. F. Mark, N. G. Gaylord, and N. M. Bikales, (eds.), Interscience Publishers, New York (1971), Vol. 15, p. 292.

37. G. J. Knight, Unpublished results.

38. G. F. L. Ehlers, K. R. Fisch, and W. R. Powell, *J. Polym. Sci. A 1* **7,** 2969 (1969).

39. L. K. McCune, *Text. Res. J.* **32,** 762 (1962).

40. J. H. Ross, *Text. Res. J.* **32,** 768 (1962).

41. L. H. Gan, P. Blais, D. J. Carlson, T. Suprunchuk, and D. M. Wiles, *J. Appl. Polym. Sci.* **19,** 69 (1975).

42. J. C. Brown and B. C. Ennis, *Text. Res. J.* **47,** 62 (1977).

43. S. L. Kwolek, U.S. Pat. 3,671,542 (1972).

44. M. Jaffe, *Polym. Preprints* **19,** 355 (1978).

45. R. A. Dine-Hart, B. J. C. Moore, and W. W. Wright, *J. Polym. Sci. B* **2,** 369 (1964).

46. Y. P. Krasnov, V. M. Savinov, L. B. Sokolov, V. I. Logunova, V. K. Belyakov, and T. A. Polyakova, *Polym. Sci. USSR* **8,** 413 (1966).

47. Y. P. Krasnov, V. I. Logunova, and L. B. Sokolov, *Polym. Sci. USSR* **8,** 2176 (1966).

48. A. T. Kalashnik, V. Y. Yefremov, N. V. Mikhailov, G. I. Kudryavtsev, A. M. Shchetinin, N. P. Panikarova, M. A. Dubrovina, and Y. P. Panksatov, *Polym. Sci. USSR A 1* **4,** 1567 (1972).

49. R. T. Conley and R. A. Gaudiana, in *The Thermal Stability of Polymers,* R. T. Conley, (ed.), Marcel Dekker, New York (1970), Vol. 1, p. 347.

50. H. Lee, D. Stoffey, and K. Neville, *New Linear Polymers*, McGraw-Hill Book Co., New York (1967).

51. J. W. Moore, Paper presented at 27th Ann. SPI Conference (1972). (Note—the text of this paper was not published in the Conference Proceedings.)

52. V. A. Sergeev, V. K. Shitikov, and V. I. Nedelkin, *Russ. Chem. Revs.* **47,** 1095 (1978).

53. A. D. Macallum, *J. Org. Chem.* **13,** 154 (1948).

54. R. W. Lenz and W. K. Carrington, *J. Polym. Sci.* **41,** 333 (1959).

55. R. W. Lenz and C. E. Handlovits, *J. Polym. Sci.* **43,** 167 (1960).

56. R. W. Lenz, C. E. Handlovits, and H. A. Smith, *J. Polym. Sci.* **58,** 351 (1962).

57. H. A. Smith, in *Encyclopedia of Polymer Science and Technology,* H. F. Mark, N. G. Gaylord, and N. M. Bikales, (eds.), Interscience Publishers, New York (1969), Vol. 10, p. 653.

58. J. T. Edmonds and H. W. Hill, U.S. Pat. 3,354,129 (1967).

59. J. N. Short and H. W. Hill, *Chem. Tech.* 481, August (1972).

60. N. S. J. Christopher, J. L. Cotter, G. J. Knight, and W. W. Wright, *J. Appl. Polym. Sci.* **12,** 863 (1968).

61. G. F. L. Ehlers, K. R. Fisch, and W. R. Powell, *J. Polym. Sci. A 1* **7,** 2955 (1969).

62. G. Montaudo, M. Pryzybylski, and H. Ringsdorf, *Makromol. Chem.* **176,** 1763 (1975).

63. G. Montaudo, M. Pryzybylski, and H. Ringsdorf, *Makromol. Chem.* **176,** 1753 (1975).

64. G. H. West and J. M. Senior, *Tribology* **6,** 269 (1973).

65. R. T. Alvarez and S. B. Driscoll, *35th Ann. SPE Tech. Conf.* **35,** 308 (1977).

66. J. T. Martin and C. W. Anderson, *11th Nat. SAMPE Tech. Conf.* **11,** 977 (1979).

67. J. T. Hartness, *25th Nat. SAMPE Symp. Exhib.* **25,** 376 (1980).

68. B. E. Jennings, M. E. B. Jones, and J. B. Rose, *J. Polym. Sci. C* **16,** 715 (1967).

69. R. N. Johnson, A. G. Farnham, R. A. Clendinning, W. F. Hale, and C. N. Merriam, *J. Polym. Sci. A 1* **5,** 2375 (1967).

70. H. A. Vogel, *J. Polym. Sci. A 1* **8,** 2035 (1970).

71. M. E. A. Cudby, R. G. Feasey, B. E. Jennings, M. E. B. Jones, and J. B. Rose, *Polymer* **6,** 589 (1965).

72. M. E. A. Cudby, R. G. Feasey, S. Gaskin, V. Kendall, and J. B. Rose, *Polymer* **9,** 265 (1968).

73. J. B. Rose, *Chem. Ind. (London)* 461 (1968).

74. A. B. Newton and J. B. Rose, *Polymer* **13,** 465 (1972).

75. T. E. Attwood, A. B. Newton, and J. B. Rose, *Br. Polym. J.* **4,** 391 (1972).

76. J. B. Rose, *Polymer* **15,** 456 (1974).

77. T. E. Attwood, D. A. Barr, R. G. Feasey, V. J. Leslie, A. B. Newton, and J. B. Rose, *Polymer* **18,** 354 (1977).

78. T. E. Attwood, D. A. Barr, T. King, A. B. Newton, and J. B. Rose, *Polymer* **18,** 359 (1977).

79. T. E. Attwood, T. King, I. D. McKenzie, and J. B. Rose, *Polymer* **18,** 365 (1977).

80. T. E. Attwood, T. King, V. J. Leslie, and J. B. Rose, *Polymer* **18,** 369 (1977).

81. V. A. Sergeyev, V. K. Shitikov, V. I. Nedelkin, and V. V. Korshak, *Polym. Sci. USSR A* **18,** 609 (1976).

82. M. G. Maximovich, S. C. Lockerby, F. E. Arnold, and G. A. Loughran, *23rd Nat. SAMPE Symp. Exhib.* **23,** 490 (1978).

83. W. F. Hale, A. G. Farnham, R. N. Johnson, and R. A. Clendinning, *J. Polym. Sci. A 1* **5,** 2399 (1967).

84. A. Davis, *Makromol. Chem.* **128,** 242 (1969).

85. L. I. Danilina, E. N. Teleshov, and A. N. Pravednikov, *Polym. Sci. USSR* **16,** 672 (1974).

86. L. I. Danilina, V. I. Muromtsev, and A. N. Pravednikov, *Polym. Sci. USSR* **17,** 2984 (1975).

87. J. R. Brown and J. H. O'Donnell, *J. Polym. Sci. B* **8,** 121 (1970).

88. B. D. Gesner and P. G. Kelleher, *J. Appl. Polym. Sci.* **12,** 1199 (1968).

89. F. Abdul-Rasonl, C. L. R. Catherall, J. S. Hargreaves, J. M. Mellor, and D. Phillips, *Eur. Polym. J.* **13**, 1019 (1977).

90. A. Davis and B. L. Diffey, *Phys. Med. Biol.* **23**, 318 (1978).

91. T. E. Bugel, *SPE J.* **24**, 52 (1968).

92. J. T. Hoggatt, *20th Nat. SAMPE Symp. Exhib.* **20**, 606 (1975).

93. R. C. Novak, NASA CR-134881 (1975).

94. R. C. Novak, NASA CR-135196 (1977).

95. D. A. Scola and M. E. Roylance, *23rd Nat. SAMPE Symp. Exhib.* **23**, 950 (1978).

96. G. E. Husman and J. T. Hartness, *24th Nat. SAMPE Symp. Exhib.* **24**, 21 (1979).

97. J. Theberge and P. Cloud, Paper presented at Conference on Elevated Temperature Performance of Thermoplastic Composites, Malvern, Pennsylvania (1979).

98. P. C. Dawson and D. J. Blundell, *Polymer* **21**, 577 (1980).

99. W. H. Bonner, U.S. Pat. 3,065,205 (1962).

100. I. Goodman, J. E. McIntyre, and W. Russell, Brit. Pat. 971,227 (1964).

101. Y. Iwakura, K. Uno, and T. Takiguchi, *J. Polym. Sci. A 1* **6**, 3345 (1968).

102. B. M. Marks, U.S. Pat. 3,442,857 (1969).

103. K. J. Dahl, Brit. Pat. 1,387,303 (1975).

104. R. A. Clendinning, A. G. Farnham, W. F. Hall, R. N. Johnson, and C. N. Merriam, *J. Polym. Sci. A 1* **5**, 2375 (1967).

105. T. E. Attwood, P. C. Dawson, J. L. Freeman, L. R. J. Hoy, J. B. Rose, and P. A. Staniland, *Polym. Prep.* **20**, 191 (1979). Also, *Polymer* **22**, 1096 (1981).

106. ICI Plastics Division Provisional Data Sheets PKPD 1 (1980), PKPD 2 (1979), PKPD 9 (1979), PKPD 12 (1979), PKPD 13 (1980), PKPD 17 (1981), and PKPD 20 (1981).

107. J. H. Golden, SCI Monograph No. **13**, 231 (1961).

108. A. S. Hay, P. Shenian, A. C. Gowan, P. F. Erhardt, W. R. Haaf, and J. E. Theberge, Phenols, oxidative polymerization, in *Encyclopedia of Polymer Science and Technology*, H. F. Mark, N. G. Gaylord, and N. M. Bikales, (eds.), Interscience Publishers, New York (1969), Vol. 10, p. 92.

109. W. F. Christopher and D. W. Fox, *Polycarbonates*, Reinhold Publishing Corp., New York (1962).

110. H. Schnell, *Chemistry and Physics of Polycarbonates*, Interscience Publishers, New York (1964).

111. L. Bottenbruch, Polycarbonates, in *Encyclopedia of Polymer Science and Technology*, H. F. Mark, N. G. Gaylord, and N. M. Bikales, (eds.), Interscience Publishers, New York (1969), Vol. 10, p. 710.

SUPPLEMENTARY BIBLIOGRAPHY

Polyphenylenes

W. Reid and D. Freitag, Oligophenyls, oligophenylenes, and polyphenyls, A class of thermally very stable compounds, *Angew Chem. Int. Ed. Engl.* **7**, 835 (1968).

G. K. Noren and J. K. Stille, Polyphenylenes, *J. Polym. Sci. D* **5**, 385 (1971).

J. G. Speight, P. Kovacic, and F. W. Koch, Synthesis and properties of polyphenyls and polyphenylenes, *J. Macromol Sci. Rev. Macromol. Chem.* **5**, 295 (1971).

V. A. Sergeev, V. K. Shitikov, and L. G. Grigoreva, Preparation and properties of oligo- and polyphenylenes with functional groups, *Russ. Chem. Rev.* **45**, 946 (1976).

D. M. Gale, Fabrication of poly (*p*-phenylene) by powder forming technique, *J. Appl. Polym. Sci.* **22**, 1955 (1978).

D. M. Gale, Properties of fabricated poly (*p*-phenylene), *J. Appl. Polym. Sci.* **22**, 1971 (1978).

Polyxylylenes

L. A. Errede, R. S. Gregorian, and J. M. Hoyt, The chemistry of xylylenes. VI. The polymerization of *p*-xylylene, *J. Amer. Chem. Soc.* **82**, 5218 (1960).

L. A. Errede and R. S. Gregorian, The chemistry of xylylenes. XIII. The problem of crystallinity and crosslinking in poly (*p*-xylylene), *J. Polym. Sci.* **60**, 21 (1962).

L. A. Errede and N. Knoll, The chemistry of xylylenes. XIV. The moldability and thermal stability of poly (*p*-xylylene) and related polymers, *J. Polym. Sci.* **60**, 33 (1962).

W. F. Gorham, Recent advances in the vapor deposition polymerization of *p*-xylylenes, *Adv. Chem. Series* **91**, 643 (1969).

M. Szwarc, Poly-para-xylylene: Its chemistry and application in coating technology, *Polym. Eng. Sci.* **16**, 473 (1976).

Polyamides

F. Dobinson and J. Preston, New high-temperature polymers. II. Ordered aromatic copolyamides containing fused and multiple ring systems, *J. Polym. Sci. A 1* **4**, 2093 (1966).

W. B. Black and J. Preston (eds.), *Symposium on High-Modulus Wholly Aromatic Fibres,* Marcel Dekker, New York (1973).

J. Preston, High performance fibers from aromatic polymers, *Org. Coat. and Plast. Preprints* **35**, (2) 160 (1975).

E. E. Magat and R. E. Morrison, Recent advances in man-made fibers, *J. Polym. Sci. Polym. Symp.* **51**, 203 (1975).

P. W. Morgan, Synthesis and properties of aromatic and extended chain polyamides, *Polym. Preprints* **17**, (1) 47 (1976).

Poly(phenylene sulphide)

J. T. Edmonds and H. W. Hill, Properties of poly (phenylene sulphide) coatings, *Polym. Preprints* **13**, (1) 603 (1972).

R. V. Jones and H. W. Hill, Polyphenylene sulphide—a new item of commerce, *Adv. Chem. Ser.* **140**, 174 (1974).

H. W. Hill and D. G. Brady, Characterization of poly (phenylene sulphide) coatings, *Org. Coat. and Plast. Preprints* **36**, (2) 363 (1976).

D. G. Brady, The crystallinity of poly (phenylene sulphide) and its effect on polymer properties, *J. Appl. Polym. Sci.* **20**, 2541 (1976).

R. T. Hawkins, Chemistry of the cure of poly (*p*-phenylene sulphide), *Macromolecules* **9**, 189 (1976).

Poly(phenylene ether sulphones)

R. N. Johnson and A. G. Farnham, Poly (arylethers) by nucleophilic aromatic substitution. III. Hydrolytic side reactions, *J. Polym. Sci. A 1* **5,** 2415 (1967).

H. Lee, D. Stoffey, and K. Neville, Aromatic polysulphones, in *New Linear Polymers,* McGraw-Hill, New York (1967), Ch. 5.

R. N. Johnson, Polysulphones, in *Encyclopedia of Polymer Science and Technology,* H. F. Mark, N. G. Gaylord, and N. M. Bikales, (eds.), Interscience Publishers, New York (1969), Vol. 11, p. 447.

W. M. Alvino, Aging behavior of polyarylsulphone films, *J. Appl. Polym. Sci.* **15,** 2521 (1971).

V. M. Laktionov and I. V. Zhuravleva, Chromatographic study of kinetics of degradation of aromatic polysulphones in vacuum, *Polym. Sci. USSR* **17,** 3232 (1975).

V.M. Laktionov, I. V. Zhuravleva, S. A. Pavlova, S. R. Rafikov, S. N. Salazkin, S. V. Vinogradova, A. A. Kulkov, and V. V. Korshak, Heat resistance of polysulphonarylates and polysulphonarylene oxides, *Polym. Sci. USSR* **18,** 379 (1976).

V. J. Leslie, J. B. Rose, G. O. Rudkin, and J. Feltzin, Polyether sulphone—a new high-temperature engineering thermoplastic, in *New Industrial Polymers,* R. D. Deanin, (ed.), ACS Symposium Series 4, American Chemical Society, Washington (1974), Ch. 6, p. 63.

POLYMERS WITH HETEROCYCLIC RINGS IN THE CHAIN

INTRODUCTION

During the late 1950s and early 1960s, requirements for thermally stable materials produced a rapid growth in new but impracticable polymers. Impracticable because, at that time, the criterion of thermal stability was narrowly interpreted as the capacity of such polymers to withstand high temperatures over long periods in air. Thus aromatic polymers, including a number of heteroaromatic systems, were produced with excellent thermo-oxidative stability, but which, because they were intractable and infusible "brick-dusts," had no practical value. Attempts to capitalize on the thermal stability of these materials involved introducing a degree of tractability into the system. In the case of highly conjugated aromatic polymers, as has already been seen, flexible (hinge) groups—linking the rings—led to increases in both molecular weight and tractability, but almost inevitably this was accompanied by a reduction in thermo-oxidative stability.

In some cases, for example the polybenzimidazoles and polyquinoxalines, introduction of the heterocyclic moiety into the polymer chain by means of cyclopolymerization improved tractability and resulted in quite high-molecular-weight, relatively soluble polymers. For the majority of structures, this improvement in properties did not apply; in such cases the most important development was the technique of "post-polymerization-cyclization." This process involved the synthesis of a high-molecular-weight, manipulable, open-chain intermediate, followed by ring closure—usually thermally induced—to the intractable heteroaromatic polymer. It was at the prepolymer stage, suitable solvents for

TABLE 5.1. Solvents for Prepolymer Synthesis

N,N-Dimethylformamide (DMF)	*N*-Methyl-2-pyrrolidone (NMP)
N,N-Dimethylacetamide (DMAC)	Pyridine
N,N-Dimethylmethoxyacetamide	Dimethyl sulphone
N-Methylcaprolactam	Hexamethylphosphoramide (HMP)
Dimethyl sulphoxide (DMSO)	Tetramethylene sulphone (TMS)
m-Cresol	*N*-Acetyl-2-pyrrolidone

which are shown in Table 5.1, that fabrication was possible. Inevitably, the nature of the process generally produced nonstructural materials such as films, fibers, coatings, and film adhesives. The post-polymerization-cyclization route led to a variety of ring-chain and fused ring heteroaromatic systems; the polyoxadiazoles, polybenzoxazoles, polythiadiazoles, polyimides, and polyimidazopyrrolones (pyrrone polymers) are typical examples.[1] Despite the successes achieved in producing fabricable heteroaromatic materials with a high degree of thermo-oxidative and chemical stability, after almost two decades of sustained research and development polyimides are virtually alone in their commercial viability. The reasons for their success probably include availability and relative cheapness of starting materials, as well as preparative routes that are capable of adaptation to changes in product requirements. A significant factor may have been that the commercially powerful Dupont Company, an early pioneer in the technological development and successful marketing of coatings, binders, and films, provided the necessary initial confidence in them.

POLYIMIDES

Introduction

There was an early recognition of the exceptionally high thermo-oxidative stability of the wholly aromatic polyimides compared with, for example, their aliphatic/aromatic counterparts. This level of stability was obviously influenced by the large component of fused rings, or rings linked only by direct carbon-carbon bonds. The rigidity of the resulting polymer chains produced materials with very high glass transitions (Tgs), which were inherently very difficult, if not impossible, to process. The commercial success of polyimides in a variety of high technology applications can be related to a balance between various methods of synthesis and retention of thermo-oxidative stability and a substantial proportion of room-temperature strength and modulus at elevated temperatures. In particular, over the past five to seven years, the compromise achieved between these factors has led to considerable improvements in ease of processing, resulting in a significant expansion in materials applications for the polyimides. From the mid-1960s onwards, a vast literature has developed covering all aspects of these polymers. Over the period 1977 to 1979, for example, the addition of almost 1,000 references has been made to the Chemical Abstracts file, and well over 100 to the NTIS (U.S. Government Reports) file. An increasing proportion of these references has dealt with improved processing techniques and applications.

Syntheses involving condensation or addition reactions have provided C- or A-type polyimides. Before the early 1970s, practical routes involved almost exclusively condensation reactions. Among these reactions was the process originally reported on by Dupont[2,3] and Westinghouse[4,5] workers in the USA and Jones *et al.*[6] in the U.K., in which the formation of a soluble intermediate (**I**) followed by a post polymerization cyclization to the intractable polymide (**II**) proved to be the predominant approach. This applied in both basic research studies and in the development of commercial high-temperature materials.

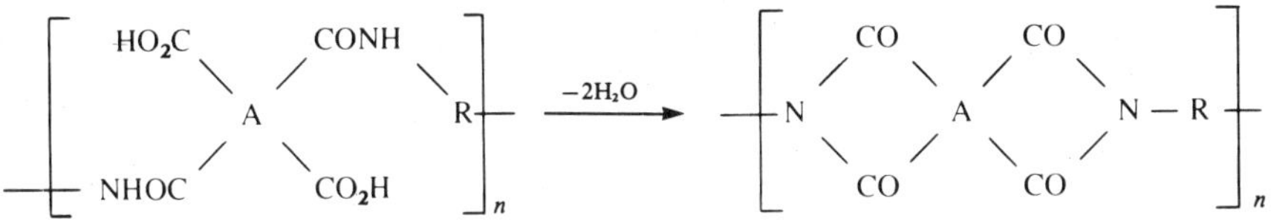

$$
\left[\begin{array}{c} HO_2C \diagdown \quad \diagup CONH \diagdown \\ A \quad\quad R \\ NHOC \diagup \quad \diagdown CO_2H \end{array}\right]_n \xrightarrow{\;-2H_2O\;} \left[\begin{array}{c} CO \quad CO \\ N \diagup \diagdown A \diagup \diagdown N-R \\ CO \quad CO \end{array}\right]_n
$$

(I)   (II)

At a relatively early stage, however, attempts to improve the processability of aromatic polyimides by increasing both solubility and tractability involved the introduction of midchain flexibility into the macromolecule. This work remained mainly of academic interest, not least because in many instances, though not all, the hinge groups lowered the stability of the system. Within the past few years a number of C-type polyimides, which are so structured that they are soluble and thermoplastic to some degree, have been developed, one or two having reached limited commercial exploitation. Even so, these "thermoplastic polyimides" have relatively high Tgs and still require measurably higher temperatures for successful processing than do conventional thermoplastics. Concurrently with these latter developments, alternative attempts to resolve the processing dilemma of the C-type polyimides prompted the introduction of A-type materials. These polyimides are formed from short, pre-imidized segments—or even monomeric reactants—that contain elements end-capped with selected unsaturated aliphatic or cycloaliphatic groups. Thermal polymerization of these tractable oligomers involving processes of chain extension and cross-linking is possible without the concomitant evolution of harmful volatiles. In certain cases a combination of condensation and addition reactions has been used in the production of polyimide materials.

Synthesis of Condensation Polyimides (Insoluble, Infusible Type)

Bogert and Renshaw[7] in 1908 reported on the intramolecular melt polycondensation (A-B polymerization) of 4-aminophthalic anhydride, or dimethyl 4-aminophthalate:

The attraction of a single monomer unit with its in-built stoichiometry led to a surprisingly limited evaluation of A-B polymerizations. Seddon,[8] for example, reexamined the polymerization of 4-aminophthalic anhydride and its hydrochloride using the two-stage solution (polyamic acid) technique. Reimschuessel *et al.*[9] produced aliphatic-linked polyimides by melt processes from β-carboxymethylcaprolactam and 4-carboxy-2-piperidone.

Melt polycondensation of monomers [(**III**)X = CO or CH(OH), R = H or CH$_3$] has produced[10,11] soluble (B-stage) prepolymers used as the binder systems in thermoset glass fiber laminates:

(III)

Major synthetic effort has concentrated on the interaction of derivatives of aromatic tetracarboxylic acids and diamines. In the early 1950s, Dupont workers, aiming at molding resins and hot-pressed films, produced moldable aliphatic/aromatic polyimides (**II**) by fusion of salt-type intermediates[12]:

(II)

e.g., A = (benzene ring) ; $R = (CH_2)_{>9}, (CH_2)_3C(CH_3)_2(CH_2)_3$

or A = (two rings joined by X) ; $R = (CH_2)_{4 \text{ or } 6}, p,p'C_6H_4OC_6H_4$

$X = O, C(CH_3)_2, CH_2$

More recent developments[13] have capitalized on the solubility of the intermediate salt in polar solvents. Copolymer (**IV**; $x = 10\text{–}70\%$, $y = 90\text{–}30\%$) produced by this process can be extruded or injection molded[14]:

(**IV**)

Early studies on wholly aromatic systems were unsuccessful; it was then claimed that aromatic diamines were insufficiently basic to form salts, although this has since been questioned.[15,16] More important, aromatic polyimides were infusible and insoluble. Soluble, fusible fluoroalkylene-linked aromatic polyimides (**V**) have been produced, however, by melt fusion of the corresponding dianhydrides and diamines[17]:

(**V**)

e.g., $x = 3, 4, 7$; $y = 3$.

The influence of structure on solubility and tractability has proved to be of preeminent importance in polyimide research and development; in this connection thermoplastic polyimides, a logical development of this approach, are referred to later. Early workers, however, concentrated their efforts on wholly aromatic systems, and the resultant polymers, unlike their aliphatic/aromatic analogs, were extremely intractable. Fortunately, it was found[2–6] that wholly aromatic systems could be prepared from dianhydrides (**VI**) and diamines (**VII**) in two stages via the soluble, fabricable, intermediate (**I**), a poly(amic acid) (PAA), followed by dehydration to product (**II**). This reaction proved to be a general route to polyimides[18]; it is still the subject of considerable research, and numerous aromatic dianhydrides and diamines incorporating a wide variety of heteroatoms and groups have been examined. Selected examples are shown in Tables 5.2 and 5.3. Nevertheless, only a very limited number have made any significant commercial impact. Among the dianhydrides, PMDA and BTDA have almost universal application, with HFDA and DEDA having rather more limited use in the U.S.A. and U.S.S.R., respectively. Commercially, application of aromatic diamines has been restricted to DADM, DAPE, MPD, and PPD.

Normally PAAs are produced by low temperature reaction of aromatic dianhydrides (**VI**) with diamines (**VII**) in an "active" polar solvent (Table 5.1), although it has been reported[19] that a solid state polymerization (560 N/m^2, 200°C) of PMDA with phenylene diamines yields DMF-soluble PAAs.

$$CO\text{-}A\text{-}CO \text{ (VI)} + H_2N-R-NH_2 \text{ (VII)} \xrightarrow[\text{polar solvent}]{<50°C} [\,HO_2C,CONH\text{-}A\text{-}NHOC,CO_2H\text{-}R\,]_n \text{ (I)}$$

A high proportion of the reports devoted to the study of this reaction have emphasized the significance of high-molecular-weight PAAs for optimum properties of the resultant polyimides. This is particularly relevant for films and coatings, but less so for laminating resins. As well as establishing an order of reactant reactivity, the early reports[2–6,18,20] singled out particularly the importance of monomer purity and stoichiometry, and the effects of concentration and solvent in achieving the optimum molecular weight of PAAs (Figures 5.1–5.3). They also demonstrated that the preferred order of addition of monomers is solid (dry) dianhydride to a solution of diamine. Unsubstituted PAAs are hydrolytically unstable, chain-scission leading to a marked reduction in viscosity (Figure 5.4).[21]

$$[\,HO_2C,CONH\text{-}A\text{-}NHOC,CO_2H\text{-}R\,]_n \text{ (I)} \xrightarrow{H_2O} [\,HO_2C,CO_2^{\ominus}\text{-}A\text{-}NHOC,CO_2H\quad H_3\overset{\oplus}{N}-R\,]_n$$

Wallach[22] showed that, using the most favorable of the reaction conditions, the maximum number average molecular weight attained ($\bar{M}_n$) was 55,000 with a molecular weight distribution ($\bar{M}_w/\bar{M}_n$) of 2.4.

TABLE 5.2. Typical Diamine Precursors to C-Type Aromatic Polyimides

Structure VII (R)	Name (abbreviation)
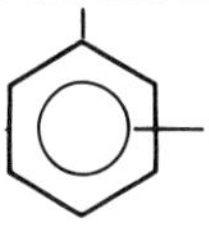	*m*- or *p*-Phenylene diamine (MPD or PPD)
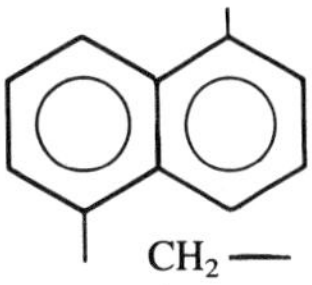	1,5-Diaminonaphthalene
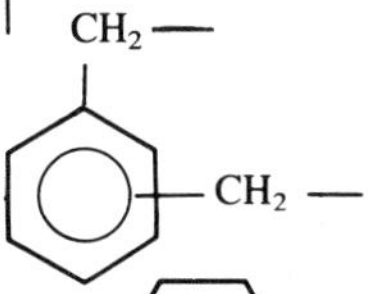	*m*- or *p*-Xylylenediamine
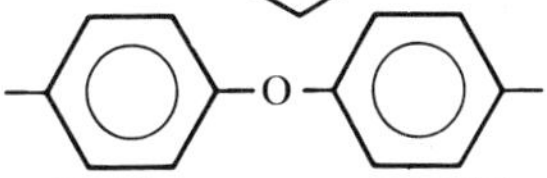	4,4′-Diaminodiphenyl ether (DAPE)
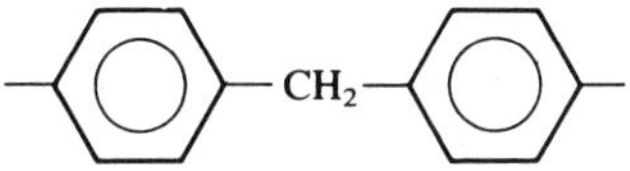	4,4′-Diaminodiphenylmethane (DADM)
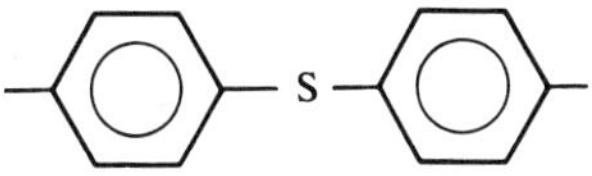	4,4′-Diaminodiphenylsulphide (TDA) (thiodianiline)
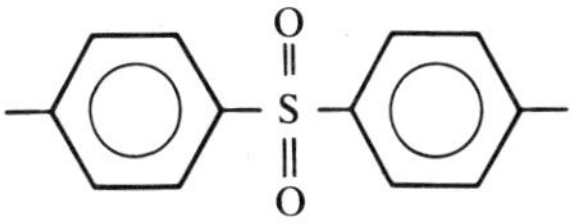	4,4′-Diaminodiphenylsulphone
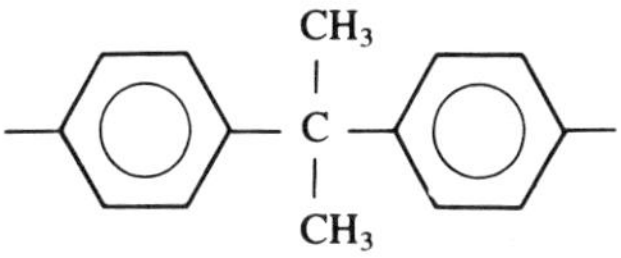	2,2-Bis (4-aminophenyl) propane
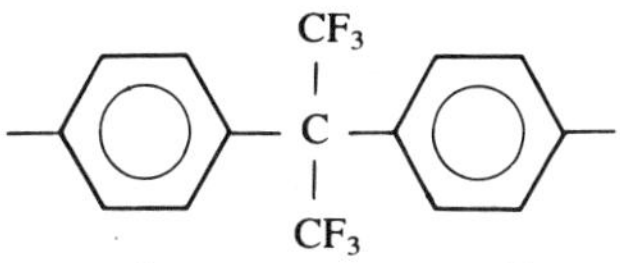	2,2-Bis (4-aminophenyl) hexafluoropropane
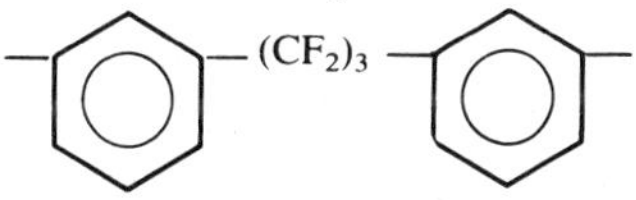	1,3-Di (3-aminophenyl) hexafluoropropane

TABLE 5.3. Typical Dianhydride Precursors to C-Type Aromatic Polyimides

Structure VI (A)	Name (abbreviation)
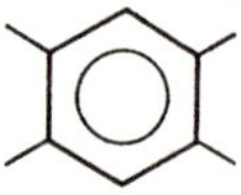	Pyromellitic dianhydride (PMDA)
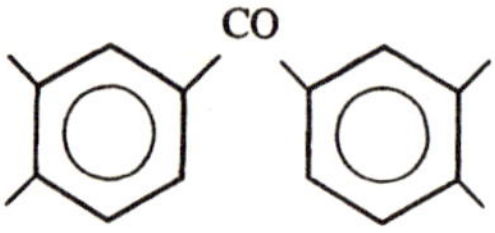	3,3',4,4'-Benzophenone tetracarboxylic dianhydride (BTDA)
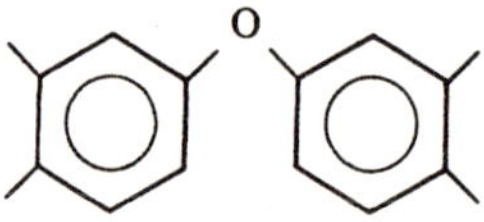	Bis (3,4-dicarboxyphenyl)ether dianhydride (DEDA)
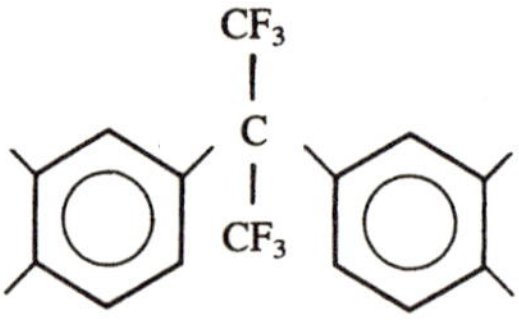	2,2-Bis (3,4-dicarboxyphenyl) hexafluoropropane dianhydride (HFDA)
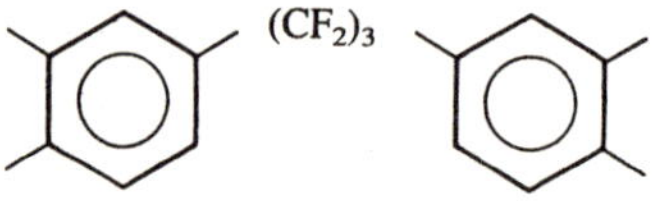	1,3-Bis (3,4-carboxyphenyl) hexafluoropropane dianhydride (DBDA)
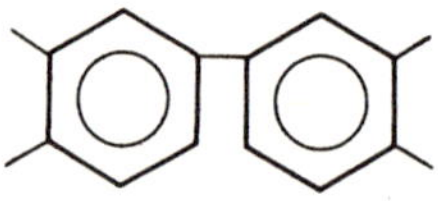	3,3',4,4'-Biphenyltetracarboxylic dianhydride and 2,2',3,3'-isomer
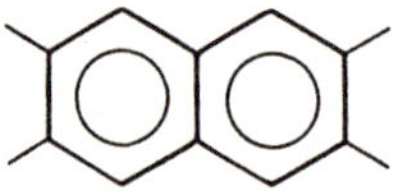	2,3,6,7 − Naphthalenetetracarboxylic dianhydride and 1,2,5,6-isomer

Optimum temperatures for the polycondensation are below 50°C; above this temperature, release of water with concomitant precipitation of polyimide results from cyclization (imidization). This together with interaction between PAA and amide solvents, or secondary amines released into solution by these solvents, has limited the development of high molecular weight.

The cyclodehydration stage is induced thermally, in steps up to 300°C, and usually[1-6,18] in an inert atmosphere or in vacuo. Less frequently chemical dehydration is employed[23-26] using agents such as aliphatic anhydrides, ketene,

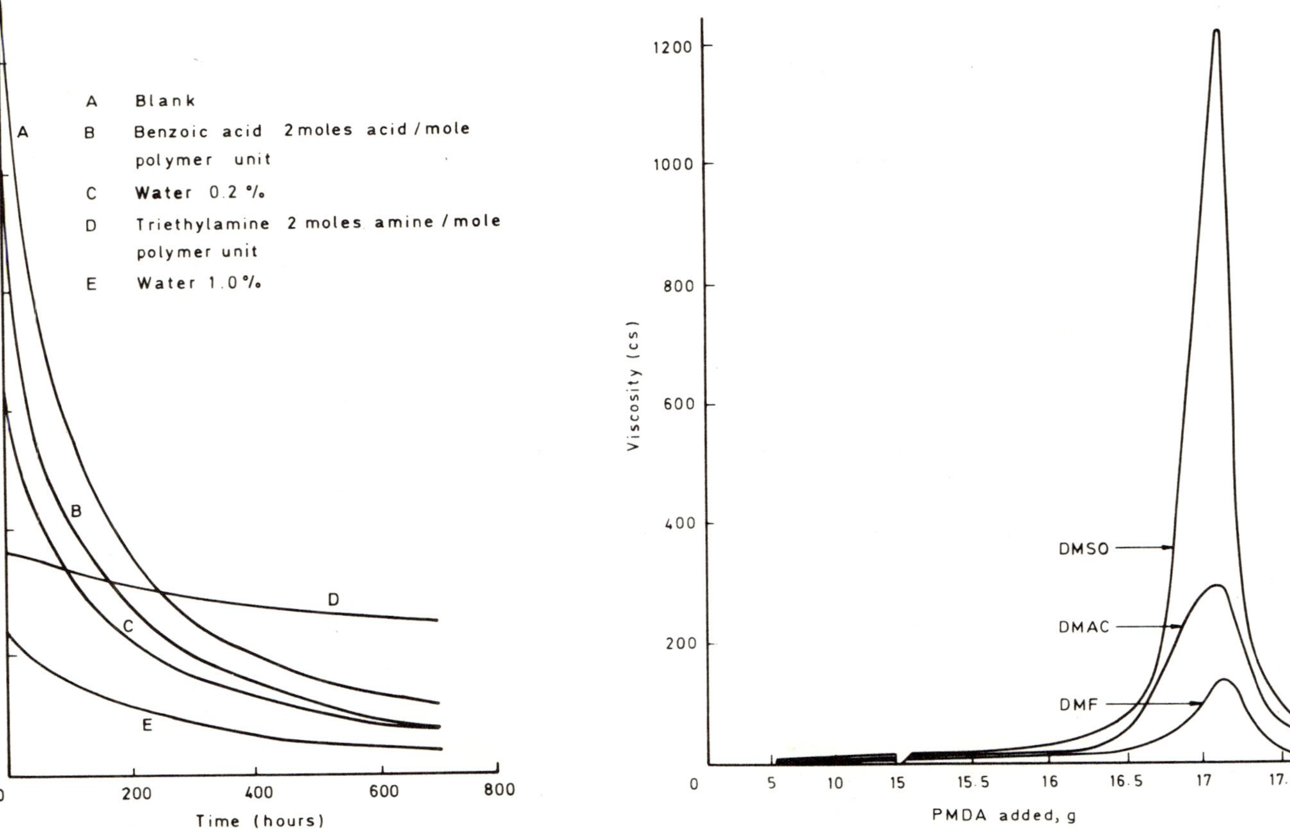

Figure 5.1.   Change in viscosity of DAPE-PMDA solutions in the presence of various additives. Reference 18b.

Figure 5.2.   Viscosity of PAA solutions as a function of amount of PMDA added to a constant amount of DADM. Reference 4.

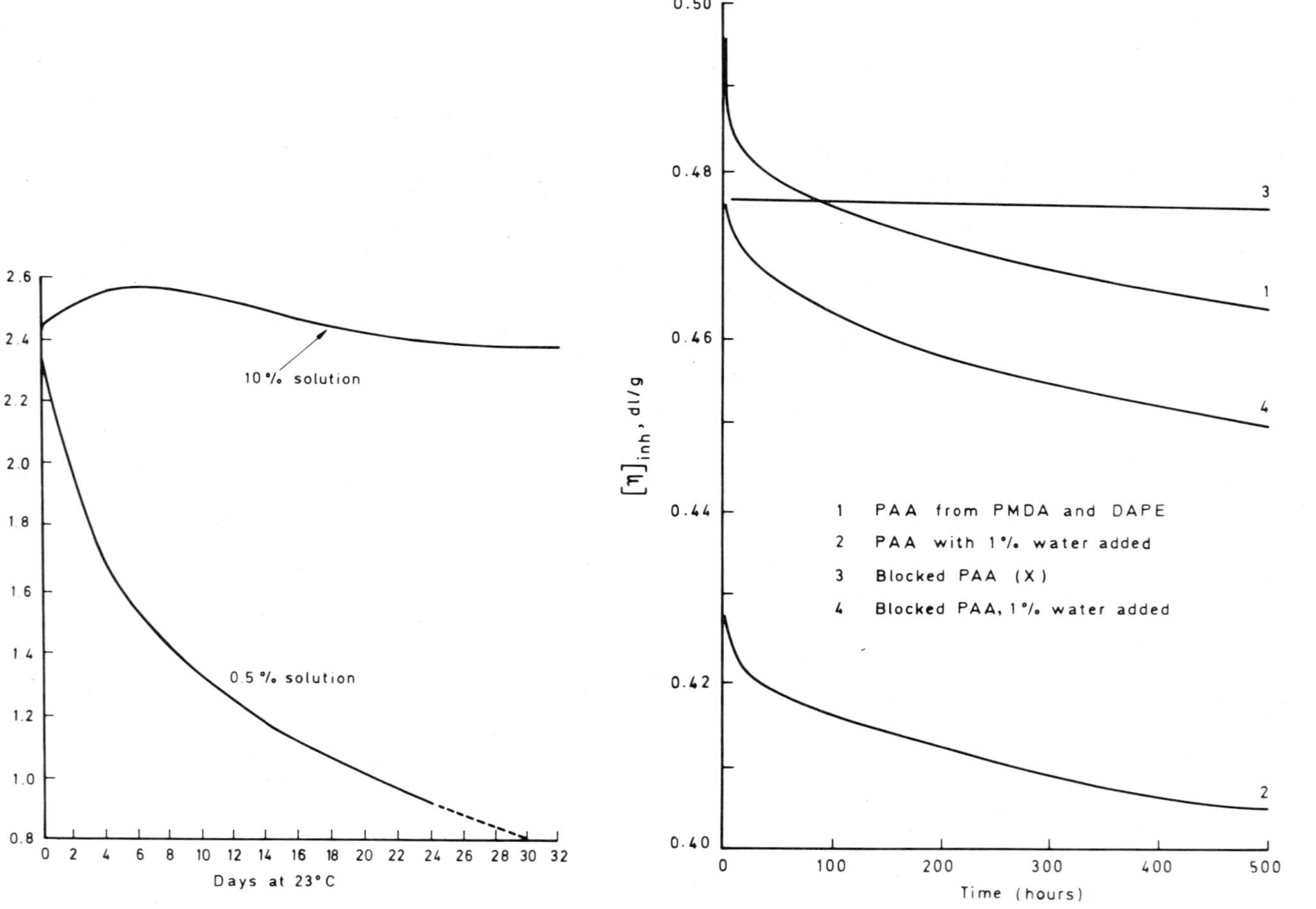

Figure 5.3. Effect of concentration of solution on stability of the PAA from PMDA and DAPE (Dimethylacetamide at 23°C). Reference 18a.

Figure 5.4. Changes of viscosity of prepolymer solutions in dimethylacetamide at 25°C. Solution concentration 0.5 g/dl. Reference 21.

N,N-dialkylcarbodiimides, or strong Lewis acids (PCl$_3$), usually in conjunction with tertiary amine catalysts.[27] Normally imidization is induced after fabrication of films, coatings, or laminates, but chemical conversions have also been extended to certain PAAs in solution.

A number of polyimides prepared by the two-stage route are soluble in various polar solvents; Soviet workers refer[28,29] to these as Cardo polyimides. They have extended their study to a variety of speciality polyimides of this type, demonstrating that such materials can be synthesized by a single-stage, high-temperature (200°C, 7–15 hours) polymerization in inert organic solvents.[30]

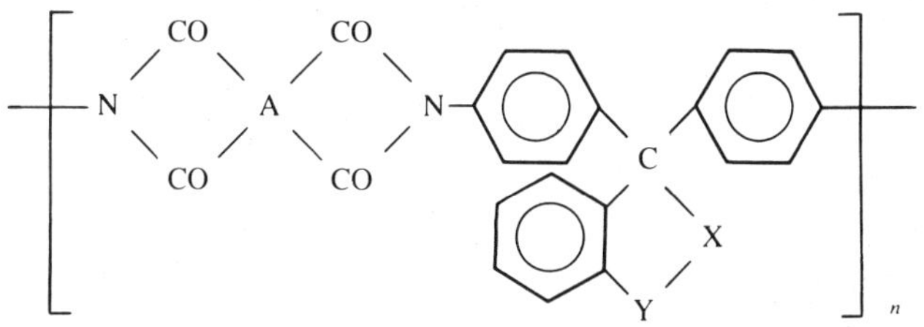

(VIII)

where A = ... or ... : X=O,CO,NH; Y=S,NH,CO.

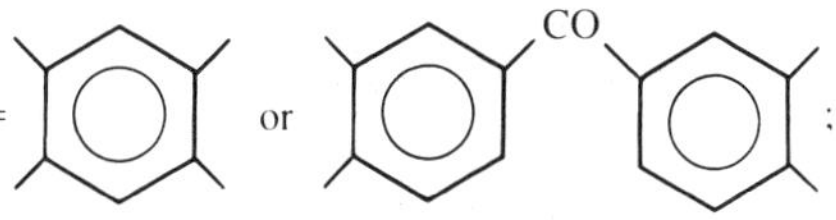

The mechanism of the initial condensation reaction between dianhydride and diamine to produce PAA has been studied spectroscopically,[31–33] and mechanisms and kinetics of the imidization process have been examined by spectroscopic,[32,34] thermal transition,[35,36] and dielectric means.[37] Rate[27] and activation energy values[38] for the thermal cyclization process have also been reported.

Attempts to improve upon the stability and processing characteristics of the PAAs, while simultaneously retaining the essential elements of the two-stage preparative route to the polyimides, have involved relatively minor modifications to either the dianhydride or diamine monomers. For example, the interaction of aromatic diamines with diester/diacid[39] or diester/diacid chloride[40] has produced poly(amide-esters), **(IX)**. Imidization of these intermediates involves elimination of an alcohol (e.g., methyl or ethyl alcohol or ethylene glycol) instead of water; in this case a marginal improvement in processing is effected by the somewhat higher temperature of cyclization in comparison to flow temperature.

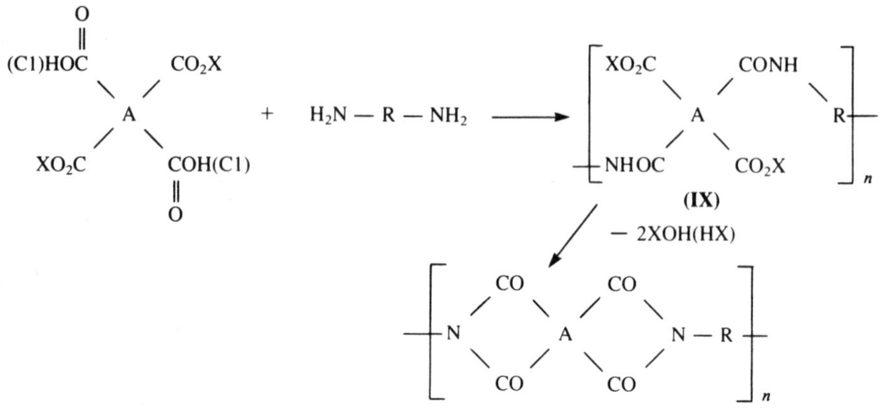

(IX)

$$- 2XOH(HX)$$

(II)

High-solids-content solutions containing monomeric reactants or low-molecular-weight oligomers (**IX**) have improved processing characteristics, particularly in laminating applications, where high flows and wetting are important. A number of commercial materials of this type, such as Skybond and its derivatives, are available. Alternative syntheses from modified dianhydrides have been reported; they also depend on blocking the elimination of water from a conventional PAA. Seddon and Reynolds[41] demonstrated that both amine and ammonium salts of PAAs are considerably more stable in aqueous solution than is the parent prepolymer in dry DMAC; thermal imidization of such salts occurs more rapidly than does that of the analogous PAA, and a number of polyimides in fabricated form were claimed. Delvig et al.[21] produced a blocked precursor amide (**X**) from the corresponding diamide/diacid chloride and DAPE; both thermal cyclization and hydrolysis, measured by changes in inherent viscosity $[\eta]_{inh}$ as a function of time (Figure 5.4), are slower in blocked prepolymer than in the analogous unsubstituted PAA.

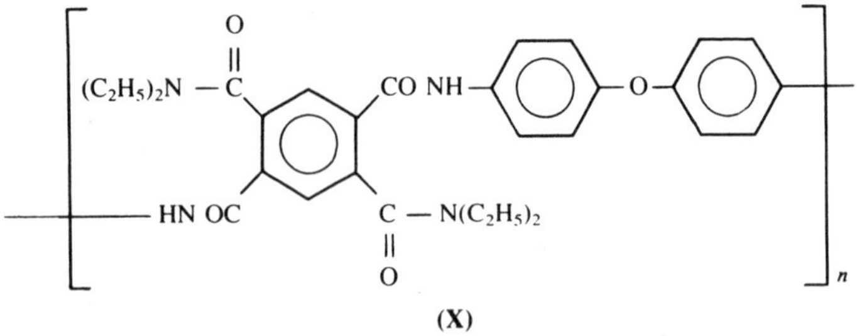

(X)

Imai[42] has also devised a route to polyimides via the blocked polyamic N-ethoxycarbonyl-amide (**XI**), which is again more resistant to thermal imidization ($>240°C/5$ hours; evolution of $C_2H_5OCONH_2$) than the analogous PAA:

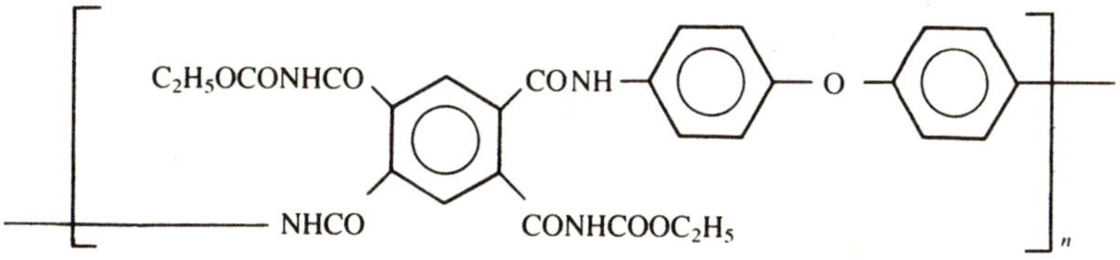

$$
\left[\begin{array}{c}
C_2H_5OCONHCO \cdots CONH{-}\text{(ring)}{-}O{-}\text{(ring)}{-} \\
{-} NHCO \qquad CONHCOOC_2H_5
\end{array} \right]_n
$$

(XI)

In another modification,[43] the interaction of dithioanhydrides with aromatic diamines in dipolar aprotic solvents leads directly to polyimides without the isolation of an intermediate.

These polyimides are claimed to be soluble in dipolar aprotic solvents and to have quite high molecular weights.

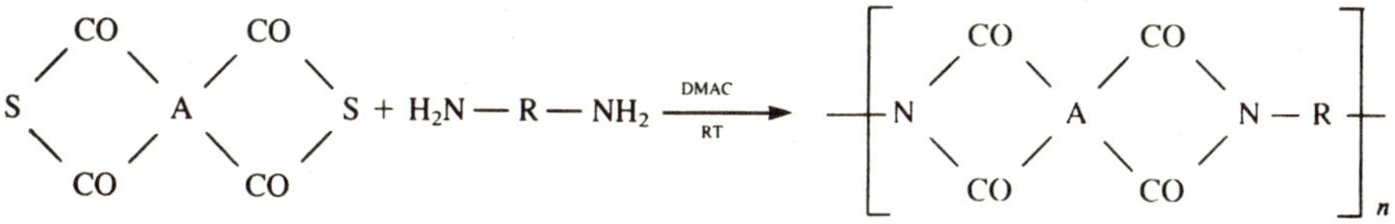

$$
S\overset{CO}{\underset{CO}{<}}A\overset{CO}{\underset{CO}{>}}S + H_2N{-}R{-}NH_2 \xrightarrow[RT]{DMAC} \left[{-}N\overset{CO}{\underset{CO}{<}}A\overset{CO}{\underset{CO}{>}}N{-}R{-} \right]_n
$$

(II)

e.g., $A =$ (ring) , (ring){-}CO{-}(ring) , (ring){-}O{-}(ring)

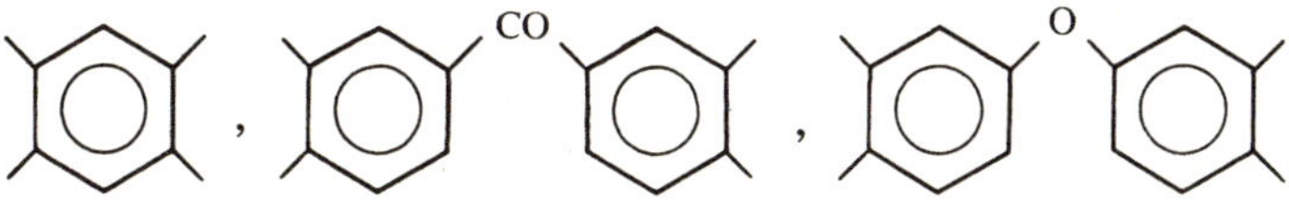

$R = p\text{-}C_6H_4,\ p,p'\text{-}C_6H_4OC_6H_4,\ p,p'\text{-}C_6H_4SO_2C_6H_4,\ p,p'\text{-}C_6H_4C(CH_3)_2C_6H_4$

An alternative to these approaches has been to prepare blocked PAAs by reaction of suitably modified diamines with conventional dianhydrides. The silylated poly(amic acid) (XII) has been produced[44] from silylated diamine and aromatic dianhydride, followed by a facile cyclization to polyimide with evolution of trimethylsilanol. Unfortunately, the intermediate (XII) is hydrolytically unstable. The principal advantage of this process is the solubility of the precursor in readily volatile solvents such as tetrahydrofuran (THF), toluene, or dioxan.

$$\underset{\textbf{(VI)}}{\text{(VI)}} \quad \text{(dianhydride)} \; + \; (CH_3)_3Si - NH - R - NH - Si(CH_3)_3 \xrightarrow[\text{RT}]{\text{THF}}$$

$$\left[\underset{\textbf{(XII)}}{\text{(XII)}} \right]_n \xrightarrow{\; - 2(CH_3)_3SiOH \;}$$

$$\left[\underset{\textbf{}}{\text{imide}} \right]_n$$

A similar, but this time commercially viable, approach involves the use of a diacetylated diamine, which leads to the evolution of acetic acid during imidization.[45,46] In this case, unlike intermediate (XII), the N-acetylated PAA (XIII) is hydrolytically stable and the temperature of cyclization is raised significantly compared to flow temperature. Volatile solvents such as acetone are used in the synthesis.

$$\underset{\textbf{(VI)}}{\text{(VI)}} \quad + \; CH_3CON - R - NCOCH_3 \longrightarrow$$

$$\left[\underset{\textbf{(XIII)}}{\text{(XIII)}} \right]_n \xrightarrow{\; - 2CH_3CO_2H \;}$$

$$\left[\underset{\textbf{(II)}}{\text{(II)}} \right]_n$$

Aromatic polyimides have also been produced from a single-stage polycondensation of dianhydride and di-isocyanate, polymerization occurring together with the evolution of carbon dioxide.

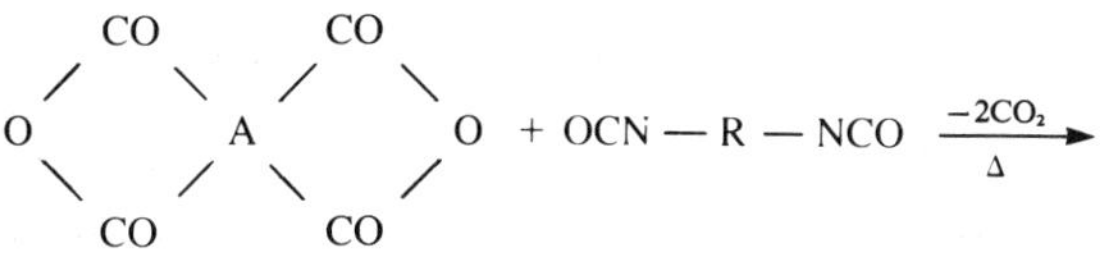

$$\text{(VI)} \quad + \quad OCN-R-NCO \xrightarrow[\Delta]{-2CO_2}$$

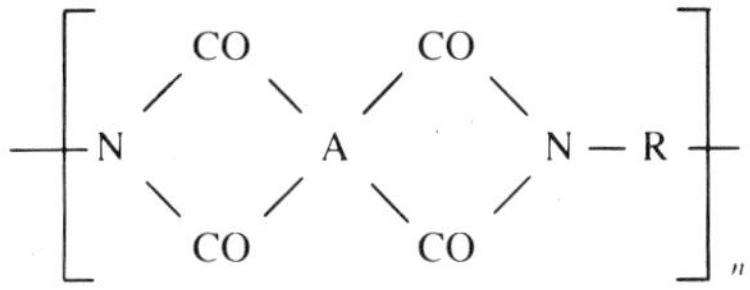

$$\text{(II)}$$

$$e.g., \quad A =$$

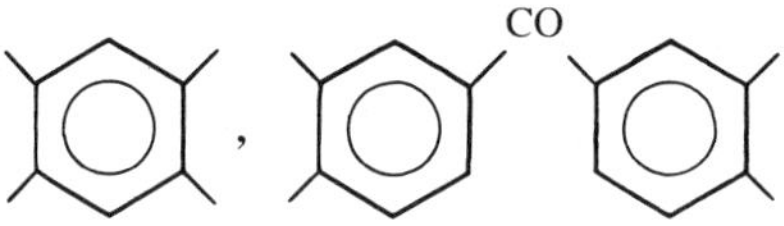

$$R = p,p'\text{-}C_6H_4OC_6H_4, \quad p,p'\text{-}C_6H_4CH_2C_6H_4$$

Conflicting interpretations of the mechanisms governing this rather complex reaction have generated variations in synthetic approach. Initial reports[47, 48] indicated the polymer-forming potential of the dianhydride/di-isocyanate reaction both in solution (dipolar aprotic solvents) and under melt fusion conditions. Subsequently it was suggested that,[49,50] with tertiary amine catalysts, the rate of polyimide formation is enhanced by the presence of water which hydrolyzes the isocyanate to amine or urea groups, either or both of which then react with anhydride to form an imide with CO_2 evolution. Alvino and Edelman (Westinghouse Corporation) proposed[51] that high-molecular-weight soluble polymers— fabricable into tough, flexible films—could be obtained by the reaction of di-isocyanate with a mixture of a dianhydride and its corresponding tetraacid in the presence of a tertiary amine catalyst. These authors stressed the significance of anhydride to acid ratio (ideally between 70:30 and 50:50), which could also be achieved by the controlled addition of water to the dianhydride; they suggested that the anhydride-acid complex might impart solubility to the growing chain, thus enhancing molecular weight. Due to the competing reaction of di-isocyanates with amide solvents at elevated temperatures, these and other reports have stressed the importance of relatively low reaction temperatures (<80°C), or the use of inert polar solvents such as NMP.

It is highly probable that the studies reported by the Upjohn Company form the basis of a patent[52] for the preparation of molding powders from BTDA and various aromatic di-isocyanates. Inherent viscosities of these polymers are low (0.2 to 0.3 dl/g), suggesting fairly low-molecular-weight materials; however, the versatile commercial resin Polyimide 2080[53] appears to have been developed from this work. An additional patent disclosure from Upjohn[54] has reported an improvement in synthesis and molecular weight of homo- and co-polyimides from dianhydrides and di-isocyanates. The reaction takes place in dry DMF at 80°C in the presence of a novel alkali metal lactamate catalyst:

$$\left[\begin{array}{c} (CH_2)_n - C{=}O \\ \diagdown \ | \\ N^{(-)} \ M^{(+)} \end{array} \right]$$

Condensation Polyimides (Insoluble, Infusible Type)— Commercial Variants

Cyclopolycondensation reactions, particularly those based on amic acid intermediates, have been responsible for a large family of polymers with carbocyclic aromatic and imide rings linked sequentially or interspersed by simple atoms or groups. These are the systems most frequently referred to above, and a limited number only, based on PMDA or BTDA, have provided the principal source of commercially available polyimide resins. In most of these the rigidity of the component polymer chains has imposed marked insolubility and intractability, and they have been rather ambiguously termed "thermoset polyimides." While it is true that in some the presence of, for example, reactive methylene or carbonyl groups may have led to relatively facile cross-linking reactions,[55–57] in most instances the solubility in relatively exotic solvents of even the most highly condensed systems strongly suggests a linear rather than a cross-linked structure. Russian workers[58] have proposed that cross-linking via the imide ring occurs above 300 to 350°C. Even so, in order to maintain long-term mechanical integrity in certain structural applications above the polymer Tg, which in the case of the wholly aromatic polyimides is frequently very high and difficult to measure,[55–57,59,60] a number of reports[61–63] have referred to the deliberate introduction of a controlled amount of cross-linking into the otherwise linear polyimide structure.

Polyimides of this general type are, or have been, available from a number of commercial sources, the most significant of which are detailed in Table 5.4. Dupont has been the most active in production and marketing. Kapton H or heat-sealable Kapton HF (FEP-coated) films and Pyre-ML grades of varnish or wire enamel are used extensively in high-temperature electrical and electronic appli-

TABLE 5.4. *Commercially Available C-Type Thermoset Resins*

Name	Manufacturer/ supplier	Form	Chemistry
Kapton[a,b]	Dupont	Film	PMDA/DAPE
Vespel	Dupont	Sintered parts	PMDA/DAPE
Pyralin[a,c] Pyre-ML	Dupont	Solution, glass fabric/roving prepregs, coatings	PMDA/DAPE
Skybond 700[a,c] Series	Monsanto	Solution	BTDA(diacid/diester)/MPD
FM-34[c]	Cyanamid	Primer solution; adhesive (glass cloth carrier)	BTDA(diacid/diester)/MPD
Meldin PI[a,c]	Dixon	Moldings and parts	BTDA(diacid/diester)MPD
QX-13[c,d]	ICI	Glass-, carbon-, and asbestos-reinforced prepregs	PMDA-BTDA/diacetylated DADM
Nolimid A380[c,d]	Rhone-Poulenc	Primer solution; adhesive (glass cloth carrier)	A-B Benzhydrol monomer (V)

[a]More than one grade supplied.
[b]Available as the polyimide.
[c]Available as PAA, substituted PAA or monomeric reactants.
[d]No longer marketed.

cations.[18,64–69] Solid resins (Vespel) are produced[70] in stock-shapes or specialized precision parts by direct forming techniques using powder metallurgy processes. Unfilled (SP-1 and 8) and graphite (SP-21 and 22), graphite/Teflon (SP-211 and 811), glass (SP-5), and MoS_2(SP-31) filled grades are used in a number of high-temperature engineering and electromechanical applications. Filled materials are particularly useful as solid lubricants.[71,72] Pyralin materials[65] are supplied in various grades; the 3550 series and 3003 are glass fabric prepregs processable by conventional vacuum bag techniques. Pyralin 3003[74,75] combines both addition and condensation chemistry and provides lower void content (<5%) with high-temperature properties. Other Pyralin products include glass fiber rovings for filament winding, conductive coatings for glass fibers, coated Nomex paper, adhesives, and sealants.[65]

Monsanto Skybond-700 materials[76] are solutions of monomer reactants (diester/diacid/diamine) that are stable for limited periods—one month at room temperature, three months at 5°C—in solvents such as ethanol, butanol, ethylene glycol, and NMP. The lower boiling solvents are particularly useful in applications where solvent removal has proved a problem. Skybond formulations have been used as varnishes, coatings, and films, but their principal use is in prepregs from which cross-linked polyimide laminates are produced by press and vacuum bag techniques.

FM-34, Cyanamid's high-temperature adhesive formulation, incorporates a resin system similar to the Skybond materials being a diethylene glycol/benzophenone-tetracarboxylate/MPD-based prepolymer, which is supplied to the user as a prepreg film incorporating aluminum powder and arsenic thioarsenate, supported on a glass carrier.[77–80] Principal applications have been in metal-to-metal bonding of stainless steel or titanium adherends. Although it has involved a different (benzhydrol-type) base resin, Rhone-Poulenc's Nolimid 380A adhesive[11, 80–82] can be used in similar fashion to FM-34. In both systems long-term bond integrity at elevated temperatures is enhanced due to a genuine cross-linking of the resin. Unfortunately, however, both commercial prepregs have been available only as PAA prepolymers, which must be stored under dry refrigerated conditions to ensure optimum bonding properties. The withdrawal of Nolimid 380A from the commercial market may have been related to this processing limitation.

QX-13 is the trade name for a material originally test marketed as a laminating resin by ICI (U.K.), but no longer available.[45,46] The synthesis (diacetyl DADM/BTDA-PMDA) and early evaluation of the system took place at the Materials Laboratories of the Royal Aircraft Establishment, Farnborough, U.K. Advantages of the system include solubility (at high solids content) in low-boiling cheap solvents such as acetone, low solution viscosity, and, unlike conventional PAAs, an extended life of the prepreg at normal temperatures. Asbestos and glass fibers were originally examined as reinforcement, but the clear potential of carbon fibers was quickly recognized.[83]

Within the group of condensation materials, it is interesting to note that a Total Materials System (TSB-130) for polyimide sandwich structures aimed particularly at aerospace applications has been made available by the Hexcel Corporation.[84] Individual items of the system include glass-reinforced laminate (F-174), film and paste adhesives (HP-954/974), structural foams (HP-901/904), and structural honeycomb (HRH-327).

Synthesis of Condensation Polyimides (Soluble, Fusible Type)

Aromatic polyimides

Using predominantly the same cyclopolycondensation routes as for the "thermosets," processable aromatic polyimides have also been developed that can be fabricated by methods typical of more conventional thermoplastics. The technique has been to so modify the chemical structure that the polyimide itself is soluble in low-boiling solvents and/or fusible at relatively low temperatures.

It is significant that in many cases processability has been increased by the introduction of relatively exotic groups midchain; that very few such systems have attained commercial recognition can probably be attributed to the relatively high cost of such precursors.

Soluble, but not necessarily fusible, materials have been produced from aromatic polyimides with large pendant groups on the repeat unit. The solubility and fusibility of the "Cardo" polymers (**VIII**), developed primarily in the U.S.S.R.[27,28] and to a lesser extent in the U.S.A.,[85] are governed both by the nature of the cyclic pendant group and the imide portion of the chain.

Further instances where large pendant cyclic groups contribute to increased room-temperature solubility have been cited,[86] but with Tgs around 400°C the polymers are not particularly thermoplastic. Pendant hydroxy or methoxy groups also convey solubility in conventional organic solvents to high-molecular-weight polypyromellitimides of the structure shown (**XIV**)[87]:

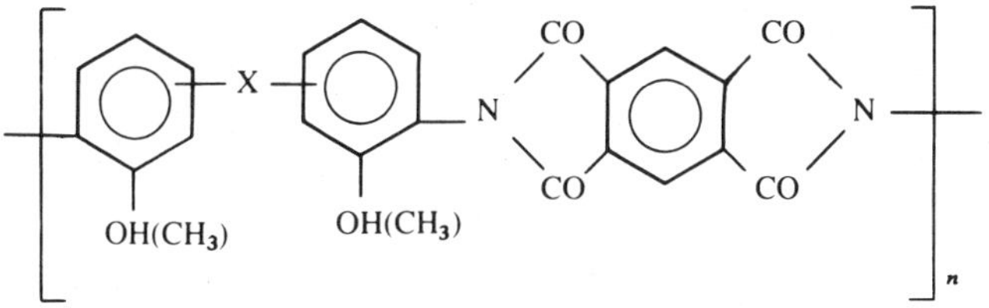

$$\left[\underset{\text{OH(CH}_3)}{\text{[benzene ring]}}-X-\underset{\text{OH(CH}_3)}{\text{[benzene ring]}}-N\underset{CO}{\overset{CO}{<}}\text{[benzene ring]}\overset{CO}{\underset{CO}{>}}N\right]_n \qquad (\textbf{XIV})$$

$$X = \text{direct bond, } CH_2, C(CH_3)_2$$

Introduction of flexible linking units into the main polyimide chain is an alternative approach. Thus, for example, silicon (siloxane) and phosphorous-modified homo- and copolymers have been developed[88–91]—the Shin-Etsu Chemical Co. markets a silicone/polyimide copolymer designated KJR 651—but, as for those systems that contain an aliphatic chain, the increase in processability has been accompanied by a reduction in thermo-oxidative stability. This problem has, however, been largely overcome by substituting stable fluoroalkylene groups midchain.[17,92] Perfluoroalkylene $[(CF_2)_n]$ and oxyperfluoroalkylene $[O(CF_2)_nO]$ "hinge" units have introduced solubility and tractability into aromatic polyimides without reducing substantially the high thermo-oxidative stability. Polymers (**XV**) are fusible and soluble in a range of solvents. This tractability is rapidly eroded if the flexible hinge groups are partially replaced; for

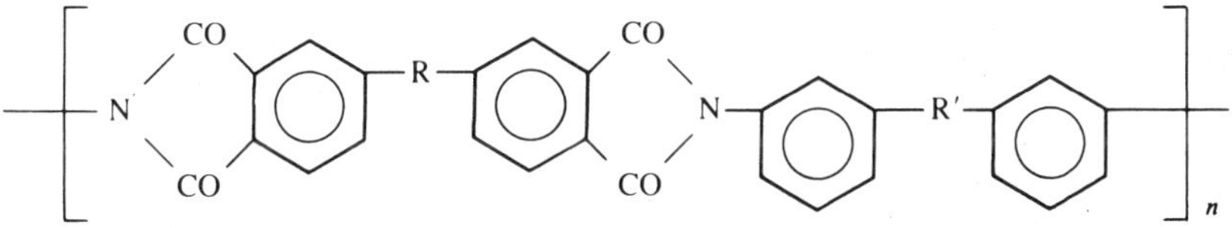

$$\left[N\overset{CO}{\underset{CO}{<}}\text{[benzene ring]}-R-\text{[benzene ring]}\overset{CO}{\underset{CO}{>}}N-\text{[benzene ring]}-R'-\text{[benzene ring]}\right]_n \qquad (\textbf{XV})$$

$$\text{e.g., } R = (CF_2)_{3,4,6,7} \; ; \; O(CF_2)_5O$$

$$R' - (CF_2)_{3,5}, \; O(CF_2\overset{CF_3}{\underset{|}{C}}FO)_x (CF_2)_5(OC\overset{CF_3}{\underset{|}{F}}CF_2)_yO \qquad [x + y = 2 \text{ or } 3]$$

example, polymer [(**XV**); R = $(CF_2)_3$, R' = O], though considerably more manipulable than most conventional aromatic polyimides (Tg ~220°C), has minimal solubility even in dipolar aprotic solvents.

For various reasons these fluoroalkylenepolyimides have remained in the laboratory development stage. Aromatic polyimides containing the hexafluoroisopropylidine group $[C(CF_3)_2]$ have, however, been marketed on a limited scale by the Dupont, Fiberite, and Hexcel Corporations. Hexafluoroisopropylidine-linked polyimides (**XVI**) were originally investigated by Rogers,[93] who demonstrated the relatively low Tg (340°C) and marked solubility of such materials in conventional solvents. The promise shown by the system generated a number of modifications, and Gibbs[94–96] in particular extended this work to produce the so-called NR-150 materials in which the diamine component is selected from a range of nonfluorinated aromatics (Table 5.5). In such cases solubility is restricted to higher boiling solvents such as DMF or NMP.

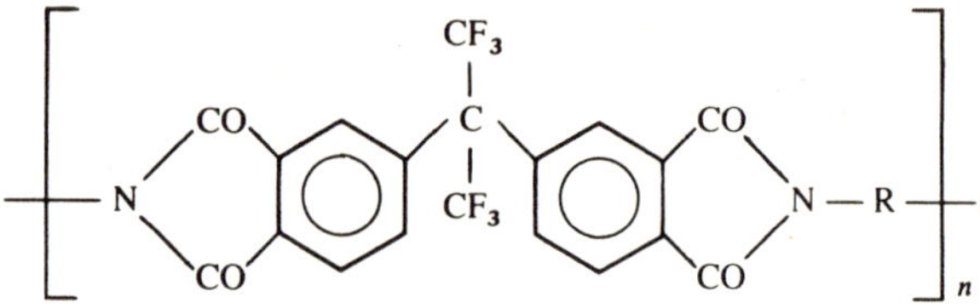

Preliminary studies were made of three systems, NR-150A (Tg ~290°C), NR-150B (Tg ~360°C), and NR-150C (Tg ~320°C), but subsequent work concentrated on materials A and B. The complete miscibility of these two has produced a continuous series of copolymers, resulting in a spectrum of resins with intermediate Tgs (Figure 5.5). NR-150 systems have been formulated as low-viscosity NMP solutions of monomeric precursors [diacid/diester (**XVII**) and aromatic diamines], which on heating produce the soluble polyimide (**XVI**). Such polymers are amorphous, noncross-linked, and they are sufficiently thermoplastic to facilitate melt flow when pressure is applied above their softening point or Tg. The majority of publications on NR-150 materials have concentrated on their evaluation in unidirectional glass or carbon fiber laminates, or as high-temperature metal-to-metal adhesives.

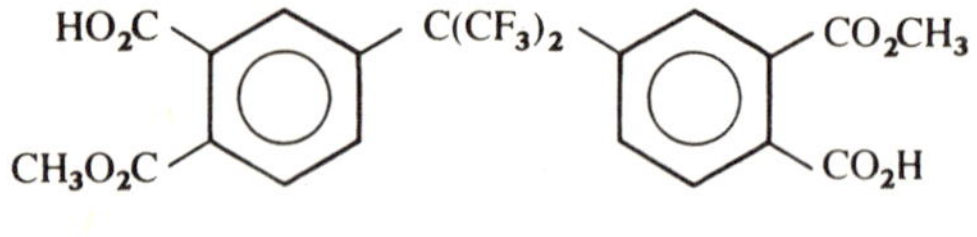

Improved processability of fully polymerized aromatic polyimides can be achieved without recourse to the relatively expensive variations described above.

TABLE 5.5. Tgs of Aromatic Polyimides of Structure XVI (Reference 95)

R	[η] inh (dl/g)	Tg (°C)
—C₆H₄—O—C₆H₄—O—C₆H₄—	0.35	229
—C₆H₄—O—C₆H₄—	0.46	285
—C₆H₄—S—C₆H₄—	0.35	283
—C₆H₄—CH₂—C₆H₄—	0.38	291
(m-phenylene)	0.41	297
(p-phenylene)	0.35	326
—C₆H₄—SO₂—C₆H₄—	0.31	336
—C₆H₄—C₆H₄— (biphenylene)	0.40	337
(naphthylene)	0.64	365

An early report by Olivier[97] of Dupont describes methods for producing melt-fabricable materials, whereby molecular weight is restricted by selective end-capping. An alternative approach, which produces high-molecular-weight, fusible products, depends on the introduction of a degree of dissymmetry into the polymer chain. Homopolyimides from traditional dianhydrides and stereoisomeric diamines (m,m' or m,p') have markedly reduced Tgs with no concomitant loss in thermo-oxidative stability.[98] Thermoplastic copolyimides have

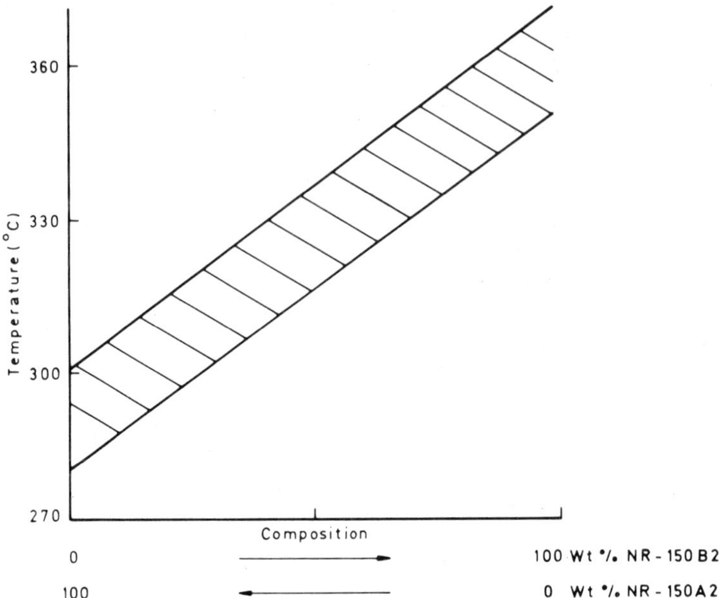

Figure 5.5.   Effect of NR-150 binder composition on glass transition temperature (Tg). Reference 120.

also[99] been produced from interaction of BTDA and two or more *m*-orientated diamines. It has been suggested that thermoplasticity in this case is related to the random distribution of repeat units in the polymer chain, which reduces the degree of crystallinity of the copolymer. None of these aromatic polyimides is, as yet, produced commercially. However, the Upjohn Polyimide 2080,[52–54] which is a fully polymerized and imidized resin used in a variety of applications (see below), incorporates certain of the features just described. It is, for example, at least in its initial form, of relatively low molecular weight and consequently highly soluble in organic solvents, particularly those of a dipolar aprotic nature. Commercial considerations have limited a full disclosure of the chemical composition of PI 2080, but it is believed to be a copolymer [(**XVIII**); x:y, 4:1] derived from BTDA and mixed aromatic di-isocyanates:

(**XVIII**)

Ciba–Geigy has recently introduced in experimental quantities a thermoplastic, soluble polyimide XU-218, which is based upon the condensation of pyromellitic dianhydride or benzophenone tetracarboxylic dianhydride with 5(6)-amino-1(4′-amino-phenyl)-1,3,3-tetramethylindane (DAPI).[100] This resin can be solvent cast into films less than 1.0 mil thick.[101]

CH$_3$

NH$_2$ — — NH$_2$

CH$_3$ CH$_3$

DAPI

Heterocyclic-Imide Copolymers

Heteroaromatic systems, other than imides, have been prepared and examined for their resistance to high temperatures. None have achieved the commercial status of the polyimides, but the majority have exhibited good thermal stability. A number of copolymer compositions have been developed in which imide and other heterocycles, e.g., oxadiazole, benzimidazole, benzoxazole, pyrrone, or quinoxaline, alternate in ordered, limited order or random sequence in the macromolecule.[102] Synthesis of these copolymers has followed conventional polyimide formation (via amic acid intermediates) from monomers containing the alternative heterocyclic,[103] or from monomers containing the preformed imide group. The latter process often involves a soluble prepolymer.[104]

It has been possible to "tailor" the properties of these copolymers towards, for example, enhanced thermal/thermo-oxidative stability or processability compared with the parent homopolyimide. However, although the comprehensive studies of Preston and Black[102,104] on ordered/limited ordered copolymers were directed specifically towards advanced fiber applications, none of the heterocyclic-imide copolymers have been marketed commercially except for an isoindoloquinazolinedione-imide material (PIQ) by the Hitachi Chemical Co. for electrical purposes.

Amide-, Ester-, or Ether-Imide Copolymers

Copolymers of this type have been developed to a large degree in parallel with the mainstream of work on aromatic polyimides. Although their elevated temperature properties are inferior, their processability and versatility of synthesis have made them commercially attractive materials. The introduction of thermoplastic aromatic polyimides has, however, somewhat diminished their significance.

Aromatic amide–imide copolymers have been prepared from precursors containing a preformed imide ring,[105,106] or by the methods described earlier[107,108] for aromatic polyimides using oligomeric diamines or dianhydrides containing amide units. The route most widely employed, via amide-(amic acid) intermediates, has involved the interaction of trimellitic anhydride (TMA), or similar anhydrides, with aromatic diamines. Numerous variations of this reaction are recorded; the cyclized product (**XIX**) from 4-chloroformylphthalic anhydride and DADM forms the basis of a range of commercial copolymers (see below):

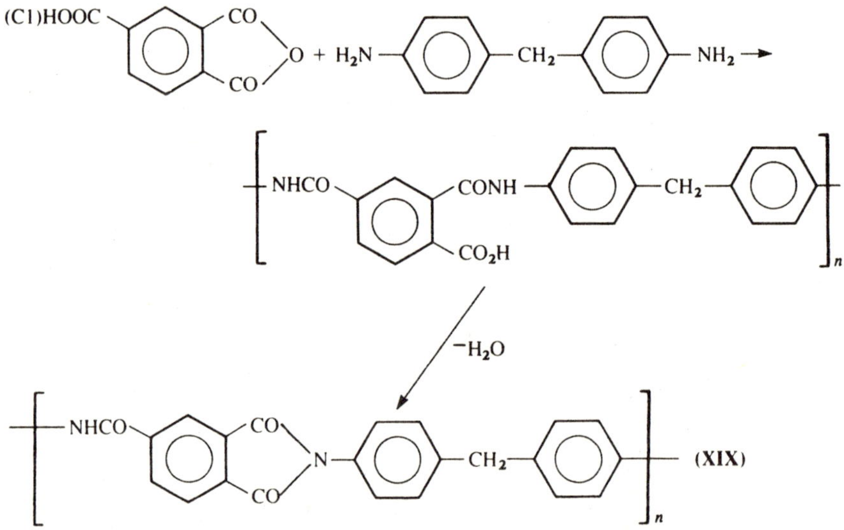

Not surprisingly, the processing problems inherent in the synthesis of polyimides by the two-stage amic acid route also occur here, although higher amide-to-imide ratios, while reducing thermal stability, to some extent minimize the problem. An alternative route, which excludes these unstable intermediates, has produced amide-imide copolymers by interaction of TMA or amide-linked dianhydrides with di-isocyanates.[109,110] A number of commercial products have been produced in this way.

Ester-imide copolymers possess superior processing characteristics—for example, flow and solubility—than the corresponding amide-imides. Essentially, they are obtained by similar methods, that is, from precursors containing ester groups (e.g., diamines, dianhydrides, and di-isocyanates) using conventional imide routes. They have also been made by interaction of aromatic dicarboxylic acids containing preformed imide groups with saturated dihydric alcohols, typically ethylene glycol.[111] This process forms the basis of several commercial materials. Poly(amide-ester-imides) have also been formulated.

Poly(ether-imides) have received less attention than other copolyimides. They are produced by conventional imide synthesis from dianhydrides, diamines, and di-isocyanates containing ether links. An interesting alternative synthesis, pioneered by the GE Co., has involved interaction of a bisphenol dianion with bis(chloro-, fluoro-, or nitrophthalimido) alkanes or arenes.[112]

While this book was in press General Electric marketed a poly(ether-imide) with the trade name Ultem.

Epoxy-Imide Copolymers

Epoxy-imide copolymers tend to bridge the gap between C- and A-type systems. They possess the ease of processing and the variety of volatileless cure reactions typical of epoxies, together with the high thermo-oxidative stability and thermomechanical strength of the aromatic polyimides. A number of structural/engineering laminating resins have been evaluated without, as yet, a commercial product resulting.

Two synthetic approaches have been described. Newey and Adler[113] and Oswitch[114] have reported the preparation of simple imido epoxies based on phthalimide, pyromellitimide, and benzophenone tetracarboxydi-imide. Resins resulting from these systems have been cured successfully with conventional amine or Lewis Acid catalysts.

A second route has involved the catalyzed direct addition of an epoxy compound to the terminal NH groups of various di-imides. Kaplan et al.,[115] as well as Soviet workers,[116] have reported on these processes and on the laminating materials produced from them.

Condensation Polyimides (Soluble, Fusible Type)— Commercial Variants

Four NR-150 products, solutions of monomeric reactants [HFDE/DAPE(A) or HFDE/mixed PPD-MPD(B)] capable of thermal cyclization, are available from Dupont. Several publications have discussed in detail[75,94–96,117–119] the application of NR-150A2 and NR-150B2 binders in combination with glass, quartz, graphite, and boron reinforcing fibers. In particular, Dupont commercial literature[120] has cited the production and processing of unidirectional glass and graphite composites using Dupont NR-150B2 binder. Fiberite (HyE 166BA and HyE 1264BB) and Hexcel (GRX 575) have also produced graphite unidirectional tape and fabric prepregs in development quantities.[121,122] NR-150A2G and B2G solutions have been examined as metal-to-metal (titanium)[120] and Kapton-film adhesives.[123]

Upjohn's PI-2080[50,53,124] is available as a fully imidized and polymerized product in two forms: as a dry molding grade or as a solution in NMP. Since no ring closure or further polymerization is necessary, these materials remain stable indefinitely under ambient conditions. PI-2080 solutions have been used in the manufacture of films, laminates (glass, boron, or carbon fibers), protective coatings, metal-to-metal adhesives (either neat or glass fiber-reinforced), fibers (wet- and dry-spun), and syntactic foams. The dry resin, neat, or in conjunction with glass, graphite, PTFE, aluminum and MoS_2 fillers, has been used in conventional compression molding applications.[125,126]

Nonstructural grades of amide- and ester-imide copolymers have been available commercially for some considerable time. The Kerimids (500 series) and Rhodeftal (101, 200, 310, 333) are marketed[127] by Rhone–Poulenc in adhesive or electrical enamel formulations; Rhodeftal 200 is also a base for thermostable paints. Similarly, Amoco's AI products (AI-10, 11, 220) are electrical grade varnishes, as are the ester-imides Terebec and Isomid marketed by Dr. Beck and Schenectady, respectively. Three grades of thermostable poly(amide-imide) fiber,[128] Kermel (203, 233, and 234), were offered in development quantities to the market by Rhone–Poulenc in the early 1970s. The very pronounced success of Dupont's Kevlar fibers appears to have robbed these materials of any real significance, and they are no longer available commercially. Amoco has also introduced a range of high performance amide-imide molding resins under the generic name Torlon.[126,129] Grades of Torlon containing various additives—TiO_2, PTFE, graphite, glass, and carbon fiber—are used in a number of electrical and structural engineering applications. A grade of glass-filled Torlon (PDX-5200), available from the LNP Corporation, is used to produce highly complex precision moldings. These materials are listed in Table 5.6.

TABLE 5.6. Commercially Available C-Type Thermoplastics

Name	Manufacturer/ supplier	Available form	Chemistry
NR-150[a,b]	Dupont Fiberite Hexcel	Solution; glass- and carbon tape prepreg	HFDE/aromatic diamines
PI-2080[a,b]	Upjohn Co.	Solution; molding resin	BTDA/mixed di-isocyanates
Torlon[a,c]	Amoco LNP Corp.	Molding resins, glass, carbon, PTFE reinforced	TMA (or derivative)/ aromatic diamine
AI[a,c]	Amoco	Solution	TMA (or derivative)/ aromatic diamine
Kerimid 500[a,c]	Rhone-Poulenc	Solution; adhesive varnish	TMA (or derivative)/ aromatic diamine
Kermel[c]	Rhone-Poulenc	Fiber	TMA (or derivative)/ aromatic diamine
Rhodeftal[a,c]	Rhone-Poulenc	Solution; varnish	TMA (or derivative)/ aromatic diamine
Terebec[d]	Dr. Beck Co.	Solution	TMA/MDA/hydroxy- terminated polyester
Isomid[d]	Schenectady	Solution	

[a]Several grades available.
[b]Polyimide.
[c]Poly(amide-imide).
[d]Poly(ester-imide).

Synthesis of Addition Polyimides

The development of high-modulus, high-strength fibers such as carbon, boron, and aromatic polyamide (Kevlar) introduced a new dimension into structural composite materials. The potential of these composites in high-temperature/high-performance applications was, however, reduced by the problems of fabrication associated with the condensation-type aromatic polyimides. The advent of A-type polyimides in which polymerization/cure occurred at reactive (unsaturated) termini of short preimidized molecular segments successfully overcame a number of the processing disadvantages inherent in the C-type polyimides. It is significant that the main applications of A-type materials have been in composites. Several addition systems have been evaluated and each has led to commercially viable materials.

From Maleimides

Imide rings can be introduced into macromolecules by taking advantage of the reactivity of the double bond in N-substituted maleimides to free radical and anionic homo- and co-polymerization.[130–132]

Polymerization of bismaleimides has been studied extensively. Copolymers with varying molecular weights and tractability have been synthesized via the Diels–Alder, 1,3-dipolar, and photosensitized addition reactions of bismaleimides (**XX**) with a number of reactive comonomers.[133–137] In some cases film-forming materials have been claimed.

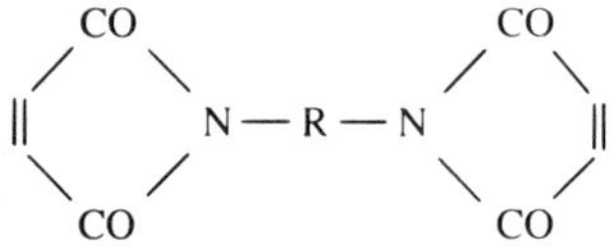

$$\text{(XX)}$$

Cross-linked polyimides have been made directly by the thermal polymerization of bismaleimides, preferably by heating them above their melting point in the presence of free radical catalysts such as dicumyl peroxide. Using a hydroquinone stabilizer, Grundschober and Sambeth[138] increased considerably the pot-life of polymerizates, and by restricting the polymerization of bismaleimide (**XX**, R = p,p'–$C_6H_4CH_2C_6H_4$) to a staged heating at 150°C, Gilwee *et al.*[139] formed a soluble resin, capable of fabrication as a glass composite. Prolonged heating at or above 150°C gave a highly cross-linked, intractable product. Increased processability has been achieved either by increasing midchain flexibility in the bismaleimide[140, 141] or by using a close-to-eutectic ternary mixture:

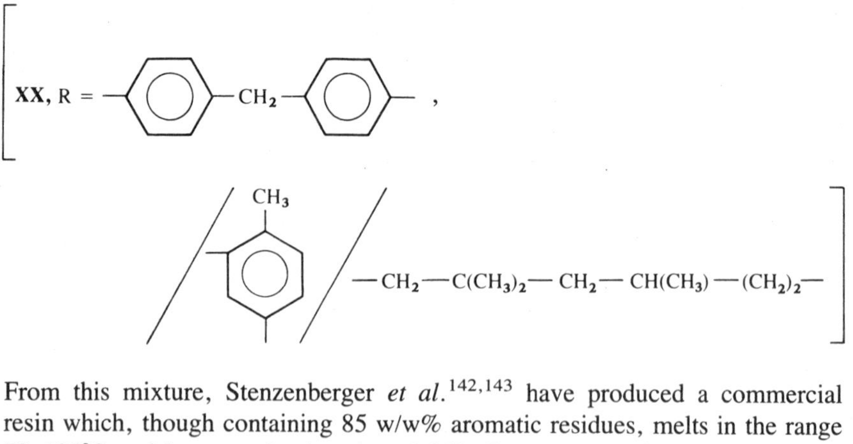

$$\mathbf{XX,\ R} = \text{...}, \quad -CH_2-C(CH_3)_2-CH_2-CH(CH_3)-(CH_2)_2-$$

From this mixture, Stenzenberger *et al.*[142,143] have produced a commercial resin which, though containing 85 w/w% aromatic residues, melts in the range 70–125°C and has a melt viscosity of 150 cP at a processing temperature of

120°C. Polymerization kinetics for the polymerization of aliphatic bismaleimides have been reported.[144]

Although high-temperature, free-radical cure of simple bismaleimides (**XX,** $R = p,p'-C_6H_4CH_2C_6H_4$) gives a tightly cross-linked and hence brittle product, studies by Kovacic and Hein[145] and Sheremeteva[146] showed that the interaction of stoichiometric mixtures of bismaleimides and aromatic diamines results in linear chain extension via a Michael addition reaction. Workers at the Rhone–Poulenc Company[147,148] observed that at elevated temperatures the mechanical properties of these linear resins were rapidly degraded. However, they established that, with nonstoichiometric mixtures of aromatic diamine and bismaleimide either in solution (NMP) or at higher temperatures in the melt, two types of reaction are possible. First, Michael addition to produce linear polymers:

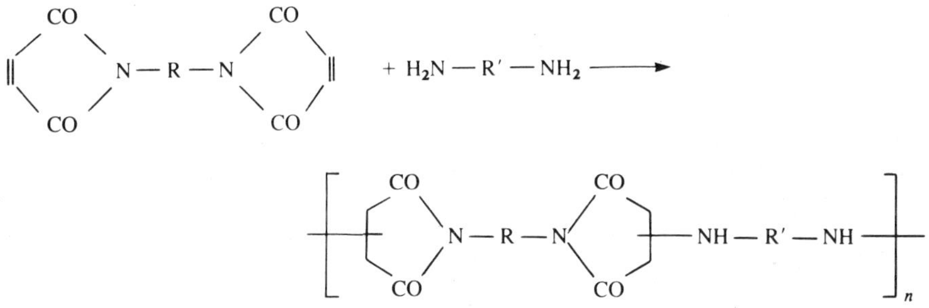

and secondly, opening of the double bond followed by a free radical cure leading to cross-linking:

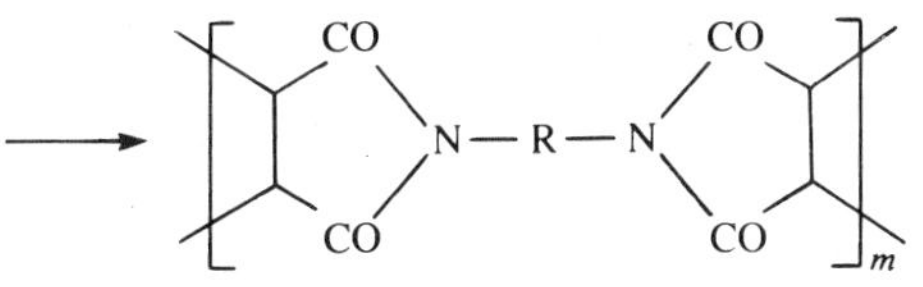

Polymerization/cure is achieved by a judicious choice of free-radical catalyst, inhibitor, and solvent.[149,150] Molding and laminating resins with a range of physical and chemical properties have been produced commercially[151] based on these reactions.

A variety of copolymer modifications to the basic bismaleimide system have been examined, mainly by Rhone–Poulenc workers. Contributions to the chemistry and technology of polyaminobismaleimides and related copolymers have also originated from the GE Co. in the U.S.A. Among these developments, Crivello and Juliano[152] have, for example, prepared alternating block copolymers (**XXI**) from polyimidothioethers ("hard segments") and thiol-terminated polysulphide elastomers ("soft segments"):

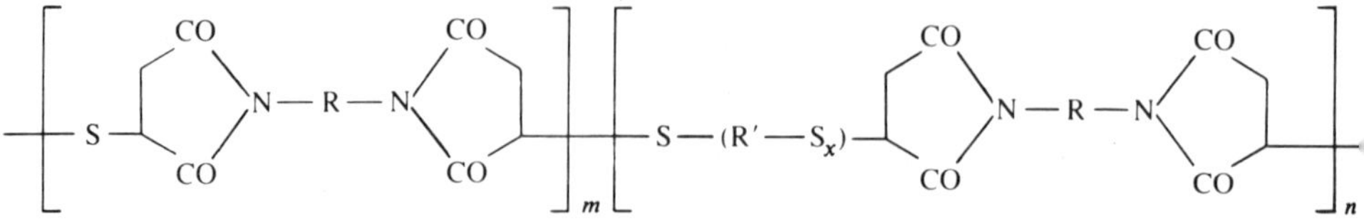

$$\left[S - \underset{CO}{\overset{CO}{\diagdown}} N - R - N \underset{CO}{\overset{CO}{\diagup}} \right]_m \left[S - (R' - S_x) - \underset{CO}{\overset{CO}{\diagdown}} N - R - N \underset{CO}{\overset{CO}{\diagup}} \right]_n$$

(XXI)

Copolymers with a wide variety of structures for R and R$'$ have been produced, and compositions with $>70\%$ polysulphide component are solvent-resistant elastomers. However, their use temperature is low due to the relative thermal instability of the polysulphide components and there have been no commercial developments. Recent publications indicate that the Soviet Union is also interested in the potential of the polyaminobismaleimides (PABMs).[153–155]

"Pyrolytic" Polyimides

Examples of A-type "pyrolytic" polyimides appeared about the same time as the first commercial PABMs. Unlike most C-type polyimides, they have been produced principally as laminating resins for use with glass, carbon, boron, and Kevlar reinforcement. It has been inevitable, therefore, that the chemistry and technological development should have been directed very specifically towards the rapid introduction of commercial products, and only a fraction of the background literature observed for C-type polyimides is available. Most reports on the subject have appeared in product-oriented literature, and much information on formulation is still commercially protected.

In their original approach, workers at TRW Systems, in close collaboration with NASA (Langley), synthesized a low-molecular-weight, soluble (40% solids in DMF) amide–acid prepolymer:

$$\left(\textbf{XXII, } R = p.p' - C_6H_4CH_2C_6H_4. \quad A = \right)$$

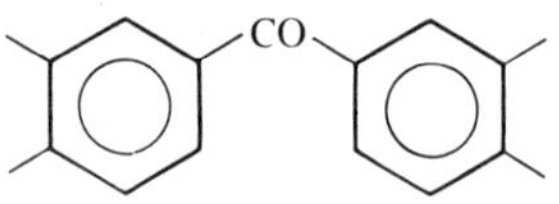

the chain ends of which were terminated or capped with norbornylene groups.[156–159] The oligomeric amic acid was isolated by various methods, most conveniently on the fiber substrate, and then advanced to the imidized state (**XXIII**), with evolution of solvent and water by heating in the temperature range 65–204°C. The low-molecular-weight oligomer (**XXIII**) melts/fuses at temperatures around 260°C and is converted into an insoluble, cross-linked product at temperatures between 275 and 350°C. It has been postulated that this pyrolytic polymerization initially involves a reverse Diels–Alder reaction yielding a maleimide prepolymer and free cyclopentadiene, which subsequently copolymerizes to the infusible, cross-linked polymer (**XXIV**). Since pyrolytic polymerization occurs without evolution of volatiles, extremely low void composites are produced.

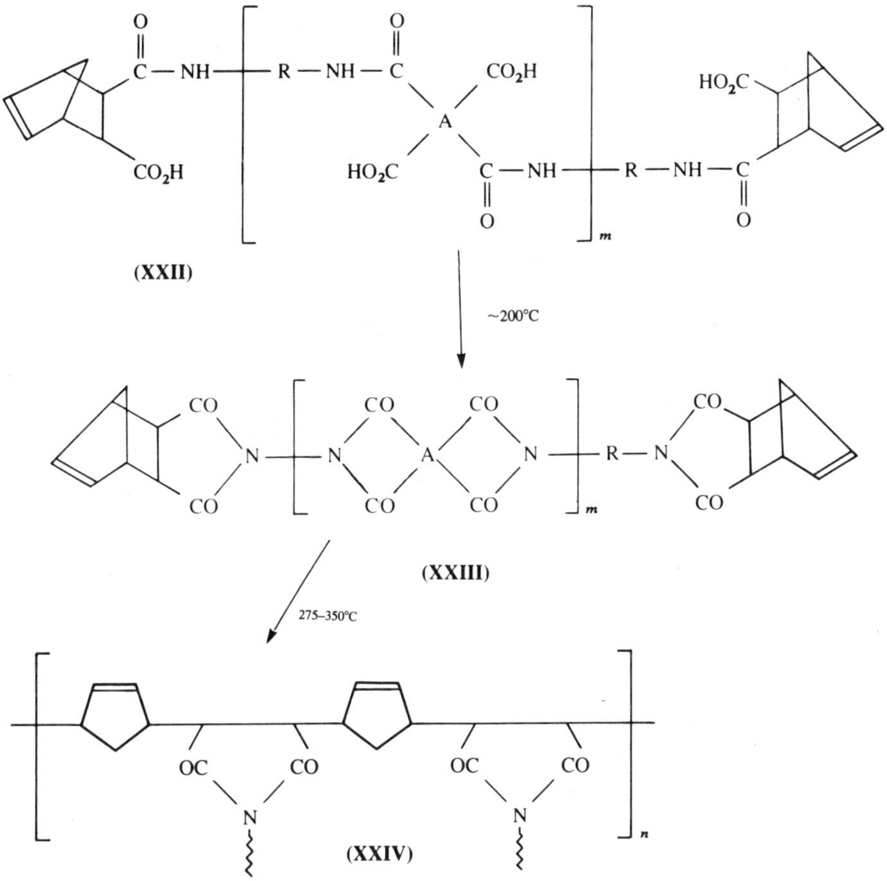

(**XXII**)

~200°C

(**XXIII**)

275–350°C

(**XXIV**)

Known by the trade name P13N [formulated molecular weight (FMW), 1300], this system is available as a DMF solution of the laminating prepolymer from which low-void-content polyimide resin matrices can be produced using short press molding cycles.[159–161] The system, however, is not without its drawbacks. In particular, an imbalance of prepolymer melting and cure temperatures has excluded its application in important autoclave molding processes. Modifications have included a change from BTDA to PMDA as main-chain dianhydride, together with a reduction in FMW ($\sim$1000) and the introduction of mixed diamines rather than a single compound. Specifically, the mixed diamines have contained thiodianiline (TDA) as one element, and it is this material that has been claimed to lower the prepolymer melt temperature and to reduce the rate of pyrolytic polymerization—vital factors in obtaining the desired improvements in processability. Modified systems of this type have been given the designations P10, P10P, P10PA, P105A, P11B, and LSU (1001–1003) by the TRW and Ciba–Geigy companies.[159–163] St. Clair[164] has also produced significant improvements in the all-round processability of the nadimide end-capped pyrolytic polyimides by the introduction of isomeric changes to the molecular structure, using meta- or meta-para-linked diamines rather than the wholly para-linked species. Several of the addition polyimides so produced have exhibited excellent potential in experimental structural composites.

The most successful approach to effective overall processability in "pyrolytic" polyimides has, however, involved the PMR[159, 165] [in situ polymerization of monomeric reactants (PMR)] resins, the route to these being rather reminiscent of the Skybond and NR-150 series of C-type materials. Instead of using end-capped amide acid prepolymers with all their obvious limitations and drawbacks, with PMR materials the reinforcing fibers (carbon, glass, or Kevlar) are impregnated with a solution containing a mixture of monomers dissolved in a low-boiling alcohol. The monomers—a dialkylester of an aromatic tetracarboxylic acid, an aromatic diamine, and a monoalkyl ester of 5-norbornene-2, 3-dicarboxylic acid—though essentially unreactive at room-temperature, polymerize on the fiber at elevated temperatures to form a polyimide by the normal pyrolytic route. A number of systems have been investigated and alterations to the chemical structure and/or stoichiometry of the monomeric reactants have shown the ability of the PMR approach to produce "tailor-made" resins with varying degrees of flow. From these extensive investigations three resin systems (Table 5.7) have been selected for more detailed evaluation—the so-called "first-generation" resins PMR-15[165–167] and LARC-160,[168] which contain the dimethyl ester of benzophenone tetracarboxylic acid (BTDE), and the "second-generation" PMR-11,[169,170] in which BTDE has been replaced by diacid/diester HFDE (**XVII**). These three systems are now under development as commercial high-temperature composites.

Despite their superior processing characteristics, the number of alicyclic rings in the cured norbornylene-terminated A-type polyimides reduces thermo-oxidative stability in comparison with the wholly aromatic C-type materials. However, materials have been developed that involve addition-type polymerization/cure reactions and yet give wholly aromatic structures. These systems, based on acetylene-terminated oligomers having molecular weights of less than 2000, were postulated to cure at temperatures above 220°C by trimerization of the acetylene groups to aromatic rings.[171–173] Subsequent work has shown that the reaction is much more complex.[174]

TABLE 5.7. First and Second Generation PMR Polyimides

Name	Monomeric reactants (moles)	FMW
	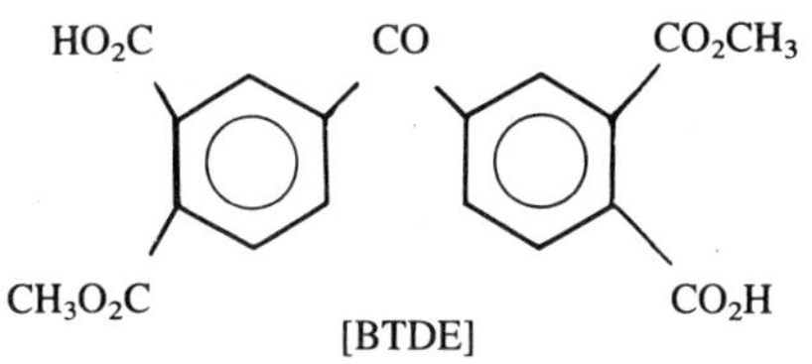 [BTDE] (2.09)	
PMR-15	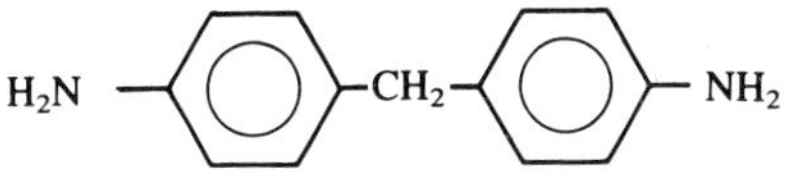 [DADM] (3.09)	1500
	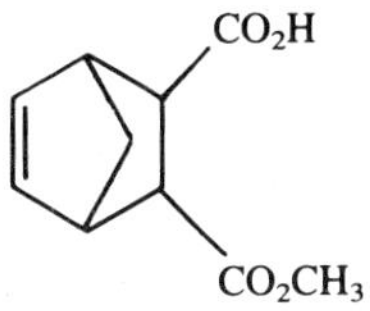 [NE] (2.0)	
	[BTDE] (0.67; 0.735)	
LARC-160	(2.56)[a]	—
	[NE] (1.22; 1.094)	
	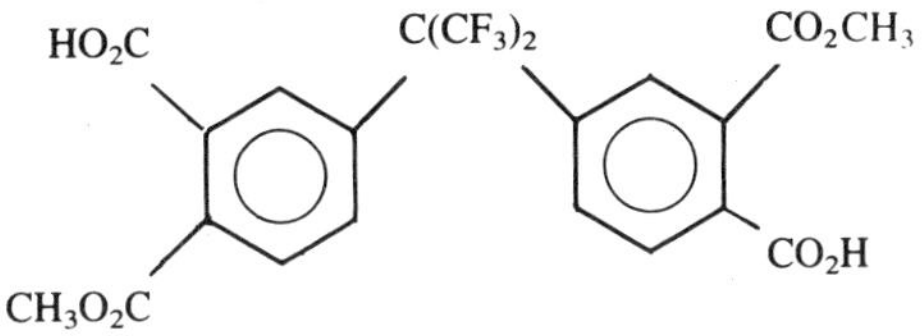 [HFDE] (1.67)	
PMR-11 	(2.67)	1260
	[NE] (2.0)	

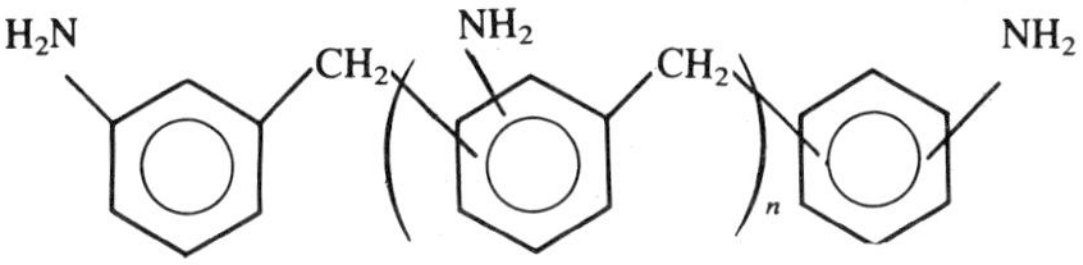

[a]Jeffamine AP22. $n = 0,1,2$, average MW − 234

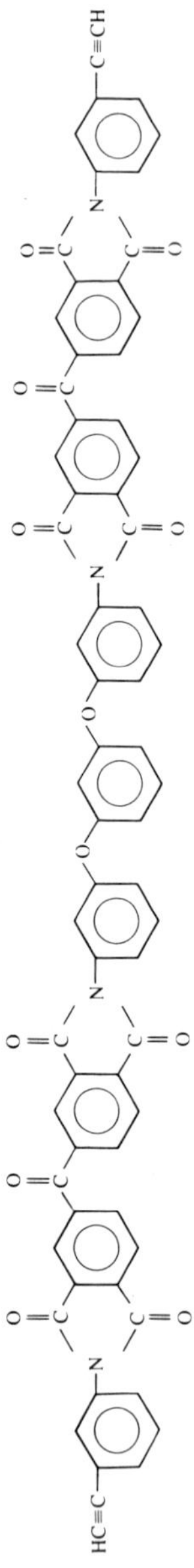

HR600

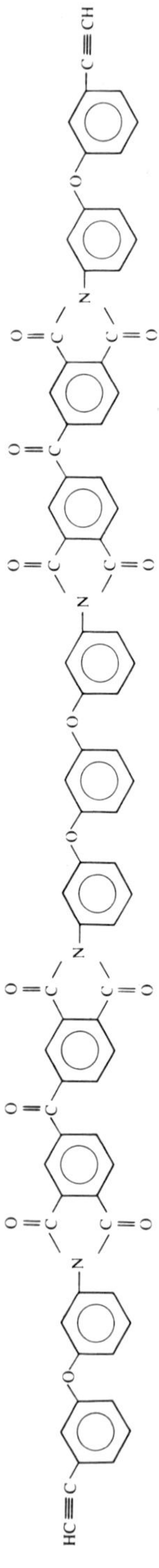

HR650

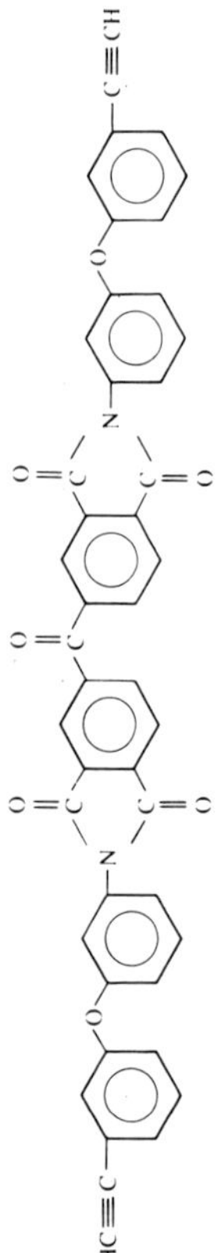

HR700

The HR-series of prepolymers shown above was developed originally by Hughes Aircraft Co. as metal-to-metal adhesives and glass/carbon fiber binders. The laminating properties of material HR-600 have been most actively examined[175] both as preimidized oligomer (in NMP) or as amic acid precursor (in acetone). HR-600 is now manufactured commercially by the Gulf Oil Chemical Co. under the trade name Thermid 600. Attempts have also been made to improve the processability of acetylene-terminated addition polymers by incorporating increased flexibility into the imide prepolymer.[176]

Addition Polyimides—Commercial Variants

Table 5.8 summarizes those A-type polyimides that have achieved a degree of commercial success. PABMs have undoubtedly received the most detailed attention, particularly from the Rhone–Poulenc organization. Their first successful products were the Kerimid 601/Kinel materials,[150,151] both manufactured from bismaleimide (**XX**) and aromatic diamine. Kerimid 601 is a thermosetting resin applied as a laminating material (NMP solution) in conjunction with glass fabric, silica, quartz, carbon, or boron fibers.[177,178] Prepreg molding has been achieved by several methods—press molding under moderate pressures, bag molding and curing under vacuum, or bag molding and curing in an autoclave. In order to achieve optimum properties, post-curing at 200°C (48 hours) or 250°C (24 hours) is necessary. Kinel molding compounds have been developed to satisfy different end uses;[151] molded parts are produced by conventional transfer (3500 series), injection (4500 series), and compression (5500 series) processes.[179] Kinel 5504, 5514, 3515/4515/5515 are glass fiber-filled resins used where mechanical properties have prime importance. Kinel 5502 (unfilled), 5505/5508 (graphite-filled), 4511/5511 (graphite/asbestos-filled), 3517/5517 (graphite/MoS_2-filled), and 3518/4518/5518 (PTFE-filled) are self-lubricating materials with low friction and wear properties and good dimensional stability under load and when heated. In the early 1970s, GE marketed Kinel and Kerimid materials under their own trade name of Gemon (2010, 2012, 3010, and Gemon L prepreg). These products are no longer commercially available.[180,181] More recently, Rhone–Poulenc has marketed the solventless composite binder resin Kerimid 353. The basis of this system[143,144]—the development was a joint Technochemie GmbH Verfahrenstechnik and Rhone–Poulenc venture—is a close-to-eutectic ternary bismaleimide mixture that can be liquified by heating to 120–130°C. Pot-life is in excess of 8 hours at 120°C and approximately 1 hour at 160°C (Figure 5.6); the low viscosity at 120°C guarantees rapid fiber impregnation. Kerimid 353 has been applied to various fiber substrates such as E-glass, Kevlar 49, and carbon in the form of rovings and filament windings. The most

TABLE 5.8. Commercially Available A-Type Thermosets

Name	Manufacturer/ supplier	Available form	Chemistry
Kinel 500[a]	Rhone–Poulenc (Rhodia)	Molding resins	Polybismaleimides (PABMs)
Kerimid[a]	Rhone–Poulenc (Rhodia)	Filament winding and laminating resins	Polybismaleimides (PABMs)
F-178	Hexcel Corp.	Filament winding and laminating resins	Polybismaleimides (PABMs)
P13N[g] P105A[g]	Ciba–Geigy Co.	Resin solution (40% solids in DMF) Molding powder	BTDA/NA/DADM
PMR-15	NASA-LEWIS Fiberite Corp.[b] Ferro Composites[c] Hexcel[d]	Solution: glass, carbon, and quartz fiber and fabric reinforced prepreg	BTDE/NE/DADM
LARC-160	NASA-LEWIS US Polymeric Fiberite Corp.[e] Hexcel[f]	Solution: Carbon fabric reinforced prepreg	BTDE/NE/Jeffamine AP22
HR-600	Gulf Oil Chemicals Co.	Molding resin; glass and carbon fiber reinforced	Acetylene terminated imide oligomer

[a]Various grades available.
[b]HMF-1133/66A carbon fiber prepreg.
[c]CPI-2237/7781 fiber reinforced prepreg.
[d]GRX-574 carbon fiber prepreg.
[e]HY-E 1678E carbon fiber prepreg.
[f]GRX-576 carbon fiber prepreg.
[g]No longer commercially available.

comprehensive information on component fabrication has, however, involved the use of three carbon fiber types, Modmor IIS, and Thornel 50S and 300. Rhone–Poulenc is now development-marketing Kerimid 711, which appears to be a variant of Kerimid 353. Hexcel also supplies[75,182] glass fabric prepregs from its F178 solventless maleimide resin; at 100% solids tack and drape properties of this resin/fiber system have allowed oven or autoclave cure, typical of standard epoxies. Finally, in this group of PABM resins, Fiberite Corporation supplies[183] a range of easily moldable materials. PI-740 and 750 are chopped-glass fiber-filled resins fabricated by compression molding techniques, while PI-730 is an unfilled resin capable of molding on conventional thermoset injection equipment.

P13N was produced and marketed by Ciba–Geigy Corporation[75,184,185] under license from TRW, Inc., as either a low-viscosity laminating solution (in

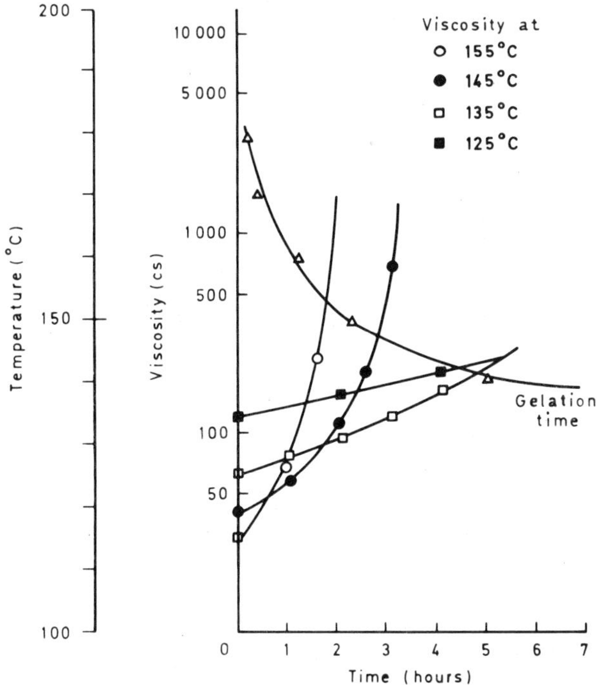

Figure 5.6.   Viscosity and gelation of Kerimid 353 at various temperatures. Reference 142.

DMF), or as neat or filled (chopped-glass fiber) molding powder.[159–161] In the varnish form, P13N is available as the amic-acid prepolymer and as such must be stored at temperatures below 5°C under dry conditions. In laminate processing, fiber-reinforced prepregs are dried at 65°C to 190°C, dependent on the drape and tack requirement, in order to remove solvent. Cyclization to the fully imidized form takes place at 190 to 205°C. Prepregs based on glass, carbon, boron, and asbestos reinforcement (resin content 25–30%) can be stored for two months or more at room temperature. Consolidation to the fully cured laminate is effected at around 300°C for 1 hour at pressures of 1.4 to 3.5 MPa. To avoid gelation prior to resin flow, very rapid heating to around 288°C in a preheated mold is imperative. Autoclave processes are not practicable. In order to improve on flow characteristics, Ciba–Geigy introduced, on an experimental basis, product P105A in which 20% of the DDM constituent was replaced by TDA. Although improved processing was achieved in this material,[75] it has now been withdrawn due, it is believed, to the unacceptable toxicity of TDA.

Alcohol solutions of monomeric reactants from both PMR-15 (BTDE/NE/DADM) and PMR-11 (HFDE/NE/PPD) have a very limited shelf-life, the limiting factors being low stability of the dimethylesters and premature precipitation of imide reaction products. Longer lifetimes are exhibited, however, when the solutions are stored under subambient temperatures. Generally it is preferable to make up fresh solutions of commercially available monomers immediately prior to fiber impregnation. Prepregs can be made by drum-winding and impregnating the selected reinforcing fiber with a solution containing up to 50 w/w% of monomer reactants so that the prepreg contains around 40 w/w% resin and 60 w/w% fiber.[75] Commercial prepreg formulations of PMR-15 are now available in limited development quantities, using glass and carbon reinforcing fibers, from the Fiberite, Ferro, and Hexcel Corporations (Table 5.8). High pressure (compression) and low pressure (vacuum bag autoclave) molding cycles are used for fabrication of composites; both processes use a maximum cure temperature of 316°C (1–2 hours) followed by post-cure at 316°C (4–16 hours). Compression molding cycles have employed high rates of heating (5 to 10°C/min) and pressures of 3.5 to 7 MPa, while in autoclave processes low heating rates (2 to 4°C/min) and pressures of 1.4 MPa or less have been used to produce essentially void-free composites. LARC-160, a PMR resin produced by NASA–Langley,[168] uses a commercially available diamine Jeffamine AP-22 (Jefferson Chemical Co.). Solventless, high-viscosity liquid resin has been produced in development quantities by the U.S. Polymeric Corporation. Carbon fiber prepregs can be prepared from the neat resin by hot-melt coating, or from a solution of this resin (in ethanol) by brush coating or immersion in the solution. Composites have been produced by compression and autoclave molding techniques as described above. Commercial prepreg formulations of LARC-160 are available in development quantities from Fiberite Corporation (Table 5.8); HMF-1176/78 has a 1000-filament woven carbon fiber reinforcement, and Hy-E 1678E is a tape prepreg consisting of 6000-filament carbon fiber yarns. Both systems are made by hot-melt impregnation with neat LARC-160 resin. Thermid 600 neat resin is available commercially[186] as a compression and transfer molding compound and as a binder lacquer (50 w/w% in NMP). The production of glass or carbon fiber-reinforced structural laminates, solid lubricants (15 w/w% MoS_2), and metal-to-metal adhesives using predominantly 6A14V titanium adherends have been described.

Thermal Characteristics

Glass Transition Temperatures

Glass transition temperature (Tg) has proved to be a key parameter in defining the thermophysical profile/processing characteristics of a number of polyimides.[75,98] Measuring techniques used have been differential scanning calorimetry (DSC), torsional braid analysis (TBA), and thermomechanical analysis (TMA). Apart from a limited number of fusible aliphatic/aromatic systems,[12] Tgs of wholly aromatic polypyromellitimides, including Dupont's H-Film, are extremely high with poorly defined dispersion maxima.[55,56,59,60,187] Although improvements have been achieved using aromatic diamines with a greater proportion of flexible links,[59] more profound effects on the level of Tg have been observed when dianhydrides (Table 5.3) such as BTDA,[55,56,59] DEDA,[188] HFDA,[93,95] and DBDA[60,189] have been used. Significant observations have been made on NR-150 and related hexafluoroisopropylidine containing resin systems[95,120] (Table 5.5, Figure 5.5), PI 2080,[50,125] and Thermid 600 and related systems.[171,176] Another important development has been the application of Tg measurements as an aid in establishing the degree of cross-linking, or cure in C- and A-type polyimides. In these systems cross-linking occurs via specifically introduced sites situated midchain or at the chain-ends. In either case the advance of Tg to higher values on exposure to higher temperatures has been interpreted[56,171,190] as a progressive advancement in the cure. In developments such as high-temperature metal-to-metal adhesives and advanced reinforced composites, where long-term strength at temperature is a priority, it has proved possible, by following Tg, to establish a link between stoichiometry, processing variables, and end-use application.

Thermal Stability

Because of their importance as thermally stable polymers, the thermal and thermo-oxidative degradation of the polyimides has been extensively studied and more than a hundred papers have been published on this topic during the last fifteen years alone. The application of thermal methods to the study of the degradation of polyimides has recently been reviewed.[191] Figure 5.7 compares the thermal stabilities of the different types in air on the basis of the weight lost in two hours at various temperatures. The superior stability of the fully aromatic polymers compared with the addition polymers containing alicyclic rings is evident. The relatively low oxidative stability of the polymer derived from the oligomers with acetylenic end-groups, which originally was thought to be fully aromatic in nature, indicates that not all the acetylenic end-groups are used up in such reactions and that residual unsaturated groups are focal points for oxidative attack. The overall activation energies for thermo-oxidative degradation of the different polyimide types derived from isothermal weight loss data are given in Table 5.9.

Low values are obtained for the degradation of the addition-type polyimides, especially for the polymer with maleimide end-groups. Synthesis of this material requires a very careful balance of free radical reactions and Michael addition reactions with diamines, and its thermo-oxidative degradation is considerably influenced by residual traces of the solvent used in its preparation.[178]

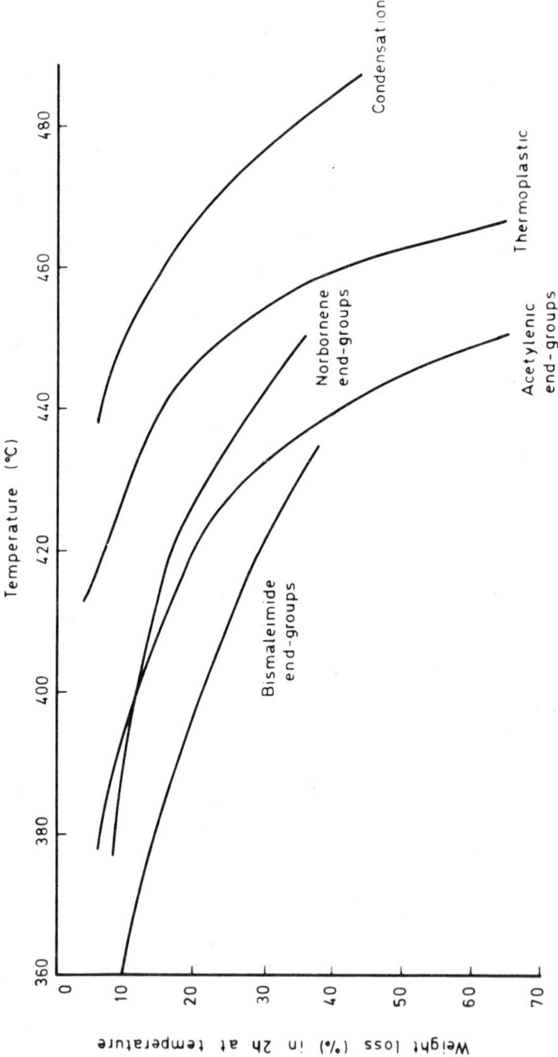

Figure 5.7. Comparison of thermal stability in air of different types of polyimide. Reference 191.

TABLE 5.9. Activation Energies for Thermo-Oxidative Degradation of Different Polyimides (Reference 191)

Polyimide type	Activation energy (kJ/mole)
Condensation—fusible (Thermoplastic)	176
Condensation—infusible	172
Addition—acetylenic end-groups	149
Addition—norbornene end-groups	143
Addition—maleimide end-groups	93

The effect of variation in chemical structure upon the thermo-oxidative stability of condensation-type polyimides has also been studied in detail,[192] and as a result the following conclusions have been drawn:

1. For polyimides derived from p-phenylene diamine and a variety of dianhydrides, the order of stability related to the dianhydride component is pyromellitic dianhydride > benzophenone-3,3',4,4'-tetracarboxylic dianhydride > 1,3-bis(3,4-dicarboxyphenyl) hexafluoropropane dianhydride > naphthalene-1,4,5,8-tetra-carboxylic dianhydride.

2. For polyimides derived from pyromellitic dianhydride and a variety of diamines, the order of stability varies as follows:

a. p-linked materials > m-linked materials.

b. p-phenylene diamine > 1,5-diaminonaphthalene $\approx$ 4,4'-diaminodiphenyl > 1,4-diaminoanthracene $\approx$ 1,6-diaminopyrene, i.e., stability decreases as the number of fused rings in the diamine increases.

c. Ring substitution in the diamine decreases stability.

d. Using diamines of structure $H_2NC_6H_4-X-C_6H_4NH_2$, stability decreases in the order X = single bond > S $\geqslant$ SO_2 > CH_2 > CO > SO $\geqslant$ O.

As would be expected, the amide-imide and ester-imide polymers are less stable than analogous polymers not containing amide or ester groups. Thermogravimetry shows a reduction in stability of 60–70°C in both cases. A recent article by Knight[193] includes weight loss and differential weight loss curves obtained at a rate of temperature rise of 2°C per minute in both air and nitrogen for a number of commercially available polyimides, i.e., Kapton, Skybond 700, Kerimid 601, Kerimid 353, Kerimid 711, PMR 15, Thermid 600, NR-150 A2, and Polyimide 2080. These curves confirm the superior stability of the fully aromatic condensation-type polyimides.

In addition to the kinetics of degradation, the volatile products of both thermal and thermo-oxidative breakdown have been the subject of numerous studies, and various mechanisms have been proposed to account for their forma-

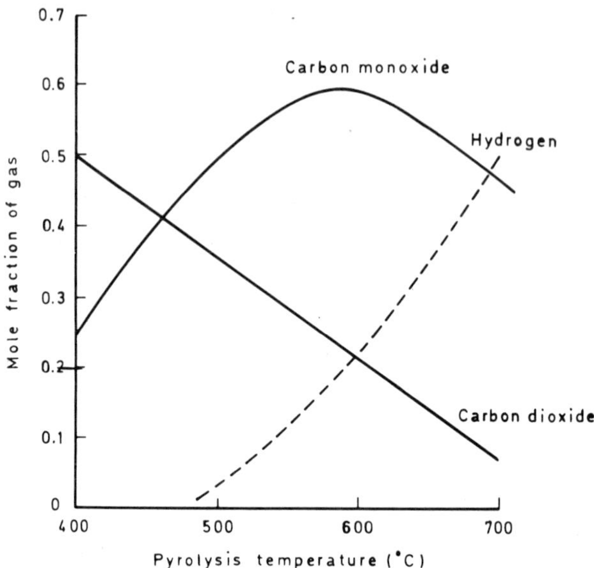

Figure 5.8. Composition of gaseous degradation products from a polyimide as a function of temperature of pyrolysis. Reference 194.

tion. The polyimide derived from 4,4'-diaminodiphenylmethane and pyromellitic dianhydride has been especially thoroughly investigated. The principal gaseous degradation products measured in the majority of experiments have been carbon monoxide and carbon dioxide. The variation in the proportion of these as a function of temperature of pyrolysis is illustrated in Figure 5.8.[194] At the lowest temperatures used carbon dioxide is the major product, but a cross-over occurs at about 450°C and above this temperature carbon monoxide is in the ascendant and the amount of carbon dioxide steadily decreases. The production of carbon monoxide from the pyromellitimide ring is easily understood, but the formation of carbon dioxide is much more a problem. Four different mechanisms have been suggested:

(i) If the imidization reaction is not complete, decarboxylation of the residual amic acid groups will yield carbon dioxide.[195]

(ii) Hydrolysis of the imide ring can regenerate the amic acid, which reacts as in (i).[197]

(iii) If isocyanates are formed as a degradation product (and there is some evidence for this), then these can react to give a carbodiimide and carbon dioxide.[196]

$$2\ R-N=C=O \longrightarrow R-N=C=N-R + CO_2$$

(iv) If a mobile imide-isoimide equilibrium exists, then carbon dioxide can arise through decomposition of the isoimide.[194]

There is some evidence to support all four of these suggestions, although a recent paper[198] claims that infrared and NMR data obtained on model compounds precludes the possibility of isoimide formation during thermal degradation of polypyromellitimides.

In the presence of oxygen, carbon monoxide and carbon dioxide are still the main volatile degradation products. To determine what proportion of these arise by pure thermal or thermo-oxidative reactions, mass spectrometric analysis has been carried out on the gases formed when a polyimide is degraded in oxygen enriched with the ^{18}O isotope.[199, 200] Table 5.10 shows the composition of the gaseous products formed during 30 minutes at 400°C. Carbon monoxide and water are produced mainly by thermal reactions, but the majority of the carbon dioxide arises via an oxidation route.

If a polyimide is pyrolyzed directly in the ion source of a mass spectrometer, then a host of other compounds are detected.[201] These comprise, in order of relative intensity, phenylene derivatives, phthalimide derivatives, and pyromellitimide derivatives, indicating extensive scission of the pyromellitimide rings.

TABLE 5.10. *Composition of the Gaseous Products from Decomposition of a Polyimide in Labeled Oxygen (Reference 200)*

Product	Amount (%)	Product	Amount (%)	Product	Amount (%)
CO	87	H_2O	60	COO	25
CO*	13	H_2O*	40	COO*	15
				CO*O*	60

$$O^* = {}^{18}O$$

A chemical study of polyimide degradation is feasible, because on treatment with hydrazine hydrate a polyimide is converted quantitatively to the diamine and the biscyclohydrazide derivative of the dianhydride from which it was originally prepared.

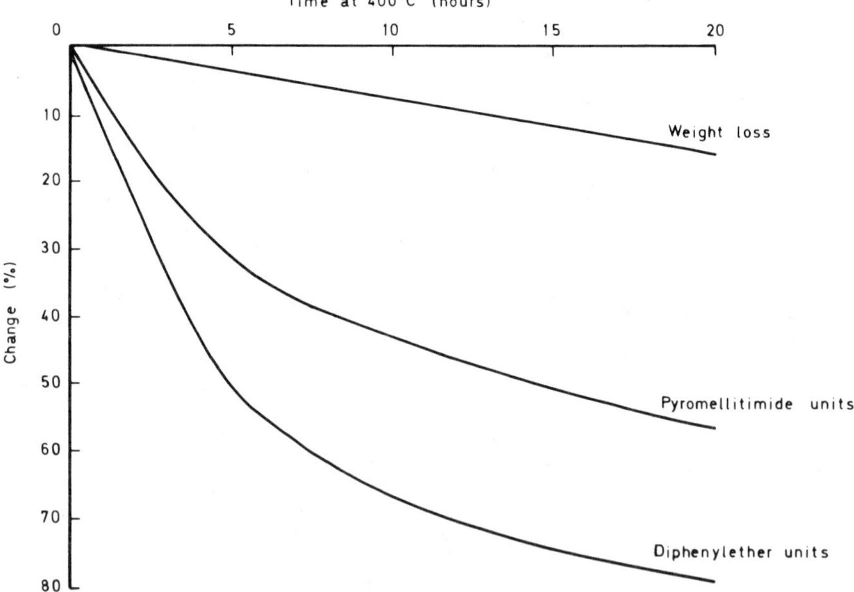

Utilizing this technique, it has been shown that almost before any weight loss is detected approximately 50% of the diphenylether units and 30% of the pyromellitimide units in the polymer have undergone structural modification[202] (see Figure 5.9). The diphenylether units cross-link either by a dehydrogenation

Figure 5.9. Change in composition of a polyimide based on pyromellitic dianhydride and diphenylether with time of aging at 400°C in air. Reference 202.

reaction or by cleavage of the ether linkages to yield phenols, which subsequently react. The pyromellitimide units undergo ring scission to give phthalimides and polycyclic structures. The thermal and thermo-oxidative degradation of polyimides is obviously a complex phenomenon. It should be stressed that slight alterations in their chemical structure can have quite marked effects upon the composition of the degradation products and their relative amounts, and hence caution should be exercised in extrapolating results from one polyimide system to another.

Elevated Temperature Properties

Since, as has already been stated, there are many polyimides that are commercially available, only a selection of the published elevated temperature data can be reproduced here. It is convenient to subdivide them into areas of application, e.g., films, moldings, wire enamels, adhesives, and laminating resins.

Polyimide films. A polyimide film with the trade name Kapton has been marketed by Dupont since 1961. It is available in two forms—Type H, polyimide alone, and Type F, in which the polyimide is coated on one or both sides with various thicknesses of fluorinated ethylene propylene (FEP) resin. The purposes of the fluorocarbon coating are to provide a heat sealable surface and to improve the chemical resistance of the film, especially to bases and concentrated acids. Table 5.11 gives some properties of 1 mil thick Kapton H film and Table 5.12 of Type F film as a function of construction. Table 5.13 shows the time required for the ultimate elongation at break of Type H film to be reduced from 70 to 1% at various temperatures.

Figures 5.10–5.13 illustrate the excellent retention of electrical properties at temperatures up to greater than 200°C.

As stated earlier, a thermoplastic, soluble polyimide film based on diaminodiphenylindane (DAPI) and either pyromellitic dianhydride or benzophe-

TABLE 5.11. *Some Properties of Kapton Type H Film (1 mil thick) (Reference 203)*

	Measured	
Property	25°C	200°C
Ultimate tensile strength (MPa)	173	117
Yield point at 3% (MPa)	69	41
Stress to produce 5% elongation (MPa)	90	59
Ultimate elongation (%)	70	90
Tensile modulus (GPa)	2.97	1.79
Dielectric constant at 1 kHz	3.5	3.0
Dissipation factor at 1 kHz	0.003	0.002
Volume resistivity (ohm cm)	10^{18}	10^{14}
Short-term dielectric strength (V/mil) at 60 Hz	7000	5600

TABLE 5.12. *Some Properties of Kapton Type F Film as a Function of Construction (Reference 204)*

Property	Nominal construction (mils)					
	1 Kapton H 0.5 FEP	1 Kapton H 1 FEP	2 Kapton H 1 FEP	2 Kapton H 2 FEP	0.5 FEP 1 Kapton H 0.5 FEP	0.5 FEP 2 Kapton H 0.5 FEP
Ultimate tensile strength (MPa)						
at 25°C	117	97	117	97	97	117
at 200°C	76	62	76	62	62	76
Yield point at 3% (MPa)						
at 25°C	50	41	50	41	41	50
at 200°C	28	21	28	21	21	28
Stress to produce 5% elongation (MPa)						
at 25°C	62	52	62	52	52	62
at 200°C	38	28	38	28	58	38
Ultimate elongation (%)						
at 25°C	75	75	90	105	75	>80
at 200°C	85	95	105	115	95	—
Tensile modulus (GPa)						
at 25°C	2.21	1.73	2.21	1.73	1.73	2.21
at 200°C	1.19	0.90	1.19	0.90	0.90	1.19

TABLE 5.13. *Time Required for Elongation at Break of Kapton H Film to Fall from 70% to 1% (Reference 205a)*

Temperature (°C)	Inert atmosphere	Oxidizing atmosphere
250	—	8 years
275	—	1 year
300	—	3 months
350	1 year	6 days
375	2 months	2 days
400	2 weeks	12 hours
425	3.5 days	5 hours
450	22 hours	2 hours

none tetracarboxylic dianhydride has recently been introduced by Ciba–Geigy.[101] Some mechanical properties of this film are given in Table 5.14.

Polyimide moldings. Molding compounds, or rather moldings, called Vespels were introduced by Dupont in 1965. A number of grades are now available containing different fillers—SP-1 is unfilled resin; SP-21 contains 15% by

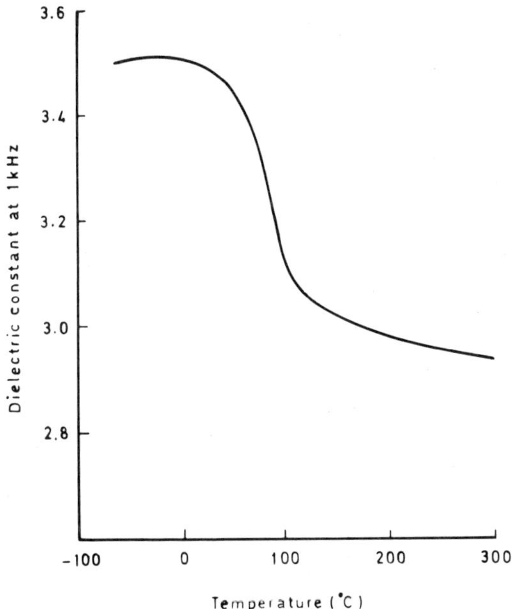

Figure 5.10. *Dielectric constant of Kapton H film (1 mil) vs. temperature. Reference 205b.*

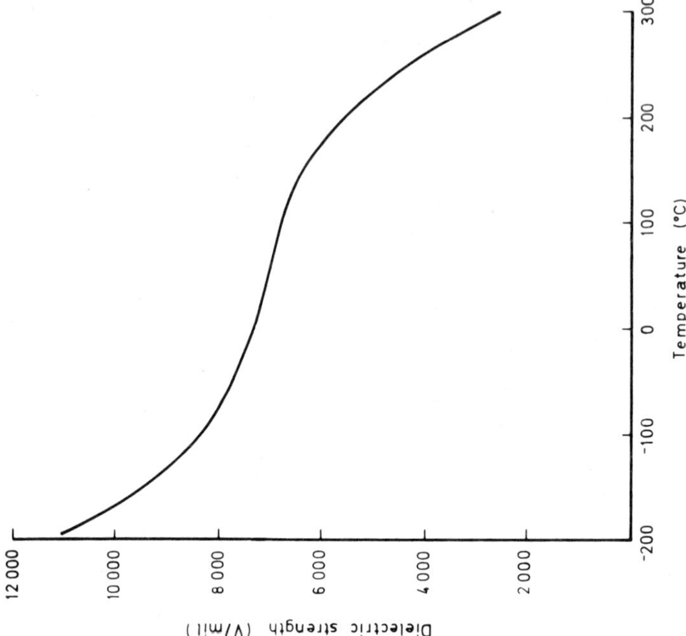

Figure 5.12. Dielectric strength of Kapton H film (1 mil) vs. temperature. Reference 205b.

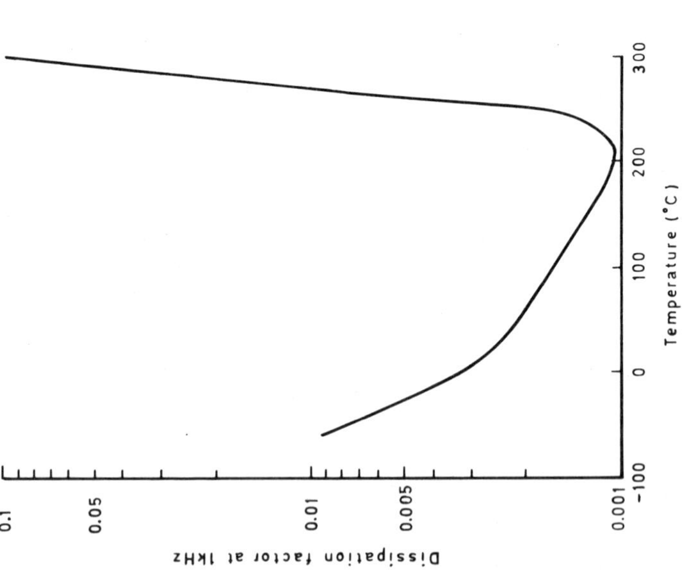

Figure 5.11. Dissipation factor of Kapton H film (1 mil) vs. temperature. Reference 205b.

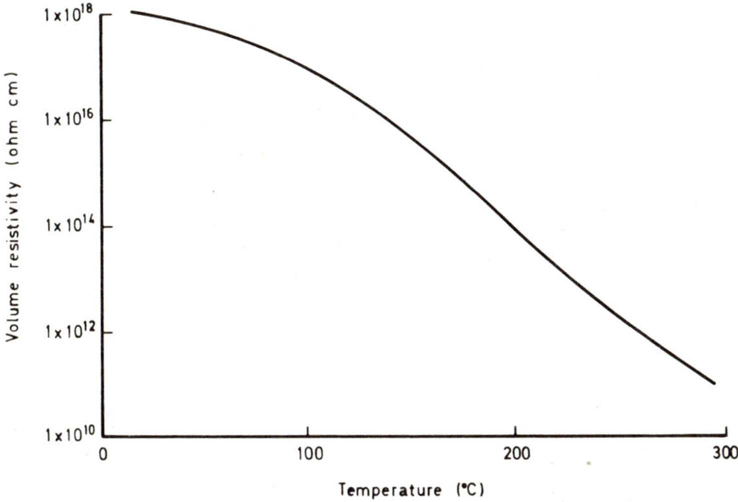

Figure 5.13. Volume resistivity of Kapton H film (1 mil) vs. temperature. Reference 205b.

TABLE 5.14. Some Mechanical Properties of DAPI-Based Polyimide Film (Reference 101)

Property	Temperature (°C)	Value
Tensile strength (MPa)	25	77
	204	36
	260	27
Tensile modulus (GPa)	25	2.48
	204	3.38
	260	3.11
Elongation at break (%)	25	10
	204	39
	260	55
Time to brittleness (hours) in air	200	>2000
	225	925
	250	250

weight graphite; SP-22 contains 40% by weight graphite; SP-31 contains 15% by weight molybdenum disulphide; SP-5 contains 42% by weight short glass fibers; and SP-211 contains 15% by weight graphite and 10% by weight polytetrafluoroethylene. A summary of the mechanical properties of some of these grades is given in Table 5.15, and estimates of the number of hours for which 50% of the initial tensile strength would be retained for continuous exposure at various temperatures are listed in Table 5.16. It should be noted that the

TABLE 5.15. Mechanical Properties of Some Vespel Compositions (Reference 206a)

Property	SP-1 Perp[a]	SP-1 Parallel[b]	SP-21 Perp	SP-21 Parallel	SP-22 Perp	SP-22 Parallel	SP-31 Perp	SP-31 Parallel	SP-5 Perp	SP-5 Parallel
Ultimate tensile strength (MPa)										
at 23°C	100	76	79	43	67	38	98	66	49	26
at 300°C	43	33	41	30	30	17	50	37	—	—
Elongation at break (%)										
at 23°C	8.0	5.0	4.5	2.0	2.8	1.8	9.4	4.9	0.9	0.5
at 300°C	3.0	2.5	3.0	1.9	1.3	0.9	6.3	3.6	—	—
Tensile modulus (GPa)										
at 23°C	3.35	3.11	4.49	3.73	5.45	4.28	3.76	3.04	9.38	6.87
at 300°C	2.07	1.90	3.11	2.52	3.04	2.14	2.17	2.10	—	—
Ultimate flexural strength (MPa)										
at 23°C	128	107	124	79	106	68	152	110	83	41
at 300°C	72	58	68	43	54	36	—	—	—	—
Ultimate flexural strain (%)										
at 23°C	12.7	5.8	7.1	3.5	3.0	2.1	8.6	5.1	1.2	1.6
at 300°C	4.9	4.1	4.0	2.4	1.9	1.6	—	—	—	—
Flexural modulus (GPa)										
at 23°C	3.17	3.04	3.93	3.48	6.21	4.35	3.68	3.42	6.90	4.28
at 300°C	1.90	1.79	2.48	2.13	3.86	2.67	—	—	—	—
Ultimate compressive strength (MPa)										
at 23°C	253[c]	311[c]	200	235	126	128	—	—	—	—
at 300°C	128[c]	133[c]	64	68	69	75	—	—	—	—
Ultimate compressive strain (%)										
at 23°C	> 50[c]	> 50[c]	33.0	38.0	17.3	18.2	—	—	—	—
at 300°C	> 50[c]	> 50[c]	14.0	13.0	11.3	11.4	—	—	—	—
Compressive modulus (GPa)										
at 23°C	3.93	3.55	4.35	3.73	4.14	4.14	—	—	—	—
at 300°C	1.90	1.62	2.07	2.00	2.62	2.97	—	—	—	—

[a] and [b] Perp and parallel mean that the specimens were tested perpendicular and parallel to the direction of molding, respectively.
[c]Indicates compressive stresses at 50% strain. The specimens did not fail.

TABLE 5.16. Estimated Times (Hours) for 50% Retention
of Tensile Strength (Reference 206a)

Temperature (°C)	SP-1	SP-2	SP-22
249	> 1000	> 1000	> 1000
300	600	> 1000	> 1000
312	400	800	> 1000
340	200	450	700
368	100	200	300
396	45	100	150
423	15	40	60
451	0.75	5	20
479	0.50	2.5	5

moldings are anisotropic in character. Figures 5.14 and 5.15 show the variation
of tensile strength and flexural modulus with temperature of test and Figure 5.16
shows the retention of tensile strength with time of aging at 300°C.

The electrical properties of the unfilled moldings are very similar to those
already quoted for polyimide film. One difference is that the dielectric strength

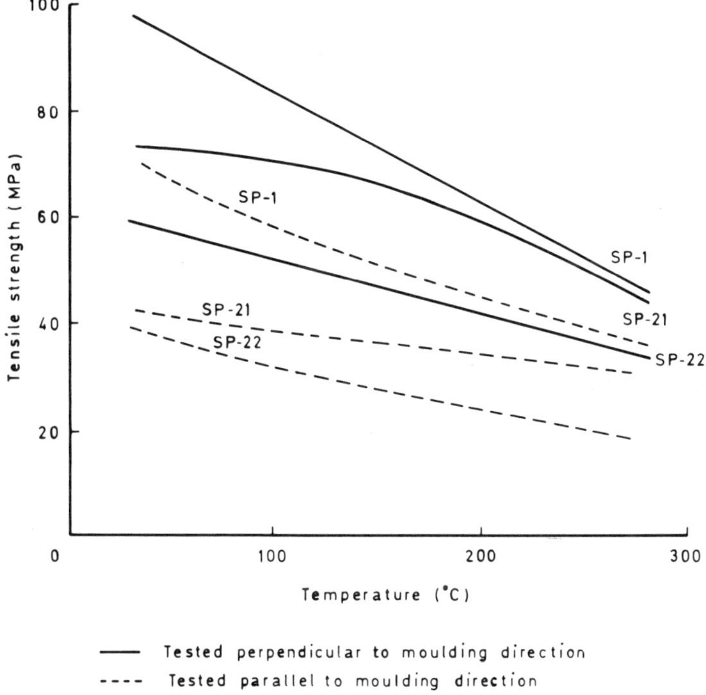

Figure 5.14. Tensile strength of polyimide moldings vs. temperature. Reference 206a.

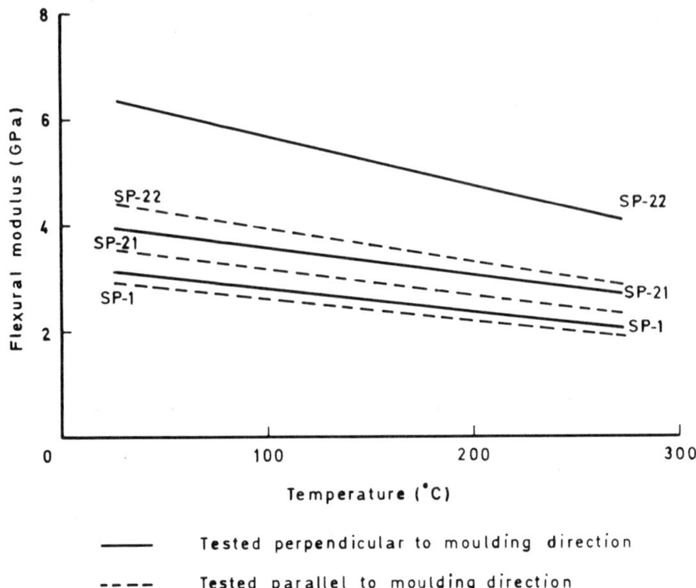

Figure 5.15. *Flexural modulus of polyimide moldings vs. temperature. Reference 206a.*

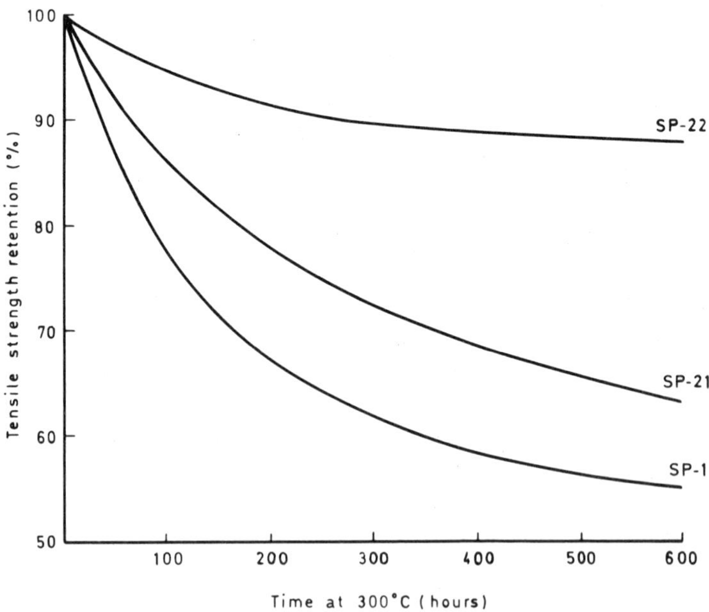

Figure 5.16. *Tensile strength retention of polyimide moldings with time of aging at 300°C. Reference 206b.*

actually rises between room temperature and 200°C by about 19% and then falls to 90% of the initial value by 300°C.

Molding powders are available based upon a thermoplastic polyimide (Upjohn 2080), upon addition-type polyimides (polyaminobismaleimides Rhone-Poulenc's Kinel), and upon amide-imide polymers (Amoco Chemical's Torlon). An indication of the elevated temperature properties of these materials is given by the data in Tables 5.17–5.21. Unfortunately a direct comparison between them is not possible because all the tests were conducted at different temperatures. It should be noted, however, that the Vespel figures above were obtained at 300°C, and the 2080, Torlon, and Kinel results are for 288°C, 260°C, and 250°C, respectively. Figure 5.17 shows the flexural strength retention of Kinel 5504 and 5514 as a function of time of aging at either 200°C or 250°C. On the basis of 50% retention of initial flexural strength (measured at room temperature), the lifetimes of Kinel 5504 are 20,000 hours at 185°C, 10,000 hours at 200°C, and 1,300 hours at 250°C.

Properties of Torlon moldings with a variety of fillers are given in Table 5.19 and of injection-molded samples containing varying amounts of chopped carbon fiber in Table 5.20.

In 1979 Rohm and Haas introduced a family of thermoplastic polyimides for injection molding trade named Kamax.[209] These are amorphous, polyaliphatic resins and consequently their heat resistance is not outstanding. The heat distortion temperature of Kamax 201, a general-purpose grade, is 130°C, and its mechanical properties are only quoted up to 120°C.

TABLE 5.17. Some Mechanical Properties of Polyimide 2080 Moldings (Reference 207)

Property	Unfilled resin	Resin + 15% graphite	Resin + 30% polytetrafluoroethylene
Tensile strength (MPa)			
at 23°C	118	72	43
at 288°C	30	—	—
Tensile modulus (GPa)			
at 23°C	2.07	1.84	1.04
at 288°C	0.97	—	—
Elongation at break (%) at 23°C	10	7	7
Flexural strength (MPa)			
at 23°C	193	95	63
at 288°C	35	28	—
Flexural modulus (GPa)			
at 23°C	3.45	3.63	2.00
at 288°C	1.11	1.42	0.63
Compressive strength (MPa) at 23°C	206	144	79
Compressive modulus (GPa) at 23°C	2.04	1.87	1.35

TABLE 5.18. Some Mechanical Properties of Kinel Moldings (Reference 151)

Property	5504[a]	5505[b]	5508[c]	5511[d]	5514[e]	5515[f]	5517[g]	5518[h]
Flexural strength (MPa)								
at 25°C	341	88	79	107	147	112	88	49
at 200°C	295	69	59	90	128	104	69	45
at 250°C	245	54	54	78	122	88	63	38
Flexural modulus (GPa)								
at 25°C	22.4	5.17	7.25	12.8	13.7	8.1	6.21	2.69
at 200°C	20.6	4.69	6.97	11.7	11.8	6.1	5.31	2.42
at 250°C	16.7	4.42	6.90	11.0	10.3	5.4	5.18	2.21
Tensile strength (MPa)								
at 25°C	186	39	32	35	44	—	39	35
at 250°C	157	29	21	26	37	—	25	25
Compressive strength								
at 25°C	224	153	108	108	235	—	138	138
at 200°C	127[i]	101	77	88	137	—	105	77

The Kinel moldings are of the following compositions:
[a]5504—contains 65% 0.25 inch long glass fiber
[b]5505—contains 25% graphite powder
[c]5508—contains 40% graphite powder
[d]5511—contains asbestos fibers and graphite powder
[e]5514—contains 50% 0.125 inch long glass fiber
[f]5515—is a modified version of 5514
[g]5517—contains molybdenum disulphide and graphite powder
[h]5518—contains polytetrafluoroethylene
[i]Measured at 250°C

Polyimide molding compounds based upon norbornene-terminated prepolymers were available for a period from TRW, Inc. and Ciba–Geigy Corp.[160] These did have a high-temperature capability and, for the sake of completeness, some of the properties of these P13N materials are listed in Table 5.21.

One application of carbon fiber-reinforced polyimide moldings is as self-lubricating bearings.[210] When molded as an outer race liner containing randomly oriented carbon fibers, the friction coefficients after run-in ranged from 0.15 ± 0.05 at 25°C to 0.05 ± 0.02 at 315°C.

Polyimide wire coatings. Wire coating insulation formulations containing polyimides, polyamideimides, and polyesterimides all became commercially available in the early 1960s; in fact the polyesterimides find their major use in this area. The polyimides have the best thermal stability with a projected lifetime of 25,000 hours at 240°C. Short-time excursions to temperatures of 300 to 400°C are permissible. Drawbacks are relatively low abrasion resistance, limited shelf-life of the enamelling solutions, and cost. The polyamideimides and polyesterimides overcome some of these disadvantages, but at the expense of a lowered temperature resistance. This is illustrated in Figures 5.18 and 5.19, which compare the service life of the different imide coatings together with those

TABLE 5.19. Some Mechanical Properties of Torlon Moldings
(Reference 129)

Property	4203[a] 4203L[b]	4301[c]	4275[d]	5030[e]	6000[f]	7030[g]
Tensile strength (MPa)						
at 23°C	186	135	125	195	146	206
at 150°C	105	73	93	137	107	142
at 260°C	52	46	44	84	28	74
Elongation at break (%)						
at 23°C	12	6	6	5	5	6
at 150°C	17	—	7	6	5	5
at 260°C	22	—	14	8	20	4
Flexural strength (MPa)						
at 23°C	212	182	175	318	212	317
at 150°C	156	141	126	231	137	218
at 260°C	75	77	48	127	48	114
Flexural modulus (GPa)						
at 23°C	4.55	6.35	7.18	11.11	7.87	17.87
at 150°C	3.59	5.04	5.93	10.49	6.35	15.11
at 260°C	2.97	4.00	4.00	8.42	3.80	14.15
Compressive strength (MPa)						
at 23°C	276	207	138	—	—	—
Compressive modulus (GPa)						
at 23°C	2.84	4.69	—	—	—	—
Shear strength (MPa)						
at 23°C	128	111	—	—	—	—

The Torlon moldings are of the following compositions:
[a]4203 contains 3% titanium dioxide
[b]4203L contains 3% titanium dioxide plus 1–2% polytetrafluoroethylene
[c]4301 contains 12% graphite powder plus 3% polytetrafluoroethylene
[d]4275 contains 20% graphite powder plus 3% polytetrafluoroethylene
[e]5030 contains 30% glass fiber plus 1% polytetrafluoroethylene
[f]6000 contains 30% mineral filler plus 1% polytetrafluoroethylene
[g]7030 contains 30% graphite fiber plus 1% polytetrafluoroethylene

of polyester [and in the one case, poly (vinylformal)] wire coatings.[211,212] (NB: The two figures are not strictly comparable as different criteria are used to assess the service life.) The superiority of the polyimide is evident, although for long-term use at lower temperatures, the polyamideimide is not greatly inferior. The projected lifetimes for the polyimide are 100,000 hours at 200°C, 10,000 hours at 260°C, and 1,000 at 315°C. Figure 5.20 shows the change in dielectric strength of Pyre ML (polyimide) when coated on glass fabric and aged at 250 and 300°C.

Polyimide adhesives. Condensation, addition, and thermoplastic polyimides and polyamideimides have all been evaluated as metal-to-metal adhesives.[79,213] The products which have been put on the market have had a rather

TABLE 5.20. Some Mechanical Properties of Graphite Fiber-Reinforced Torlon Moldings (Reference 208)

Property	Temperature (°C)	XG25	XG30	XG40	XG50
Carbon fiber content (%)	—	25	30	40	50
Tensile strength (MPa)	23	173	166	169	166
	150	124	121	119	104
	260	13	12	12	8
Tensile modulus (GPa)	23	19.3	19.8	29.3	30.6
	150	10.2	13.0	21.9	25.7
	260	1.57	1.64	1.68	1.76
Elongation at break (%)	23	2.9	3.3	3.7	3.0
	150	5.2	4.8	4.7	5.5
	260	7.2	7.1	6.9	5.5
Flexural strength (MPa)	23	271	276	311	299
	150	150	208	230	237
	260	66	65	60	60
Flexural modulus (GPa)	23	16.9	18.0	23.5	24.0
	150	11.5	17.6	21.9	22.7
	260	6.1	7.7	7.9	9.6

TABLE 5.21. Some Mechanical Properties of P13N Moldings (Reference 160)

Property	Unfilled resin	50% Mineral-filled	40% Glass fiber-filled
Flexural strength (MPa)			
at 25°C	69–83	69–97	179–207
at 260°C	41–55	62–83	152–179
at 288°C	35–41	41–55	138–166
Flexural modulus (GPa)			
at 25°C	3.17–3.38	8.28–9.66	13.80–16.56
at 260°C	2.07–2.28	—	11.73–13.11
at 288°C	1.93–2.07	4.83–6.90	10.35–11.73
Tensile strength (MPa)			
at 25°C	50	54	99
at 260°C	39	34	64
Tensile modulus (GPa)			
at 25°C	3.84	9.04	15.04
at 260°C	2.39	5.87	11.52
Elongation at break (%)			
at 25°C	1.4	0.6	0.8
at 260°C	1.8	0.8	0.6
Compressive strength (MPa)			
at 25°C	258	235	225
at 260°C	—	119	150
Compressive modulus (GPa)			
at 25°C	2.89	3.46	4.44
at 260°C	—	2.59	3.39

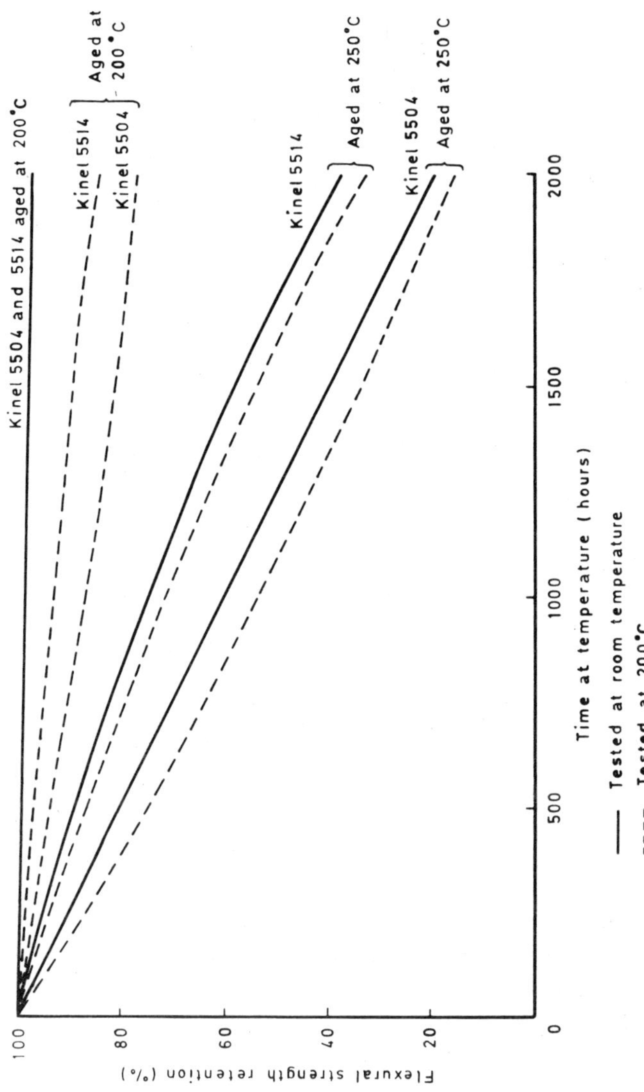

Figure 5.17. Flexural strength retention of Kinel moldings after aging at 200 and 250°C. Reference 151.

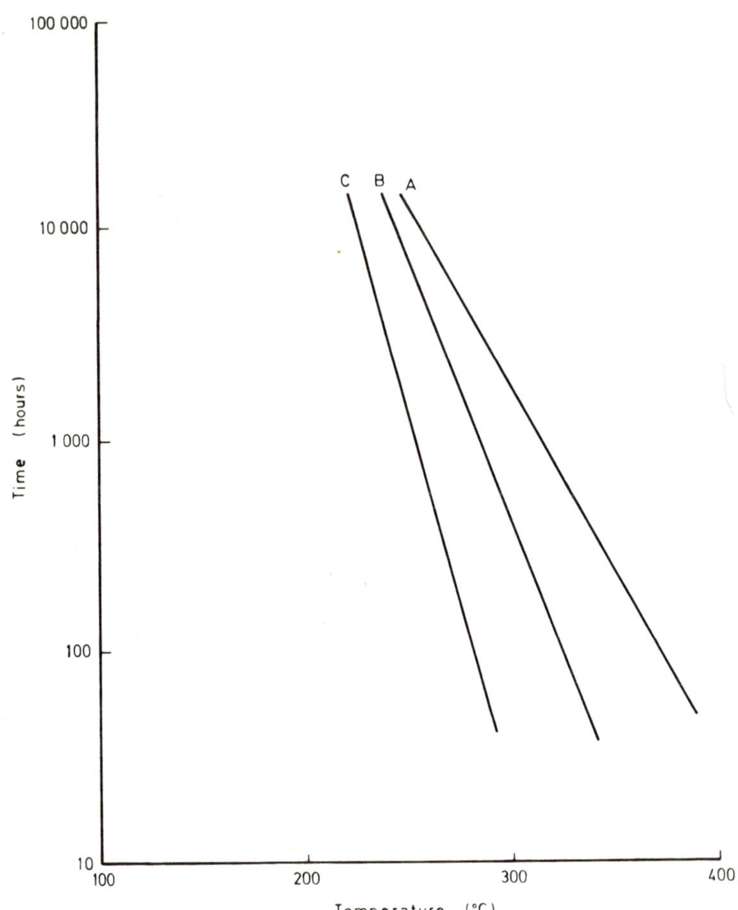

Figure 5.18. Comparison of service life of wire-coating enamels. (A) polyimide; (B) polyamideimide; (C) polyester. Reference 211.

checkered history and, as was indicated previously, some have subsequently been withdrawn. In the case of condensation products, careful attention must be paid to the glue-line temperature and to the applied pressure to ensure that volatile products evolved during the cure reaction do not result in a porous glue-line with low mechanical strength. Normally, a solution of the intermediate polyamic acid is applied to a glass cloth carrier, the bulk of the solvent removed, and the material partially cured. Completion of cure is carried out under pressure at high temperature—typically 0.5 MPa at 316°C for 90 minutes. A post-cure of some hours at 300–350°C is necessary for development of optimum properties. It has also been observed with condensation-type polyimide adhesives that the solvent

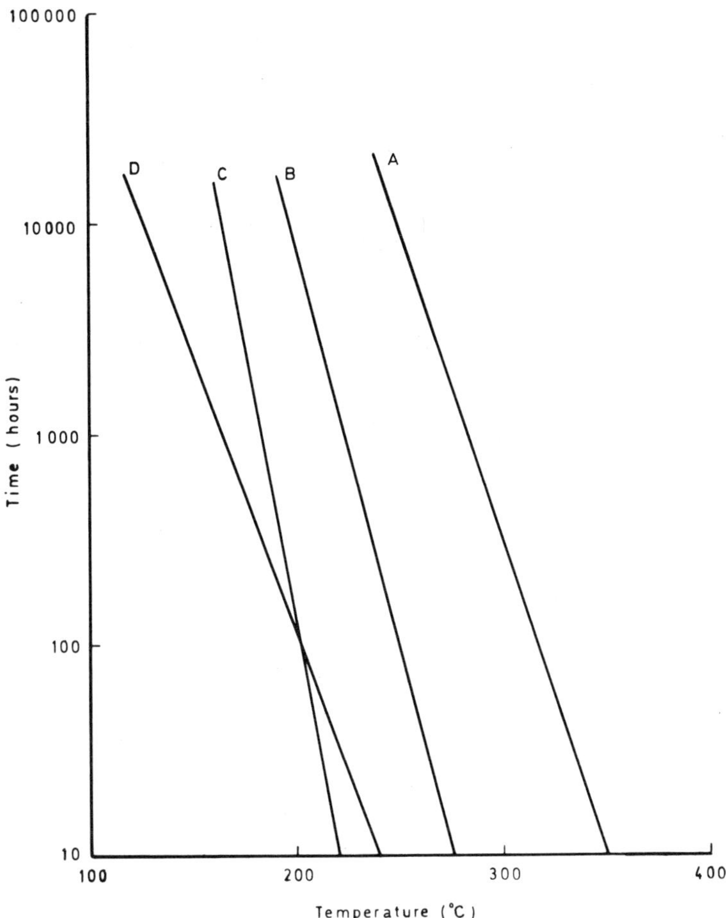

Figure 5.19. Comparison of service life of wire-coating enamels (ASTM D2307). (A) polyimide; (B) polyesterimide; (C) polyester; (D) poly(vinylformal). Reference 212.

used plays an important role in the performance of the resulting adhesive bond.[214,215] If (2-methoxyethyl) ether is used instead of N,N′-dimethyl-formamide, then bond strengths are almost doubled. This is attributed to enhanced wetting of the substrate surfaces by the ether. It goes without saying that the correct metal surface pretreatment is essential for the long-term durability of adhesive joints.

Most commercial polyimide adhesives are supported on glass cloth carriers, and all contain antioxidants, generally arsenic compounds, whose purpose is to prolong service life at elevated temperatures. The improvement resulting from

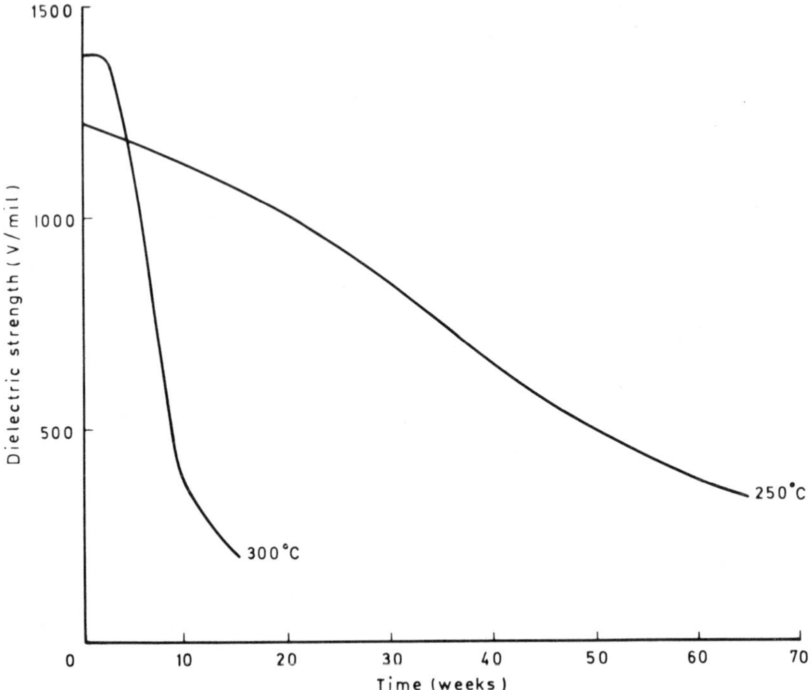

Figure 5.20.   Effect of time at 250 and 300°C on the dielectric strength of Pyre ML-coated glass fabric (0.003 in thick). Reference 211.

the incorporation of 2% by weight of arsenic thioarsenate into a polyimide adhesive is illustrated in Figure 5.21. Aluminum powder is also often added to adhesive formulations to reduce thermal expansion stresses and to act as a reinforcing agent. The first commercially available (condensation) polyimide adhesive was American Cyanamid's FM-34. Detailed results on its properties are given in the paper by Pike, Novak, and DeCrescente,[78] who used it to bond titanium alloys and BORSIC/aluminum. They found that vacuum fabrication procedures are necessary to produce low-void-content bonds. Table 5.22 gives the results of lap-shear tests with time at temperature for stainless steel and titanium joints, and Figure 5.22 shows the very good strength retention with time of aging of titanium joints. These are American Cyanamid data taken from a brochure on FM-34 adhesive film. Nolimid A380 was marketed by Rhone–Poulenc for a period but has recently been withdrawn. On the basis of 6.9 MPa as the failure criterion for titanium bonds, it was estimated[11] that its useful lifetimes were 9,000 hours at 300°C, 500 hours at 340°C, 100 hours at 380°C, 25 hours at 400°C, and one hour at 450°C. Data for comparison with those of FM-34 are given in Table 5.23.

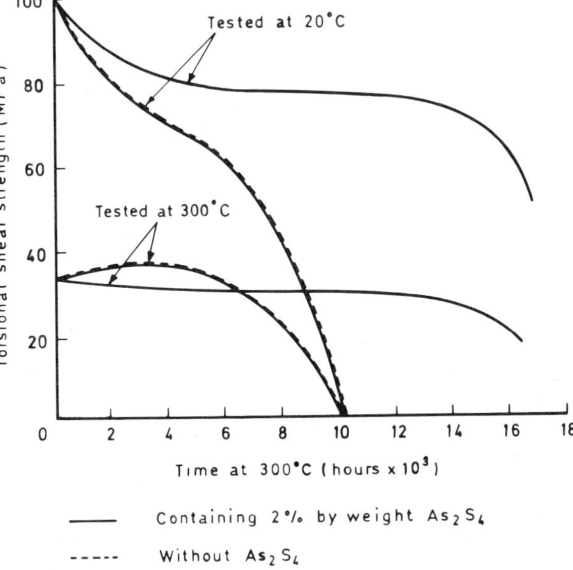

Figure 5.21. Influence of arsenic thioarsenate on the strength retention at 300°C of steel joints bonded with polyimide adhesive.

TABLE 5.22. Heat Aging Data for FM-34 Bonded Lap-Shear Joints

Adherend	Exposure conditions	Lap-shear strength (MPa) at	
		25°C	260°C
Stainless steel	As made	29	16
	After 192 hours at 316°C	22	17
	500 hours at 316°C	18	15
	1000 hours at 316°C	14	12
Titanium	As made	25	12
	After 1000 hours at 260°C	20	14
	2000 hours at 260°C	17	14
	5000 hours at 260°C	17	13
	As made	26	13
	After 192 hours at 316°C	12	15
	500 hours at 316°C	14	16
	1000 hours at 316°C	15	15
	2000 hours at 316°C	7	9

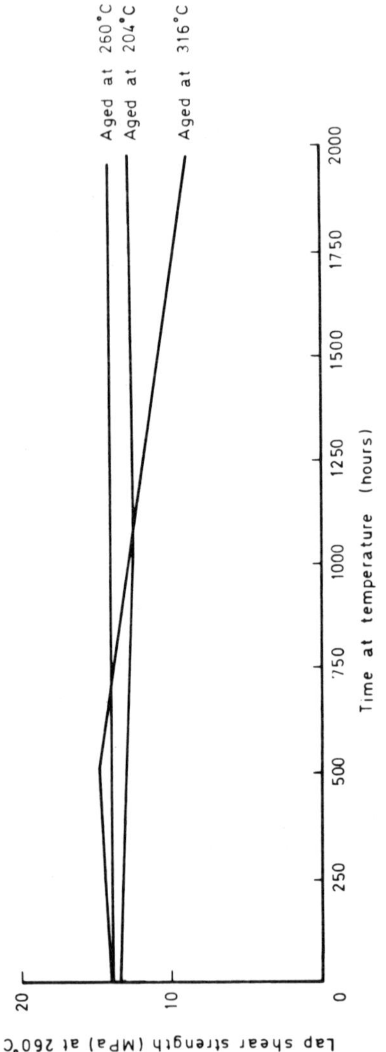

Figure 5.22. Strength retention of titanium lap-shear joints bonded with FM-34 with time at different temperatures.

TABLE 5.23. *Heat Aging Data for Nolimid A380 Bonded Titanium Lap-Shear Joints (Reference 11)*

Exposure temperature(°C)	Test temperature (°C)	Lap-shear strength (MPa) after time at temperature of (hours)				
		0	500	4000	8000	12000
260	25	20	22	24	23	22
260	260	18	19	20	20	19
300	25	20	16	15	9	6
300	300	16	15	14	11	6
300	350	12	11	—	—	—

The torsional shear strength retentions of FM-34 and Nolimid A380 bonded steel joints have been compared with those of an experimental fluoropolyimide adhesive[63] at 250 and 300°C, and the results are illustrated in Figure 5.23. The superiority of the fluoropolyimide is evident, but cost considerations have precluded its commercial development.

Because of difficulties encountered with the condensation-type polyimides, e.g., volatile evolution and hence void formation, long and involved processing cycles, inherent brittleness and poor crack resistance, considerable effort has

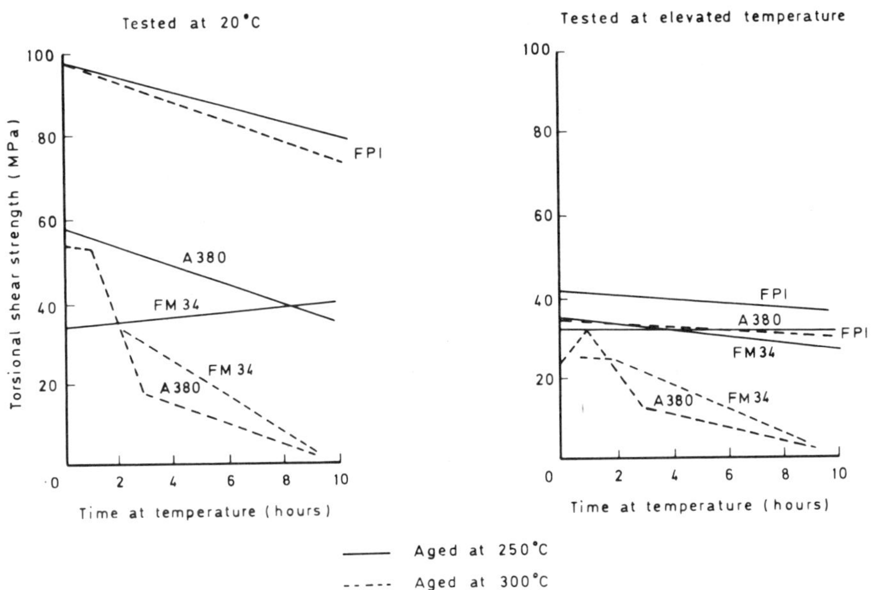

Figure 5.23. *Comparison of torsional strength retention at elevated temperatures of stainless steel bonded with polyimide adhesives.*

TABLE 5.24. *Lap-Shear Strengths of Titanium/NR Polyimide Joints*

Adhesive type	Test temperature (°C)	Lap-shear strength (MPa)
NR-150 A2G	23	34
	260	21
NR-150 B2G	23	26
	316	10
NR-056X	23	24
	316	15

been devoted to the formulation of thermoplastic polyimide adhesive systems. Dupont has produced a number of these designated NR-150 A2G, NR-150 B2G, and NR-056X.[96,119,216] Although bonding conditions have probably not been optimized, lap-shear strengths have been obtained for titanium joints as listed in Table 5.24.

Maximum use temperatures of 260°C and 343°C are claimed for the A2G and B2G materials, respectively.

Some attention has also been given to developing adhesive systems based on the acetylene-terminated polyimide prepolymers.[173] Lap-shear strengths have been obtained for titanium joints of 25 MPa at ambient temperature and 18 MPa at 260°C. After 1,000 hours at temperature, strength retentions of 100, 80, and 30–40% have been measured at 150, 230, and 260°C, respectively.

Amide-imide adhesives are commercially available from Amoco (AI-1030 and AI-1137) and from Rhone–Poulenc (Kerimid 500). An indication of their properties taken from product bulletins is given in Tables 5.25 and 5.26.

Polyimide laminating resins. This has recently been the area of maximum activity and the one that has seen the marketing of the greatest number and greatest variety of polyimides. As a consequence, very many papers have been published on the use of polyimides in composites, specifically with carbon fibers

TABLE 5.25. *Lap-Shear Strengths of Metal/AI Adhesive Joints*

Metal	Test condition	Lap-shear strengths (MPa) at RT for adhesive	
		AI-1030	AI-1137
Aluminum	As made	24	24
	After 500 hours at 260°C	17	18
Stainless steel	As made	25	28
	After 500 hours at 260°C	19	20
Titanium	As made	28	28
	After 500 hours at 260°C	15	16

TABLE 5.26. Lap-Shear Strengths of Stainless
Steel/Kerimid 500 Adhesive Joints

Test condition	Lap-shear strength (MPa) at RT
As made	25
After 1000 hours at 250°C	17
After 2500 hours at 250°C	12
After 4000 hours at 250°C	10
After 100 hours at 300°C	10

as reinforcement. To summarize the whole of the available data would produce a monograph in its own right. We have, therefore, attempted to compare and contrast in general terms the processing of the different types of polyimide and the level of properties that can be expected from carbon fiber-reinforced composites made from them. This topic has recently been reviewed,[75, 217] and what follows is largely based upon these reviews. Tables 5.27 and 5.28 show the conditions necessary for the processing by press and autoclave techniques of carbon fiber prepreg made from the different polyimides. The data are taken from the schedules recommended by the manufacturers of the resins. To facilitate comparison, salient features of the press cure conditions are summarized in Table 5.29, listing the resins in ascending order of temperature necessary for cure. The temperatures required differ by as much as 250°C and the pressures by 16 MPa. The condensation- and addition-type polyimides are fabricated at the lowest temperatures and pressures, but in the main require extended and stepwise post-cure cycles. The thermoplastic polyimides need high press temperatures and in some cases high pressures, but no post-cure cycle is necessary. Some of the systems have not yet been developed in a form suitable for vacuum bag/autoclave procedures.

Tables 5.30 and 5.31 give the available figures for the mechanical properties of the unreinforced polyimide resins at elevated temperatures. Points to note are the relatively lower tensile strengths and elongations at break of the addition polymers; the high compressive strength and modulus of the condensation polymers; the high flexural modulus of the acetylenic end-capped material; and the high flexural strength and tensile modulus of the thermoplastic variants.

Table 5.32 compares the initial room temperature mechanical properties of unidirectional composites made from the different polyimides and the different types of carbon fiber. The figures quoted are the arithmetical means of data taken from Refs. 83, 162, and 218 to 236. Variation from these means can be as much as ±15%, though it should be realized that the results cover a span of seven years, during which time, due to improvements in manufacturing processes, the properties of the carbon fibers themselves had to some extent altered. It can be seen that differences in resin properties are not markedly reflected in the proper-

TABLE 5.27. *Comparison of Processing of Prepreg Based on Carbon Fiber and Polyimide Resin—Condensation and Thermoplastic Types (Reference 217)*

	Condensation type	Thermoplastic type 1 (NR 150)	Thermoplastic type 2 (Upjohn 2080)
Solvent used	N-methylpyrrolidone (NMP)	Ethanol/NMP	NMP
Solids content (%)	68–72	48–54	~20
Viscosity (Pa.s)	3–7	1.5–3.5	~10
Resin pick-up on fiber (%)	40–45	50–55	
Volatile content (%)	9–13	14–16	
Flow (%)	10–23		
"Precure"	—	Heat to 200°C at 2°C/min under full vacuum Hold at 200°C for 30 min	Heat at 250°C for 30 min to reduce solvent content to <4%
Press cure	Preheat press to 205°C Apply contact pressure for 1 min	Preheat press to 427°C	Heat press to 350°C "bumping" repeatedly
	Then apply 1.7 MPa for 30 min	Apply 17.3 MPa for 10 min	Apply 2–3.5 MPa for 60 min
	Cool to below 60°C before pressure release	Cool to below Tg under a pressure $\not>$1.4 MPa	Cool to below 250°C before pressure release
Post-cure	4 hours at 205°C + 4 hours at 260°C + 4 hours at 315°C + 4 hours at 345°C	None	None
Vacuum bag/autoclave cure	Apply full vacuum Heat to 120°C at 1–2°C/min Apply 0.7 MPa Heat to 175°C at 1–2°C/min Hold for 60 min	Apply full vacuum Heat to 200°C at 2°C/min Hold for 30 min Heat to 343°C over 2 hours Apply 1.4 MPa Hold for 60 min	—
	Cool under vacuum and pressure	Cool under vacuum and pressure	
Post-cure	As above	None	—

TABLE 5.28. *Comparison of Processing of Prepreg Based on Carbon Fiber and Polyimide Resin—Addition Types (Reference 217)*

	Norbornene end-capped	PMR Type	Bismaleimide end-capped	Acetylenic end-capped
Solvent used	Dimethyl formamide	Methanol	N-methylpyrrolidone (NMP)	NMP
Solids content (%)	38–42	~50	45–55	~50
Viscosity (Pa.s)	0.2–0.3	$\sim 2 \times 10^{-2}$	0.2–0.8	
Resin pick-up on fiber	25–30		30–50	
Volatile content (%)	<2		7–9	
Flow (%)			20–40	
"Precure"	—	Heat at 205°C for 2h	—	—
Press cure	Preheat press to 315°C	Preheat press to 232°C Apply contact pressure for 10 min	Preheat press to 120°C	Preheat press to 250°C Apply contact pressure for 1 min "Bump" 3 times in 30 sec
	Apply pressure of 1.4 MPa	Apply pressure of 6.9 MPa Heat to 315°C over 20 min	Apply pressure of ≥ 1.4 MPa Heat to 180°C	Apply pressure of 1.4 MPa
	Hold for 60 min Can remove hot	Hold for 60 min Cool under pressure	Hold for 60 min Can remove hot	Hold for 60–120 min Cool to below 95°C before pressure release
Post-cure	Not necessary, but advisable for maximum properties	16 hours at 343°C	24 hours at 250°C or 48 hours at 200°C	Heat to 230°C Then from 230–345°C at 15°C/h Hold at 345°C for 4h Heat from 345–400°C at 75°C/h Hold at 400°C for 4h Cool slowly to 95°C
Vacuum bag/autoclave	Not recommended	—	Apply full vacuum Place in autoclave at 120°C Heat to 180°C over 1h	—
Post-cure	—	—	As above	—

TABLE 5.29. *Comparison of Salient Features of Press Cure (Reference 217)*

	Temperature	Pressure	Time	Post-cure	
Polyimide type	(°C)	(MPa)	(min)	Temperature (°C)	Time (hours)
Bismaleimide	120–180	1.4	60	250	24
Condensation	205	1.7	30 + cool	205–345	16
Acetylenic	250	1.4	60–120 + cool	230–400	22
PMR	232–315	6.9	90 + cool	343	16
Norbornene	315	1.4	60	Not necessary	
Thermoplastic 2	250–350	2.0–3.5	60 + cool	None	
Thermoplastic 1	427	17.3	10 + cool	None	

TABLE 5.30. *Comparison of Unreinforced Polyimide Properties (Reference 217)*

			Thermoplastic	
Property	Temperature (°C)	Condensation	Type 1 (NR150)	Type 2 (2080)
Tensile strength (MPa)	25	76–100	110	118
	260	—	—	30
	288	—	—	28
	300	33–43	—	—
	316	—	31	—
Tensile modulus (GPa)	25	3.10–3.35	4.00–4.66	1.30
	260	—	—	—
	288	—	—	0.67
	300	1.90–2.07	—	—
	316	—	1.04	—
Compressive strength (MPa)	25	253–310	—	206
	300	128–133	—	—
Compressive modulus (GPa)	25	3.93	—	2.04
	300	1.90	—	—
Flexural strength (MPa)	25	107–128	117	119
	200	—	—	124
	250	—	—	55
	288	—	—	35
	300	58–72	—	—
Flexural modulus (GPa)	25	3.03–3.17	3.80–4.17	3.32
	200		2.42	2.00
	250		1.93	1.59
	288		1.66	1.11
	300	1.79–1.89	—	—
Impact strength (J/cm) Izod notched		0.38–0.76	0.43	0.38
Elongation at break (%)	25	5–8	6	10
	300	3	[316°C] 65	
Heat distortion temperature (°C)		357	—	270–280
Water pick-up (% in 24 hours)		0.32	—	0.6

TABLE 5.31. *Comparison of Unreinforced Polyimide Properties (Reference 217)*

Property	Temperature (°C)	Addition		
		Norbornene/PMR	Bismaleimide	Acetylenic
Tensile strength (MPa)	25	48–83	48–59	83
	260	39	—	—
Tensile modulus (GPa)	25	3.80	—	3.93
	260	2.39	—	—
Compressive strength (MPa)	25	255	197	173
Compressive modulus(GPa)	25	2.89	—	—
Flexural strength (MPa)	25	76	128	131
	200	—	83	—
	250	48	57	—
	260	41–55	—	—
	288	35–41	—	—
	316	—	—	29
Flexural modulus (GPa)	25	3.17–3.38	3.80	4.49
	200	—	3.17	—
	250	2.17	2.76	—
	260	2.07–2.28	—	—
	288	1.93–2.07	—	—
Elongation at break (%)	25	1.4–2.5	<1	2
	260	1.8	—	—
Heat distortion temperature (°C)		>300	—	195–210
Water pick-up (% in 24 hours)		0.4	—	[1000 hours] < 1

ties of the carbon fiber composites. A similar conclusion has been reached by Petker,[237] although the values he cites for interlaminar shear strength and flexural strength for composites of zero void-content are some 30–50% greater than the figures in Table 5.32.

In Table 5.33 are listed the percentage retentions of interlaminar shear strength, flexural strength, and flexural modulus for composites after aging for different times at different temperatures, the test measurements being made either at room temperature or at the aging temperature itself. Under either test condition, at least 80% of the initial properties are retained up to 200°C for periods of 1,500 hours. Above this temperature loss in properties becomes progressively greater, the lifetimes of composites based upon 50% retention of flexural strength being 1,820 hours at 232°C, 1,350 hours at 300°C, and 470 hours at 316°C. These figures should be regarded as maximum values, as in service, the effects of absorbed moisture and applied stress must be considered in addition to thermo-oxidative aging.

TABLE 5.32. Comparison of Properties of Carbon Fiber/Polyimide Resin Unidirectional Laminates (Reference 217)

Polyimide type	Carbon fiber type	Interlaminar shear strength (MPa)	Flexural strength (MPa)	Flexural modulus (GPa)
Condensation	High modulus	49	710	163
	High strength	72	1340	115
	Type III or A	66	1420	108
Addition—norbornene	High modulus	52	—	—
	High strength	78	1430	100
Addition—bismaleimide	High strength	85	1410	120
Addition—acetylenic	High strength	83	1350	104
	Type III or A	116	1190	108
Thermoplastic type I	High modulus	51	870	145
(NR 150)	High strength	61	1240	108
	Type III or A	88	1390	104

Table 5.34 gives an indication of what can currently be achieved in respect of interlaminar shear strength of polyimide/carbon fiber composites as a function of time and temperature.[238] Marginal improvements may be possible in these figures as a result of improved process control and technology, but major advances are unlikely. Poly(amideimides) have also been used as laminating resins and an illustration of their behavior at elevated temperatures is given in Figure 5.24, which shows the flexural strength retention of style 181E glass cloth laminates made with Amoco AI-1030 as a function of time at temperature.[239]

Polyimide foams. A limited amount of work has been put into the development of polyimide foams. In 1970 Monsanto made available four precursor powders, designated RI-7271-01, RI-7271-06, RI-7271-12, and RI-7271-18, based on the Skybond system.[240] The numbers -01, -06, -12, and -18 indicate the approximate density in lbs/ft^3 of the foams that result from heat curing the powders. Table 5.35 gives the compressive strengths of the denser of these foams at room and elevated temperatures. They have very low flame spread and smoke production characteristics.

A lightweight syntactic foam has been produced from a mixture of equal weights of ICI's QX-13 resin and silica microballoons (Eccospheres) by processing at 300°C under moderate pressures.[241] The foam has a density of 610 kg/m^3, a flexural strength of 13.8 MPa, a flexural modulus of 2.1 GPa, and a compressive strength of 31 MPa. No elevated property data are available.

While this book was in production International Harvesters marketed a polyimide foam under the trade name Solimide.

Polyimide fibers. Some effort has also been devoted to the production of polyimide fibers. Success has been limited, but Figures 5.25 and 5.26 indicate the level of property retention at elevated temperatures, which has been achieved

TABLE 5.33. *Retention on Aging at Elevated Temperatures of Properties of Carbon Fiber/Polyimide Resin Unidirectional Laminates (Reference 217)*

Polyimide type	Aging time (hours)	Aging temperature (°C)	Test temperature (°C)	% Retention of initial		
				ILSS	Flexural strength	Flexural modulus
Condensation	25	200	25	99	100	100
	100	200	25	98	100	98
	1000	200	25	81	100	99
	25	300	25	100	100	99
	100	300	25	98	99	98
	1000	300	25	61	76	82
	100	316	25	—	100	—
	140	316	25	—	90	—
	200	316	25	—	97	—
	250	316	25	—	94	—
	325	316	25	—	100	—
	400	316	25	—	86	—
	25	200	200	—	91	—
	100	200	200	—	88	—
	1000	200	200	—	97	—
	1500	200	200	—	89	—
	0.5	232	232	76	82	100
	500	232	232	70	82	100
	1000	232	232	65	66	93
	2000	232	232	61	45	83
	25	300	300	—	85	—
	100	300	300	—	78	—
	1000	300	300	—	53	—
	1800	300	300	—	42	—
	100	316	316	—	71	—
	140	316	316	—	90	—
	200	316	316	—	71	—
	250	316	316	—	73	—
	325	316	316	—	64	—
	400	316	316	—	55	—
Addition–norbornene	125	260	260	52	—	—
	500	260	260	52	—	—
	1000	260	260	57	—	—
	2000	260	260	52	—	—
	125	288	288	51	—	—
	500	288	288	49	—	—
	1000	288	288	38	—	—
	2000	288	288	21	—	—
	125	316	316	48	—	—
	500	316	316	41	—	—
Addition–bismaleimide	100	232	25	100	100	—
	500	232	25	92	83	—
	100	232	232	19	36	—
	500	232	232	35	30	—

continued

TABLE 5.33 (continued). Retention on Aging at Elevated Temperatures of Properties of Carbon Fiber/Polyimide Resin Unidirectional Laminates (Reference 217)

	Aging time (hours)	Aging temperature (°C)	Test temperature (°C)	% Retention of initial		
Polyimide type				ILSS	Flexural strength	Flexural modulus
Addition–	500	316	25	—	72	—
acetylenic	1000	316	25	—	45	—
	125	316	316	100	83	90
	500	316	316	—	88	—
	1000	316	316	—	56	—
Thermoplastic–	5000	260	260	84	76	86
type I (NR150)	1850	288	288	80	72	100
	3000	316	316	80	66	90
	500	343	343	78	60	81

TABLE 5.34. Currently Achievable Levels of Performance with Polyimide/Carbon Fiber Composites (Reference 217)

Polyimide type	Interlaminar shear strength (MPa)	Temperature (°C)	Time (h)
Condensation	83	260	1000
Addition–PMR	83	260	1000
Addition–bismaleimide	83	230	1000
Condensation	55	350	10
Condensation	55	320	1000
Thermoplastic	55	320	1000
Addition–PMR	55	320	500
Addition–PMR	55	290	1000
Thermoplastic	55	260	50000
Addition–PMR	55	260	10000
Additon–bismaleimide	55	260	1000
Thermoplastic	28	350	10
Addition–PMR	28	350	10
Addition–PMR	28	320	1000
Condensation	28	290	10000
Thermoplastic	28	290	10000
Addition–PMR	28	290	10000
Thermoplastic	28	230	50000
Addition–PMR	28	230	50000
Addition–bismaleimide	28	230	10000

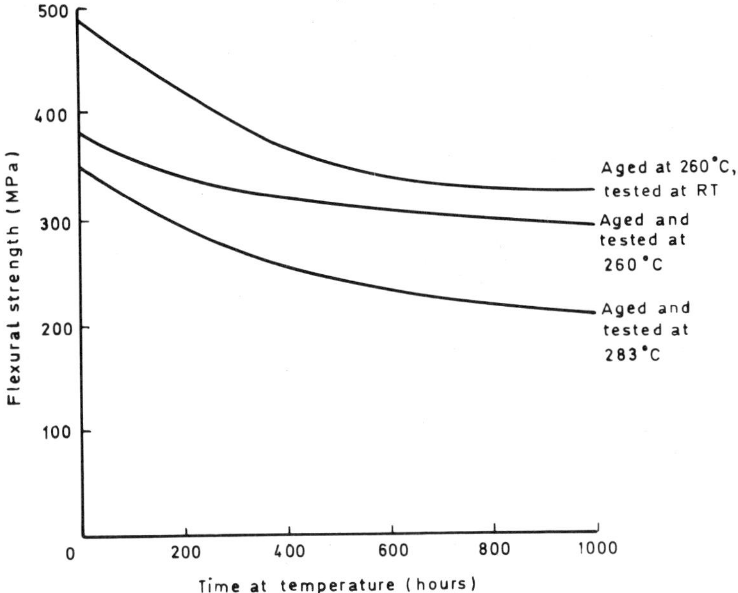

Figure 5.24.   Flexural strength retention of glass/AI-1030 polyamideimide laminates aged at 260 and 283°C. Reference 239.

with fibers made by the wet spinning of Upjohn's Polyimide 2080.[242] Depending upon draw ratio, the initial properties of the fibers are in the following ranges: tenacity, 2.2–3.2 g/den; modulus, 65–88 g/den; and elongation at break, 28–37%. Earlier work by Irwin and Sweeney[243] cited lifetimes for polyimide fibers (time for tenacity to fall to half its original value) of 750 hours at 283°C, 280 hours at 333°C, and 10 hours at 400°C.

TABLE 5.35. Compressive Strengths of Polyimide Foams at Room and Elevated Temperatures (Reference 240)

Property	Temperature (°C)	RI-7271-06	RI-7271-12	RI-7271-18
Density (Kg/m³)		107	218	328
Pressure (KPa) required for				
5% compression	RT	97	649	2450
10% compression	RT	145	932	4209
25% compression	RT	276	1622	7489
Pressure (KPa) required for				
10% compression at	93	138	828	3726
10% compression at	204	138	690	3312
10% compression at	232	83	621	1656
10% compression at	288	69	242	1518
10% compression at	316	14	14	69

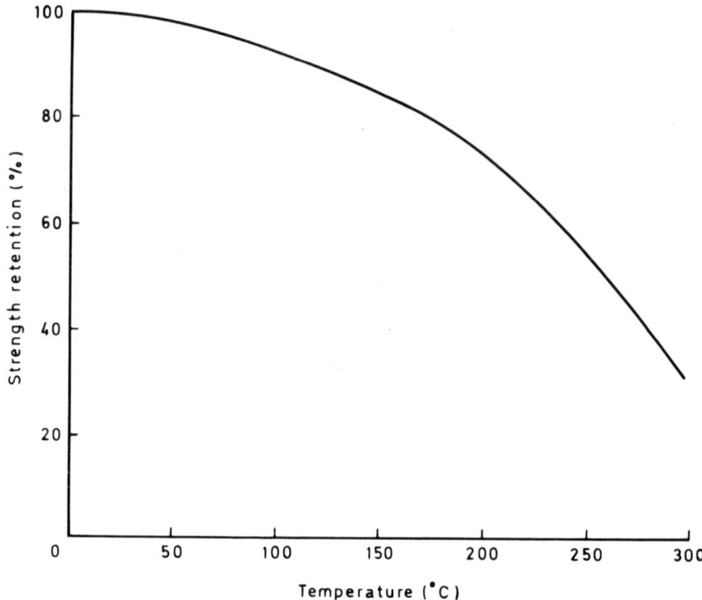

Figure 5.25. Strength retention of polyimide fibers as a function of temperature. Reference 242.

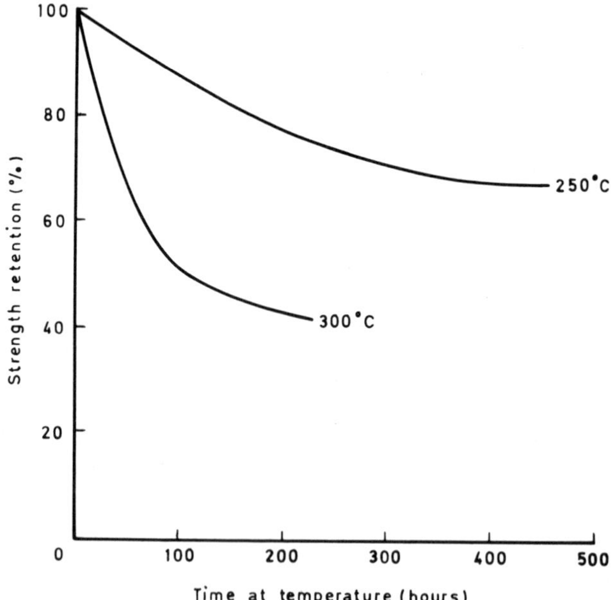

Figure 5.26. Strength retention of polyimide fibers as a function of time at temperature. Reference 242.

POLYBENZIMIDAZOLES

Introduction

Reference has already been made to the fact that, apart from the polyimides (PIs), the only viable commercial materials emerging from the polyheteroaromatics have been the polybenzimidazoles (PBIs). They have not been very substantial competitors to the PIs, for in spite of a number of possible applications, effectively the only real commercial success, and that is still at the development stage, is in the area of high-temperature/flame-resistant fibers and fabrics.[244–246] It is quite clear from the number and scope of publications that this relative failure cannot be attributed in any sense to inadequate research in the field. During the early 1960s to early 1970s, a flood of papers from both Western and Soviet sources, comparable in number to those on PIs, described the synthesis and properties of the PBIs. The volume of work since that time has considerably decreased; some research is still being reported, especially in the U.S.S.R., but there is now only a minor interest in the West. A number of reasons have been advanced (see earlier) for the commercial success of the PIs in comparison to the PBIs. For example, the flexibility of PI synthesis has not been paralleled in the PBIs. Although a number of alternative routes to PBIs exist, a major obstacle to their development has been the dependence on the use of aromatic tetraamines as comonomers in synthesis. These tetraamines have proved difficult and expensive to obtain in the reproducibly pure form necessary for successful polymer-forming processes. Furthermore, there has been considerable anxiety with regard to the carcinogenic activity of all of them, but in particular to that of the most widely used 3,3′-diaminobenzidine (DAB), and very stringent precautions have been necessary in its use.

Preparation

In the Melt

Polymers containing the unsubstituted heterocyclic ring pendant to the main chain were the first polyimidazoles to be produced and to find application.[247,248] Research into these polymers, particularly in the context of model enzyme systems, is still active. High-temperature, in-chain heterocyclic group polybenzimidazoles (**XXV**, m=4, 6, 8; R = direct bond, O, CH$_2$), which like the PIs originated in the Dupont laboratories, were first obtained[249] from the melt polycondensation of aromatic tetraamines (**XXVI**) with aliphatic dicarboxylic acid.

$$HO_2C-(CH_2)_m-CO_2H \;+\; \xrightarrow{-2H_2O}$$

(XXVI)

(XXV)

As distinct from these AA–BB-type polycondensations, melt-processable aliphatic A–B-type polybenzimidazoles (**XXVII**) have also been produced[250] from melt polymerizations.

$$XO_2C(CH_2)_m \xrightarrow[Ar]{230-250°C}$$

(XXVII)

$X = H$ or CH_3; $m = 2$ and 3

The most important route to the wholly aromatic PBIs (**XXVIII** and **XXIX**) has also involved melt polycondensation using aromatic tetraamines (**XXVI** and **XXX**) and aromatic diacids or diesters. While the application of aromatic diacids has been severely restricted[251] due to their facile decarboxylation at high temperatures, Marvel *et al.*[251–253] in their early work found that aromatic dicarboxylic esters, particularly the diphenyl esters, were very suitable monomers. A very large proportion of the subsequent synthetic approaches to aromatic PBIs have used tetraamine/dicarboxylic acid diphenyl ester monomer systems. A–B-type PBIs (**XXXI**) have been obtained from the corresponding monomers.[251, 254]

(XXVI)

$+ XO_2C—R—CO_2X \xrightarrow[-2XOH]{\Delta}$

(XXVIII)

(XXX)

(XXIX)

$X = OH, OCH_3, NH_2, CN$
$R = H, C_6H_5$

(XXXI)

With the favored diphenyl esters, the evolution of phenol is considered by some workers to produce a beneficial plasticizing effect; certain lower alkyl esters on the other hand induce side reactions that reduce the effectiveness of the polycondensation.

Early studies[251–253] of the melt polycondensation of aromatic monomers used a simple single-stage process of heating between 200 and 350°C. Above 220°C melt fusion occurred, with evolution of phenol and water, and the resolidified high-melting polymeric product was then heated for up to 6 hours at 350–370°C. Subsequently, a more sophisticated procedure[255–257] was developed, which involved the comelting (1–2 hours at 220°C) of monomers under inert conditions, removal of residual volatiles in vacuo, powdering of the high softening polymer at room temperature, and completion of polymerization (solid-state) by heating at 350–385°C. A majority of the all-ring-containing PBIs produced by this process, which was designed to eliminate thermo-oxidative cross-linking, are insoluble and intractable. A notable exception, poly[2,2′-(*m*-phenylene)-5,5′-bibenzimidazole] (**XXVIII,** R = direct bond; R′ = *m* C_6H_4), proved to be noncrystalline. It is soluble in dipolar aprotic solvents from which

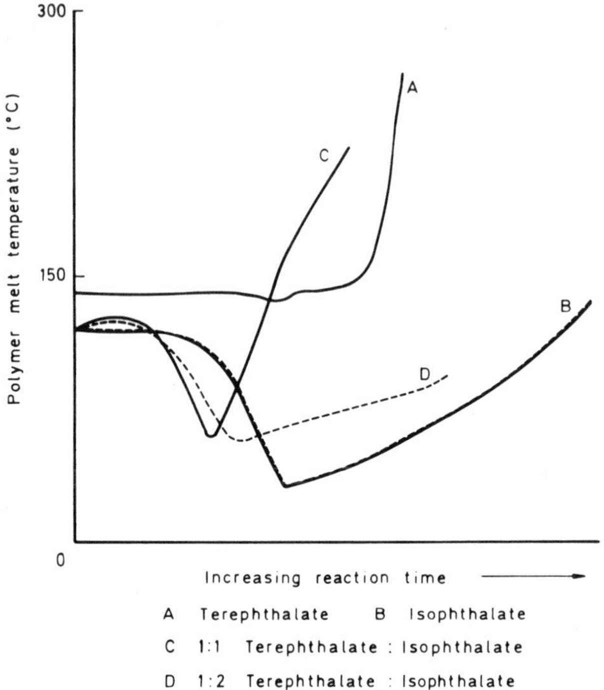

Figure 5.27. Variation of polymer melt temperature with reaction time for the condensation of DAB and diphenylesters under isothermal conditions. Reference 258.

films and fibers can be made; as a result this system has formed the basis for PBI materials development. For composite and adhesive resin formulations, high flow and wetting are important characteristics, and here the pre-polymer—probably a dimer or tetramer—has been exploited; polymerization is completed on the substrate. It is interesting to compare (Figure 5.27) the very pronounced decrease in prepolymer melting temperature with increasing reaction time shown by this polymer and related copolymers from diphenylisophthalate with the rapid increase for the rigid crystalline system based on the wholly para-linked polymer.[258] Various mechanisms have been proposed[259–261] for the formation of aromatic PBIs by the synthetic processes outlined above.

Wrasidlo and Levine,[260] from a kinetic study, have suggested that condensation proceeds via a Schiff's-base (syn) intermediate, which subsequently loses phenol to form the (linear) PBI:

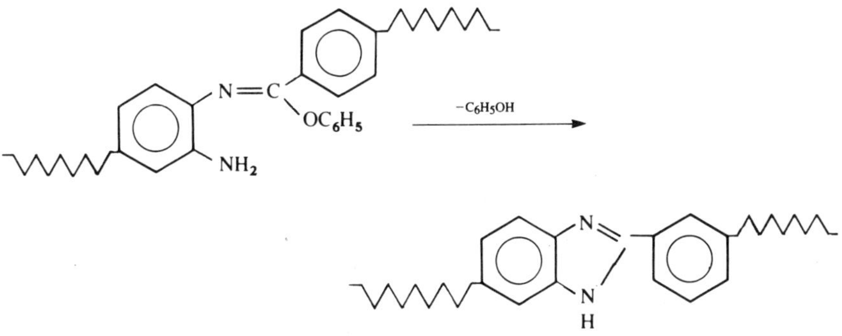

Using principally analytical and infrared/mass spectral data, other workers[261] have found that phenol is evolved before complete dehydration. They propose a poly(amino–amide)/poly(hydroxybenzimidazoline) intermediate, which above 260–280°C eliminates water with formation of the PBI. This mechanism suggests that the PBI prepolymer may contain both amino–amide/hydroxybenzimidazoline and benzimidazole structures which, in the solid-state reaction above 350°C, undergo complete conversion to PBI.

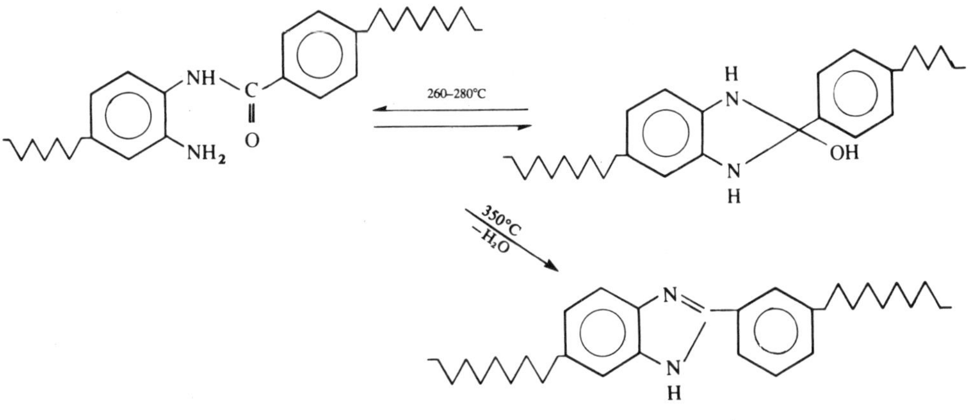

Alternative melt-condensation routes to PBIs from aromatic tetraamines have been investigated. Aromatic dinitriles[262] (in the presence of ammonium salts of inorganic acids), acid chlorides,[251] and phthalic anhydride[252] or poly(isophthalic anhydride)[263] have all yielded reasonably high-molecular-weight products. Gray *et al.*[264] developed a relatively low-melt-temperature polymerization route from DAB and 1,4-diacetylbenzene. The two-stage process involved formation of a high-molecular-weight Schiff-base/benzimidazoline prepolymer at 250°C, which was subsequently cyclized at 300°C with evolution of methane.

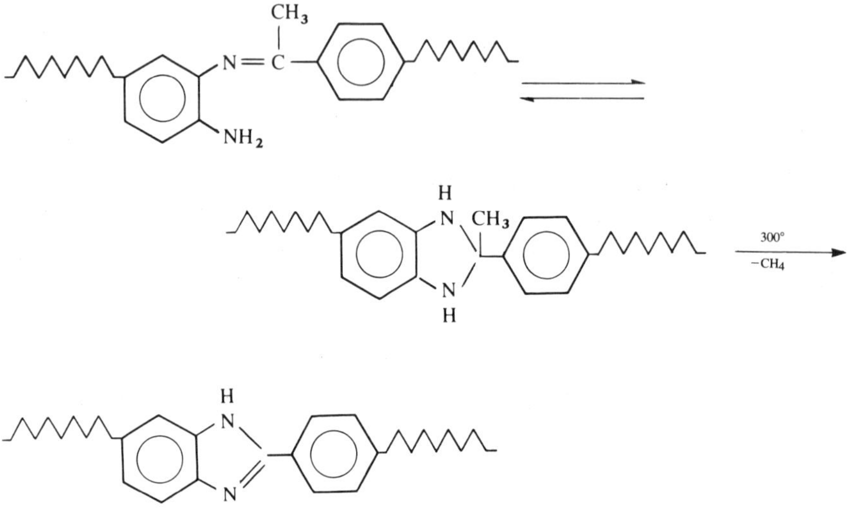

The most significant, practical alternative to the use of diphenylisophthalate with DAB proved to be that of isophthalamide.[254,265] With this material, a tractable low-molecular-weight prepolymer was formed by partial reaction (Figure 5.28) of a stoichiometric ratio of monomers. The full course of the polymerization could be followed by measurement of ammonia evolution, and the prepolymers remained fusible up to 93% of theoretical reaction compared with 60–70% for diphenylisophthalate-derived materials.

In Solution

The polycondensation of various tetraamines with dicarboxylic acids and acid derivatives has been carried out using several high-boiling solvents.[266–271] The obvious aim behind this technique, which has not been achieved in practice, has been to obtain solutions which could be used directly in prepreg formation. In many instances, however, the process has produced very low-molecular-weight PBIs, and often cyclization to the benzimidazole structure has been incomplete. Attempts have been made to overcome these problems either by using an initial solution polymerization when prepolymer is formed, followed by a final solid-state polymerization at 300°C[272–275] or, alternatively, by producing a prepolymer in the melt at 200–220°C, followed by completion of polymerization in the high-boiling-point solvent.[276–279] Limited work has

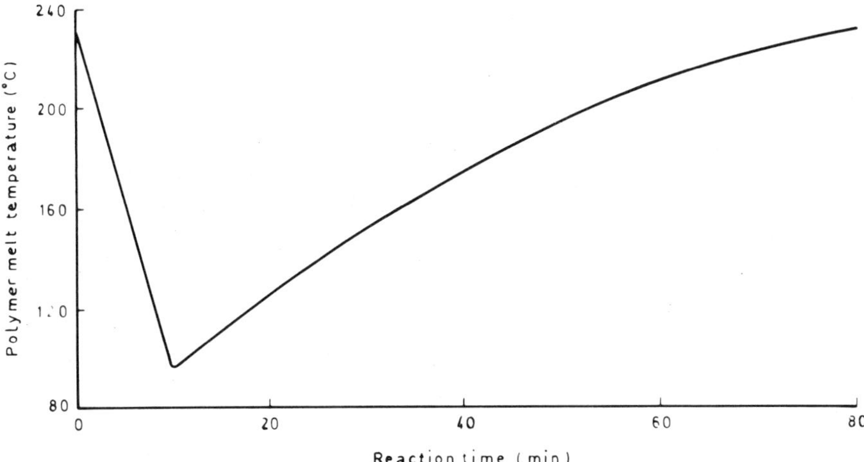

Figure 5.28. *Variation of polymer melt temperature with reaction time for condensation of DAB and isophthalamide at 273–277°C. Reference 265.*

been reported on the polymerization of aromatic dinitriles with diamines; in dipolar aprotic solvents, intermediate polyamidines are produced[280] which are subsequently cyclized, in the presence of N HCl/CH$_3$OH, to low-molecular-weight PBIs.

Alkoxide-catalyzed reaction of phthalonitrile with aromatic tetraamines in methoxyethanol solution has also produced[281] low-molecular-weight PBIs. Relatively high-molecular-weight PBIs have been made by a two-stage synthesis from aromatic dicarboxylic acid chlorides and substituted tetraamines. The initial low-temperature polycondensation in polar solvents produces[267, 273–275, 282] poly(aminoamides), which can be thermally or chemically cyclized in solution, or film, or after isolation of solid polymer by solid-state polymerization. This method has proved especially useful in the preparation of N-substituted PBIs.

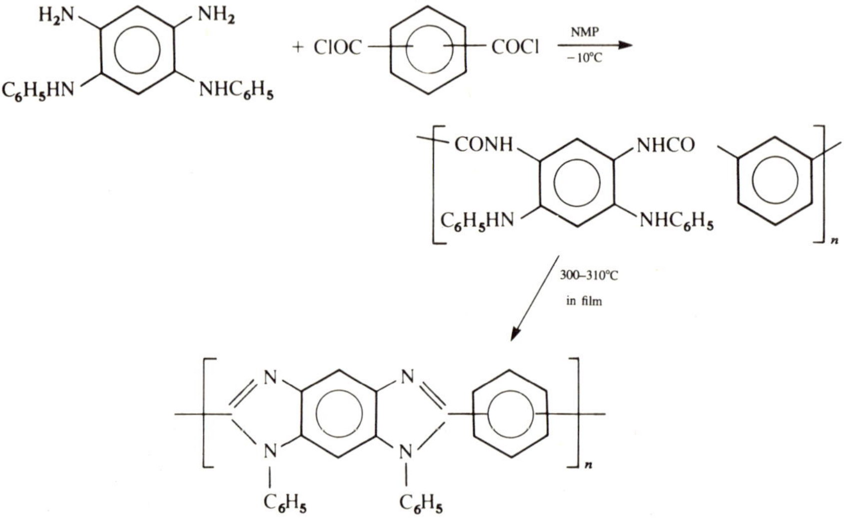

Korshak *et al.*[283, 284] have also recently described the preparation of PBIs by the reductive polyheterocyclization at 180–200°C of poly(*o*-nitro) amides which were produced by a low-temperature technique similar to the one used with the poly(aminoamides) above. High-molecular-weight polyamidoximes (**XXXII**, R = CH_2 or O) have been[285] chemically cyclized in benzene/pyridine solution by *p*-toluene-sulphonylchloride to the corresponding PBI, and reaction of aniline with polybenzoxazoles in polyphosphoric acid (see below) has also produced[286] N-substituted PBIs.

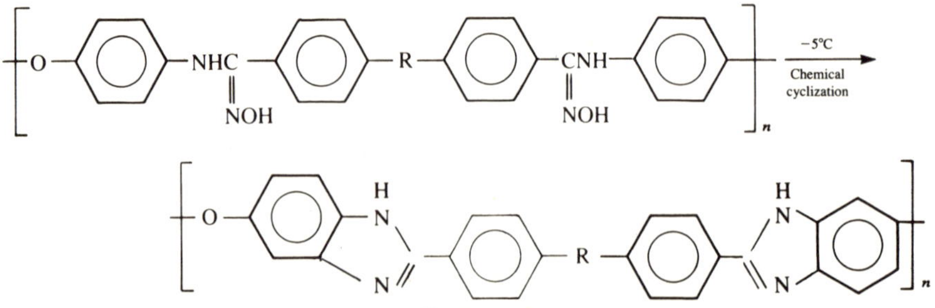

(**XXXII**)

The most widely used solvents for the production of PBIs have undoubtedly been polyphosphoric acid (PPA) or polyphosphoric acid esters; indeed, PPA has been shown to be a suitable solvent/catalyst for a number of polyheterocyclization systems.[287] The method was first reported by Iwakura *et al.*,[288, 289] but has since been used by a large number of workers[290–295]; its main benefit is that the oxidation sensitive tetraamines can be replaced by the more stable tetrahydrochloride salt. This process, which applies to the preparation of the fully cyclized and polymerized PBI and not the prepolymer, has been used particularly with aromatic dicarboxylic acids/dimethylesters, diamides, and dinitriles. Neither diphenylesters nor diacid chlorides are successful in the process. Some drawbacks have been observed with the use of PPA. It has been suggested that, even after lengthy polymerization at 140°C, cyclization is often incomplete and, also, that the complete removal of inorganic contaminants from the final polymer is difficult and not always successful.

Less conventional monomers have also been used to a limited extent in solution polycondensations. Aromatic PBIs have been produced, directly, by reaction of tetraamines with dialdehydes or dialdehyde adducts[296, 297] in polar solvents. Using low-temperature polycondensation techniques, Neuse[298] isolated the intermediate polySchiff's bases (**XXXIII**), which were solution cast as films and then thermally cyclized.

$$\left[CH=N \bigcirc\bigcirc NH_2, \quad H_2N \bigcirc\bigcirc N=CH \bigcirc \right]_n \xrightarrow[\text{Film–form}]{60\text{–}110°C}$$

$$\left[\underset{H}{N}\diagdown\diagup N \bigcirc\bigcirc N \diagup\diagdown \underset{H}{N} \bigcirc \right]_n$$

(**XXXIII**)

Korshak *et al.*[299] have reported on the preparation of aliphatic/aromatic PBIs from tetraamines and imidic acids/imino esters and orthoesters. A recent report by Dudgeon and Vogl[300] suggests that the interaction of tetraamines with bisorthoesters (**XXXIV**) in polar solvents (particularly DMSO) at 100°C produces high-molecular-weight PBIs in quantitative yield. This reaction appears to be significant for producing PBI fiber systems.

+ (MO)$_3$C — R′ — C(OM)$_3$ $\xrightarrow[\text{DMSO}]{100°C}$

(XXXIV)

R = O,CO, direct bond; R′ = *m* and *p*-C$_6$H$_4$, direct bond

Polymer Types and Materials Applications

The wide variety of structural modifications prepared in the condensation-type polyimides is also observed in the PBIs. Very few of these structures have been incorporated into practical products, however. Overwhelmingly, research has been merely one element of a wider study throughout the heteroaromatic polymers designed to relate structure with processability (solubility/tractability) and thermal/thermo-oxidative stability. Table 5.36 lists several of the most frequently used tetraamines, of which DAB, TAB, TADM, and TADE, in the order shown, have been employed the most frequently. In general the number of available dicarboxylic acids, or acid derivatives, is largely responsible for the diversity of PBI structures. Following the original work of Brinker and Robinson,[249] further studies have been made of the alkylene-linked PBIs (**XXV**; m = 2, 4, 6, 7, 8, 11) and **XXIX; R′** = (CH$_2$)$_4$ derived from aliphatic dicarboxylic acids or esters.[251,252,288,293,301,302] These experiments have been aimed mainly at the determination of solubility/tractability relationships in such systems. For example, the effect of the length of alkylene chain on polymer stiffness (Tg) has been studied in some detail by Tsur *et al.*[295] The only polymer of this type to receive much attention as a practical material has been[303–305] poly(2,2′-octamethylene-5,5′-bibenzimidazole) (**XXV;** m = 8, R = direct bond). Problems associated with the synthesis (HF evolution), which have also led to low-

TABLE 5.36. *Aromatic Tetraamine/Substituted Tetraamine PBI Monomers*

Structure	Name
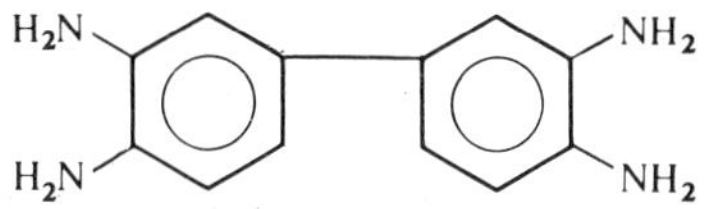 (structure: 3,3′-diaminobenzidine)	3,3′-Diaminobenzidine (DAB)
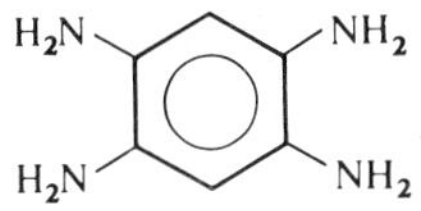 (structure: 1,2,4,5-tetraaminobenzene)	1,2,4,5-Tetraaminobenzene (TAB)
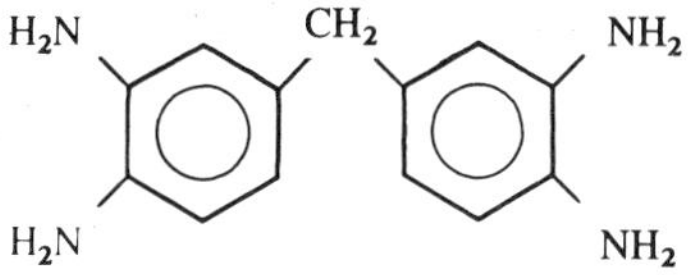 (structure: 3,3′,4,4′-tetraaminodiphenylmethane)	3,3′,4,4′-Tetraaminodiphenylmethane (TADM)
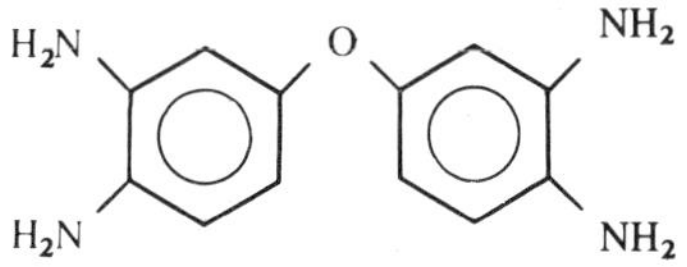 (structure: 3,3′,4,4′-tetraaminodiphenylether)	3,3′4,4′-Tetraaminodiphenylether (TADE)
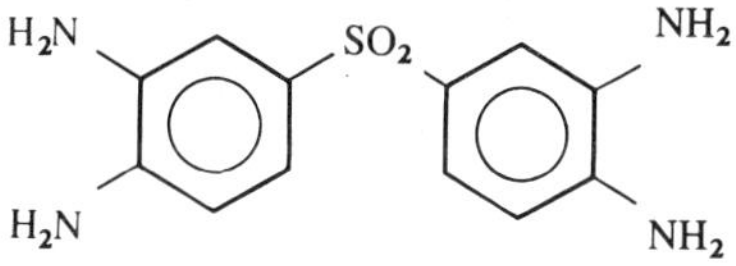 (structure: 3,3′,4,4′-tetraaminodiphenylsulphone)	3,3′,4,4′-Tetraaminodiphenylsulphone (TADS)
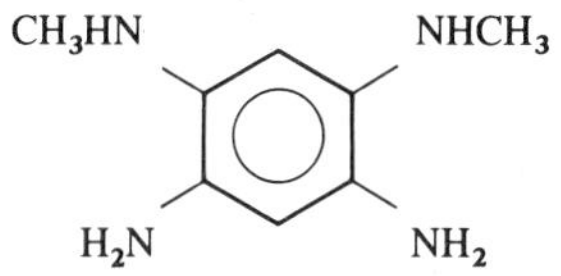 (structure: 1,3-bis(methylamino)-4,6-diaminobenzene)	1,3-Bis(methylamino)-4,6-diaminobenzene
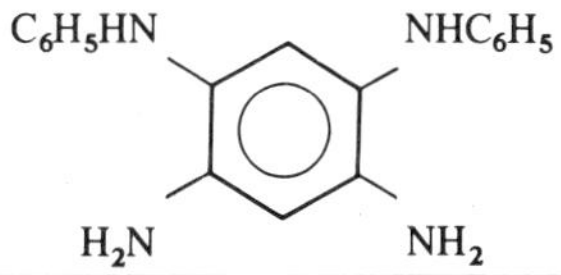 (structure: 1,3-dianilino-4,6-diaminobenzene)	1,3-Dianilino-4,6-diaminobenzene

molecular-weight, brittle, insoluble products, have made[306–308] the fluoroalkylene-linked PBIs [**XXVIII**, R = direct bond, R′ = $(CF_2)_3$ and $(CF_2)_6$] considerably less attractive than other fluoroalkylene-linked heteroaromatic polymers such as the polyperfluoroalkylene-*s*-triazines.

From the outset the wholly aromatic PBIs based on DAB and TAB have generated considerable interest, although a limited study has also been made using the tetraamines from naphthalene[307] and anthracene.[292] Very many aromatic dicarboxylic acid/acid derivatives have been utilized in the polymerization and these have been fully discussed in a number of reviews (see the supplementary bibliography). It is noteworthy that most of the wholly aromatic PBIs are crystalline, insoluble (to a great extent), and intractable, the exceptions[258, 309] being the unsymmetrical systems (**XXVIII**, R = direct bond;

$$R' = \quad \text{and} \quad).$$

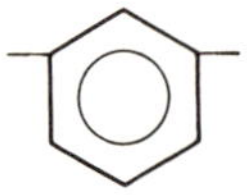

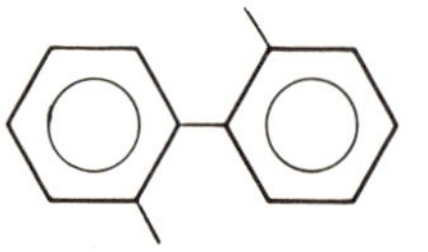

Aiming for useful polymers which combine processability at relatively low temperatures with strength and rigidity at high temperatures, PBIs derived from blends of *m*- and *p*-linked systems (see Figure 5.27) have been developed.[258, 294] Attempts to establish controlled cross-linking in PBIs, involving blends of precursor acid esters containing trifunctionality (diphenyl trimesate), were not successful.[258] Increased solubility/tractability in wholly aromatic PBIs has resulted from the introduction of appropriate substituents in the benzene and imidazole rings, although the replacement of imido hydrogen in the imidazole moiety was primarily aimed at improving the thermo-oxidative stability (see below). Direct introduction of methyl[310] or carboranyl[311] groups in polymer

$$(\textbf{XXXV}, \ R = CH_3, \ \text{—} \ C \text{———} CH)$$
$$B_{10}H_{10}$$

has resulted in partial N-alkylation, but by using tetraamines such as 1,3-bis(methylamino)-4,6-diaminobenzene and 1,3-dianilino-4,6-diaminobenzene (see Table 5.36) the quantitative introduction of both N-Me and N-Ph has been possible and more representative thermal and thermo-oxidative stabilization achieved.

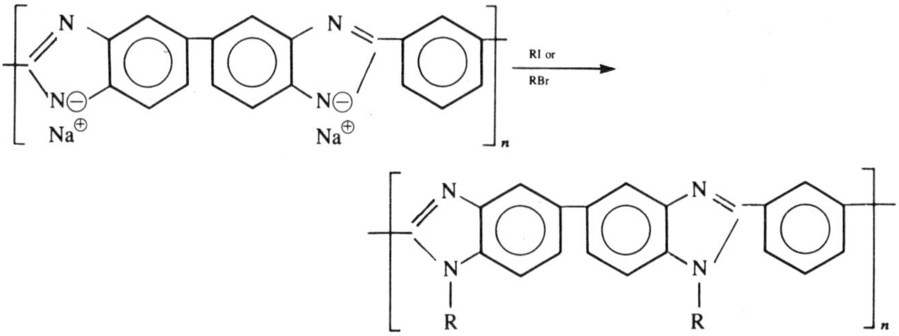

(XXXVI)

Soviet workers have improved the processability of aromatic PBIs (**XXXVI**) by introduction of methyl groups into both benzene and imidazole rings,[313–316] as well as by introducing[291,317,318] particularly bulky polar side groups into the systems (**XXXVII**, R = direct bond, O, CH_2; X =

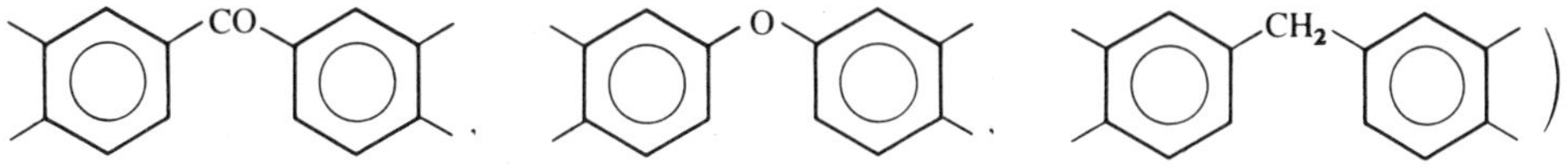

and (**XXXVIII**).

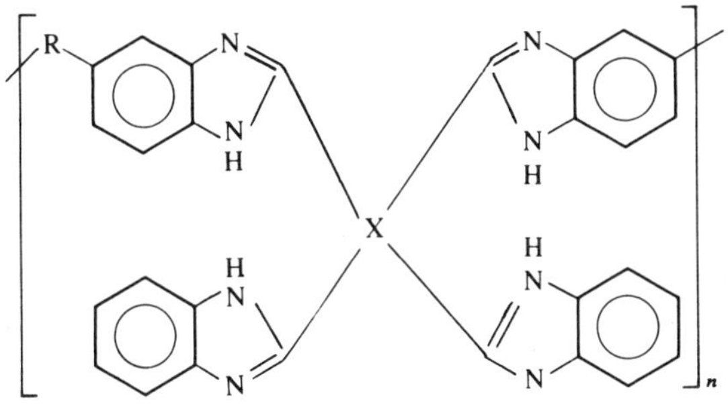

(XXXVII)

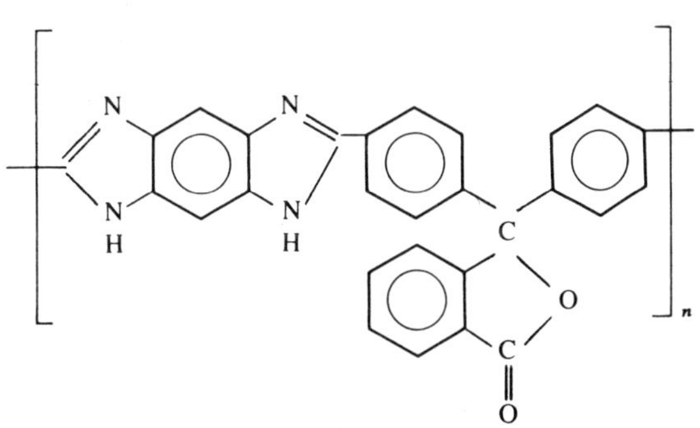

(XXXVIII)

Alternatives to the all-ring-containing main-chain PBIs have also been developed in attempts to obtain a compromise between solubility/tractability and thermal/thermo-oxidative stability. Heteroatoms or simple groups such as oxygen, methylene, or sulphone have been incorporated[274,279,283,284,288,295,300,319–327] via the tetraamine (e.g., TADM, TADE, or TADS) and/or diacid/diester monomer. The effect of more complex linking groups in aliphatic/aromatic and wholly aromatic PBIs such as amide,[277,328–336] ester,[302,337] siloxane,[338–340] silane,[341] phosphorus (phosphine oxide),[342–345] adamantane,[346] phenylene-F-trimethylene,[347,348] and phenylene-F-isopropylidine[349] has also been investigated. Despite some early synthetic difficulties,[311,350] both *m*- and *p*-carborane species have been successfully introduced[351,352] into the polymer chain of (**XXVIII**, R = direct bond, O, CH_2, SO_2; R′ = *m*- and *p*- $CB_{10}H_{10}C-$). Boron has also been introduced[342, 353] into the analogous polybenzoborimidazolines (**XXXIX**).

(**XXXIX**)

R = direct bond, CH_2; R′ = *p*- and *m*-C_6H_4;

In addition, PBI copolymers containing alternative heteroaromatic systems such as imide,[102,103,354–358] 1,3,4-oxadiazole,[102,103,359] benzoxazole,[360] quinoxaline,[266,361] *s*-triazine,[362] imidazopyrrolone,[363] and pyridine[251,364] have been produced.

Thermal Stability

A very high proportion of the publications on polybenzimidazoles have included at least an element of work devoted to their thermal characterization, particularly to a study of thermal degradation in inert or oxidizing conditions. In this respect most comparability and kinetic studies have employed thermogravimetry and, to a much lesser extent, differential thermal analysis (DTA) and torsional braid analysis (TBA). Elucidation of the mechanism of degradation has most frequently involved gas chromatography (GC) and infrared (IR) and mass (MS) spectrometry, often in conjunction with thermogravimetry. Electron spin resonance (ESR) techniques have also been used.

Although the measurements of thermochemical transition phenomena have featured less with the polybenzimidazoles than with the polyimides, nevertheless, some useful information has been produced. Gillham,[365] using TBA, has investigated low temperature ($-70°C$) transition behavior as well as glass transitions (Tgs) in some systems.[309,366] Overall, Tg has not been used consistently in structure-property correlations as in the polyimides, although occasionally[250,302,313,323,340,367] Tgs measured by differential scanning calorimetry (DSC) have been included in overall property data. However, Tsur *et al.*,[295] using Tg—measured by thermomechanical analysis (TMA)—as a key parameter, selected the 3:1 copolymer of isophthalic and meta-phenylenediacetic acid with DAB from a number of possible aromatic systems as having the best combination of oxidative resistance and processing characteristics. The increase of Tg in some of these processable aliphatic-aromatic polybenzimidazoles, observed after repeated heating cycles, was linked with decreased mobility of the polymer chains caused by cross-linking at temperatures $<350°C$.

An indication of the likely overall thermal stability of polymers based on the benzimidazole system has come from measurement of the thermal decomposition temperatures (T_D) of fairly simple model compounds.[368,369] In some instances, DTA has been used to determine transitions which could be related to degradative processes; most often, though, the method has been used to characterize melting or softening points of PBIs. DTA analysis has also been of value in following the course of cyclization reactions from prepolymers to fully aromaticized polybenzimidazoles.

Thermogravimetry has been, by far, the most widely used tool to determine the level of thermal stability. Both dynamic and isothermal heating have been used under inert, vacuum, air, and pure oxygen conditions. The polyalkylenebenzimidazoles, for example, (**XXV,** m = 4 or 8; R = direct bond) prepared by AA–BB polycondensations, are stable both in nitrogen and air up to 350–380°C; above 380–400°C rapid degradation occurs[251,288,301–303] (Figure 5.29). In air, unlike nitrogen, there is virtually no residue above 600–700°C. Poly[2,5-ethylenebenzimidazole (**XXVII,** m = 2)] produced by A–B polycondensation has[250] a similar temperature for initial decomposition, although this varies, as does the temperature for 50% weight loss (510–540°C), with the degree of polyheterocyclization and the reduced viscosity of the sample. Significant variations observed from laboratory to laboratory in the thermal stability of comparable polymer systems throughout the range of polybenzimidazoles may have resulted from measurements, like these, on nonstandard materials. With the poly(amide benzimidazoles) [**XL,** R = direct bond, O; Z = $(CH_2)_4$, $(CH_2)_8$, m-C_6H_4; x = 5 or 6], significant degradation begins at 350, 310, and 260°C for m-C_6H_4, $(CH_2)_4$, and $(CH_2)_8$ systems, respectively.[334, 335]

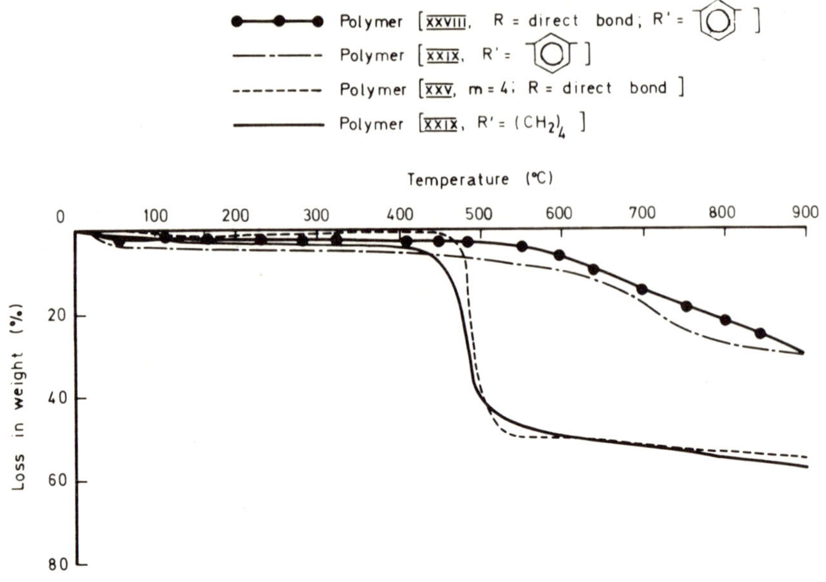

Figure 5.29. Thermogravimetry of aliphatic- and aromatic-linked PBI's in nitrogen. (Heating rate 2.5°C/min) Reference 251.

(XL)

Surprisingly, the perfluoroalkylene-linked polybenzimidazoles [**XXVIII**, R = direct bond; R′ = (CF$_2$)$_3$, CF$_2$)$_6$] in both inert and oxygen-containing atmospheres show[306–308] relatively poor stability (Figure 5.30). It has been suggested that the ease of rupture of the imidazole N-H bond is responsible for the evolution of hydrogen fluoride in these systems. In polybenzimidazoles [**XXVIII**, R = direct bond; R′ = p,p'C$_6$H$_4$(CF$_2$)$_{5\ or\ 6}$C$_6$H$_4$], where perfluoroalkylene groups are shielded from the imidazole ring by a paraphenylene unit, initial degradation temperatures in air (350–380°C) are similar to the alkylene-linked systems.[347,348] Thermal stabilities nearer to those of the wholly aromatic PBIs have been claimed[349] for systems [**XXVIII**, R = R′ = C(CF$_3$)$_2$].

For the predominantly aromatic PBIs, the approach has been either to maximize thermal/thermo-oxidative stability, largely ignoring processing fac-

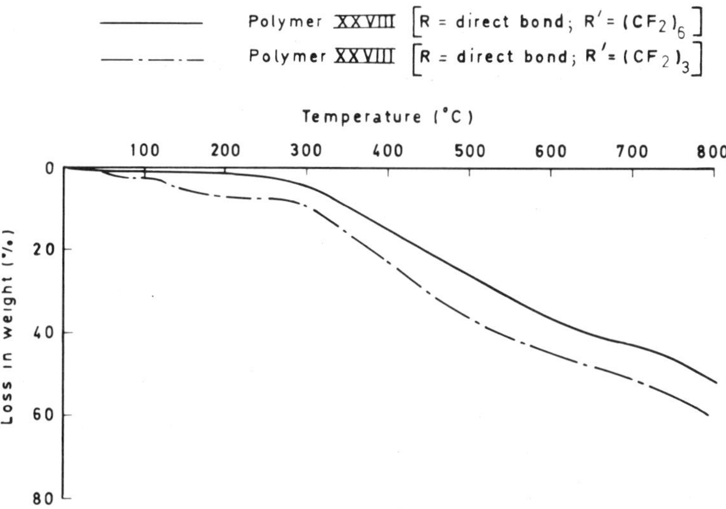

Figure 5.30. Thermogravimetry of F-alkylene-linked PBI's in nitrogen. (Heating rate 2.5°C/min) Reference 307.

tors, or, alternatively, to introduce a level of tractability while retaining the highest possible stability. Replacement of alkylene by arylene linking groups has a considerable effect on the stability (Figure 5.29); Wrasidlo[370] has suggested that, overall, substitution of aliphatic by aromatic units increases the temperature of initial weight loss by around 150°C. In their early studies on the entirely ring-substituted PBIs, Marvel and Vogel[251,252] established (Table 5.37) that variation of the aromatic linking units had only a marginal effect on stability in inert or oxidizing atmospheres. They demonstrated that most of the polymers show no change in properties on heating at 550°C and less than 5% weight loss after several hours at 600°C in nitrogen. Two of the systems [**XXVIII,** R = direct bond; R' = m- or o-C₆H₄] combine a level of tractability with high thermal stability. In particular, the meta-linked system has attracted considerable interest and, apart from the original work by Marvel, several other investigations[213,274,284,288,294,300,370–378] have employed thermogravimetry and DTA to determine thermal stability. The appreciably lower stability under oxidizing conditions has been confirmed by a number of these studies. Since the source of this instability has been considered to be the lability of the imidazole hydrogen, the effect of substitution has been examined. Introduction of N-Me produces very little change in either thermal or thermo-oxidative stability although processability is improved.[312–314] Substitution of the imidazole hydrogen by phenyl in polymers (**XXVIII,** R = direct bond; R' = m-C₆H₄) and (**XXXI**) has little effect on the level of thermal stability under inert or oxidizing

TABLE 5.37. Thermal Stability of Wholly Ring Containing PBIs (References 251, 252)

Benzimidazole repeat unit	Linking group	Wt loss (%) after 1 hour at					Total wt loss (%)
		400°C	450°C	500°C	550°C	600°C	
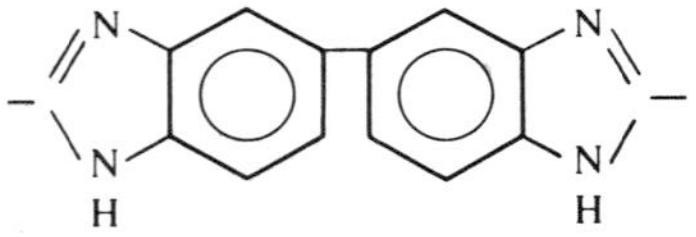 (benzimidazole structure)	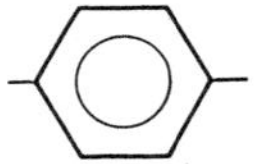 (structure) [a]	1.0	1.0	0	1.7	1.0	4.7
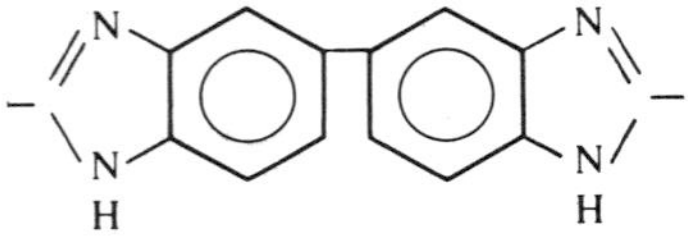 (benzimidazole structure)	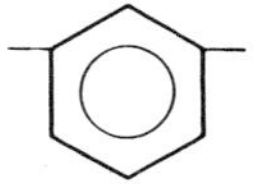 (structure) [a]	0.6	0	0.4	1.3	2.2	4.5
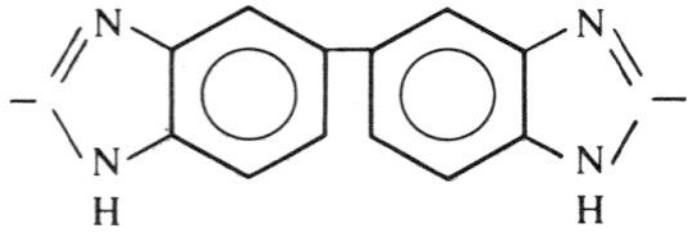 (benzimidazole structure)	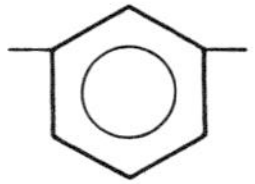 (structure) [b]	0	0	0.3	3.9	3.7	7.9
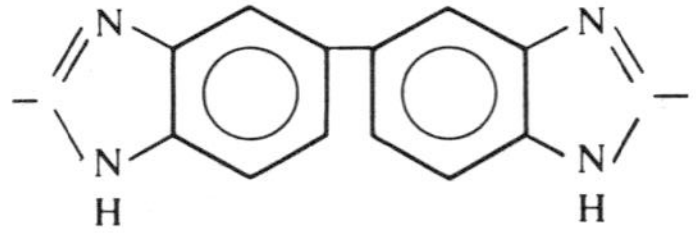 (benzimidazole structure)	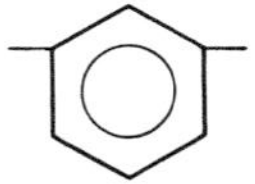 (structure) [c]	0	1.7	5.2	7.6	9.0	22.5

(bibenzimidazole structure)	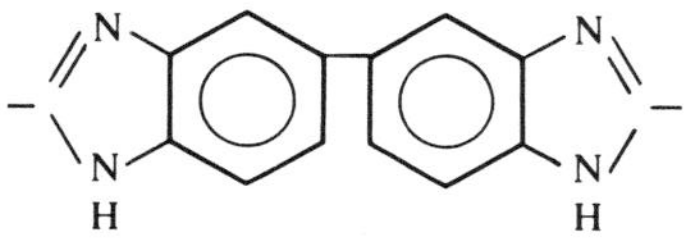 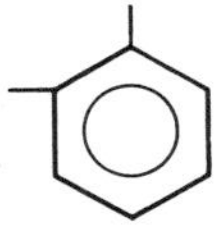 (benzene ring) [a]	0	0.4	0.4	3.7	—	4.5 (to 550°C)
(bibenzimidazole structure)	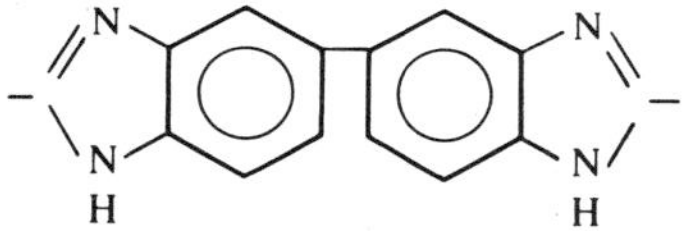 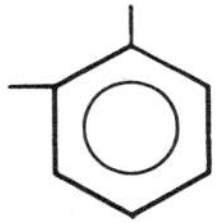 (benzene ring) [c]	0	1.5	7.0	7.6	—	16.1 (to 550°C)
(bibenzimidazole structure)	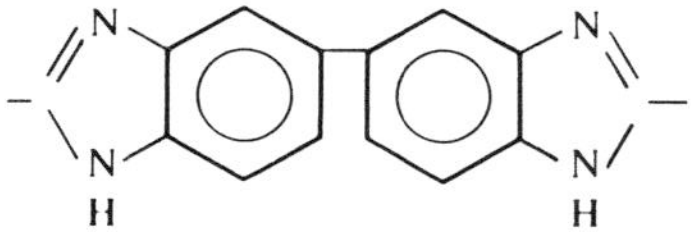 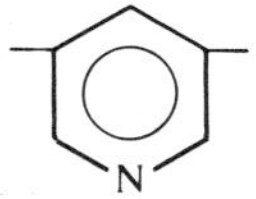 (pyridine ring) [a]	0.2	0.8	0.5	1.4	2.7	5.6
(bibenzimidazole structure)	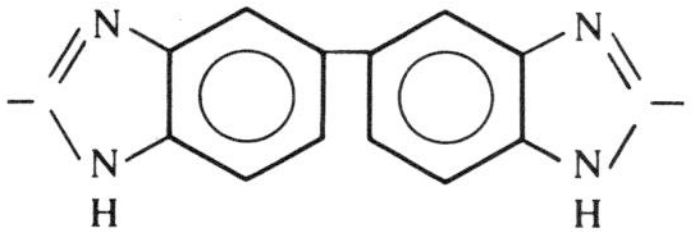 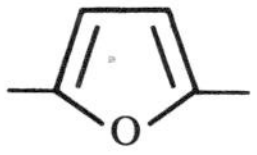 (furan ring) [a]	1.4	1.7	2.6	2.3	2.0	10.0

continued

TABLE 5.37 (continued). Thermal Stability of Wholly Ring Containing PBIs (References 251, 252)

Benzimidazole repeat unit	Linking group	Wt loss (%) after 1 hour at					Total wt loss (%)
		400°C	450°C	500°C	550°C	600°C	
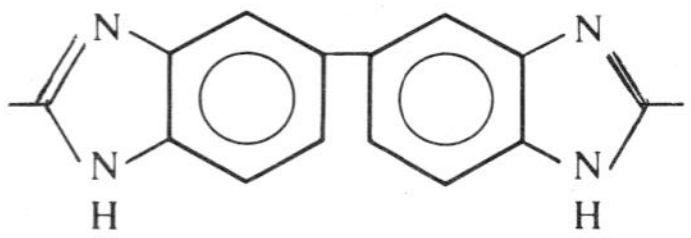 (benzimidazole structure)	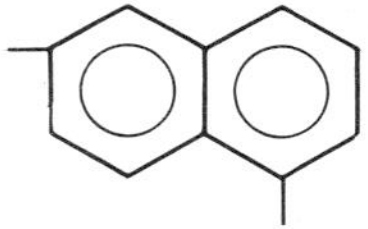 (naphthalene linking group) [a]	0.4	0.4	0.8	1.2	3.7	6.5
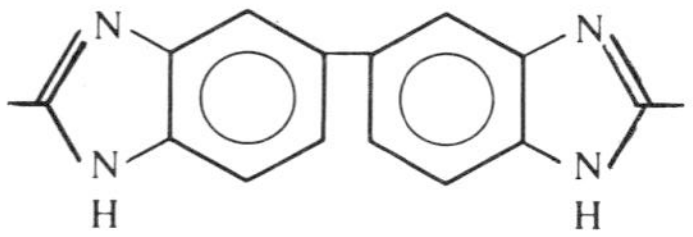 (benzimidazole structure)	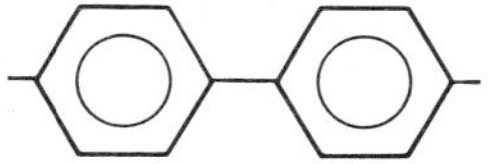 (biphenyl linking group) [a]	0.3	0	0.8	0.3	2.1	3.5
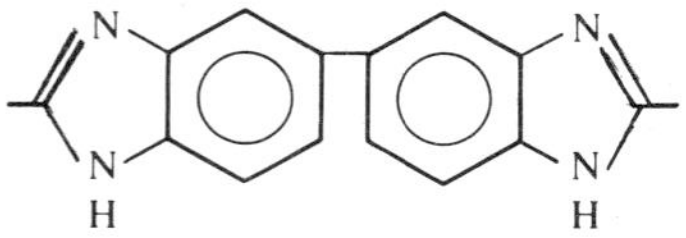 (benzimidazole structure)	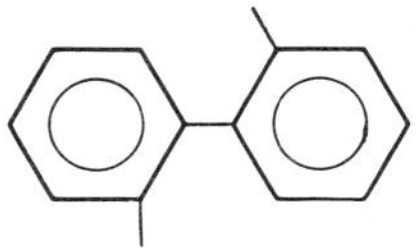 (2,2'-dimethylbiphenyl linking group) [a]	0	0.5	8.0	8.5	0.5	17.5
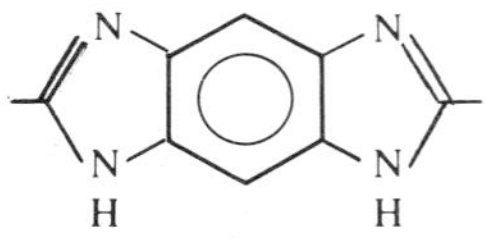 (benzimidazole structure)	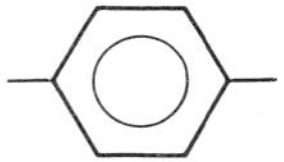 (phenylene linking group) [a]	2.8	0.5	1.0	1.9	4.0	10.2

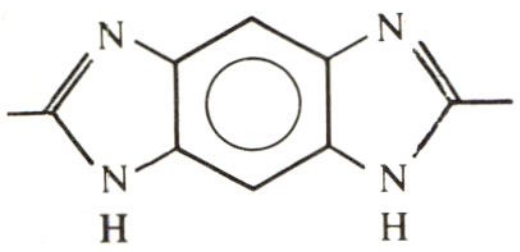	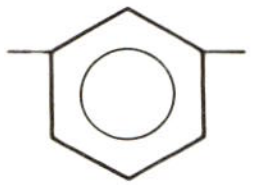 [a]	0.7	1.4	0.3	1.4	1.4	5.2
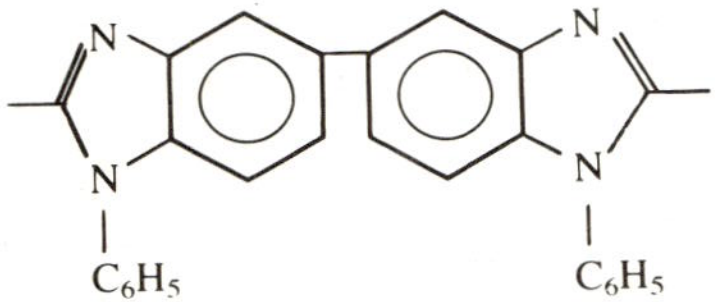	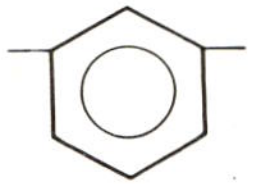 [a]	0	0.4	1.6	6.0	—	8.0 (to 550°C)
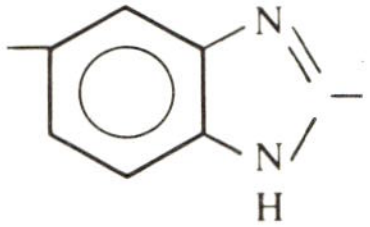 [a]		1.1	0.4	0.4	0.4	5.0	7.3

[a]Under nitrogen.
[b]In vacuo.
[c]In air.

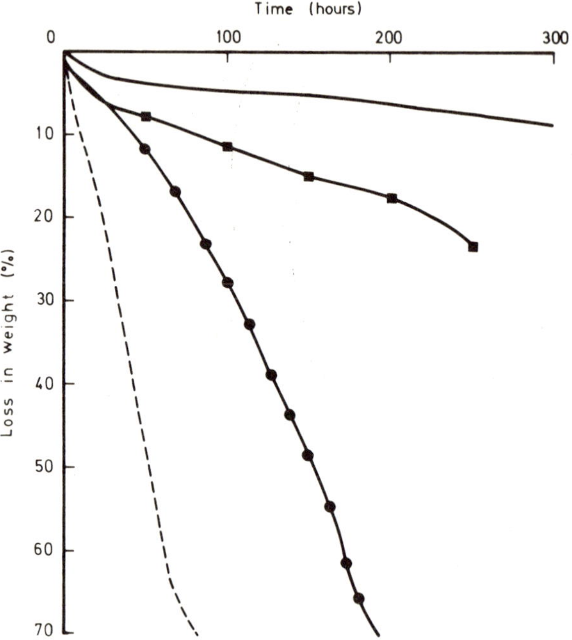

Figure 5.31. Isothermal weight loss curves for PBI and N-phenyl PBI in air. Reference 373.

conditions as measured by dynamic weight loss.[252,274,282,287,379] However, determination of isothermal weight loss shows[254,370,373,378-380] the phenyl-substituted PBIs to be markedly more stable under oxidizing conditions (Figure 5.31). Data produced by Ehlers[381] does not corroborate these findings, however, and it has been suggested that the improved oxidative stability observed under isothermal conditions may have been caused by cross-linking processes occurring during prolonged heating.

In spite of the original evidence to the contrary, some variation in stability due to different substitution patterns in aromatic linking groups has been observed. It has been claimed, for example, that the paraphenylene-linked system is more stable under oxidizing conditions than the corresponding ortho- or meta-analogs,[284, 294, 332, 382] although the absolute level of this stabilization is in

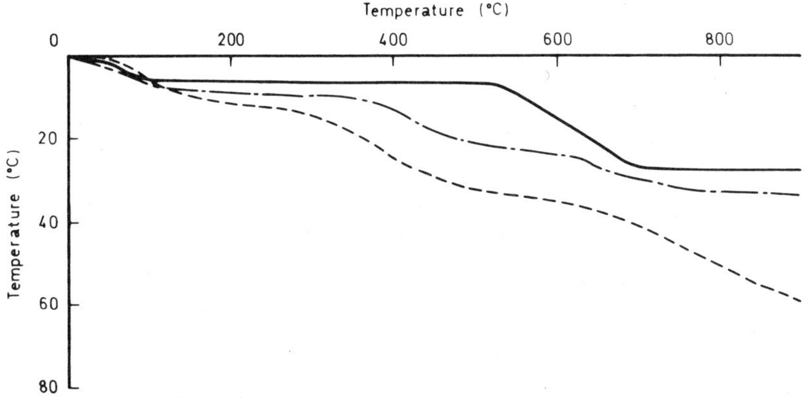

Figure 5.32. Thermogravimetry of isomerically substituted naphthylene PBIs in nitrogen. (Heating rate 2.5°C/min) Reference 307.

dispute and PBIs from DAB and isomeric diphenyl naphthalenedicarboxylates[307] have different levels of stability (Figure 5.32). Although earlier work[370] pointed to a greater stability of A-B type polymers compared with AA-BB systems (approximately 100°C in temperature of initial weight loss), more recently Ehlers[381] has claimed that there is in fact no significant difference in the stability of the structures shown below:

Aromatic PBIs with flexible linking groups such as $-O-$[274,300, 319,320,323,369,383–385] and $-SO_2-$[274,283,326] in either the amine or acid moiety, or $-CH_2-$[323,381–386] in the amine moiety, are marginally less thermally or thermo-oxidatively stable than the all-ring systems. Their decomposition temperatures in inert atmospheres or air range between 450 and 550°C. To varying degrees they are more tractable and soluble. Depending to some extent on their structure, aromatic poly(amide-benzimidazoles)[329–332] combine a relatively high level of processability with adequate thermal stability in nitrogen (initial weight loss at 420°C, 10% loss at 500°C). In air, however, weight loss is between 25 and 50% at 500°C. Several of the benzimidazole-heteroarylene copolymers containing 1,3,4-oxadiazole,[102,103,359] quinoxaline,[266,361] imide,[102,103,354–358] benzoxazole,[360] or imidazole[307] nuclei exhibit thermal and thermal oxidative stabilities at least equivalent (>500°C in N_2, 480–500°C in air) to the phenylene-linked homopolymers. In some cases, moreover, processing characteristics are improved. The introduction of siloxane,[338–340] silane,[341] and phosphine oxide[342–345] units into aromatic PBIs [**XXVIII**] has had a marked effect on tractability and solubility.

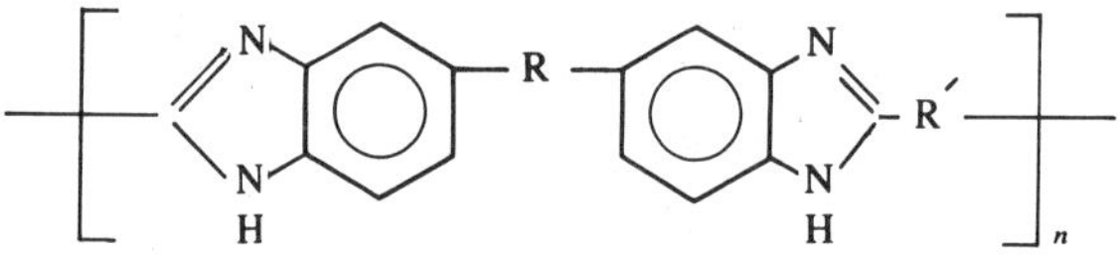

$$\text{(XXVIII)}$$

(a) R = direct bond; R′ = $(CH_2)_m$—Si(CH_3)_2—$\left(O-Si(CH_3)_2\right)_y$—$(CH_2)_m$— ($m$ = 2 or 3),

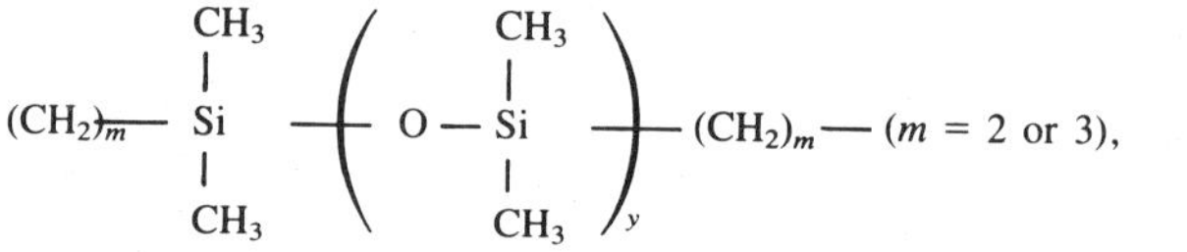

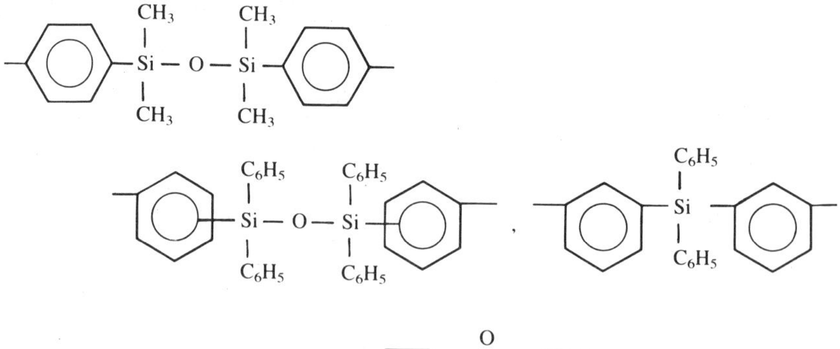

(b) R = direct bond, O, CH_2 ; R′ =

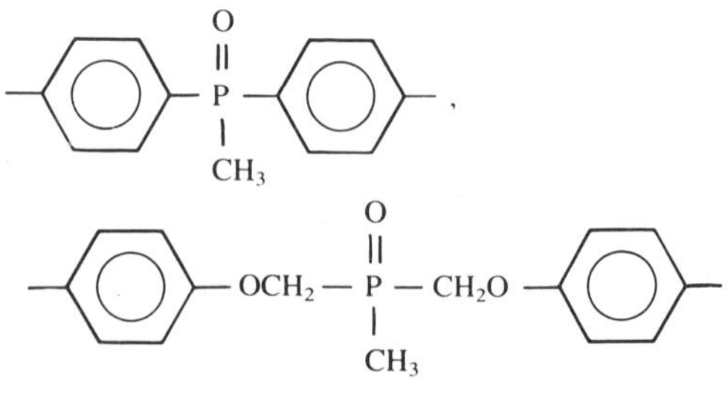

Although dynamic experiments indicate little weight loss in nitrogen or air below 500°C for the fully aromatic systems, both siloxane- and silane-linked copolymers show initial degradation in nitrogen at 350°C and in air at 300°C under isothermal heating conditions. PBIs containing the dimethylsiloxy group[339] begin to lose weight at temperatures around 260°C and decompose rapidly above 400°C under thermo-oxidative conditions. Similarly MeP(O)-modified polymers suffer extensive degradation above 400°C, though they show a high resistance to ignition in contact with a flame.[345] The introduction of intrinsically more stable heterogroups such as adamantane[346] and carborane[351,352] into the PBI chain leads to high decomposition temperatures (~600°C in nitrogen), but this is allied to poorer processing properties. Surprisingly, introduction[307] of the ferrocenyl group into PBI (**XXVIII, R** = direct bond;

$$R' = \text{(ferrocenyl group: two cyclopentadienyl rings linked by Fe)}$$

reduces the thermal stability (in nitrogen); weight losses of 20 and 25% at 400 and 500°C, respectively, were recorded.

Comparisons of the thermal and thermo-oxidative stability of the PBI system with those of other heteroaromatic polymers based on dynamic weight loss results have been made by a number of authors.[370–373,376,377,381,387,388] (Some typical results are shown in Figures 5.33 and 5.34.) Although there are divergences in the conclusions reached, taking as the criterion the temperatures for initial weight loss, or the temperatures for onset of major decomposition, the relative order of stability under inert conditions is: Polybenzothiazoles ~ polybenzimidazoles ≥ polyimides > polybenzoxazoles > polyquinoxalines.

The relative order of stability under oxidizing conditions is somewhat different: Polybenzothiazoles > polybenzoxazoles ≥ polyimides > polybenzimidazoles > polyquinoxalines. Isothermal weight-loss curves measured at 371°C in air (Figure 5.35) give yet another order: Polyimide > polybenzoxazole > polyquinoxaline > polybenzothiazole > poly-N-phenylbenzimidazole > polybenzimidazole. This last order is the most realistic from a practical use standpoint.

Research into the actual mechanisms of degradation have been much more limited. Three early studies[375,389,390] involved the application of combined thermogravimetry and mass spectrometry to the identification of pyrolysis products from the thermal degradation of poly(2,2′-m-phenylene-5,5′bibenzimidazole) (**XXVIII, R** = direct bond; R′ = m-C$_6$H$_4$) over the temperature range 500–900°C. Major products identified were hydrogen, hydrogen cyanide, ammonia, and water, together with lesser amounts of carbon monoxide and methane. Trace amounts of carbon dioxide, benzene, benzonitrile, aniline,

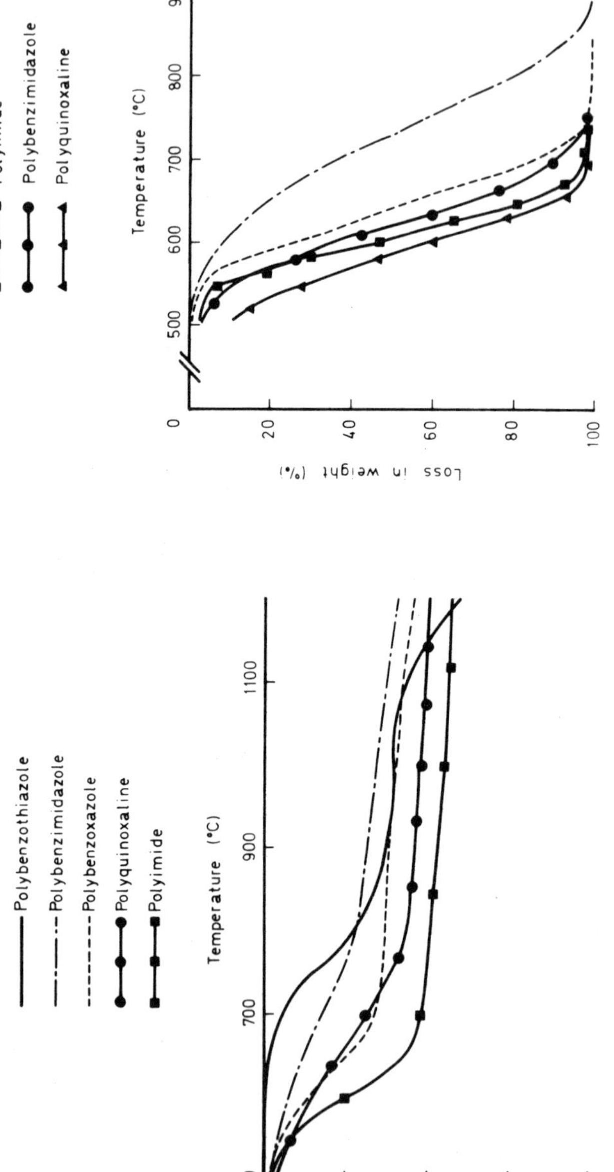

Figure 5.34. Thermogravimetry of heteroaromatic polymers in air. (Heating rate 6.6°C/min) Reference 370.

Figure 5.33. Thermogravimetry of heteroaromatic polymers in helium. (Heating rate 6.6°C/min) Reference 370.

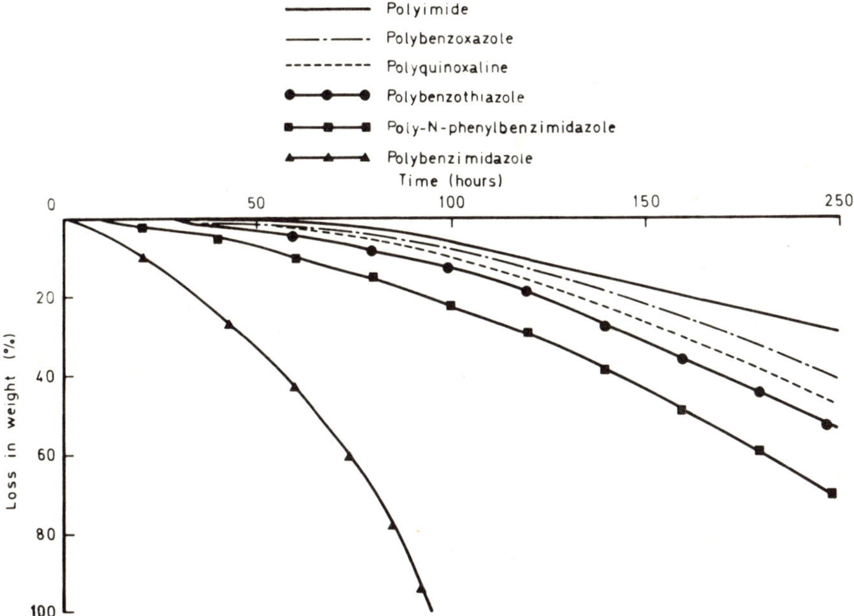

Figure 5.35. Isothermal weight loss curves of heteroaromatic polymers at 371°C in air. Reference 213.

phenol, aromatic diamines, and phthalonitrile were also detected. In the most comprehensive of these studies, Schulman and Lochte[375] suggested that the low level of carbon-containing volatiles is probably a result of the benzenoid-ring structures forming a residue which amounts to 70–80% of the initial polymer weight. To account for the major volatile products, these workers postulated a three-stage degradative mechanism involving hydrolytic ring opening (<500°C), stepwise homolytic sequences (550–650°C) involving formation and coupling of radicals, and (700–800°C) decomposition of products formed at lower temperatures. Ehlers et al.[389] have since put forward similar mechanisms for the thermal degradation of several aromatic polybenzimidazoles. An alternative mechanism of thermal degradation for poly(2,2'-m-phenylene-5,5'-bibenzimidazole) proposed by Schulman and Lochte,[375] which also accounts for the major pyrolysis products, involves the direct homolytic cleavage of the imidazole ring to form benzyne and nitrene intermediates. Soviet workers have also studied[378] the thermal degradation of poly(2,2'-p-phenylene-5,5'-bibenzimidazole) (**XXVIII**, R = direct bond; R' = p-C_6H_4) and the N-phenylated analog using thermogravimetry coupled to GC/TLC and IR/UV analytical techniques. The spectra of solid and gaseous pyrolysis products from these polymers are similar to those reported for poly(2,2'-m-phenylene-5,5'bibenzimidazole). Rode et al.[378] suggested that in both polymers degrada-

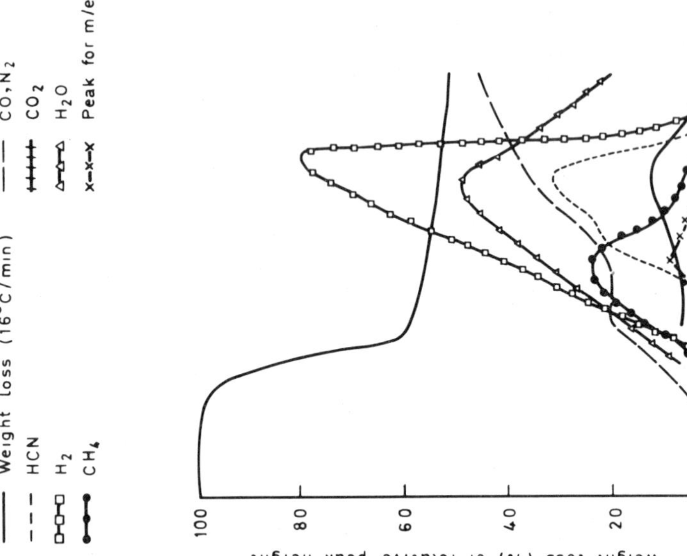

Figure 5.36(b). Thermogravimetry and mass spectrometric analysis of volatile products from thermal degradation of PBI from m-phenylene diacetic acid and 3,3',4,4'-diaminobenzidine. Reference 383.

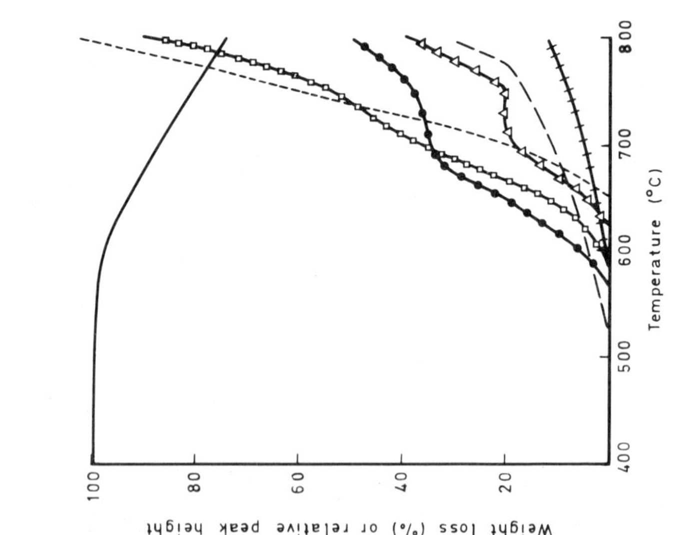

Figure 5.36(a). Thermogravimetry and mass spectrometric analysis of volatile products from thermal degradation of PBI from isophthalic acid and 3,3'-diaminobenzidine. Reference 383.

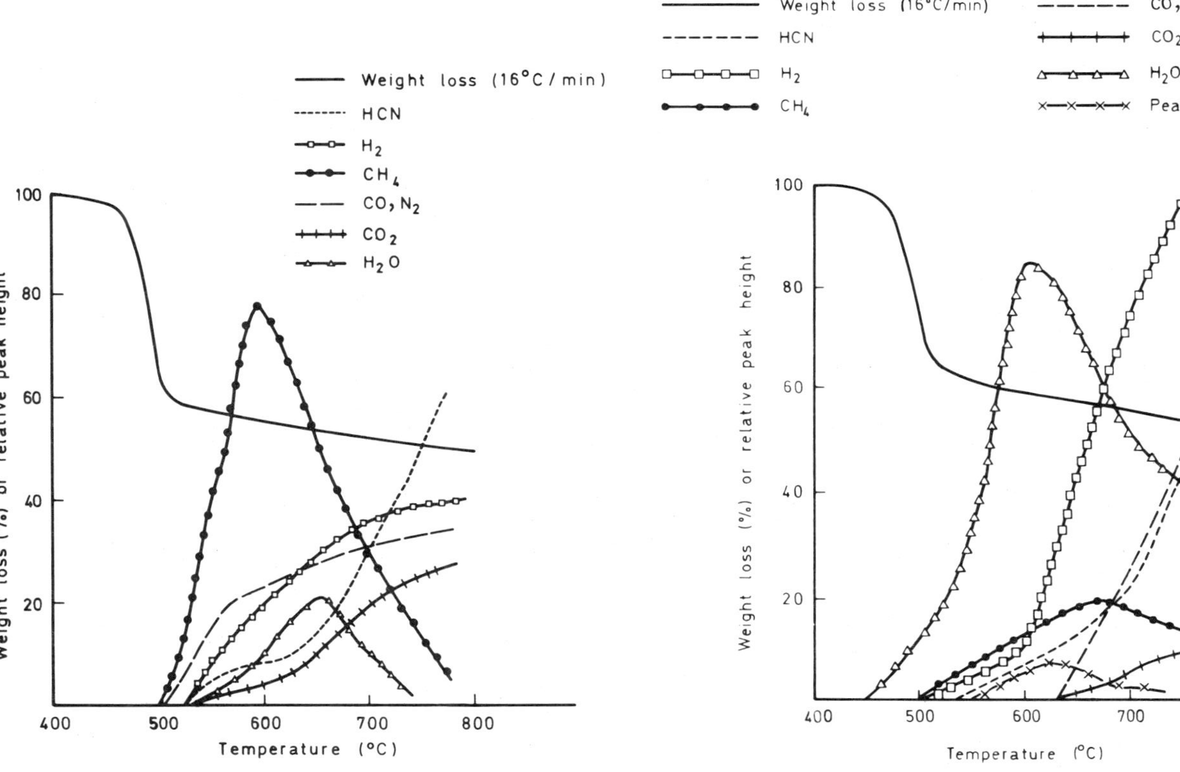

Figure 5.36(c). Thermogravimetry and mass spectrometric analysis of volatile products from thermal degradation of PBI from succinic acid and 3,3'-diaminobenzidine. Reference 383.

Figure 5.36(d). Thermogravimetry and mass spectrometric analysis of volatile products from thermal degradation of PBI from m-phenylene diacetic acid and 3,3',4,4'-tetraaminodiphenylether. Reference 383.

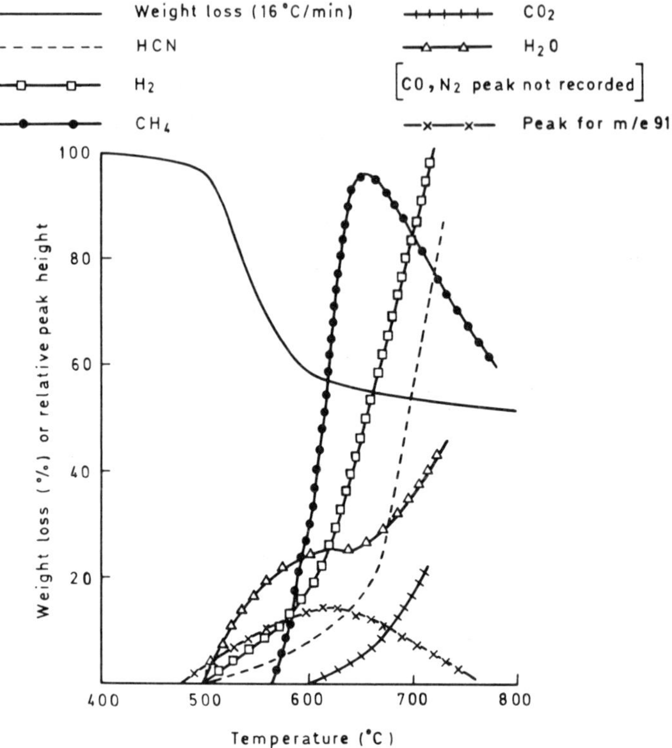

Figure 5.36(e). Thermogravimetry and mass spectrometric analysis of volatile products from thermal degradation of PBI from isophthalic acid and 3,3′,4,4′-tetraaminodiphenylmethane. Reference 383.

tion occurs by homolytic rupture of the benzimidazole rings. However, with the poly-N-phenylbenzimidazole, early formation of benzene indicates that this process is preceded by rupture of the N-phenyl bond. A complementary kinetic analysis showed that the thermal degradation of both polymers takes place by a first-order reaction, effective activation energy for the nonphenylated and phenylated polymers being 189 and 178 kJ/mole, respectively.

Tsur *et al.*[383] studied the thermal degradation in argon of poly(2,2′-*m*-phenylene-5,5′-bibenzimidazole) using thermogravimetry/mass spectrometry. The evolution of the major gaseous pyrolysates begins at the temperature ($\sim$ 600°C) of initial weight loss [Figure 5.36(a)]; composition of the products, with the exception of ammonia whose presence was questioned, corresponds with earlier findings. The investigation was extended to methylene-containing PBIs (**XXVIII**, R = direct bond, O, CH₂; R′ = *m*-CH₂C₆H₄CH₂-*m*, *p*-CH₂-C₆H₄CH₂-*p*) and (**XXV**, m = 2, 4, 6, 8; R = direct bond). Typically [Figure 5.36(b)], degradation of polymer (**XXVIII**, R = direct bond; R′ = *m*-CH₂C₆-

H$_4$CH$_2$-*m*) exhibits a two-stage decomposition; initially, a sharp weight loss (40%) occurs around 560°C associated with a solid sublimate, followed by slower weight loss at higher temperatures when gaseous products associated with a more profound degradation of imidazole rings are detected. Mass spectral analysis of the sublimate indicates a mixture of fractions derived from CH$_2$ bond-breaking, the most abundant fragment (m/e 234) being the bibenzimidazole unit:

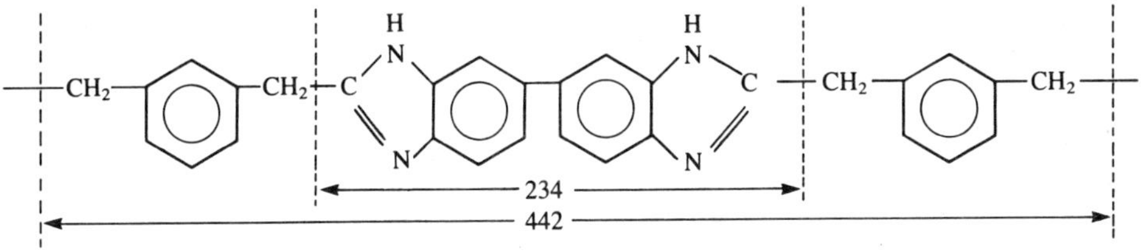

Polymers with aliphatic chains in the acid moiety [Figure 5.36(c)] lose up to 80% weight (500°C) as sublimate, the most abundant mass fragment (m/e 262) indicating cleavage of the weak CH$_2$—CH$_2$ bond:

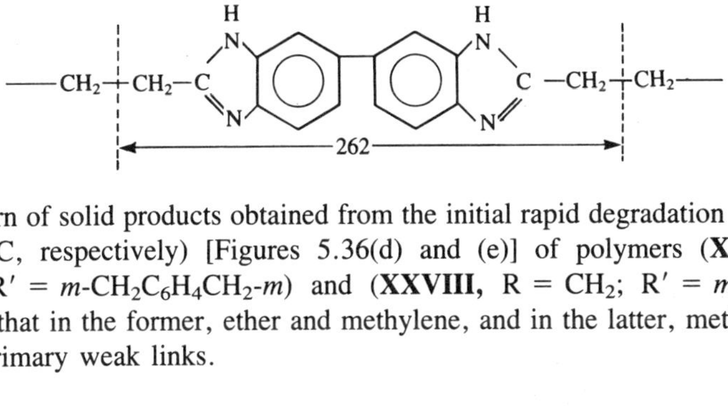

The pattern of solid products obtained from the initial rapid degradation (at 450 and 520°C, respectively) [Figures 5.36(d) and (e)] of polymers (**XXVIII,** R = 0, R′ = *m*-CH$_2$C$_6$H$_4$CH$_2$-*m*) and (**XXVIII,** R = CH$_2$; R′ = *m*-C$_6$H$_4$) indicates that in the former, ether and methylene, and in the latter, methylene, are the primary weak links.

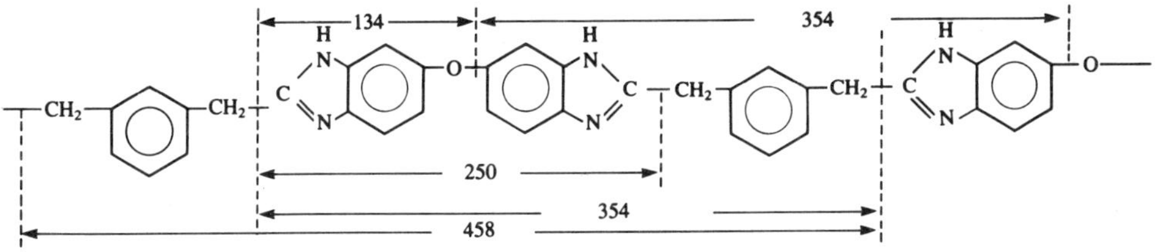

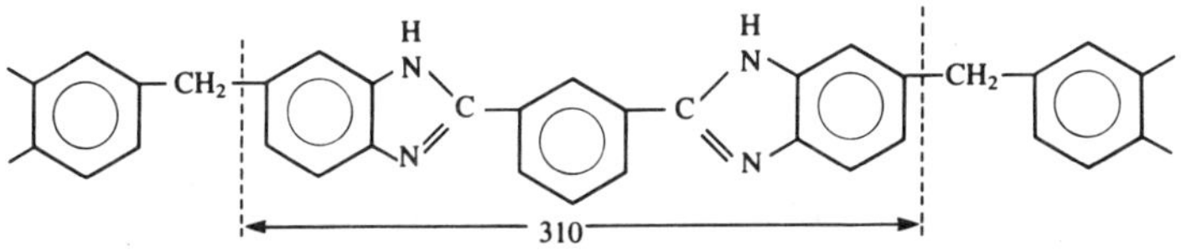

Mechanistic studies of the thermo-oxidative degradation of poly(2,2'-*m*-phenylene-5,5'-bibenzimidazole), based on IR and GC analysis of volatile products, have been carried out principally by Conley and co-workers[391,392] at Wright State University, Ohio. The oxidative pyrolysis of polybenzimidazole (**XXVIII**, R = direct bond; R' = *m*-C$_6$H$_4$), as well as that of related model compounds, reveals a very close similarity in the types of degradation product—water, cyanogen, nitriles, carbon dioxide, and carbon monoxide—to those produced under inert conditions. Coupling these results with previous reports on the enhanced stability of the N-phenylated polymer analog to isothermal aging in air, it was concluded that the N-H bond of the imidazole ring is highly susceptible to both thermal- and oxidatively- induced rupture. The scission could involve either hydrogen abstraction by oxygen, or homolytic cleavage of the bond with formation of the same radical species. Under oxidative conditions it was postulated that degradation occurs either by straightforward attack of oxygen on the benzimidazole unit to form an oxygenated structure (XLI) with subsequent formation and decomposition of intermediates, or by thermal generation of free radical intermediates (**XLII**), which react with oxygen to give species (**XLIII**), which then degrade:

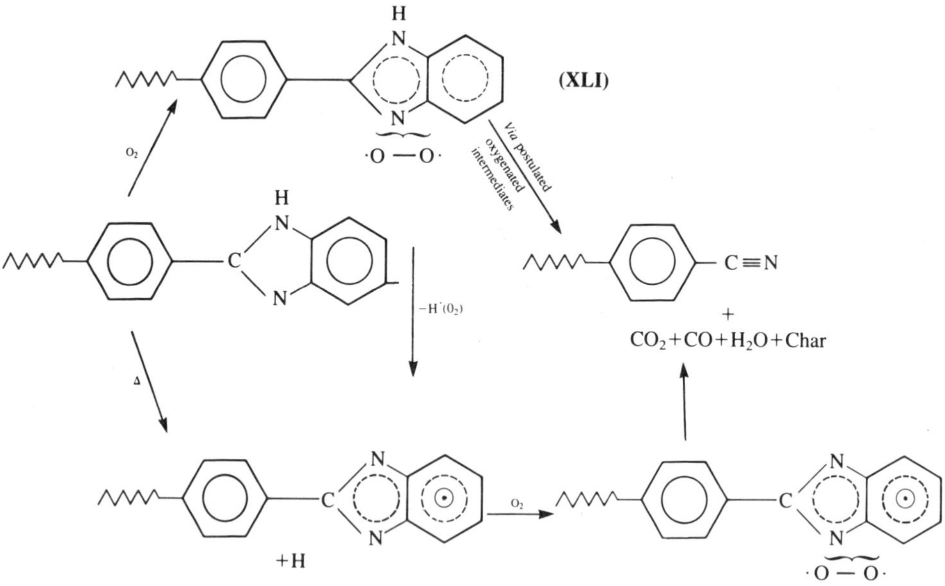

The spectrum of degradation products is adequately explained by either reaction sequence. Ultimately, however, strong support for the route involving free radical formation and subsequent oxidation came from an observation [393] that functional end-groups play a significant role in polybenzimidazole oxidative degradation. The carboxylic acid function, through decarboxylation, and the primary amine, through removal of amine hydrogen, were believed to serve as free radical sources for removal of the imidazole hydrogen atom and formation of the relatively stable benzimidazole radical.

Tsur *et al.*,[383] using thermogravimetry/mass spectrometry techniques, confirmed that during oxidative degradation of poly(2,2′-*m*-phenylene-5,5′-bibenzimidazole) maximum weight loss corresponds to a maximum in the production of volatile pyrolysis products (CO_2 and H_2O).

Soviet workers have also examined[274,378] the thermo-oxidative degradation of a series of polybenzimidazoles (**XXVIII,** R = direct bond, O, CH_2; R′ = *m*-C_6H_4, *p*-C_6H_4) and their N-phenylated analogs. These studies confirmed the greater long-term oxidative stability of N-phenylated materials and led to the suggestion that oxygen participates mainly in the oxidation of secondary products of degradation. Belyakov *et al.*[394] have also compared the thermal and thermo-oxidative stability of a number of heteroaromatic polymers —polypyromellitimides, polybenzimidazoles, polyimidazopyrrolones, polybenzoxazoles, and poly-1,3,4-oxadiazoles—containing various in-chain stable linking units

Using kinetic data based on weight loss and production of CO_2 and CO, they concluded that at lower temperatures (350°C) the polyoxadiazoles are the most resistant to thermal oxidation, while at higher temperatures (>400°C) the polyimides and polybenzoxazoles are most stable. Introduction of the electron-withdrawing SO_2 unit into the chain results in increases in stability of all five heterocyclic systems.

Elevated Temperature Properties

There are four applications for PBIs that have received a degree of commercial or semicommercial development. These are laminates/filament winding resins, adhesives, fibers, and foams. All of these applications have used the poly(2,2′-*m*-phenylene-5,5′-bibenzimidazole) (**XXVIII;** R = direct bond, R′ = *m*-C_6H_4) system. The production of laminating resins and adhesives is

TABLE 5.38. Low Temperature Strength of PBI Resin (Reference 395)

Test temperature (°C)	Adhesive tensile shear strength[a] (MPa)	Laminates		
		Flexural strength[b] (MPa)	Modulus[b] (GPa)	Impact strength[c] cN/m Notch
23	26.2	689.5	28.3	17
−170	33.2	1248	48.3	—
−190	—	—	—	17
−252	39.2	1213.5	45.5	—

[a]17-7PH stainless steel adherends.
[b]AF-994 fabric, HTS finish, 1581 style.
[c]Molding.

very similar. They are manufactured using essentially the method of processing described earlier (p.259), which involves the formation of a low-molecular-weight soluble/tractable prepolymer that can readily be applied from solution or melt onto the fibrous reinforcement or cloth support. Cure (solid-state polymerization) of the prepreg occurs at high temperature in the laminate or in-situ in the adhesive bond. As a result of its pioneering work, sponsored by the U.S.A.F. (AFML), the Whittaker Corporation (Narmco Division) produced the Imidite range of resins, which has formed the basis of both laminating and adhesive formulations.

A feature of the polymer (**XXVIII**; R = direct bond, R′ = m-C$_6$H$_4$) which is of major importance both in laminate and adhesive applications is that, despite a Tg > 400°C, the resin remains tough[395] throughout a range that extends down to cryogenic temperatures (Table 5.38).

It has been suggested that features inherent in the polymer such as low density and strong interchain forces (H-bonding), coupled with low-temperature relaxation mechanisms, are important factors governing the observed properties.

Composites

The commercial glass fiber-reinforced laminate made from a resin system based on poly(2,2′-m-phenylene-5,5′-bibenzimidazole) has been designated[396–399] Imidite 1850; other development codes are AF-R-100 and AF-R-151. Some mention has been made[400,401] of the reinforcement of PBI resin with carbon or other high strength fibers or whiskers. However, the main development of the system took place prior to the proliferation of work on high strength/high modulus carbon and organic fibers—an area of fundamental importance in the development of polyimide laminating materials—and the vast majority of reports refer to glass reinforcement.

Although PBI laminating resins have been produced by solution (DMAC or NMP) coating methods, the deleterious effects of combined volatiles and solvent on the quality of the laminates has prompted the more general use of hot-melt coating techniques. Even so, considerable problems have arisen in optimizing laminating procedures and the high (371°C) cure temperatures have caused difficulties with tooling materials and release agents. Most importantly, despite the absence of solvent, the necessary use of prepolymer in the hot-melt process has resulted in high void levels due to the entrapment of volatiles (phenol and water). The deleterious effects of these volatiles on laminate properties have been reduced by surface treatment of the glass. Although very reasonable flexural strengths at ambient and high temperatures have been achieved, the initial composites had unacceptably low (15–20%) resin contents. To improve on this, processing conditions were developed[399,400,402] involving an initial cure of 3 min at 371°C followed by 3 hours at 371°C and 1.4 MPa pressure, coupled with effective venting of volatiles using a semipermeable barrier/breather of Teflon/glass fabric, which retains 30–40% resin in the composite. This is followed by a post-cure under inert conditions of 24 hours each at 315, 345, and 371°C and 8 hours at 400, 425, and 450°C. These procedures have been followed for both press and autoclave cures and have been developed for commercial production of Imidite 1850 laminates.

Excellent room-temperature strengths have been achieved[399–404] with these PBI laminates: values of 650–800 MPa, 380–450 MPa, and 520–600 MPa, for flexural, compressive, and tensile strengths, respectively, and 30–40 GPa for all three moduli. Reference has already been made[395] to the surprisingly good cryogenic properties of PBI laminates. Very impressive strengths and moduli are also exhibited at high temperatures on short-time aging (Figure 5.37). Indeed, of the high-temperature polymers the PBIs together with the polybenzothiazoles probably show the greatest strengths over short periods at elevated temperatures. With longer time aging, however, a marked deterioration in properties is observed (Figures 5.38–5.41), due in large part to the oxidative degradation of the base PBI resin. Although the PBI glass laminates have considerable advantages in strength retention over some other heat-resistant laminates (e.g., phenolics)[405] this loss in long-term strength is very marked in comparison with the main competitor, the polyimides (Table 5.39).[406] Spain and Ray[373] have suggested maximum temperature capabilities for PI, PBT, and PBI glass laminates of 377, 538, and > 650°C, respectively, for 10 minutes at temperature. After exposure for longer periods (200 hours), the order—357, 340, and 315°C—is reversed. This short-time high temperature advantage of the PBI laminates almost certainly prompted their development for use in ablative heat shields.[401,407]

Some work[399] has been reported on filament winding with PBI resin using hot-melt and solution-coated roving, and data for standard NOL test rings are given in Table 5.40. Unlike press and autoclave cure of laminates, filament-wound materials must be cured and post-cured in an inert atmosphere. Values for both hoop tensile and horizontal shear strengths as a function of temperature and

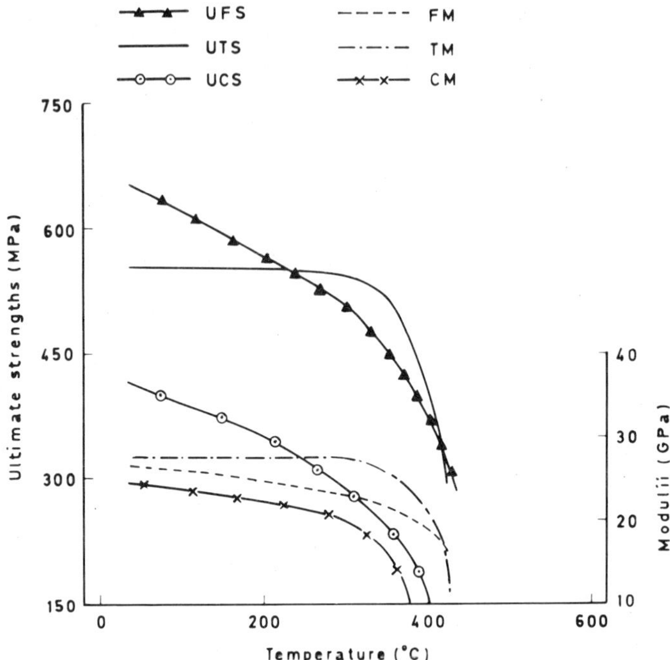

Figure 5.37. Effect of temperature on the levels of ultimate strengths [flexural (UFS), tensile (UTS), compressive (UCS)] and moduli [flexural (FM), tensile (TM), compressive (CM)] for Imidite 1850 laminates. Reference 397.

time for several high-temperature systems are shown in Table 5.41. Once again, the high initial strengths of PBI composites at elevated temperatures are rapidly lost on long-term aging, a situation not observed with polyimide materials. Electrical properties, substantially the same as those of competitive high-temperature materials—relatively flat loss tangent and dielectric constant curves versus temperature—have been obtained[403,404] for PBI laminates subjected both to short-time exposure at temperatures up to 600°C and for longer times (300 hours) to 316°C. There is no evidence, at least after these relatively short-term exposures, of a similar degradation of electrical properties on aging at high temperature as is apparent in PBIs mechanical properties.

Imidite 1850 materials suitable for laminate production were supplied by Narmco in the form of "B-Stage" (partially cured) glass fabric prepreg. In order to prevent hydrolytic degradation and/or premature gelation the fabric had to be transported and stored in tightly sealed, moisture-proof packing under refrigerated conditions. The problems associated with the marketing of this prepreg material, together with the many difficulties encountered in fabricating laminates with reproducible properties, were ultimately significant factors in preventing PBI materials from making a lasting impact in the advanced materials field.

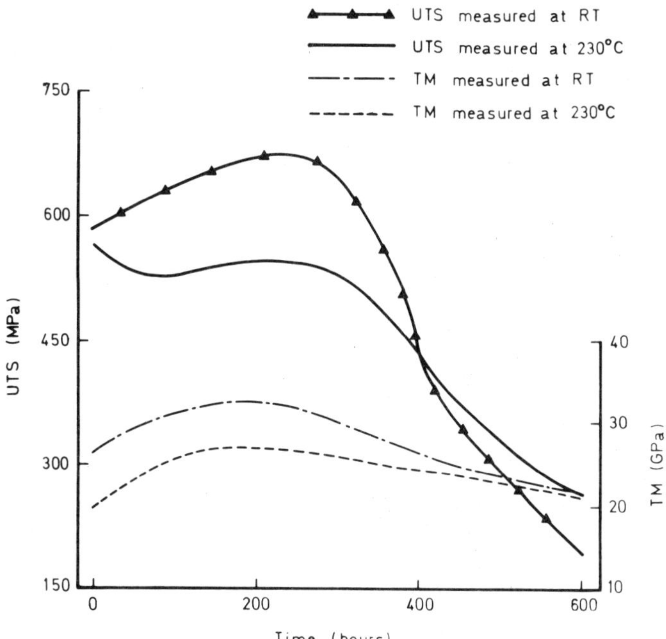

Figure 5.38. *Ultimate tensile strength (UTS) and tensile modulus (TM) of Imidite 1850 as a function of time at 230°C. Reference 399.*

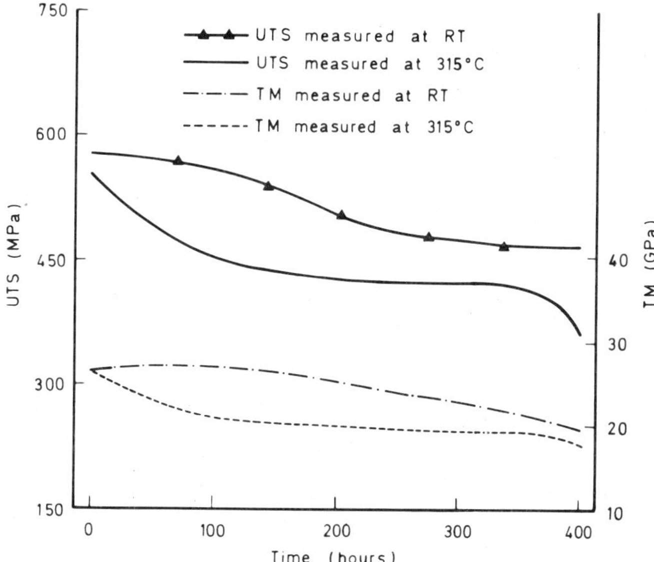

Figure 5.39. *Ultimate tensile strength (UTS) and tensile modulus (TM) of Imidite 1850 as a function of time at 315°C. Reference 399.*

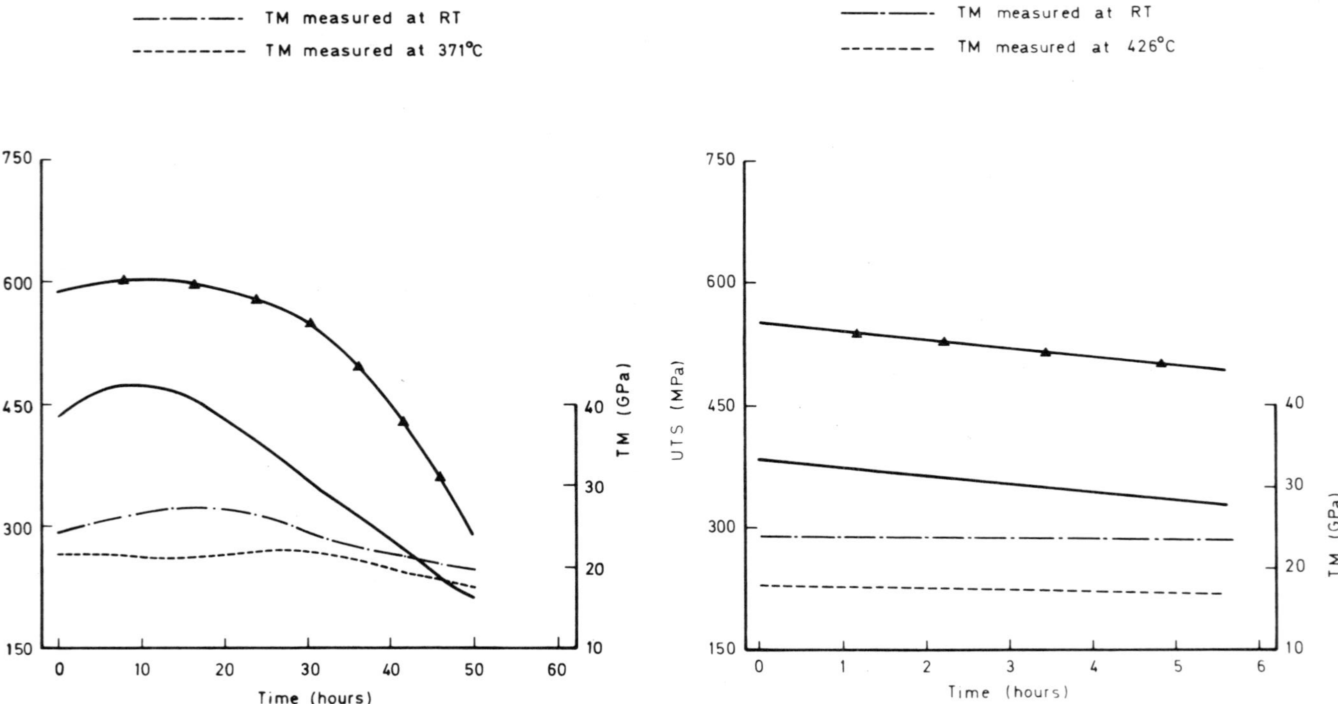

Figure 5.40. Ultimate tensile strength (UTS) and tensile modulus (TM) of Imidite 1850 as a function of time at 371°C. Reference 399.

Figure 5.41. Ultimate tensile strength (UTS) and tensile modulus (TM) of Imidite 1850 as a function of time at 426°C. Reference 399.

TABLE 5.39. *Flexural Strength of High-Temperature Resin/Glass Laminates (Reference 406)*

Test temp (°C)	Aging conditions	Flexural strength (MPa)				
		Polyimide	Amide–imide copolymer	Epoxy	Silicone	PBI
25	None	300	310	483	269	780
25	2 hours in boiling water	290	—	—	—	513
300	0.5 hours at 300°C	228	152	152	138	514
300	100 hours at 300°C	217	152	145	131	510
300	200 hours at 300°C	192	145	138	131	345
300	500 hours at 300°C	148	124	131	117	104
300	1000 hours at 300°C	114	55	—	—	55
300	1500 hours at 300°C	84	0.0	—	—	—
300	2000 hours at 300°C	95	0.0	—	—	—
320	0.5 hours at 320°C	179	128	—	—	773
320	100 hours at 320°C	198	152	—	—	324
320	200 hours at 320°C	163	124	—	—	83
320	500 hours at 320°C	94	69	—	—	0.0
320	1000 hours at 320°C	20	0.0	—	—	—
360	0.5 hours at 360°C	108	—	—	—	615
360	16 hours at 360°C	128	—	—	—	83
360	34 hours at 360°C	90	—	—	—	43
425	0.5 hours at 425°C	108	—	—	—	461

TABLE 5.40. *Mechanical Strengths of Filament-Wound NOL Rings of S-994 PBI Glass Fiber Composites (Reference 399)*

Coating technique	Split disk tension (MPa)		RT Tensile modulus (GPa)
	RT	371°C	
Solution	889	698	59
Hot-melt	1289	961	70

TABLE 5.41. *Hoop Tensile and Horizontal Shear Strengths of Filament-Wound Glass Composites (Reference 406)*

Resin (Content %)	Hoop tensile strength (MPa) at		Horizontal shear strength (MPa) at			
	30°C	300°C (0.5 hours)	−30°C	−130°C (0.5 hours)	300°C (0.5 hours)	360°C (0.5 hours)
Polyimide (16.0)	1275	1183	31.7	25.5	13.7	6.8
Polybenzimidazole (13.6)	1103	827	42.0	—	—	9.2
Epoxy (19.0)	1181	769	53.8	37.6	6.9	1.1
Silicone (24.2)	964	618	21.4	9.6	3.2	2.6

Adhesives

As with their laminating resin, Narmco's Commercial PBI adhesive system (Imidite 850)[408] was made from polymer **XXVIII** (R = direct bond, R′ = m-C_6H_4). Soviet workers have reported[409] that polymer **XXV** (n = 8, R = direct bond) can also be used effectively as a film adhesive for bonding various metals, e.g., Stainless Steel and Duralumin.

Original U.S.A.F.-sponsored work examined[404] a number of possible high-temperature adhesives based on DAB and aromatic diphenyl esters. The relative thermoplasticity of the system made from DAB and diphenylisophthalate (AF-A-100) is decreased (see Figure 5.27) by using[410] blends of diphenylisophthalate and terephthalate (AF-R-121 and AF-R-131). The copious evolution of volatiles (phenol and water) from the low-molecular-weight prepolymer was as great a problem in adhesive technology (causing voidy glue-lines) as in the corresponding laminates. By using formulations based on DAB and isophthalamides the level of volatiles can be considerably reduced—to as low as 4% in a well-advanced prepolymer—but the high vapor pressure of the ammonia evolved (204 atm at 370°C) compared with 14 atm for phenol and water largely negates any benefits. Despite the volume of research and development devoted to alternative formulations,[410–414] Narmco's Imidite 850 adhesive was, therefore, based on the development of the AF-A-100 system.

In many ways, PBI adhesives have proved to have a more complex but more flexible technology than the comparable laminating systems. They can be used in film, solution, or tape (glass fiber substrate) form.[415] Imidite 850 itself was supplied as a resin on a glass cloth (112-style) carrier. Resin impregnation of the carrier involved either solution coating using pyridine as solvent or a hot-melt process, which resulted in higher solids contents. The highest strengths are obtained when supported film is used in conjunction with a primer, preferably a polyimide-based material. Incorporation of aluminum, or in certain cases amorphous boron powder, as filler improves the properties of the adhesive and, as in other high-temperature systems, addition of inorganic arsenic compounds such as arsenic thioarsenate (As_2S_4) substantially reduces the oxidative degradation of these adhesives on steel adherends.[410] The enhanced degradation of PBI materials in contact with steel surfaces has been studied by Flom *et al.*[416] Decomposition profiles (using mass spectrometry) indicate that the acceleration of polymer degradation by iron is due to oxidative processes involving iron oxides rather than to catalytic effect of the metal. It was shown that aluminum, vapor deposited onto the steel surface, eliminates the degradative effects of the iron oxides.

Imidite 850 and related PBI adhesives were investigated[410–413,417–419] with stainless steel (17-7 PH), titanium (Ti-8Al-1Mo-1V, Ti-6Al-4V), aluminum alloy (2219-T81), and beryllium adherends. Flat plate and honeycomb[420] structures were examined. With PBI adhesives, as with other high-temperature adhe-

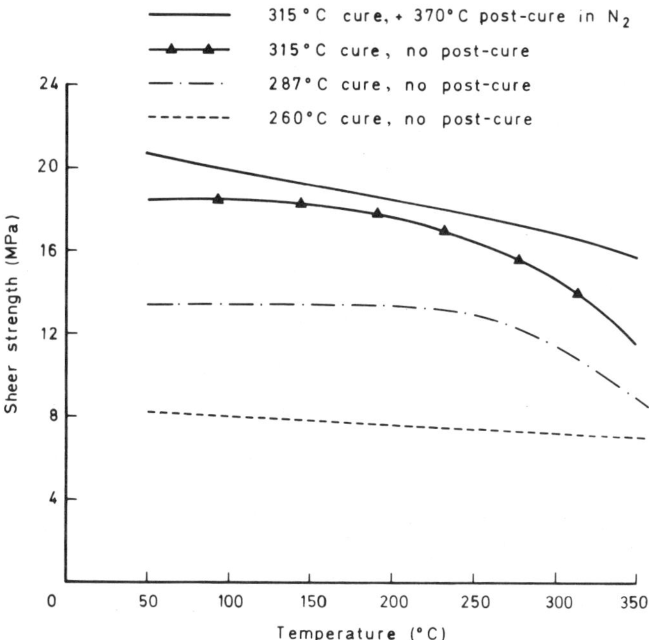

Figure 5.42.   Effect of cure conditions on the performance of Imidite 850. Reference 397.

sive systems, surface preparation of the adherends is an important factor in performance, particularly in long-term aging. Considerable efforts have been made to match surface preparation (sand blasting, solvent degreasing, acid etching) to the metallic adherend. PBI adhesives require high curing temperatures (1 hour at 220°C followed by 1 hour at 315°C) under pressure (1.4 MPa) and, except for the titanium alloys, this is ideally followed by a post-cure under inert conditions (1 hour at 400°C). Changes in cure cycle can result in substantial variations in the performance of PBI-bonded systems (Figure 5.42).[397] Hill[412] has suggested that in bonding stainless steel as little as 70 kPa of applied pressure is required for small area specimens. Adhesive testing of PBI materials has included both metal-to-metal and honeycomb sandwich structures, and this has been carried out under a range of temperature and aging conditions. Specific test procedures have included lap shear, T-peel and climbing drum peel, edgewise compression, and flatwise tensile.

Like the cast resins and laminates, PBI adhesive formulations are capable of functioning for short periods at high temperatures, but properties fall off fairly rapidly on more prolonged aging. Comparisons of the shear strengths of titanium joints as a function of temperature (10 min exposure at each temperature) and as a function of time of aging in air at 260°C of three advanced adhesives, including PBI, are made in Figures 5.43 and 5.44.[411] The elevated temperature perform-

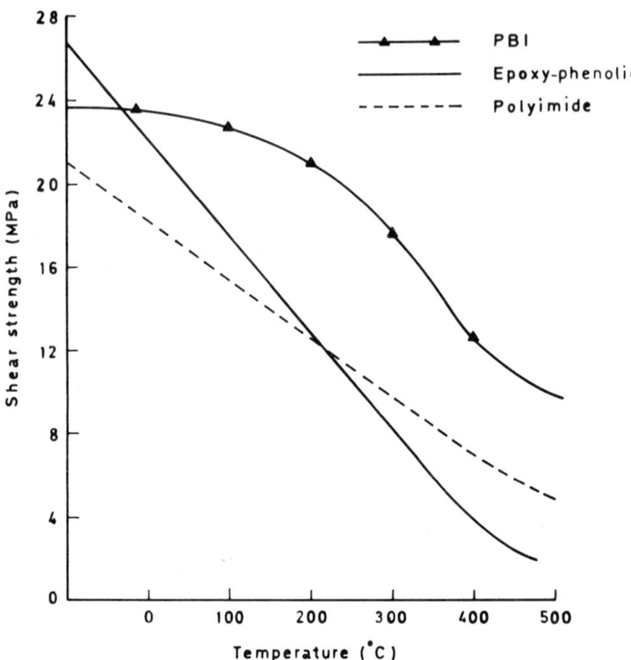

Figure 5.43.   Effect of temperature on the shear strength of high-temperature adhesives. Reference 411.

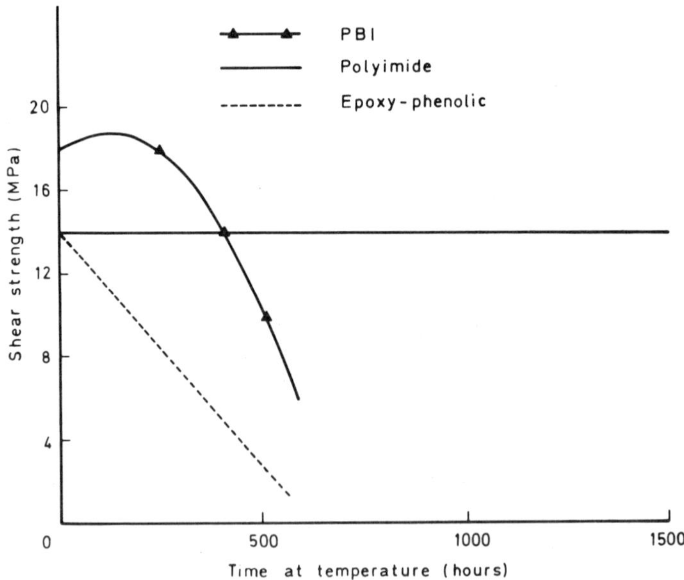

Figure 5.44.   Effect of aging in air at 260°C on the shear strength of high-temperature adhesives. Reference 411.

TABLE 5.42. *Lap Shear Strengths of PBI (AF-A-121)-Bonded Titanium and Stainless Steel Adherends after Short-Term Exposure to Elevated Temperatures (Reference 410)*

| Test temp (°C) | Joint strength (MPa) | | | | | |
| | 10 min soak at temp | | | 1 hour soak at temp | | |
	Ti-6Al-4V	Ti-8Al-1Mo-1V	17-7 PH St St	Ti-6Al-4V	Ti-8Al-1Mo-1V	17-7 PH St St
23	26.6	23.0	25.3	—	—	—
149	26.6	19.8	20.8	22.0	21.4	22.0
204	27.5	21.6	21.6	23.2	22.4	23.3
260	25.0	19.8	19.9	22.8	21.6	21.1
316	18.3	19.5	18.3	18.9	18.1	20.1
372	15.0	12.2	18.1	16.8	10.3	17.8
426	10.2	8.2	13.1	10.6	7.2	13.1
483	8.3	7.8	10.9	10.1	4.4	8.5
583	9.4	9.5	9.6	3.3	1.4	3.7

ance of PBI-bonded titanium and stainless steel reflects closely this situation.[410–412] In all cases (Tables 5.42 and 5.43), while initial lap shear strengths are high, particularly for stainless steel joints, after short-time (10 min) exposure at temperature, there are significant reductions in tensile shear strengths with time at temperature. This reduction in strength is most marked at temperatures above 260°C and has been directly attributed to thermo-oxidative degradation of the adhesive.

In the application of PBI adhesives to the bonding of beryllium structures, it has been noted that limited aging (24 hours at 370°C) results in higher strength retention[410–412,418–419] at exposure temperature than is the case with either stainless steel or titanium. Even so, practically total strength loss is observed (Figure 5.45) after 50 hours at 316°C. Although the strength retention of stainless

TABLE 5.43. *Lap Shear Strengths of PBI (AF-A-121)-Bonded Titanium and Stainless Steel Adherends after Long-Term Exposure to Elevated Temperature (Reference 410)*

| Time at temp (hours) | Joint strength (MPa) | | | | | |
| | Aging at 316°C | | | Aging at 360°C | | |
	Ti-6Al-4V	Ti-8Al-1Mo-1V	17-7 PH Stainless steel	Ti-6Al-4V	Ti-8Al-1Mo-1V	17-7 PH Stainless steel
2	—	—	—	15.1	13.4	18.1
5	—	—	—	14.5	13.5	17.1
10	21.4	21.2	21.2	13.3	13.1	15.2
15	—	—	—	11.7	11.3	13.6
24	—	—	—	11.4	8.4	10.7
25	19.2	17.9	20.6	—	—	—
50	10.3	12.8	17.7	—	—	—
100	7.6	7.2	9.7	—	—	—
150	10.0	7.2	2.3	—	—	—
200	2.4	3.0	0.0	—	—	—

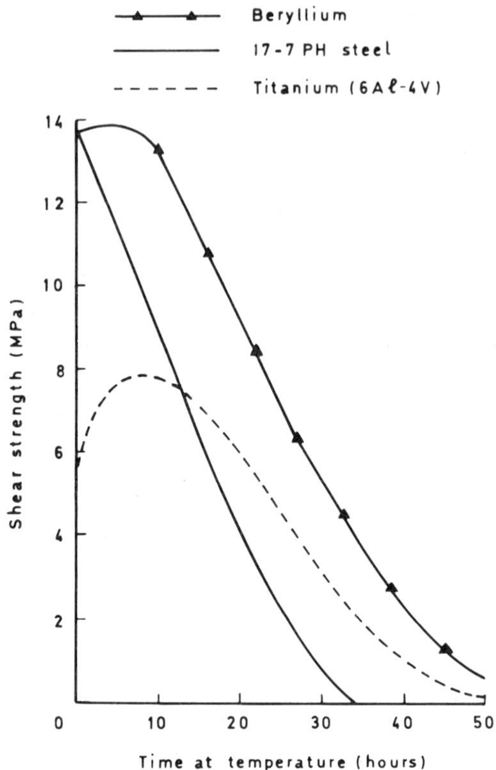

Figure 5.45.   Effect of aging at 316°C on lap shear strength of metals bonded with Imidite 850. Reference 412.

steel-bonded joints reveals (Table 5.44) PBIs' aging properties at high temperature to be superior to that of commercial epoxy-phenolic material, it is significantly inferior to that of a polyimide adhesive.[421] A comparison of the environmental resistance of these bonded joints reveals (Table 5.45) the order of stability to be epoxy-phenolic > polyimide > PBI. Undoubtedly the most severe environmental effects—reduction in room temperature strength of 60%—are observed with PBI-bonded titanium alloy joints on exposure to high humidity for 30 days. It is important, however, to reiterate the excellent performance of PBI-bonded titanium and stainless steel adherends (lap shear and peel strengths) in the cryogenic region. In particular, it has been shown that fatigue strengths at −195°C are superior to those at room temperature.[395]

Fibers

A substantial part of the early research effort into the application of aromatic heterocyclic polymers involved their examination as high-temperature fibers. In

TABLE 5.44. Comparison of Lap Shear Strengths at Elevated Temperatures of Stainless Steel Bonded with High Temperature Adhesives (Reference 421)

Conditions	Joint strength (MPa)		
	Polyimide	PBI	Epoxy-phenolic
at 23°C	19.3	27.6	> 27.6
at 288°C after			
1 hour at 288°C	11.0	24.1	14.5
100 hours at 288°C	13.1	17.2	0
200 hours at 288°C	9.6	7.6	0
500 hours at 288°C	9.0	4.8	—
1000 hours at 288°	9.6	0	—
at 371°C after			
1 hour at 371°C	9.0	13.8	6.2
10 hours at 371°C	7.6	13.8	0
24 hours at 371°C	8.3	6.2	0
60 hours at 371°C	7.6	0	—

particular, assessments were made of the polyimides, polybenzimidazoles, poly-1,3,4-oxadiazoles, poly-N-phenyltriazoles, poly(imidazopyrrolones), as well as a number of heterocyclic-heterocyclic and heterocyclic-amide copolymers. In the Soviet Union several of these systems have been used in a number of advanced applications, but in the West only PBI, and to a limited extent BBB, fibers have been developed to a semicommercial position. Undoubtedly the successful introduction of the high-temperature aromatic polyamide fibers—initially Nomex (HT-1) and subsequently the Kevlar (Aramid) range—has reduced the development potential of the heterocyclic systems.

The original work (USAF-AFML) on PBI fibers from poly(2,2'-*m*-phenylene-5,5'-bibenzimidazole) established the basic technology of fabrication.[422–424] Results from high-temperature testing of these fibers reflect base

TABLE 5.45. Environmental and Chemical Resistance of Stainless Steel Adherends Bonded with High Temperature Adhesives (Reference 421)

Conditions	Joint strength (MPa)		
	Polyimide	PBI	Epoxy-phenolic
at 23°C after			
30 days in RT water	14.5	13.1	17.9
30 days 43°C/100% RH	13.1	13.1	20.0
30 days 40°C/5% salt spray	13.8	13.1	
7 days JP-4 jet fuel	15.9	26.9	22.1
7 days isopropyl alcohol	17.9	22.8	> 27.6
7 days hydraulic oil	15.9	26.9	> 27.6
7 days oil (70/30 isooctane/ toluene)	20.0	27.6	> 27.6

resin properties in that they show somewhat lower thermal stability (in air) at 300°C than aromatic polyamides. Stability to long-term aging at temperatures above 300°C, though better than that of aromatic amides, is inferior to that of some of the other experimental heteroaromatic fibers referred to above. Furthermore, PBI fibers are subject to photo-oxidative degradation under ambient conditions. Despite these drawbacks, PBI fibers exhibit excellent textile properties, and their tolerance to short-time exposure at extreme temperatures while retaining a large proportion of their initial mechanical strength has encouraged a continued examination of the system. It was envisaged, for example, that these properties would be ideal in military applications such as decelerators or brake parachutes.

From the mid-1960s onwards, further major development of PBI fiber technology has been continued jointly by the Celanese Corporation and AFML. They have reexamined[256,257,425] processing variables and scale-up polymerization conditions so as to achieve the maximum polymer molecular weight (inherent viscosities $\geq$ 1.0 dl/g) commensurate with solubility in solvents such as DMAC, NMP, and DMSO. Detailed examinations have been made of the properties of these solutions. Significant improvements have been made in the processes and conditions for wet and dry spinning from these and even more exotic solvents.[256,426] Postspinning processes of orientation and crystallization of the predominantly amorphous fiber are most important[256] in the achievement of optimum properties, as are processes for stabilizing the fibers against ultraviolet (uv) degradation.[427] Fibers have been dyed with conventional dyestuffs[428] using a variety of techniques and this has also provided protection against photo-oxidation.

Continuous filament or staple forms of PBI fiber produced by the methods described have properties (Table 5.46) typical for a working fiber.[256,426,429,430] The effect of temperature on the tenacity of oriented versus crystalline fibers is shown in Figure 5.46.[422] In contrast to most other fibers PBI has a high moisture regain [>13%; cf., 0% (glass), 2% (acrylics), 5% (Nomex)] at 65% RH and 21°C; this has considerable significance in respect to the wearability of PBI fabrics.[431–433]

TABLE 5.46. PBI Fiber Properties (Reference 256)

	Tenacity (cN/tex)	Elongation (%)	Modulus (N/tex)	Yarn tex
Undrawn	15[a]–20[b]	115[a]–138[b]	3.9[a]	44
Drawn	43.3[a]–51.3[b]	24[a]–19[b]	9.3[a]–12.4[b]	22

[a] η_{inh} = 0.72 dl/g
[b] η_{inh} = 1.2 dl/g

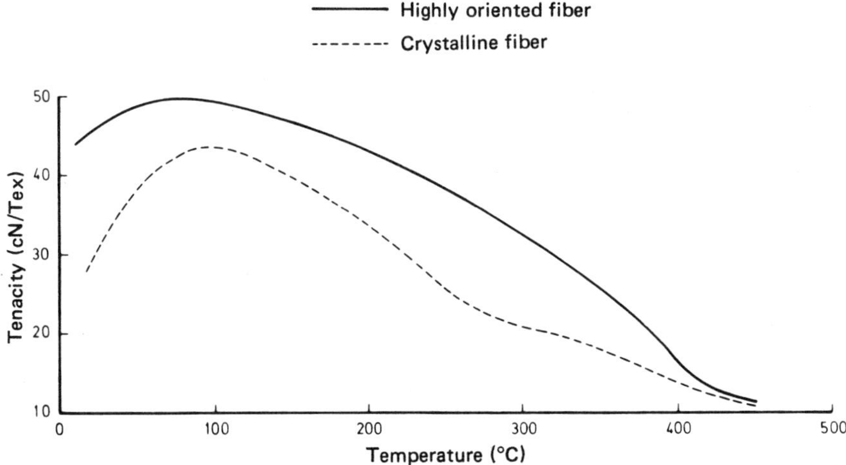

Figure 5.46. Variation of tenacity of PBI fiber with temperature. Reference 422.

Spun yarn has been woven into fabrics; such materials have shown considerably superior dimensional stability and abrasion resistance than more conventional glass, Aramid, and Kynol high-temperature fabrics.[434]

An application for advanced fibers that increased in importance during the later development phase of PBI was that of nonflammable materials suitable for air and spacecraft crew protection. Probably the most significant property of PBI fiber has turned out to be its almost complete nonflammability in air.[246,429,430,433,435,436] A quantitative assessment of fiber or fabric flammability is given by the Critical Oxygen Concentration (COC) or Limiting Oxygen Index (LOI), which are measurements of the concentration of oxygen necessary before the yarn or fabric will ignite.

Comparison of relative values (Table 5.47) for a number of materials indicates PBI to be outstanding in this respect.[437] Moreover, PBI flight suits of comparable weight provide significantly greater protection to high intensity fuel-fires than do aramid materials.[246,433,435,436] Although nonflammability is a key factor, low smoke generation at high heat flux and nontoxic gas evolution—at least up to 600°C—are also of primary importance in the application of PBIs in the hazardous environments of air- or spacecraft.[436]

PBI fiber in the form of continuous yarn has been converted into high strength/high modulus carbon fibers.[438–441] Independent workers have reported the rapid graphitization of precursor fibers that had been stabilized by air oxidation. In its ease of conversion to carbon fibers (in air), PBI exhibits unique properties. The carbon fibers produced by this process are highly oriented with tensile strengths of 1.1–1.7 GPa and initial tensile moduli of 296.5–372.3 GPa.

TABLE 5.47. Limiting Oxygen Index of Various Fabrics (Reference 433)

Material	LOI [BW][a] (% O_2)	LOI [TW][b] (% O_2)
Polyoxymethylene	12.2	13.6
Cotton	12.8	16.8
Epoxy	13.2	19.0
Cellulose acetate	14.6	15.8
Polypropylene	15.3	15.8
Polyacrylonitrile	15.3	17.0
Poly(ethyl-co-methacrylate)	15.4	17.6
Polystyrene	15.4	17.8
Poly(ethylene terephthalate)	15.5	18.5
Nylon 6, 6	15.5	20.5
Nomex	17.0	26.0
Durette 400X (aromatic polyamide)	18.0	36.0
Kynol (cross-linked phenolic)	18.5	29.0
Kapton	18.5	35.0
PVC (unfilled)	19.5	32.0
PBI	28.5	48.0

[a]Ignition from the bottom.
[b]Ignition from the top.

Foams

Low-molecular-weight prepolymers, typical of those used in adhesives and laminates, based on the system (**XXVII**, R = direct bond, R^1 = m-C_6H_4) have been converted[442–445] into high-temperature PBI foams with a performance capability above 500°C. Foam formation from the low-melting prepolymer is carried out under nitrogen (> 103 kPa) with the temperature rising from 180–280°C at 2°C per minute, followed by a cure at 470–530°C. Blown-foams having densities in the range 25–80 Kg/m^3 and syntactic foams produced from blends of prepolymer and fibrous fillers (carbon or PBI) or microspheres, with densities of 400–500 Kg/m^3, have been produced by these processes. Syntactic materials have been marketed as Imidite SA and Imidite PC by the Whittaker Corporation.[444]

Mechanical properties of syntactic foams are given[443] in Table 5.48. The effect of temperature on compressive properties and dimensional recovery of low density PBI foams is shown in Figures 5.47 and 5.48.[442]

The thermal performance of PBI foams has been determined by measuring the thermal efficiency, thermal conductivity, flame spread, ease of ignition, and smoke evolution. Both thermal efficiency and conductivity are comparable with, or better than, those of other organic foams such as polyimide or silicone elastomer.[442,445] Like asbestos, PBI foam exhibits zero flame spread, and lev-

TABLE 5.48. *Mechanical Properties of PBI Syntactic Foam[a] (Reference 443)*

Test	Result (MPa)		
	at 24° C	316°C	540°C
Compressive strength	26.2	20.5	5.4
Compressive modulus	1662	248.2	—
Tensile strength	9.0	8.8	2.0
Tensile modulus	752	730.8	—

[a] ρ = 400–500 Kg/m^3.

els of flammability (LOI) and smoke evolution that are similar to those of PBI fiber. PBI foam is effectively a nonflammable material.

Although PBI-cured foams can be easily machined into complex shapes, ultrastrong, rigid moldings have also been produced by shredding the foams and molding the shreds into the desired shape using a bis(maleimide) adhesive.

PBI syntactic foams (containing chopped carbon or silica fiber and phenolic or silica microspheres) have been examined as candidate ablative heat shield ma-

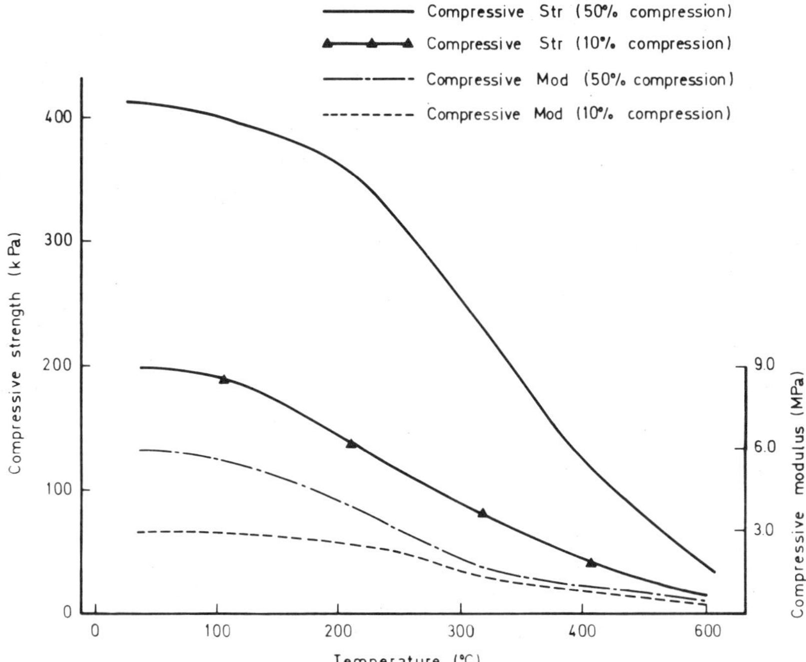

Figure 5.47. *Effect of temperature on the compressive strength and modulus of PBI low density (43 Kg/m^3) foam. Reference 442.*

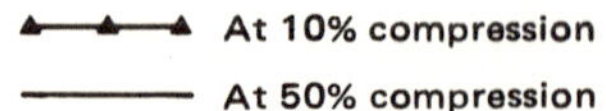

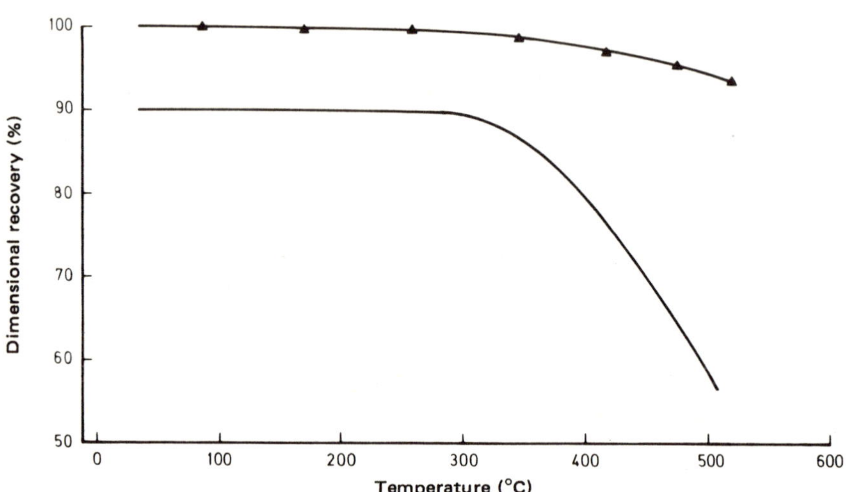

Figure 5.48.   Effect of temperature on the dimensional recovery of PBI low density (43 Kg/m³) foam. Reference 442.

terials. Under conventional ablative tests (heating rate 800 W/cm^2 and surface pressure 2.5 atm) it has been shown[406] that the linear PBI composite foam has a lower mass-loss rate than comparable ($\rho = 500$ kg/m^3) phenolic-nylon material. Using thermally cross-linked (prolonged heating at 700°C) or chemically cross-linked material, ablation performance is considerably improved.

SUMMARY

1. Despite the high level of thermal and chemical stability exhibited by the majority of the heteroaromatic polymers, only the polyimides, and to a very limited extent the polybenzimidazoles, have had a significant commercial impact.

2. It is suggested that, in addition to their excellent thermal and environmental stability, the primary reason for the commercial success of polyimides has been the wide choice of viable synthetic routes. The nature and diversity of these routes has resulted in the systematic development of polyimides from largely condensation (thermoset) materials with limited application to condensation (thermoplastic) and addition systems. These latter polymers have shown considerably greater potential in the field of structural composites. In contrast to

the position with the polyimides, only one effective synthetic route has been found for the polybenzimidazoles.

3. Overall, condensation (thermoset) polyimides have provided the highest levels of thermal and thermo-oxidative stability in experimental polymers, resins, and fabricated articles. In general, as the systems have changed from those highly condensed, wholly aromatic polymers to thermoplastic and addition types, so the stability has decreased. The latter materials, however, have offered considerable advantages in processability with, in many instances, only modest sacrifices in the stability of the end-product. Polybenzimidazoles, although showing comparable or even possibly slightly superior thermal stability in inert atmospheres to the polyimides, are more susceptible to thermo-oxidative degradation at temperatures above 300°C.

4. Applications of polyimides have ranged from laminating resins, moldings, molding powders, films, coatings, varnishes, adhesives, and foams to fibers. In contrast, although the polybenzimidazoles have been evaluated in many of the above applications, and laminating resin and adhesive formulations showed some commercial prospects in the late 1960s to early 1970s, only PBI fibers and possibly foams appear now to have limited commercial potential. It may well be that the very success of the polyimides has reduced the development potential of the less versatile polybenzimidazoles.

5. Polyimide materials have found most outlets in the aerospace, electronic, nuclear power, and automotive industries. Typical uses have included seals, bearings, gear wheels, piston and back-up rings, and disk brakes. In advanced gas turbine engines polyimide binders have been used in combination with graphite reinforcement in cowls, exhaust nozzle flaps, and compressor blades. Applications for composite structures using polyimide resins reinforced with carbon and Kevlar fibers have been reported in missiles and spacecraft. Electrical outlets have included insulators, radomes, printed circuit boards, slot liners, wire and cable insulation. As adhesives, polyimides have been used to bond titanium (metal-to-metal) and titanium-to-honeycomb.

Polybenzimidazole (PBI) fibers are now under limited evaluation in high-flame-resistant and low-smoke-toxicity fabrics for military and specialized civilian uses.

REFERENCES

1. J. P. Critchley, A review of the poly(azoles), in *Progress in Polymer Science*, A. D. Jenkins, (ed.), Pergamon Press Ltd., Oxford (1970), Vol. 2, Ch. 2.
2. E. I. du Pont de Nemours, French Patent No. 1,239,491 (1960); Australian Patent No. 58,424 (1960).
3. A. L. Endrey, Canadian Patent No. 659,328 (1963); U.S. Patent Nos. 3,179,631 and 3,179,633 (1965).

4. G. W. Bower and L. W. Frost, *J. Polym. Sci. A* **1,** 3135 (1963).

5. L. W. Frost and I. Kesse, *J. Appl. Polym. Sci.* **8,** 1039 (1964).

6. J. Idris Jones, F. W. Ochynski, and F. A. Rackley, *Chem. and Ind.* 1686 (1962).

7. T. M. Bogert and R. R. Renshaw, *J. Amer. Chem. Soc.* **30,** 1140 (1908).

8. J. D. Seddon, U.K. Patent No. 1,192,001 (1970); 1,228,364 (1971).

9. H. K. Reimschuessel, K. P. Klein, and G. J. Schmitt, *Macromolecules* **2,** 567 (1969).

10. G. Rabilloud, B. Sillion, and G. De Gaudemaris, French Patent No. 1,601,091 (1971).

11. F. P. Darmory, *Adhes. Age* **17,** (3) 22 (1974).

12. W. M. Edwards and I. M. Robinson, U.S. Patent No. 2,710,853 (1955); 2,867,609 (1959); 2,900,369 (1959).

13. I. Tobushi, K. Tayaki, and R. Oda, *J. Chem. Soc. (Japan) (Ind. Chem. Sec.)* **67,** 1081 (1964).

14. German (FDR), Ministry of Defence, U.K. Patent No. 1,076,947 (1967).

15. V. L. Bell and N. T. Wakelyn, *Diss. Abstr. Int. B* **35,** 2207 (1974).

16. A. V. Samokhvalov, K. N. Vlasova, L. A. Gardeeva, and L. S. Bublik, *Plast. Massy* **9,** 15 (1971).

17. J. P. Critchley, P. A. Grattan, M. A. White, and J. S. Pippett, *J. Polym. Sci. A 1* **10,** 1789 (1972).

18. C. E. Sroog, A. L. Endrey, S. V. Abramo, C. E. Berr, W. M. Edwards, and K. L. Olivier, *J. Polym. Sci. A* **3,** 1373 (1965); C. E. Sroog, Polyimides, in *Encyclopedia of Polymer Science and Technology,* H. F. Mark, N. G. Gaylord, and N. M. Bikales, (eds.), (1969), Vol. 11, p. 249.

19. M. Prince and I. Honyak, *J. Polym. Sci. B* **4,** 601, (1966).

20. R. A. Dine-Hart and W. W. Wright, *J. Appl. Polym. Sci.* **11,** 609 (1967).

21. P. Delvigs, Li-Chen Hsu, and T. Serafini, *J. Polym. Sci. B* **8,** 29 (1970).

22. M. L. Wallach, *J. Polym. Sci. A* **2,** 653 (1967).

23. A. L. Endrey, U.S. Patent No. 3,179,630 (1965).

24. J. A. Kreuz, U.S. Patent No. 3,541,057 (1970).

25. E. R. Hoegger, U.S. Patent No. 3,342,774 (1967).

26. R. J. Angels, U.S. Patent No. 3,282,898 (1966), 3,345,342 (1967).

27. J. A. Kruez, A. L. Endrey, F. P. Gray, and C. E. Sroog, *J. Polym. Sci. A* **1,** 2607 (1966).

28. V. V. Korshak, S. V. Vinogradova, Y. S. Vygod'skii, S. M. Pavlova, and L. V. Boyko, *Izvest. Akad. Nauk S.S.S.R. Ser. Khim.* **10,** 2267 (1967).

29. S. V. Vinogradova, G. L. Slonimskii, Y. S. Vygod'skii, A. A. Askadskii, A. I. Mzhel'skii, N. A. Churchina, and V. V. Korshak, *Polym. Sci. U.S.S.R A* **11,** 3098 (1969).

30. S. V. Vinogradova and Y. S. Vygod'skii, *Russ. Chem. Rev.* **42,** 551 (1973).

31. M. H. Leucheux and A. D. Bondert, *J. Polym. Sci.* **48,** 405 (1960).

32. W. Wrasidlo, P. Hergenrother, and H. Levine, *Polym. Preprints* **5,** 141 (1964).

33. A. Ya Ardashnikov, E. I. Kardash, B. V. Kotov, and A. N. Pravednikov, *Dokl. Akad. Nauk S.S.S.R.* **164,** 1293 (1966).

34. A. S. Teleshova, E. M. Teleshov, and A. N. Pravednikov, *Polym. Sci. U.S.S.R.* **13,** 2593 (1971).

35. E. Sacher and D. G. Sedor, *J. Polym. Sci.* **12,** 629 (1974).

36. Yu N. Sazanov, L. V. Krasilnikova, and L. M. Scherbavava, *Eur. Polym. J.* **11,** 801 (1975).

37. G. A. Lusheikin, V. V. Surova, B. L. Gringnt, M. L. Dobrokhotova, L. N. Emelyanova, and V. D. Voroblev, *Sov. Plast.* **8,** 31 (1972).

38. L. A. Lauis, M. I. Bessonov, and F. S. Florinskii, *Polym. Sci. U.S.S.R.* **13,** 2257 (1971).

39. E. Lavin, A. H. Markhart, and R. E. Kass, French Patent No. 1,360,488 (1963).

40. E. I. du Pont de Nemours, French Patent No. 1,399,077 (1965).

41. R. J. W. Reynolds and J. D. Seddon, *J. Polym. Sci.* **23,** 45 (1968).

42. Y. Imai, *J. Polym. Sci. B* **8,** 555 (1970).

43. Y. Imai and K. Kojima, *J. Polym. Sci. A* **1**, 2091 (1972).
44. E. M. Boldebuck and J. K. Klebe, U.S. Patent No. 3,303,157 (1967).
45. R. A. Dine-Hart and W. W. Wright, Reinforced Plastics Group Conference, New and Improved Resin Systems, London (1973), Paper 7.
46. D. R. Dixon, J. B. Rose, and C. N. Turton, U.S. Patent No. 3,832,330 (1974).
47. Societe Rhodiacata, Netherlands Patent No. 6,690,214 (1967).
48. R. A. Meyers, *J. Polym. Sci. A* **1**, 2757 (1969).
49. P. S. Carleton, W. J. Farrissey, and J. S. Rose, *J. Appl. Polym. Sci.* **16**, 2983 (1972).
50. W. J. Farrissey, L. M. Alberino, and A. A. R. Sayigh, *J. Elast. Plast.* **7**, 285 (1975).
51. W. M. Alvino and L. E. Edelman, *J. Appl. Polym. Sci.* **19**, 2961 (1975).
52. The Upjohn Company, U.S. Patent No. 3,708,458 (1973).
53. K. W. Rausch, W. J. Farrissey, and A. A. R. Sayigh, *Proc. 28th Ann. Tech. Conf. Reinf. Plast./Comp. Inst. Soc. Plast. Ind.* (1973), Sect. 11-E.
54. B. K. Onder, U.S. Patent No. 4,094,864 (1978).
55. S. L. Cooper, A. D. Mair, and A. V. Tobolsky, *Textile Res. J.* **35**, 1110 (1965).
56. R. E. Coulehan and T. L. Pickering, *Polym. Preprints* **12**, 305 (1971).
57. V. L. Bell, B. L. Stump, and H. Gager, *J. Polym. Sci., Polym. Chem. Ed.* **14**, 2275 (1976).
58. A. V. Sidorovich and Ye. V. Kushinskii, *Polym. Sci. U.S.S.R. A* **10**, 1627 (1968).
59. J. K. Gillham and H. C. Gillham, *Polym. Eng. Sci.* **13**, 447 (1973).
60. J. M. Barton and J. P. Critchley, *Polymer* **11**, 212 (1970).
61. G. M. Bower, J. H. Freeman, E. J. Traynor, L. W. Frost, H. A. Burgman, and C. R. Ruffing, *J. Polym. Sci. A 1* **6**, 877 (1968).
62. E. I. du Pont de Nemours, U.S. Patent No. 3,492,270 (1970); 3,533,997 (1970).
63. J. P. Critchley, U.K. Patent No. 1,373,904 (1974); 1,457,191 (1976); also unpublished work.
64. Dupont, Product Bulletin A-44627, -46069, -48710.
65. Anon., *Materials Eng.* **2**, 69 (1974).
66. J. F. Heacock and C. E. Berr, *SPE Trans.* **4**, 105 (1965).
67. B. W. Melvin and D. J. Parish, *Insulation* **9**, 120 (1965).
68. J. R. Cannizzaro, *Solid State Tech.* **11**, 31 (1969).
69. Anon., *Br. Plast.* **41**, 69 (February, 1968).
70. Dupont, Product Bulletin A-52349, -74598, -83708, E-06470, -17624, -19546.
71. Anon., *Plastics* **33**, 372 (1968).
72. J. D. Hensel, *Plast. Eng.* **33**, 20 (1977).
73. Dupont, Product Bulletin A-62113, -64675.
74. M. W. Westley, *19th Nat. SAMPE Symp.* **19**, 135 (1974).
75. L. E. Lorensen, *SAMPE Q.* **6**, 1 (1975).
76. The Monsanto Company, Skybond 700 Series, Tech. Bull. No. 6234-6239.
77. American Cyanamid (Bloomingdale Dept.), FM-34 Product Bulletin (1971).
78. R. A. Pike, R. C. Novak, and M. A. DeCrescente, *Proc. 30th Ann. Tech. Conf. Reinf. Plast./Comp. Inst. Soc. Plast. Ind.* (1975), Sect. 18-G.
79. J. L. Cotter, *Rev. High Temp. Mater.* **3** (4), 277 (1978).
80. Anon., *Mater. Eng.* **5**, 73 (1976).
81. F. P. Darmory, *5th Nat. SAMPE Tech. Conf.* **5**, 415 (1973).
82. Anon., *SAMPE J.* **10**, 120 (1974).
83. L. N. Phillips and K. F. Rogers, *Composites* **1**, 286 (1970).
84. Hexcel Aerospace Company, Data Bulletins 123, 9500, 9501, 9510, 9520, 3100.
85. F. P. Darmory and M. DiBenedetto, U.S. Patent No. 3,705,869 (1972); 3,748,338 (1973).
86. F. W. Harris, W. A. Feld, and L. H. Lanier, *J. Polym. Sci. B* **13**, 283 (1975).
87. G. I. Timofeyeva, S. A. Pavlova, G. M. Tseitlin, V. I. Azevov, V. I. Skoortsova, and V. V. Korshak, *Polym. Sci. U.S.S.R.* **13**, 2637 (1971).

88. D. E. Green, U.S. Patent No. 3,338,859 (1967).

89. W. A. Fessler, U.S. Patent No. 3,736,290 (1972).

90. N. A. Adrova, M. M. Koton, and L. K. Prokhorova, *Dokl. Akad. Nauk S.S.S.R.* **177,** 579 (1976).

91. A. Rie, French Patent No. 1,529,727 (1968).

92. J. A. Webster, J. M. Butler, and T. J. Morrow, *Polym. Preprints* **13,** 612 (1972).

93. F. E. Rogers, U.S. Patent No. 3,356,648 (1967).

94. H. H. Gibbs, *Proc. 28th Ann. Tech. Conf. Reinf. Plast./Comp. Inst. Soc. Plast. Ind.* (1973), Sect. 2-D.

95. H. H. Gibbs and C. V. Breder, *Polym. Preprints* **15,** 775 (1974).

96. H. H. Gibbs, *7th Nat. SAMPE Tech. Conf.* **7,** 244 (1975); *21st Nat. SAMPE Symp.* **21,** 592 (1976).

97. K. L. Olivier, U.S. Patent No. 3,234,181 (1966).

98. V. L. Bell, U.S. Patent No. 4,094,862 (1978).

99. L. Acle, U.S. Patent No. 3,726,834 (1973).

100. J. H. Bateman, W. Geresey, and D. S. Neiditch, *Org. Coat. and Plast. Chem. Preprints* **35,** (2) 77 (1975).

101. W. Collins and T. Villani, Paper presented at Conference on Thermosets versus High Performance Thermoplastics, New Haven, Connecticut, October, 1977.

102. J. Preston and W. B. Black, *Appl. Polym. Symp.* **21,** 61 (1973).

103. L. W. Frost, G. M. Bower, J. H. Freeman, H. A. Burgman, E. J. Traynor, and C. R. Ruffing, *J. Polym. Sci. A 1* **6,** 215 (1968).

104. J. Preston, W. DeWinter, and W. B. Black, *J. Polym. Sci. A 1* **10,** 1377 (1972).

105. W. Wrasidlo and J. M. Augl, *J. Polym. Sci. A 1* **7,** 321 (1969).

106. F. Hayano and K. Komoto, *J. Polym. Sci. A 1* **10,** 1263 (1972).

107. L. W. Frost and G. M. Bower, U.S. Patent No. 3,179,635 (1965).

108. F. F. Holub, U.S. Patent No. 3,277,043 (1966).

109. J. H. Incremona and D. Strugar, U.S. Patent No. 3,852,106 (1974).

110. M. W. Alvino and L. W. Frost, *J. Polym. Sci. A 1* **8,** 2209 (1971).

111. R. E. Van Strian and J. R. Erzner, U.S. Patent No. 3,377,321 (1968).

112. J. G. Wirth and D. R. Heath, U.S. Patent No. 3,838,097 (1974).

113. H. A. Newey and H. V. Adler, U.S. Patent No. 3,692,705 (1972).

114. S. Oswitch, *Composites* **5,** 55 (1974).

115. S. L. Kaplan, E. L. Mitch, and A. Katzakian, *Proc. 30th Ann. Tech. Conf. Reinf. Plast./ Comp. Inst. Soc. Plast. Ind.* (1975), Sect. 19-A.

116. M. F. Sorokin, I. M. Kochnov, and L. G. Shode, *Plast. Massy* **2,** 47 (1974).

117. H. H. Gibbs, *10th Nat. SAMPE Tech. Conf.* **10,** 211 (1978); *23rd Nat. SAMPE Symp.* **23,** 806 (1978); *24th Nat. SAMPE Symp.* **24,** 11 (1979).

118. H. H. Gibbs and J. R. Ness, *SAMPE J.* **15,** 11 (1979).

119. P. S. Blatz, *Adhes. Age* **21,** 39 (1978).

120. Dupont, NR-150 Polyimide Precursor Solutions, Product Bulletin E-08227(3/77).

121. Fiberite Corporation, NR-150B2/Unidirectional Type Prepreg Bulletin (4/6/78).

122. Hexcel Corporation, Hexstrand Advanced Composite Materials from Hexcel, May 1979.

123. A. K. St. Clair, N. S. Slemp, and T. L. St. Clair, *23rd Nat SAMPE Symp.* **23,** 113 (1978).

124. Anon., *Product Eng.* **4,** 46 (1973).

125. The Upjohn Company, Donald S. Gilmore Laboratories, DSG Reports and Tech. Bulletins.

126. D. J. Goldwasser and E. P. Otocka, *Org. Coat. and Plast. Chem. Preprints* **35,** (2) 198 (1975).

127. Rhone-Poulenc Co., Product Bulletins 6463/6473 and Data Sheets from Dept. of Thermostable Polymers.

128. J. Chambion, *Chem. Ind. Genie Chim.* **106,** 453 (1973).
129. Amoco Chemical Corp., *Torlon Engineering Resins,* Tech. Bulletins and Data Sheets (TB-47 and AT-C/1978).
130. I. E. Coleman and J. A. Conrady, *J. Polym. Sci.* **38,** 241 (1953).
131. P. O. Tawney, R. H. Snyder, R. P. Conger, K. A. Leibbrand, C. H. Stiteler, and A. R. Williams, *J. Org. Chem.* **26,** 15 (1961).
132. R. C. P. Cubben, *Polymer* **6,** 419 (1965).
133. J. K. Stille and R. Morgan, *J. Polym. Sci. A* **3,** 2379 (1965).
134. F. W. Harris and S. O. Norris, *J. Polym. Sci., Polym. Chem. Ed.* **11,** 2143 (1973).
135. R. Oda, K. Takagi, and T. Shono, *J. Polym. Sci. B* **7,** 11 (1969).
136. Y. Musa and M. P. Stevens, *J. Polym. Sci. A 1* **10,** 319 (1972).
137. F. C. DeSchryver, W. J. Feast, and G. Smets, *J. Polym. Sci. A 1* **8,** 1939 (1970).
138. F. Grundschober and J. Sambeth, U.S. Patent No. 3,533,996 (1970).
139. W. J. Gilwee, R. W. Rosser, and J. A. Parker, *18th Nat. SAMPE Symp.* **18,** 284 (1973).
140. G. T. Kwiatkowski, L. M. Robeson, G. L. Brode, and A. W. Bedwin, *J. Polym. Sci., Polym. Chem. Ed.* **13,** 961 (1975).
141. F. F. Holub and C. M. Emerick, Canadian Patent No. 959,494 (1974).
142. H. D. Stenzenberger, *Appl. Polym. Symp.* **22,** 77 (1973) and **31,** 91 (1977).
143. G. Menges, A. Harnier, and H. D. Stenzenberger, *Kunststoffe* **63,** 911 (1973).
144. D. O. Hummel, K. U. Heinen, H. D. Stenzenberger, and H. Siesler, *J. Appl. Polym. Sci.* **18,** 2015 (1974).
145. P. Kovacic and R. W. Hein, *J. Amer. Chem. Soc.* **81,** 1187 (1959).
146. T. V. Sheremeteva, *J. Polym. Sci. C* **5,** 1631 (1967).
147. M. Bargain, A. Combet, and P. Grosjean, U.K. Patent No. 1,280,846 (1972); 1,280,847 (1972).
148. M. Gruffaz, German Patent No. 2,251,065 (1973).
149. M. A. J. Mallet, *Mod. Plast.* **50,** 78 (1975).
150. M. A. J. Mallet and F. P. Darmory, Polyaminobismaleimides, in *New Industrial Polymers,* D. Deanin, (ed.), Marcel Dekker, N.Y. (1975), Ch. 9, p. 112.
151. F. P. Darmory, Processable polyimides, in *New Industrial Polymers,* D. Deanin, (ed.), Marcel Dekker, N. Y. (1975), Ch. 10, p. 124.
152. J. V. Crivello and P. C. Juliano, *J. Polym. Sci., Polym. Chem. Ed.* **13,** 1819 (1973).
153. A. A. Berlin, B. I. Liogonkii, and B. I. Zapadinskii, Soviet Patent No. 584014 (1979).
154. A. M. Kandybko, A. M. Kotukhova, A. P. Suslov, V. S. Volkov, and S. A. Dolmatov, *Plast. Massy* **10,** 43 (1978).
155. Yu. A. Mikhaylin and Ye. B. Trostyanskaya, *Plast. Massy* **10,** 18 (1978).
156. H. R. Lubowitz, U.S. Patent No. 3,528,950 (1970); 3,781,240 (1973); 3,827,927 (1974).
157. H. R. Lubowitz, *Polym. Preprints* **12,** 561 (1971).
158. H. R. Lubowitz, W. P. Kendrick, J. F. Jones, R. S. Thorpe, and E. A. Burns, U.S. Patent No. 3,647,529 (1972).
159. T. T. Serafini and P. Delvigs, *Appl. Polym. Symp.* **22,** 89 (1973).
160. S. L. Kaplan, D. Helfand, and S. S. Hirsh, *Proc. 27th Ann. Tech. Conf. Reinf. Plast./Comp. Inst. Soc. Plast. Ind.* (1972), Sect. 2-A.
161. S. L. Kaplan and S. H. Hirsh, Polyimides, chemistry, processing, properties, in *New Industrial Polymers,* R. D. Deanin, (ed.), Marcel Dekker, N.Y. (1975), Ch. 8, p. 100.
162. R. W. Vaughan, R. J. Jones, and E. A. Burns, *Proc. 28th Ann. Tech. Conf. Reinf. Plast./Comp. Inst. Soc. Plast. Ind.* (1973), Sect. 17-E.
163. S. L. Kaplan and C. W. Snyder, *Proc. 29th Ann. Tech. Conf. Reinf. Plast./Comp. Inst. Soc. Plast. Ind.* (1974), Sect. 11-B.

164. A. K. St. Clair and T. L. St. Clair, *Org. Coat. and Plast. Chem. Preprints* **35**, 77 (1975).
165. T. T. Serafini, P. Delvigs, and G. R. Lightsey, *J. Appl. Polym. Sci.* **16**, 905 (1972); U.S. Patent No. 3,745,149 (1973).
166. T. T. Serafini, R. D. Vannucci, and W. B. Alston, *Proc. 30th Ann. Tech. Conf. Reinf. Plast./ Comp. Inst. Soc. Plast. Ind.* (1975), Sect. 14-E.
167. M. P. Hanson and C. C. Chamis, *Proc. 29th Ann. Tech. Conf., Reinf. Plast./Comp. Inst. Soc. Plast. Ind.* (1974), Sect. 16-C.
168. T. L. St. Clair and R. A. Jewell, *23rd Nat. SAMPE Symp.* **23**, 520 (1978).
169. T. T. Serafini, R. D. Vannucci, and W. B. Alston, NASA TM 71894 (1976).
170. W. E. Winters and P. J. Cavano, *10th Nat. SAMPE Tech. Conf.* **10**, 661 (1978).
171. N. Bilow, A. L. Landis, and T. J. Aponyi, *20th Nat. SAMPE Symp.* **20**, 618 (1975).
172. N. Bilow and A. L. Landis, *8th Nat. SAMPE Tech. Conf.* **8**, 94 (1976).
173. R. H. Boschan, A. L. Landis, and T. J. Aponyi, *21st Nat. SAMPE Symp.* **21**, 356 (1976).
174. M. D. Sefcik, E. O. Stejskal, R. A. McKay, and J. Schaefer, *Macromolecules* **12**, 423 (1979).
175. T. J. Aponyi, C. B. Delano, J. D. Dodson, R. J. Milligan, and J. M. Hurst, *23rd Nat. SAMPE Symp.* **23**, 763 (1978).
176. N. Bilow, L. B. Keller, A. L. Landis, R. H. Boschan, and A. A. Castillo, *23rd Nat. SAMPE Symp.* **23**, 791 (1978).
177. Rhone–Poulenc, Kerimid 601, Product Report 6413.
178. F. P. Darmory, *18th Nat. SAMPE Symp.* **18**, 317 (1973); *SAMPE J.* **10**, 33 (1974).
179. Rhone-Poulenc, Kinel Moulding Compounds, Product Reports CFH 723/923.
180. Anon., *Mater. Eng.* **6**, 33 (1970).
181. J. M. Witzel, R. J. Jablonski, and D. Kruh, *Chem. Tech.* **7**, 441 (1972).
182. Hexcel Corp., F-178 Polyimide, Product Bull. (1/15/78).
183. Fiberite Corp., Polyimide Molding Compounds, Product Data Sheet.
184. Ciba–Geigy Corp., P13N, Polyimide Laminating Varnish Product Bulletin.
185. F. P. Darmory, S. S. Hirsch, and S. L. Kaplan, *SAMPE J.* **1**, 9 (1972).
186. Gulf Oil Co., Thermid 600 Polyimide Resins, Product Bull. (1978).
187. J. K. Gillham, K. D. Hallock, and S. J. Stadnicki, *J. Appl. Polym. Sci.* **16**, 2595 (1972).
188. S. V. Vinogradova, V. V. Korshak, and Y. S. Vygodskii, *Vysokomol. Soed. A* **8**, 809 (1966).
189. J. P. Critchley and Mary A. White, *J. Polym. Sci. A* **1**, 1809 (1972).
190. P. Delvigs, *Polymer Eng. Sci.* **16**, 323 (1976).
191. W. W. Wright, Application of thermal methods to the study of the degradation of polyimides, in *Developments in Polymer Degradation–3*, N. Grassie, (ed.), Applied Science Publishers, London (1981), Ch. 1.
192. R. A. Dine–Hart and W. W. Wright, *Makromol. Chem.* **153**, 237 (1972).
193. G. J. Knight, High-temperature properties of thermally stable resins, in *Developments in Reinforced Plastics–1*, G. Pritchard, (ed)., Applied Science Publishers, London (1980), Ch. 6.
194. F. P. Gay and C. E. Berr, *J. Polym. Sci. A 1* **6**, 1935 (1968).
195. A. S. Teleshova, E. N. Teleshov, and A. N. Pravednikov, *Polym. Sci., U.S.S.R.* **A13**, 2593 (1971).
196. G. F. L. Ehlers, K. R. Fisch, and W. R. Powell, *J. Polym. Sci. A 1* **8**, 3511 (1970).
197. Z. A. Kabilov, T. M. Muinov, L. A. Shibaev, Y. N. Sazanov, L. N. Korzhavin, and N. R. Prokopchuk, *Thermochimica Acta* **28**, 333 (1979).
198. J. Zurakowska-Orszagh, T. Chreptowicz, A. Orzeszko, and J. Kaminski, *Eur. Polym. J.* **15**, 409 (1979).
199. L. C. Scala and W. M. Hickam, *J. Appl. Polym. Sci.* **9**, 245 (1965).
200. B. M. Kovarskii, N. G. Annenkova, V. V. Guryanova, and A. B. Blyumenfeld, *Polym. Sci. U.S.S.R.* **A15**, 2783 (1973).
201. H. J. Dussel, H. Rosen, and D. O. Hummel, *Makromol. Chem.* **177**, 2343 (1976).
202. R. A. Dine–Hart, D. B. V. Parker, and W. W. Wright, *Br. Polym. J.* **3**, 226 (1971).

203. Dupont, Product Bulletin H-1B.
204. Dupont, Product Bulletin F-1A.
205. Dupont, Product Bulletin H-2; Dupont, Product Bulletin H-4.
206. Dupont, Vespel Design Handbook; Dupont, Vespel Precision Parts.
207. Upjohn, Highlights: Polyimide 2080, Report No. 5.
208. Y. T. Chen, *23rd Nat. SAMPE Symp. Exhib.* **23,** 826 (1978).
209. Anon., *Plast. Technol.* **25,** 23 (1979).
210. H. E. Sliney and T. P. Jacobson, *J. Amer. Soc. Lubrication Eng.* **31,** 609 (1975).
211. H. Lee, D. Stoffey, and K. Neville, *New Linear Polymers,* McGraw-Hill, N.Y. (1967), p. 194.
212. W. Grunsteidl, *Kunststoffe* **58,** 739 (1968).
213. P. M. Hergenrother, *SAMPE Q.* **3,** 1 (1971).
214. T. St. Clair and D. J. Progar, *Polym. Preprints* **16,** 538 (1975).
215. D. J. Progar and T. St. Clair, *7th Nat. SAMPE Tech. Conf.* **7,** 126 (1975).
216. P. S. Blatz, NASA Contractor Report 3017, (July 1978).
217. J. P. Critchley and W. W. Wright, *Rev. High Temp. Mater.* **4,** 107 (1979).
218. C. E. Browning and J. A. Marshall, *J. Composite Mater.* **4,** 390 (1970).
219. R. A. Pike and M. A. DeCresente, *Proc. 26th Ann. Tech. Conf. Reinf. Plast./Comp. Inst. Soc. Plast. Ind.* (1971), Sect. 13-D.
220. M. P. Hanson and T. T. Serafini, NASA TN-D-6604 (1971).
221. R. J. Jones, R. W. Vaughan, and E. A. Burns, NASA CR-72984 (1971).
222. R. Kollmansberger and E. Birchfield, MCAIR 72-009 (1972).
223. P. J. Cavano, R. J. Jones, and R. W. Vaughan, NASA CR-72983 (1972).
224. N. Sung and F. J. McGarry, AMMRC-CTR-73-31 (1973).
225. W. Wolkowitz and L. M. Poveromo, *28th Ann. Tech. Conf. Reinf. Plast./Comp. Inst. Soc. Plast. Ind.* (1973), Sect. 15-C.
226. T. T. Serafini and P. Delvigs, NASA TM X-2798 (1973).
227. H. Petrovicki, H. D. Stenzenberger, and A. Harnier, *32nd SPE Ann. Tech. Conf.* **32,** 88 (1974).
228. W. F. Wennhold, CASD-NSC-74-001 (1974).
229. R. T. Alvarez and F. P. Darmory, *20th Nat. SAMPE Symp.* **20,** 253 (1975).
230. R. W. Vaughan, R. J. Jones, M. K. O'Rell, and T. V. Roszhart, *20th Nat. SAMPE Symp.* **20,** 365 (1975).
231. W. E. Winters and T. T. Serafini, *20th Nat. SAMPE Symp.* **20,** 629 (1975).
232. T. L. St. Clair and R. A. Jewell, *8th Nat. SAMPE Tech. Conf.* **8,** 82 (1976).
233. P. J. Cavano and W. E. Winters, NASA CR-135113 (1976).
234. B. L. Lee and F. J. McGarry, *AMMRC-CTR-76-10* (1976).
235. N-H. Sung and F. J. McGarry, *Polym. Eng. Sci.* **16,** 426 (1976).
236. J. S. Jones, *21st Nat. SAMPE Symp.* **21,** 438 (1976).
237. I. Petker, *21st Nat. SAMPE Symp.* **21,** 37 (1976).
238. NASA CR-132685 (1975).
239. Amoco Bulletin HT 6a.
240. I. Serlin, A. H. Markart, and E. Lavin, *Proc. 25th Ann. Tech. Conf. Reinf. Plast./Comp. Div. Soc. Plast. Ind.* (1970), Sect. 19-A.
241. M. C. Cray, Paper presented at Plastics Institute RPG Conference, Reinforced plastics in aerospace applications, London, 5–6 April, 1973.
242. Upjohn, *Highlights: Polyimide Fibers,* Report No. 3.
243. R. S. Irwin and W. Sweeney, *J. Polym. Sci. C* **19,** 43 (1967).
244. J. Veghte, A. R. Marko, and C. Wilson, AD Report No. 754935/1974.
245. J. H. Ross, *Textile Res. J.* **32,** (9) 768 (1969).
246. J. H. Ross and R. M. Stanton, *Appl. Polym. Symp.* **21,** 109 (1973).

247. J. Meisenheimer and B. Wieger, *J. Pract. Chem.* **15,** 152 (1938).
248. C. G. Overberger, B. Kosters, and T. St. Pierre, *J. Polym. Sci. A 1* **5,** 1987 (1967).
249. K. C. Brinker and I. M. Robinson, U.S. Patent No. 2,895,948 (1959).
250. S. W. Shalaby, R. L. Lapinski, and E. A. Turi, *J. Polym. Sci., Polym. Chem. Ed.* **12,** 2891 (1974).
251. H. Vogel and C. S. Marvel, *J. Polym. Sci.* **50,** 511 (1961).
252. H. Vogel and C. S. Marvel, *J. Polym. Sci. A* **1,** 1531 (1963).
253. C. S. Marvel and H. A. Vogel, U.S. Patent No. 3,174,947 (1965).
254. H. H. Levine and R. D. Stacy, AFML-TR-65-350 (1966), p. 45.
255. A. B. Conciatori and E. C. Chenevey, *Macromol. Syn.* **3,** 24 (1968).
256. A. B. Conciatori, E. C. Chenevey, T. C. Bohrer, and A. E. Prince, *J. Polym. Sci. C* **19,** 49 (1967).
257. L. P. Suffredini, U.S. Patent No. 3,340,325 (1967).
258. H. H. Levine, C. B. Delano, and K. J. Kjoller, *Polym. Preprints* **5,** 160 (1964).
259. A. A. Izyneev, V. V. Korshak, T. M. Frunze, and V. V. Kurashov, *Izv. Akad. Nauk S.S.S.R., Otdel., Khim. Nauk* **10,** 1828 (1963).
260. W. Wrasidlo and H. H. Levine, *J. Polym. Sci. A* **2,** 4995 (1964).
261. D. N. Gray and G. P. Schulman, *J. Macromol. Sci. A* **1,** 395 (1967).
262. Y. Ohfiyu and T. Eguchi, U.S. Patent No. 3,655,632 (1972).
263. G. Rabilloud, B. Sillion, and G. De Gaudemaris, French Patent 1,526,830 (1968).
264. D. N. Gray, L. L. Rouch, and E. I. Strauss, *Macromolecules* **1,** 478 (1968).
265. H. H. Levine, *AFML-TR-64-365* **1** (1964), Pt. 1.
266. B. Durif-Varambon, B. Sillion, and G. De Gaudemaris, *C.R. Acad. Sci. (Paris), Ser. C,* **267,** 471 (1968).
267. B. Sillion and G. De Gaudemaris, *Bull. Soc. Chim. (France)* **12,** 4452 (1970).
268. A. R. Gerber, H. Kokelenberg, and C. S. Marvel, *J. Polym. Sci.* **11,** 1703 (1973). *J. Polym. Sci. A 1* **8,** 3199 (1970).
269. G. Rabilloud, B. Sillion, and G. De Gaudemaris, French Patent 1,492,776 (1967).
270. R. Pense and C. S. Marvel, *J. Polym. Sci. A 1* **8,** 3189 (1970).
271. F. L. Hedberg and C. S. Marvel, *J. Polym. Sci., Polym. Chem. Ed.* **12,** 1823 (1974).
272. Y. Iwakura and K. Uno, Japanese Patent 70,18675 (1970).
273. V. V. Korshak, A. L. Rusanov, and R. D. Katsarava, *Izv. Akad. Nauk S.S.S.R. Ser. Khim.* **2,** 480 (1969).
274. V. V. Korshak, A. L. Rusanov, D. S. Tugushi, and G. M. Cherkasova, *Macromolecules* **5,** 807 (1972).
275. S. Hara, M. Seo, M. Uchida, T. Yoshida, Y. Yoshida, and Y. Imai, U.S. Patent No. 3,518,234 (1970).
276. A. E. Prince, U.S. Patent No. 3,551,389 (1970).
277. G. Rabilloud, B. Sillion, and G. De Gaudemaris, French Patent No. 1,530,445 (1968).
278. E. C. Chenevey and A. B. Conciatori, U.S. Patent No. 3,433,772 (1969).
279. A. A. Izyneev, V. P. Maznrevskii, and V. V. Korshak, *Polym. Sci. U.S.S.R.* **17,** 1374 (1975).
280. R. Brand, M. Bruma, R. Kellman, and C. S. Marvel, *J. Polym. Sci., Polym. Chem. Ed.* **16,** 2275 (1978).
281. D. I. Packham, J. D. Davies, and H. M. Paisley, *Polymer* **10,** 923 (1969).
282. D. S. Tugushi, V. V. Korshak, A. L. Rusanov, V. G. Danilov, G. M. Cherkasova, and G. M. Tseitlin, *Polym. Sci. U.S.S.R.* **15,** 1087 (1973).
283. V. V. Korshak, A. L. Rusanov, I. M. Gverdtsiteli, D. S. Tugushi, and A. S. Shubashvili, *Dokl. Akad. Nauk S.S.S.R.* **240,** 346 (1978).
284. V. V. Korshak, I. M. Gverdtsiteli, L. G. Kipiani, D. S. Tugushi, and A. L. Rusanov, *Polym. Sci. U.S.S.R.* **21,** 133 (1979).
285. C. N. Zellner and H. W. Steinman, U.S. Patent No. 3,729,453 (1973).

286. V. V. Korshak, E. S. Krongauz, and A. V. D'yachenko, U.S.S.R. Patent No. 228,941 (1968). *Izobret. Prom. Obraztsy, Tovarnye Znaki.* **45,** (32) 75 (1968).
287. N. Yoda and M. Kurihara, *J. Polym. Sci., Pt. D (Macromol. Revs.)* **5,** 109 (1971).
288. Y. Iwakura, K. Uno, and Y. Imai, *J. Polym. Sci. A* **2,** 2605 (1964); *Makromol. Chem.* **77,** 33 (1964).
289. Y. Iwakura, Y. Imai, S. Inone, and K. Uno, *Makromol. Chem.* **95,** 236 (1966).
290. V. V. Korshak, A. L. Rusanov, O. M. Zurlova, Ts. G. Iremashvili, S. N. Leonteva, and R. D. Katsarava, U.S.S.R. Patent No. 399,515 (1973). *Otkrytiya, Izobret., Prom. Obraztsy, Tovarnye, Znaki* **50,** (39) 64 (1973).
291. V. V. Korshak, E. S. Krongauz, A. L. Rusanov, and A. P. Travnikova, *Polym. Sci. U.S.S.R.* **16,** 38 (1974).
292. H. Kokelenberg and C. S. Marvel, *J. Polym. Sci. A 1* **8,** 3235 (1970).
293. I. Matei, Gh. Mandrie, V. Taranu, and I. A. Schneider, *Rev. Roum. Chim.* **12,** 715 (1967).
294. I. K. Varma and Veena, *J. Polym. Sci., Polym. Chem. Ed.* **14,** 973 (1976); *J. Macromol. Sci. Chem. A* **11,** 845 (1977).
295. Y. Tsur, H. H. Levine, and M. Levy, *J. Polym. Sci., Polym. Chem. Ed.* **12,** 1515 (1974).
296. M. Saga, T. Shono, and K. Shinra, *Kogyo Kagaku Zasshi* **70,** 585 (1967).
297. T. Shono, M. Izumi, S. Matsumura, and N. Asano, Japanese Patent No. 68,15995 (1968).
298. E. W. Neuse, *Chem. and Ind. (London)* **7,** 315 (1975).
299. V. V. Korshak, S. V. Vinogradova, and V. A. Pancratov, *Vysok. Soed. B* **13,** 550 (1971).
300. C. D. Dudgeon and O. Vogl, *J. Polym. Sci., Polym. Chem. Ed.* **16,** 1831 (1978).
301. V. V. Korshak, T. M. Frunze, V. V. Kurachev, and G. P. Lopatina, *Polym. Sci. U.S.S.R.* **6,** 1379 (1964).
302. T. M. Frunze, V. V. Korshak, and A. A. Izyneev, *Izvest. Akad. Nauk S.S.S.R. Ser. Khim.* **11,** 2104 (1964).
303. B. M. Ginzburg, N. V. Mikhailova, V. N. Nikitin, A. V. Sidorovich, Sh. Tuichiev, and S. Ya. Frenkel, *Polym. Sci. U.S.S.R.* **9,** 2694 (1967).
304. S. Ya. Frenkel and B. M. Ginzburg, *J. Polym. Sci. C* **22,** 813 (1967).
305. N. A. Adrova, S. Ya. Frenkel, M. M. Koton, L. N. Korzhavin, L. M. Pyrkov, L. A. Laius, T. P. Pushkina, E. M. Moskvina, and B. M. Ginzburg, U.S.S.R. Patent No. 202,430 (1967).
306. A. Ya. Yakubovich, G. G. Rozantsev, G. I. Braz, and V. P. Bazov, *Polym. Sci. U.S.S.R.* **6,** 922 (1964).
307. L. Plummer and C. S. Marvel, *J. Polym. Sci. A* **2,** 2559 (1964).
308. D. Burmeister, M. Sander, and K. H. Bergert, *Makromol. Chem.* **89,** 199 (1965).
309. J. K. Gillham, *Science* **139,** 494 (1963).
310. C. S. Marvel, ML-TDR-64-39 **V,** 85 (1968).
311. J. Green and N. Mayes, *J. Macromol. Sci. Chem. A* **1,** 135 (1967).
312. K. Mitsuhashi and C. S. Marvel, *J. Polym. Sci. A* **3,** 1661 (1965).
313. V. V. Korshak, M. M. Teplyakov, and R. D. Fedorova, *Polym. Sci. U.S.S.R. A* **11,** 2117 (1970).
314. V. V. Korshak, M. M. Teplayakov, and R. D. Fedorova, *J. Polym. Sci. A 1* **9,** 1027 (1971).
315. N. N. Voznesenskaya, V. I. Berendyaev, B. V. Kotov, V. S. Voishchev, and A. N. Pravednikov, *Vysok. Soed. B* **16,** 114 (1974).
316. N. V. Karayakin, I. B. Rabinovich, V. V. Korshak, A. L. Rusanov, D. S. Tugushi, A. N. Mochalov, and V. N. Sapozhnikov, *Polym. Sci. U.S.S.R.* **16,** 795 (1974).
317. V. V. Korshak, A. A. Izyneev, V. P. Mazurevskii, V. D. Vorob'ev, M. V. Cherkasov, and A. D. Markov, U.S.S.R. Patent No. 491,667 (1975). *Otkrytiya Izobret. Prom. Obraztsy, Tovarnye Znaki* **52,** 68 (1975).
318. A. A. Izyneev, V. P. Mazurevskii, A. D. Markov, and V. V. Korshak, *Dokl. Akad. Nauk S.S.S.R.* **231,** 1126 (1976).
319. R. T. Foster and C. S. Marvel, *J. Polym. Sci. A* **3,** 417 (1965).

320. T. M. Frunze, V. V. Korshak, A. A. Izyneev, and V. V. Kevaelev, *Polym. Sci. U.S.S.R.* **7**, 313 (1965).
321. N. A. Adrova, M. M. Koton, and L. K. Prokharova, *Dokl. Akad. Nauk. S.S.S.R.* **166**, 9 (1966).
322. V. V. Korshak, G. M. Tseitlin, G. M. Cherkasova, and N. A. Berezkina, *Izv. Akad. Nauk S.S.S.R., Ser. Khim* **9**, 2143 (1968).
323. V. V. Korshak, V. M. Mamedov, and G. E. Golubkov, *Polym. Sci. U.S.S.R.* **14**, 2718 (1972).
324. V. V. Korshak, I. F. Manucharova, T. M. Frunze, and V. V. Kurashev, *Polym. Sci. U.S.S.R.* **6**, 1540 (1964).
325. R. Salle, B. Sillion, and G. De Gaudemaris, *Makromol. Chem.* **135**, 279 (1970).
326. T. V. L. Narayan and C. S. Marvel, *J. Polym. Sci. A* **5**, 1113 (1967).
327. V. V. Korshak, V. V. Izyneev, A. M. Egorov, N. Sh. Aldarova, and A. D. Markov, U.S.S.R. Patent No. 350,805 (1972). *Otkrytiya. Izobret. Prom. Obraztsy, Tovarnye Znaki* **49**, 82 (1972).
328. V. V. Korshak, T. M. Frunze, A. A. Izneev, and T. N. Shishkina, *Polym. Sci. U.S.S.R.* **6**, 992 (1964).
329. Y. Iwakura, K. Uno, Y. Imai, and M. Fukui, *Makromol. Chem.* **77**, 41 (1964).
330. G. Rabilloud, B. Sillion, and G. De Gaudemaris, *Bull. Soc. Chim. (France)* **3**, 932 (1966).
331. J. Horikawa, S. Tammoto, and R. Oda, *Chem. High Polym. Japan* **24**, 501 (1967).
332. Gh. Mandrie and I. Zugravescu, *Rev. Roum. Chim.* **16**, 1419 (1971).
333. T. M. Kiseleva, M. M. Koton, N. P. Kuznetsov, Yu. N. Sazanov, and S. N. Nikolaeva, *Vysok. Soed. B* **15**, 514 (1973).
334. A. A. Iznyeev, Zh. P. Mazurevskay, and V. V. Korshak, *Vysok. Soed. B* **17**, 373 (1975).
335. V. V. Korshak, A. A. Izyneev, and I. S. Novak, *Vysok. Soed. B* **17**, 229 (1975).
336. S. Hara and M. Senoo, Japanese Patent No. 74,26520 (1974).
337. O. L. Herberts, French Patent No. 1,491,016 (1967).
338. J. E. Mulvaney and C. S. Marvel, *J. Polym. Sci.* **50**, 541 (1961).
339. T. Nakajia and C. S. Marvel, *J. Polym. Sci. A 1* **7**, 1295 (1969).
340. L. W. Breed and J. C. Wiley, *J. Polym. Sci., Polym. Chem. Ed.* **14**, 83 (1976).
341. H. N. Kovacs, A. D. Delman, and B. B. Simms, *J. Polym. Sci. A 1* **6**, 2103 (1968).
342. V. I. Kasatochkin, V. V. Korshak, V. V. Kurashev, Z. S. Smutkina, T. M. Frunze, and T. M. Khrenkova, *Polym. Sci. U.S.S.R.* **7**, 1267, (1965).
343. M. L. Nielsen, U.S. Patent No. 3,330,805 (1967).
344. S. Kon-ya and M. Yokayama, *Kogaku Daigaku Kenkyu Hogoku* **33**, 66 (1973).
345. H. Sivriev and G. Borissov, *Eur. Polym. J.* **13**, 25 (1977).
346. S. Moon, A. L. Schwartz, and J. K. Hecht, *J. Polym. Sci. A 1* **8**, 3665 (1970).
347. A. Ya. Yakubovich, R. M. Gitina, E. L. Zaitseva, G. S. Markova, and A. P. Simonov, *Polym. Sci. U.S.S.R.* **12**, 2854 (1970).
348. R. M. Gitina, E. L. Zaitseva, and A. Ya. Yakubovich, *Russ. Chem. Rev.* **40**, 679 (1971).
349. V. V. Korshak, G. M. Tseitlin, M. S. Ustinova, M. V. Cherkasov, V. D. Vorob'ev, and B. R. Livshits, U.S.S.R. Patent No. 527,453 (1976). *Otkrytiya, Izobret., Prom. Obraztsy, Tovarnye Znaki* **53**(33), 78 (1976).
350. M. B. Sheratte, *Nucl. Sci. Abstr.* **21**, 38848 (1967).
351. V. V. Korshak, V. V. Vagin, A. A. Izyneev, and N. I. Bekasova, *Polym. Sci. U.S.S.R.* **17**, 2181 (1975).
352. V. V. Korshak, L. D. Radnaeva, L. G. Komarova, N. I. Bekasova, and A. A. Izyneev, *Polym. Sci. U.S.S.R.* **18**, 2954 (1976).
353. J. E. Mulvaney, J. J. Bloomfield, and C. S. Marvel, *J. Polym. Sci.* **62**, 59 (1962).
354. T. Kurosaki and P. R. Young, *J. Polym. Sci. C* **23**, 57 (1966) (published 1968).
355. V. V. Korshak, G. M. Tseitlin, G. M. Cherkasova, and A. L. Rusanov, U.S.S.R. Patent No. 267,065 (1969). *Otkrytiya. Izobret. Prom. Obraztsy, Tovarnye Znaki* **47**, 90 (1970).

356. N. P. Kuznetsov, M. I. Bessanov, T. M. Kiseleva, and M. M. Koton, *Polym. Sci. U.S.S.R.* **14**, 2282 (1972).
357. N. A. Adrova, V. N. Bagal, A. M. Dribnova, I. Ya. Kvitko, M. M. Koton, N. P. Kuznetsov, and F. S. Florinskii, *Vysok. Soed. B* **15**, 509 (1973).
358. M. M. Koton, T. M. Kiseleva, and S. N. Nikolaeva, *Vysok. Soed. B* **17**, 18 (1975).
359. B. M. Culbertson and S. Dietz, *J. Polym. Sci. B* **3**, 247 (1968).
360. V. V. Korshak, A. A. Izyneev, G. M. Tseitlin, and A. I. Pavlov, U.S.S.R. Patent No. 221,278 (1969). *Otkrytiya, Izobret., Prom. Obraztsy, Tovarnye Znaki* **46**, 184 (1969).
361. K. Nagakubo, F. Akutsu, and I. Kunio, Japanese Patent No. 76,53676 (1976).
362. R. J. Kray, R. Seltzer, and R. A. E. Winter, *Org. Coat. and Plast. Chem. Preprints* **31**(1), 569 (1971).
363. N. Dogoshi, S. Toyama, K. Ikeda, M. Kurihara, and N. Yoda, Japanese Patent No. 69,20111 (1969).
364. A. H. Gerber, *J. Polym. Sci., Polym. Chem. Ed.* **11**, 1703 (1973).
365. J. K. Gillham, *Polym. Preprints* **7**, 513 (1966).
366. J. K. Gillham, *Polym. Eng. Sci.* **7**, 225 (1967).
367. F. D. Trischler, K. J. Kjoller, and H. H. Levine, *J. Appl. Polym. Sci.* **11**, 1325 (1967).
368. I. B. Johns, E. A. McElhill, and E. O. Smith, *J. Chem. Eng. Data* **6**, 87 (1961).
369. V. V. Korshak, V. V. Rode, and N. M. Kotsoeva, *Polym. Sci. U.S.S.R.* **17**, 1418 (1975).
370. W. Wrasidlo, Paper presented at Polymer Conference on High Temperature Polymers, Wayne State University, 1968.
371. D. A. Bochvar, I. V. Stankevich, O. B. Tomilin, E. S. Krongauz, A. P. Travnikova, and V. V. Korshak, *Dokl. Akad. Nauk S.S.S.R.* **209**, 1097 (1973).
372. W. Wrasidlo and R. Empey, *J. Polym. Sci. A 1* **15**, 1513 (1967).
373. R. G. Spain and J. D. Ray, *Soc. Plast. Eng. (SPE) Conf. on Stability of Plastics*, Washington D.C. (1967), pp. E1–E14.
374. W. W. Wright and R. Phillips, *J. Polym. Sci. B* **2**, 47 (1964).
375. G. P. Shulman and W. Lochte, *J. Macromol. Sci. A 1* (3), 413 (1967).
376. G. F. L. Ehlers and K. R. Fisch, Int. Symp. Polym. Characterization, *Appl. Poly. Symp.* **8**, 171 (1969). *Proc. 3rd Anal. Conf.* **3**, 187 (1971), published 1972.
377. P. M. Hergenrother, *Proc. High Temp. Polym. Symp.*, Western Regional A.C.S. Meeting, Los Angeles (November 1965), p. D1.
378. V. V. Rode, N. M. Kotsoeva, G. M. Cherkasova, D. S. Tugushi, G. M. Tseitlin, A. L. Rusanov, and V. V. Korshak, *Polym. Sci. U.S.S.R.* **12**, 2103 (1970).
379. N. N. Voznesenskaya, V. L. Berendyaev, B. V. Kotov, V. S. Voischev, and A. N. Pravednikov, *Vysok. Soed. B* **16**, 114 (1974).
380. H. H. Levine, N. P. Loire, and C. B. Delano, AFML Report No. AFML-TR-67-23 (1967).
381. G. F. L. Ehlers, AFML Report No. AFML-TR-74-177 (1974).
382. V. I. Kasatochkin, V. V. Korshak, V. V. Kurachev, Z. S. Smutkina, T. M. Frunze, and T. M. Khrenkova, *Dokl. Akad. Nauk S.S.S.R.* **159**, 843 (1964).
383. Y. Tsur, Y. L. Freilich, and M. Levy, *J. Polym. Sci., Polym. Chem. Ed.* **12**, 1531 (1974).
384. R. Yokata, I. Mita, and H. Kambe, *Kob. Kagaku.* **29**, 428 (1972).
385. V. V. Korshak, N. M. Kotsoeva, and V. V. Rode, *Dokl. Akad. Nauk S.S.S.R.* **209**, 356 (1973).
386. T. M. Frunze, V. V. Korshak, A. A. Izyneev, and V. V. Kurachev, *Polym. Sci. U.S.S.R.* **7**, 308 (1965).
387. W. W. Wright, in *Degradation and Stabilization of Polymers*, G. Geuskens, (ed.), Applied Science Publishers, London (1975), p. 43.
388. J. Preston and W. B. Black, *J. Polym. Sci. A 1* **5**, 2429 (1967).
389. G. F. L. Ehlers, K. R. Fisch, and W. R. Powell, AFML Report No. AFML-TR-70-63 (1970). *J. Polym. Sci. A* **1**, 2931 (1969).

390. H. L. Freeman, AFML Report No. AFML-TDR-64-274 (1965).

391. R. A. Gauchana and R. T. Conley, *J. Macromol. Sci., Chem.* A **4,** 441 and 1599 (1970).

392. R. T. Conley, S. Ghosh, and J. J. Kane, *Org. Coat and Plast. Chem. Preprints* **31**(1), 151 (1971).

393. G. Dubey and J. J. Kane, *Org. Coat and Plast. Chem. Preprints* **30**(2), 187 (1970).

394. V. K. Belyakov, I. V. Belyakova, M. V. Kozlova, P. A. Okunev, and O. G. Tarakanov, *Polym. Sci. U.S.S.R.* **15,** 2990 (1973).

395. R. B. Gosnell and H. H. Levine, *J. Macromol. Sci. Chem.* A **3,** 1381 (1969).

396. Anon. Narmco Imidite 850 Data Sheet.

397. R. Poet, *Proc. 20th Ann. Tech. Conf. Reinf. Plast./Comp. Inst. Soc. Plast. Ind.* (1965), Sect. 6-E.

398. H. H. Levine, Report ASD-TDR-63-396 (1962), pp. 56–58 .

399. R. Reed and T. J. Reinhart, *Proc. 21st Ann. Tech. Conf. Reinf. Plast./Comp. Inst. Soc. Plast. Ind.* (1966), Sect. 7-B.

400. J. D. Ray, *Proc. 24th Ann. Tech. Conf. Reinf. Plast. Comp. Inst. Soc. Plast. Ind.* (1969), Sec. 15-E.

401. E. B. Bahnsen and L. E. Shoff, *12th Nat. SAMPE Symp. Exhib.* **12,** 1 (1967).

402. F. J. Riel and S. Litvak, *12th Nat. SAMPE Symp. Exhib.* **12,** 15 (1967).

403. H. A. Mackay, *Mod. Plast.* **1,** 149 (1966).

404. R. Reed and R. Hidde, Report AFML-TR-136 (1965) AFML-TR-65-146, Vols. 1 and 2 (1965).

405. E. L. Strauss, *Polym. Eng. Sci.* **61,** 24 (1966).

406. A. H. Frazer, *High Temperature Resistant Polymers,* Wiley-Interscience, New York (1968), pp. 290–292.

407. R. R. Dickey, J. H. Lundell, and J. A. Parker, *J. Macromol. Sci. Chem.* A **3,** 573 (1969).

408. Whittaker Corp., Narmco Division, Imidite 850 Adhesive Product Information (July 1964).

409. V. V. Korshak, T. M. Frunze, and A, A. Izyneev, *Ref. Zhur. Khim.,* Abstr. No. 125,412 (1967).

410. S. Litvak, *Adhes. Age* **11**(1), 17 (1968); **11**(2), 24 (1968).

411. P. Giuliani, *Les Hauts Polymères Thermostables,* Publications de L'Institut Français du Pétrole, Paris (1971), p. 79.

412. J. R. Hill, *Adhes. Age* **9** (8), 32 (1966).

413. H. H. Levine, *9th Nat. SAMPE Symp.* **9,** V-1 (1965).

414. R. B. Krieger and R. E. Politi, *9th Nat. SAMPE Symp.* **9,** V-4 (1965).

415. H. A. King, *Adhes. Age* **15**(2), 22 (1972).

416. D. G. Flom, A. L. Speece, and G. A. Schmidt, *Polym. Chem. Preprints* **8**(2), 1190 (1967).

417. S. Y. Yoshino, M. A. Nadler, and D. H. Richter, *Adhes. Age* **10**(8), 26 (1967).

418. R. Kanter and S. Litvak, *Adhes. Age* **12**(11), 24 (1969).

419. T. J. Reinhart and R. Hidde, *Appl. Polym. Symp.* **3,** 299 (1966).

420. S. V. Messineo, Hawthorne, and S. Y. Yoshimo, U.S. Patent No. 3,549,468 (1970).

421. S. B. Twiss, *Appl. Polym. Symp.* **3,** 455 (1966).

422. R. O. Denyes, Report AFML-TR-66-167, Vols. I and II (1966).

423. J. H. Ross, *Amer. Dyestuff Reporter* **51** (20), 29 (1962).

424. J. H. Ross, *Text. Res. J.* **32,** 768 (1962).

425. R. Spain and L. G. Picklesimer, *Text. Res. J.* **36,** 619 (1966).

426. J. G. Santangelo, U.S. Patent No. 3,441,640 (1964).

427. R. Dauskys, U.S. Patent No. 3,856,549 (1974).

428. K. A. Rinehart, E. J. Powers, G. W. Calundann, and C. P. Driscoll, Report ASD-TR-73-49 (1974).

429. R. W. Singleton, *Appl. Polym. Symp.* **9,** 133 (1969).

430. R. W. Singleton, H. O. Noether, and J. F. Tracy, *J. Polym. Sci.* C **19,** 65 (1967).

431. R. M. Stanton and R. S. Schulman, Report AFML-TR-72-139 (1972).
432. J. H. Ross, R. S. Schulman, and R. M. Stanton, *Text. Res. J.* **4**, 146 (1971).
433. L. R. Belohlav, *Angew. Makromol. Chem.* **40/41**, 465 (1974).
434. R. M. Stanton, Report ASD-TR-73-27 (1973).
435. J. H. Ross, *19th Nat. SAMPE Symp.* **19**, 166 (1974).
436. R. H. Jackson, *Text. Res. J.* **48**, 314 (1978).
437. W. D. Freeston, J. S. Panto, L. Barish, and M. M. Schoppee, Report AFML-TR-70-267 (1970).
438. H. M. Ezekiel, *Org. Coat and Plast. Chem. Preprints* **31**(1), 415 (1971).
439. H. M. Ezekiel and R. G. Spain, U.S. Patent No. 3,528,774 (1967).
440. D. E. Stuetz, *Org. Coat. and Plast. Chem Preprints* **31**(1), 389 (1971).
441. H. Yokata, K. Kobayashi, and J. Harikawa, Japanese Patent No. 74,54629 (1974).
442. D. A. Kourtides, J. A. Parker, and C. L. Segal, *Soc. Plast. Eng.* **32**, 580 (1974).
443. C. L. Segal, *19th Nat. SAMPE Symp.* **19**, 51 (1974).
444. Whittaker Corp., Narmco Division, *Prod. Eng.* **38**(18), 95 (1967).
445. R. Reed and J. Feher, AIAA/ASME Struct., *Struct. Dgn. Mater. Conf., Collect. Tech. Pap. Mater.* **9** (1968), pp. 1–6.

SUPPLEMENTARY BIBLIOGRAPHY

Polyimides

Aromatic imide polymers for high-temperature adhesives, H. A. Burgman, J. H. Freeman, L. W. Frost, G. M. Bower, E. J. Traynor, and C. R. Ruffing, *J. Appl. Polym. Sci.* **12**, 805 (1968).
Effect of long-term oxidation at 200–300°C on six types of aromatic amide and imide resins, L. C. Scala, W. M. Hickam, and I. Marschik, *J. Appl. Polym. Sci.* **12**, 2339 (1968).
Heat resistant polyamides and polyimides for electrical insulation, J. A. Almouli and P. F. Bruins, *Plast. Elec. Insul.* 155 (1968).
Polyimides. A new class of heat resistant polymers, N.A. Androva, M. I. Bessonov, L. A. Laius, and A. P. Rudakov, Israel Program for Scientific Translations, Jerusalem (1969).
Polyimide manufacture, M. W. Ranney, *Chem. Progr. Rev.*, Vol. 54, Noyes Data Corp., New Jersey (1969).
Polyimides, J. Preston, in *Encyclopedia of Chemical Technology,* Suppl. Vol., 2nd ed., H. F. Mark, J. J. McKetta, and D. F. Othmer, (eds.), Interscience Publishers, New York (1971).
Engineers guide to polyimide plastics, R. J. Falian, *Mater. Eng.* **74**(2), 26 (1971).
Chemistry and kinetics of polyimide degradation, C. Arnold and L. K. Burgman, *Ind. Eng. Chem. Prod. Res. Develop.* **11**, 322 (1972).
Future prospects in the field of aromatic polyimides (polyarimides) and their derivatives, M. M. Koton, *Polym. Sci. U.S.S.R.* **16**, 1383 (1974).
Mechanism of formation and degradation of polyimides, P. P. Nechaev, Y. S. Vygodskii, G. E. Zaikov, and S. V. Vinogradova, *Polym. Sci. U.S.S.R.* **18**, 1903 (1976).
Polyimides, C. E. Sroog, *J. Polym. Sci. Macromol. Rev.* **11**, 161 (1976).
When, where and how to use polymides, J. D. Hensel, *Plastics Eng.* **33** (10), 20 (1977).
Polyimides take the heat, J. A. Vaccari, *Prod. Eng.* **49**, 41 (1978).
How to process the new thermoplastic polyimides, Anon., *Plast. Technol.* **25** (8), 23 (1979).
Soluble polyimides, S. V. Vinogradova, Y. S. Vygodskii, V. V. Korshak, and T. N. Spirins, *Acta. Polym.* **30**, 3 (1979).

Polybenzimidazoles

Polybenzimidazoles. A review, J. R. Hall and D. W. Levi, Plastec Report No. 28 (1966).

Polybenzimidazoles, H. H. Levine, in *Encyclopedia of Polymer Science and Technology*, H. F. Mark, N. Gaylord, and N. M. Bikales, (eds.), Interscience Publishers, New York (1969), Vol. 11, p. 188.

Polybenzimidazoles, J. R. Leal, *Mod. Plast.* **52**(8), 60 (1975).

General

Heteroatom ring containing polymers, A. D. Delman, *J. Macromol. Sci. (Rev. Macromol. Chem.) C* **2**, 153 (1968).

Thermal stability of polyheteroarylenes, E. M. Bondarenko, V. V. Rode, and V. V. Korshak, *Usp. Khim. Fiz. Polim.* 206 (1970).

Synthesis methods and properties of polyazoles, V. V. Korshak and M. M. Teplyakov, *J. Macromol. Sci. (Rev. Macromol. Chem.) C* **5**, 409 (1971).

The heat resistance of polyheteroarylenes, V. K. Belyakov, I. V. Belyakova, S. S. Medved, A. F. Yerin, V. A. Kosobutskii, V. M. Savinov, and L. B. Sokolov, *Polym. Sci. U.S.S.R.* **13**, 1956 (1971).

New synthetic routes to high temperature polymers by cyclopolycondensation reactions, N. Yoda, M. Kurihara, and N. Dokoshi, *Progr. Polym. Sci. Japan* **4**, 1 (1972).

New aspects of polycyclization, E. S. Krongauz, *Russ. Chem. Rev.* **42**, 857 (1973).

Thermally stable hetcrocyclic polymers as adhesives, M. G. Maximovich, in *Handbook of Adhesives Bonding*, C. V. Cagel, (ed.), McGraw-Hill, New York (1973), p. 5/1.

Synthesis of polymers by polycyclotrimerization, V. A. Sergeev, V. K. Shitikov, and V. A. Pankratov, *Russ. Chem. Rev.* **48**, 79 (1979).

Thermally stable polymers: Polyoxadiazoles, polyoxadiazole-N-oxides, polythiazoles and polythiadiazoles, P. E. Cassidy and N. C. Fawcett, *J. Macromol. Sci. (Rev. Macromol. Chem.) C* **17**, 209 (1979).

6

SILICON-CONTAINING POLYMERS—SILICONES

INTRODUCTION

The polyorganosiloxanes, or as they are more commonly called, the silicones, may be defined as polymers with backbones comprising alternate silicon and oxygen atoms in which the silicon atoms are also linked to organic groups. The structures concerned are

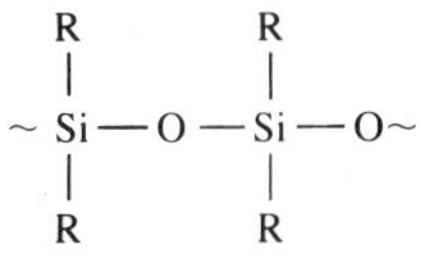

$$
\begin{array}{ccccc}
 & R & & R & \\
 & | & & | & \\
\sim & Si-O-Si-O\sim & \text{and} & \\
 & | & & | & \\
 & R & & R &
\end{array}
$$

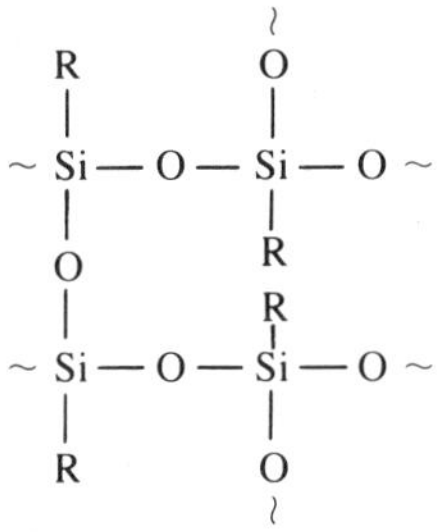

and both the linear and network silicones have found technological applications. They are attractive commercially because, although the products made from them have relatively low mechanical properties at room temperature, the strength falls off very much more slowly at elevated temperatures than is the case with many other systems.

The first organosilicon compounds were prepared by Friedel and Crafts in 1863,[1] but the most detailed study was that of F. S. Kipping at Nottingham University between 1900 and 1940.[2] The name silicone was given by Kipping because it was thought at first that the compounds had structures similar to the organic ketones and, accordingly, they were called silicoketones, or silicones for

323

short. Kipping's work was concerned almost exclusively with nonpolymeric compounds, and commercial interest in silicon polymers only developed in the 1930s during searches for heat-resistant electrical insulating materials. Research was carried out by Hyde and Sullivan at the Corning Glass Works (U.S.A.), McGregor and co-workers at the Mellon Institute, and Rochow, Patnode, and Marshall at the General Electric Co. (U.S.A.). Simultaneously, work progressed in the U.S.S.R. under Andrianov.

SYNTHESIS OF ORGANO-SILICON COMPOUNDS

Of the many synthetic methods for obtaining organo-silicon compounds, precursors for the silicones, only a few are important commercially. These include:

1. The Grignard process;
2. The direct process;
3. Silane-olefin addition reactions;
4. Aromatic silylation;
5. Redistribution reactions.

The Grignard reaction is the classical method for preparing organosilanes and was the first commercially important synthetic route.

$$RMgCl + SiCl_4 \longrightarrow RSiCl_3 + MgCl_2$$
$$RMgCl + RSiCl_3 \longrightarrow R_2SiCl_2 + MgCl_2$$
$$RMgCl + R_2SiCl_2 \longrightarrow R_3SiCl + MgCl_2$$
$$RMgCl + R_3SiCl \longrightarrow R_4Si + MgCl_2$$

The reaction is useful because of its versatility in converting chlorosilane intermediates into other alkylated organo-silane materials, for example:

$$CH_3SiCl_3 + C_6H_5MgBr \longrightarrow Cl-\underset{\underset{\textstyle C_6H_5}{|}}{\overset{\overset{\textstyle CH_3}{|}}{Si}}-Cl + MgBrCl$$

However, the most important commercial method for making organochlorosilanes is the direct process discovered by Rochow.[3] In this reaction, alkyl or aryl halides are reacted directly with elemental silicon in the presence of catalysts.

$$2RCl + Si \xrightarrow[250-280°C]{Cu} R_2SiCl_2$$

TABLE 6.1. *Composition of Crude Product from the Direct Process (methyl chloride and silicon) (Reference 4)*

Compound		Boiling point (°C)	Content (%)
Dimethyldichlorosilane	$(CH_3)_2SiCl_2$	70	75
Methyltrichlorosilane	CH_3SiCl_3	66	10
Trimethylchlorosilane	$(CH_3)_3SiCl$	58	4
Methyldichlorosilane	CH_3SiHCl_2	41	6
Silicon tetrachloride	$SiCl_4$	58	
Tetramethylsilane	$(CH_3)_4Si$	26	In small
Trichlorosilane	$SiHCl_3$	32	amounts
High-boiling residue (disilanes)		100–200	

In practice, other reactions occur simultaneously and a mixture of products is obtained. The composition of the reaction products from methyl chloride is shown in Table 6.1, together with the boiling points of the various components. The mixture may be separated by distillation.

In the olefin addition process, compounds containing SiH bonds are added to olefins.

$$\geq Si - H + R_2C = CR_2 \longrightarrow R_2CH - CR_2 - Si \leq$$

The process may also be applied to alkynes to give unsaturated silanes.

$$\geq Si - H + RC \equiv CR \longrightarrow RCH = CR - Si \leq$$

The reaction is usually performed in the presence of catalysts which influence both the rate of the reaction and whether the silicon becomes attached to the α or β carbon atom. An example of a commercially important reaction is the addition of 3,3,3-trifluoroprop-1-ene to a silane.

$$CH_3SiHCl_2 + CH_2{=}CH - CF_3 \longrightarrow \underset{\displaystyle CH_2 - CH_2 - CF_3}{\overset{\displaystyle CH_3}{Cl - Si - Cl}}$$

Cyanosiloxanes may be prepared similarly by reacting an alkenyl nitrile with dichloromethyl silane.

$$CH_3SiHCl_2 + CH_2{=}CH - CN \longrightarrow \underset{\displaystyle CH_2 - CH_2 - CN}{\overset{\displaystyle CH_3}{Cl - Si - Cl}}$$

The phenyl group attached to silicon also imparts important properties to silicone elastomers. The corresponding chlorosilanes are generally made either by the direct process between chlorobenzene and silicon or by aromatic silylation. This latter reaction will take place at 400°C, or, in the presence of Friedel–Crafts catalysts, lower temperatures may be used.

$$C_6H_6 + HSiCl_3 \xrightarrow[\text{AlCl}_3]{\text{Heat}} C_6H_5SiCl_3 + H_2$$

Under appropriate conditions various substituents on the silicon atom can be exchanged for one another. This provides a convenient method of converting by-product chlorosilanes into more useful materials. A typical example, which is of practical interest, is the combination of a mixture of trimethylchlorosilane and methyltrichlorosilane to form dimethyldichlorosilane.

$$(CH_3)_3SiCl + CH_3SiCl_3 \rightleftharpoons 2(CH_3)_2SiCl_2$$

The reaction is carried out at 200–400°C in the presence of aluminum chloride. These silane compounds are then reacted further to obtain polymeric materials. For convenience the synthesis and properties of elastomers are examined separately from the preparation of thermosetting resins.

ELASTOMERS

Synthesis

The further conversion of the chlorosilanes to the cyclic siloxanes and silanol precursors of silicone high polymers involves hydrolysis and condensation. The functionality of the chlorosilane and the conditions used for hydrolysis have an influence on the structure of the siloxane that is obtained. Monomeric silanols are extremely reactive and condense readily with elimination of water. Dihydroxydimethylsilane is so reactive that it is extremely difficult to prepare as such, as it condenses readily to form polymers with silanol end-groups. Aromatic silanols are less reactive and diphenylsilanediol may be easily prepared.

The hydrolysis of dimethyldichlorosilane leads to a mixture of linear and cyclic polymers.

$$(CH_3)_2\,SiCl_2 \longrightarrow$$

$$HO-\underset{\underset{CH_3}{|}}{\overset{\overset{CH_3}{|}}{Si}}-O-\left[\underset{\underset{CH_3}{|}}{\overset{\overset{CH_3}{|}}{Si}}-O-\right]_n\underset{\underset{CH_3}{|}}{\overset{\overset{CH_3}{|}}{Si}}-OH$$

$$\left[\underset{\underset{CH_3}{|}}{\overset{\overset{CH_3}{|}}{Si}}-O-\right]_n$$

$$n = 3,4,5\ldots$$

Polydimethylsiloxane of sufficiently high molecular weight that it can be used for making good quality silicone elastomers cannot be made by direct hydrolysis of dimethyldichlorosilane; traces of monofunctional impurities result in a lowering of molecular weight, while trifunctional impurities lead to branching. However, it is possible to obtain very pure octamethylcyclotetrasiloxane from which satisfactory gums can be prepared.

In a typical process dimethyldichlorosilane is dissolved in ether and the solution is mixed with an excess of water. Cyclic polymers make up about 50% of the reaction product, of which octamethylcyclotetrasiloxane is the main component. This tetramer has a boiling point of 175°C and is readily purified by distillation; it may then be polymerized by heating at 150–200°C with a trace of sodium hydroxide and a very small amount of monofunctional material to control the molecular weight. The product is a highly viscous gum with no elastic properties. Dimethyl silicone gums are normally cured by heating with organic peroxides such as benzoyl peroxide and 2,4-dichlorobenzoyl peroxide at temperatures of 110–175°C. This curing reaction presumably gives rise to dimethylene crosslinks. Pressure needs to be applied during cure to prevent volatilization of the decomposition products of the peroxides. After press cure, these decomposition products may be safely removed by high-temperature-oven post-cure.

A gum stock containing up to 0.5% of vinyl groups is much easier to vulcanize; it needs less peroxide for the cure and forms a vulcanizate that is more resistant to the rearrangements that cause reversion and high compression set. As a consequence, nearly all commercial gums now contain vinyl-modified polymers. Other common modifications are the incorporation of phenyl, trifluoropropyl, or cyanoalkyl groups. High-molecular-weight gums of these substituted siloxanes may be prepared in a similar way to polydimethylsiloxane, for example, by reacting cyanoalkylcyclosiloxane with octamethylcyclotetrasiloxane in the presence of sodium hydroxide catalyst.[5]

A dimethyl silicone rubber tends to be stiff below −50°C; 5 to 10% of phenyl groups extends the service range to below −90°C.[6] The addition of trifluoropropyl ($-CH_2-CH_2-CF_3$), cyanoethyl, or cyanopropyl groups[7] gives silicone rubbers with better oil and solvent resistance.

Room Temperature Vulcanizing Silicones

As mentioned above, silicone rubbers may be prepared from polydimethyl siloxane gums by oxidative cross-linking using organic peroxides. Such cures require the application of heat and pressure and the systems are the so-called High Temperature Vulcanization (HTV) silicones. It is also possible to obtain cross-linked materials by the condensation of linear polymers with hydroxyl or hydrolyzable terminal groups. Such reactions take place at room temperature and give rise to the Room Temperature Vulcanization (RTV) silicones.[4]

The RTV silicones are supplied in two forms, either as a two-part mixture, where a curing agent and/or catalyst are added prior to use, or as compounds where condensation reactions take place on exposure to moisture. The base silicone polymer is of a lower molecular weight than the materials used for HTV, but otherwise they are chemically the same, either wholly polydimethyl siloxane or with some of the methyl groups replaced by phenyl or trifluoropropyl groups.

There are several different curing reactions used for the two pack RTV rubbers. A commonly used system consists of a polysiloxane terminated with hydroxyl groups, together with an alkoxysilane as cross-linking agent, which may

$$4\ HO-\left[\begin{array}{c} R \\ | \\ Si-O \\ | \\ R \end{array}\right]_n H\ +\ R'O-\overset{\displaystyle OR'}{\underset{\displaystyle OR'}{\overset{|}{\underset{|}{Si}}}}-OR'$$

$$\Big\downarrow\ \text{Sn salt as catalyst}$$

$$O-\left[\begin{array}{c} R \\ | \\ Si-O \\ | \\ R \end{array}\right]_n$$

$$-\left[\begin{array}{c} R \\ | \\ O-Si \\ | \\ R \end{array}\right]_n O-Si-O-\left[\begin{array}{c} R \\ | \\ Si-O \\ | \\ R \end{array}\right]_n\ +\ R'OH$$

$$O-\left[\begin{array}{c} R \\ | \\ Si-O \\ | \\ R \end{array}\right]_n$$

be an ester of orthosilicic or polysilicic acids. The cure reaction is catalyzed by metallic salts such as stannous octoate. The preferred cross-linking agents are the methoxy and ethoxysilanes; hence cure involves the evolution of methanol or ethanol.

Siloxanes containing Si–H bonds may also be used as cross-linking agents, in which case the reaction with the siloxanol group liberates hydrogen. Here again metal salts of carboxylic acids are used as catalysts.

$$
\text{HO} \left[\begin{array}{c} R \\ | \\ \text{Si} - \text{O} \\ | \\ R \end{array} \right]_n \!\!\! H + H - \overset{|}{\underset{|}{\text{Si}}} - R
$$

$$
R - \overset{|}{\underset{|}{\text{Si}}} - H \;+\; H \left[\begin{array}{c} R \\ | \\ \text{O} - \text{Si} \\ | \\ R \end{array} \right]_n \!\!\! \text{OH} \xrightarrow{\text{catalyst}}
$$

$$
\text{HO} \left[\begin{array}{c} R \\ | \\ \text{Si} - \text{O} \\ | \\ R \end{array} \right]_n \!\!\! H + H - \overset{|}{\underset{|}{\text{Si}}} - R
$$

The addition reaction of silanes with vinyl groups may also be used as a low-temperature cure reaction for polysiloxanes. In this case, siloxanes containing vinyl end-groups can be made to react with siloxanes containing Si–H groups in the presence of a suitable catalyst. Using chloroplatinic acid or other platinum compounds, the reaction occurs at temperatures of less than 100°C. The slightly higher cure temperature for these systems can be advantageous, as it gives a longer pot-life at room temperature than that of the other RTV types.

Where the finished articles are to be subjected to high temperatures in use, it is best to give the material a post-cure at the maximum use temperature. The temperature needs to be raised slowly to this value, otherwise the siloxane may depolymerize.

The one-pack systems rely on the ingress of moisture to bring about the cross-linking reaction. The first systems of this type were made by reacting polydimethylsiloxane-α, ω-diols with methyltriacetoxysilane; the tetraacetoxy-polydimethylsiloxane so formed can be stored in the absence of moisture. In contact with atmospheric humidity vulcanization takes place by hydrolytic cleavage of the acetoxy groups to form silanol groups. The condensation of the silanol with the remaining acetoxy groups takes place with evolution of acetic acid.

$$\text{(chemical scheme: acetoxysilane crosslinking of silanol-terminated polysiloxane)}$$

Vulcanization begins at the surface and proceeds towards the interior as water vapor diffuses in and acetic acid diffuses out; consequently, thick sections are slow to vulcanize completely. Magnesium oxide added to the mix will absorb the acid and aid the cross-linking of quite thick layers.

Reinforcement of Elastomers

Rubbers are seldom used without the incorporation of other materials, some of them in large amounts. The silicone rubbers are no exception to this rule. The common reinforcing additive found in many rubber formulations, carbon black, is, however, seldom used with silicones; it interferes with the peroxide cure and does not provide the same degree of reinforcement that it does in other systems. The reinforcing filler most frequently used with silicones is fume silica,[6,8] ob-

TABLE 6.2. *Properties of Polymethylvinylsiloxane Vulcanizates (Reference 9)*

Silicone gum	100 parts	100 parts	100 parts	100 parts
Highly active silica	40 parts	—	—	—
Medium activity silica	—	43 parts	—	—
Kieselguhr	—	—	50 parts	100 parts
Benzoyl peroxide	1.25 parts	1.25 parts	1.25 parts	1.25 parts
Mechanical properties after-post cure of 15 hours at 200°C				
Tensile strength (MPa)	6.4–7.4	5.4–6.4	2.5–3.4	3.9–4.9
Elongation at break (%)	250–300	300–350	200–250	150–200
Shore A hardness	60±5	55±5	40±5	60±5
Compression set after 22 hours at 175°C %	30–40	20–30	35–45	40–50
Density (g/cc)	1.17	1.18	1.23	1.40
After aging 14 days at 200°C				
Tensile strength (MPa)	5.4–6.4	5.4–6.4	2.5–3.4	3.9–4.9
Elongation at break (%)	200–250	250–300	150–200	100–150
Shore A hardness	65±5	60±5	45±5	65±5

tained by burning a silicon-containing compound at high temperatures. Extending fillers are also used and, as silicone elastomers are very thermally stable, the fillers also need to show a high thermal stability in order not to detract from that of the polymer. Examples of the more common extending fillers are ground quartz, diatomaceous earth, zinc oxide, titanium dioxide (both rutile and anatase), calcium carbonate, and iron oxides. Table 6.2 gives some examples of the mechanical properties achieved with different fillers and also shows the effects of aging in air at 200°C.[9]

The exact nature of the polymer-filler interaction is in some dispute, but it probably involves a combination of mechanical keying to a rough surface, hydrogen bonding, Van der Waals forces, and also some covalent bonding resulting from interaction of silanol end-groups on the polymer chain with hydroxy groups on the surface of the silica. It has also been suggested that some bonds may be formed during the peroxide cure.[10]

The fume silicas present a mixing problem in that an interaction occurs between the filler and the polymer to such an extent that the material has the appearance of being partly vulcanized; this effect is termed "crepe hardening." Apart from causing problems in milling, crepe hardening also results in variation of the properties of the final vulcanized material as a function of the time of storage of the unvulcanized stock. Many compounded stocks, therefore, have to be aged for a few weeks before being processed and vulcanized.[11–14] It is possible to reduce the effect by the use of additives, commonly silanediols, or substituted diols such as diphenyl silanediol or pinacoxy dimethyl silane.[15,16] Alternatively, silicas that have been pretreated with chlorosilanes may be used. Addition of these silica fillers results in good mechanical properties as well as excellent thermal aging characteristics.[17]

Thermal Stability of Elastomers

A detailed review of the thermal behavior of polymers, including the polysiloxanes, has been written by Wright and Lee[18] and, more recently, a review of the thermo-oxidative stability of siloxane elastomers has been published by Goldovskii and Kuzminskii.[19] Overall, the silicones can be regarded as providing materials with good high-temperature resistance, but the exact application needs to be carefully considered before the degree of stability can be assessed because of reversion problems that can arise under certain conditions.

Figures 6.1 to 6.4 serve to illustrate the thermal stability that can be achieved by polysiloxanes. These curves were obtained by thermogravimetric analysis in air and nitrogen of some unfilled polysiloxane vulcanizates (rate of temperature rise 2°C/min). The compositions of the polymers examined are shown in Table 6.3.

From the figures it can be seen that samples 1 and 2 behave very similarly, decomposition starting in air at about 300°C and in inert atmosphere at 330°C. In air, a cross-linking reaction takes place that results in 60% by weight of residue, even when the temperature is taken above 800°C. The phenyl-containing polymer has a similar stability to that of sample 1 in air up to 20% weight loss, but in nitrogen it is markedly less stable. It will also be noticed that sample 3 gives a much smaller residue in air, only about 10%. The fluorosilicone, sample 4, is the least stable in air, although it does give a residue of about 30%. Its stability in nitrogen is comparable with that of sample 3. The results are compared in Table 6.4, which gives the temperatures for 5, 10, and 50% weight loss in both air and nitrogen.

Isothermal weight loss experiments produce similar results. The weight losses after two hours at temperature are plotted against temperature in Figures 6.5 and 6.6, which show the relative order of stability of the four vulcanized polymers based on the first 10% weight loss to be, in air, polymethylphenyl > polydimethyl > polymethylvinyl > polymethyltrifluoropropylsiloxane, and in inert atmosphere, polydimethyl > polymethylvinyl > polymethylphenyl > polymethyltrifluoropropylsiloxane.

The activation energies for breakdown of the polymers calculated from the isothermal weight loss data are given in Table 6.5.

The stability of the silicones may be attributed to the high bond strengths found in the polymers, Table 6.6.

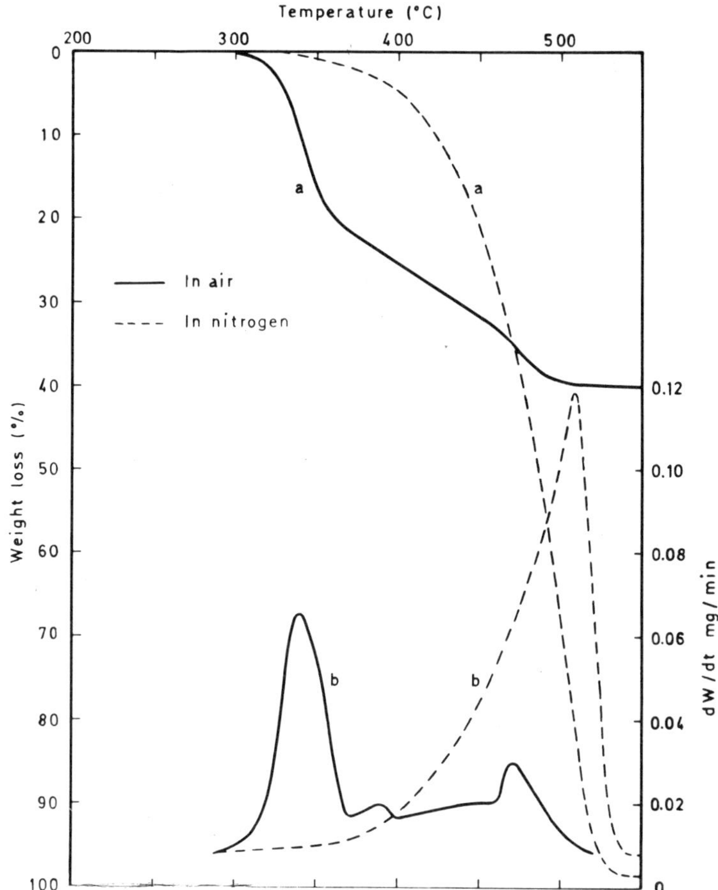

Figure 6.1. *Polydimethylsiloxane vulcanizate; (a) weight loss and (b) rate of weight loss dW/dt. (Heating rate 2°C/min)*

However, when heated at temperatures $\geq 200°C$ in vacuum or inert atmosphere, pure linear polydimethylsiloxanes degrade by rupture of the Si-O bonds with the formation of cyclic volatile materials,[20, 21] this process being called "reversion."

Grassie and MacFarlane have shown that more than 99% of the volatile degradation products are members of the series of cyclic oligomers, trimer, tetramer, pentamer, etc. The proportion of these products does not vary significantly with temperature or extent of degradation in the ranges 275–405°C and 0–80% degradation, respectively. Average values for the products are given in Table 6.7.

It is thought that these cyclic products are formed by reaction of silanol groups in, or at the ends of, the chain with siloxane bonds in the chain.[21–23]

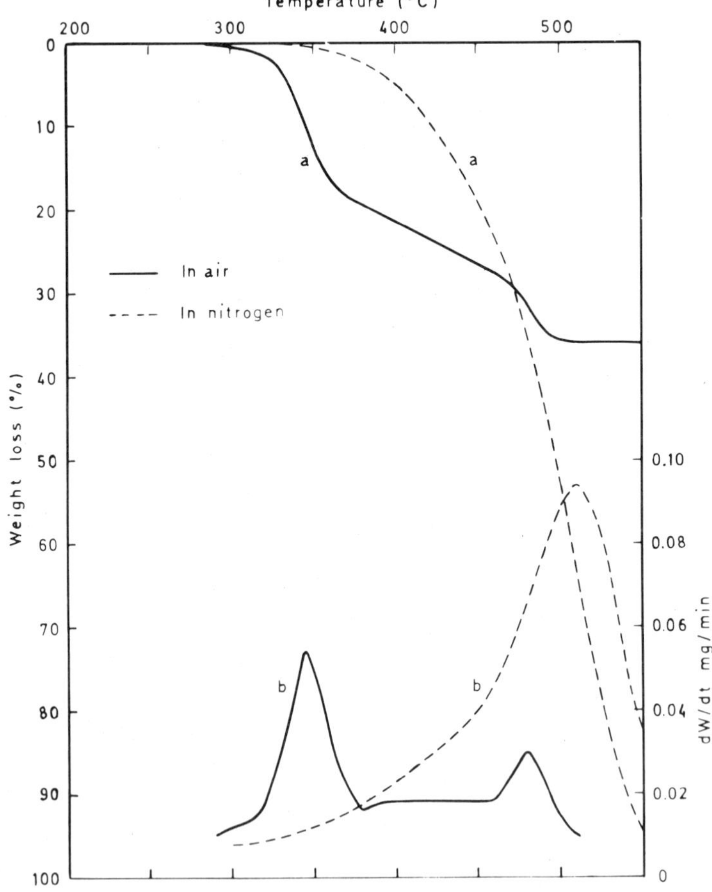

Figure 6.2.   Polymethylvinylsiloxane vulcanizate; (a) weight loss and (b) rate of weight loss dW/dt. (Heating rate 2°C/min)

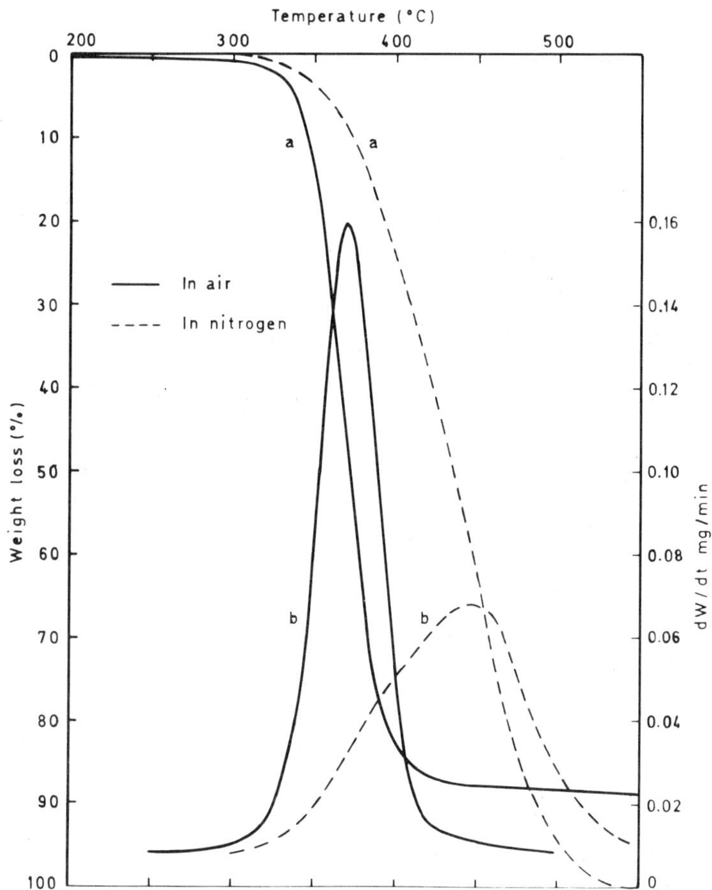

Figure 6.3. Polymethylphenylsiloxane vulcanizate; (a) weight loss and (b) rate of weight loss dW/dt. (Heating rate 2°C/min)

There is no evidence of charring, suggesting that the Si-C and C-H linkages are stable at these temperatures (275–405°C). It is known that branched-chain polysiloxanes are stable at temperatures in excess of 400°C and that tetramethylsilane vapor is stable to above 600°C, confirming the stability of the Si-C bond in nonoxidizing conditions. Grassie *et al.* [24–26] have also examined the thermal degradation of phenyl-substituted polymethylsiloxanes and have shown them to be more thermally stable on a weight loss basis than polydimethylsiloxane (contrast the results of Figure 6.6 for vulcanizates), the stability decreasing with increasing molecular weight and with replacement of terminal OH by Me_3SiO groups. The main degradation products comprise a

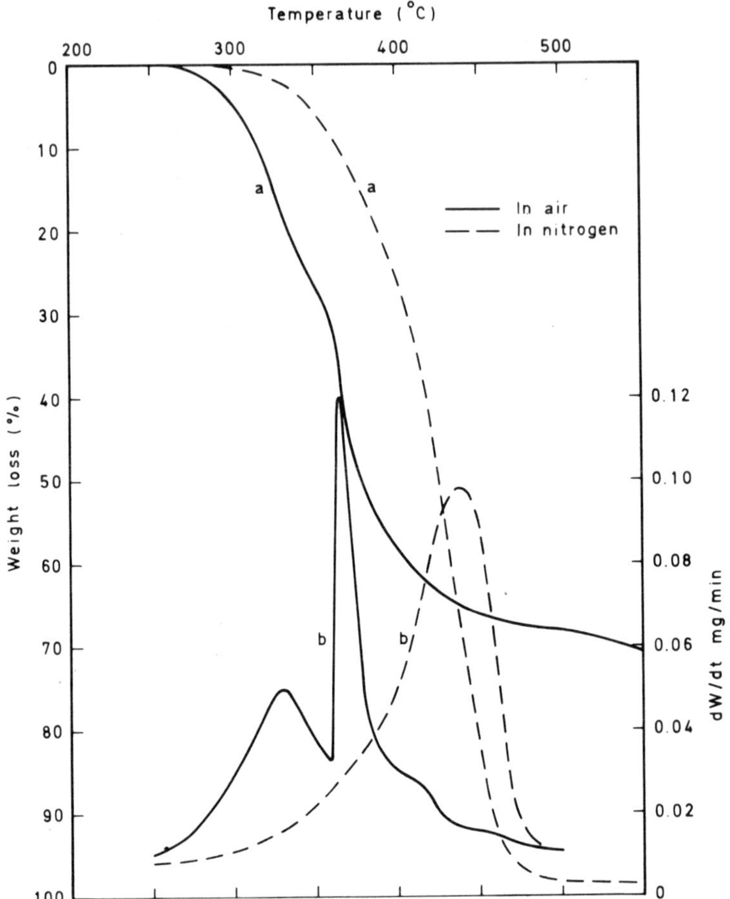

Figure 6.4. Polymethyltrifluoropropylsiloxane vulcanizate; (a) weight loss and (b) rate of weight loss, dW/dt. (Heating rate 2°C/min)

mixture of all possible stereoisomeric cyclic trimers and tetramers together with small amounts of the cyclic pentamer and the two cyclic oligomers.

TABLE 6.3. *Composition and Cure of the Polymers Used in the Thermogravimetric Experiments Illustrated in Figures 6.1 to 6.4*

Sample	Polymer	Trade name	Cross-linking agent	Proportions	Cure
1	Polydimethyl-siloxane	ICI E300	2,4-dichloro-benzoyl peroxide	E300 100 parts	Press cure 1 hour at 160°C
				PDS50 3 parts	Post-cure 24 hours at 250°C
2	Polymethyl-vinylsiloxane	ICI E302	2,5-dimethyl-2,5-di-*t*-butylperoxy hexane	E302 100 parts	Press cure 1 hour at 160°C
				Varox 2 parts	Post-cure 24 hours at 250°C
3	Polymethyl-phenylsiloxane	ICI E350	2,5-dimethyl-2,5-di-*t*-butylperoxy hexane	E350 100 parts	Press cure 1 hour at 160°C
				Varox 8 parts	Post-cure 24 hours at 250°
4	Polymethyl-trifluoro-propylsiloxane	Dow Corning LS 420	2,5-dimethyl-2,5-di-*t*-butylperoxy hexane	LS420 100 parts	Press cure 1 hour at 160°C
				Varox 4 parts	Post-cure 24 hours at 250°C

TABLE 6.4. *Temperatures of 5%, 10%, and 50% Weight Loss of Vulcanized Silicone Polymers Heated in Air and Nitrogen at a Rate of Temperature Rise of 2°C/min*

	Temperature (°C) for a weight loss in					
	Air of			Nitrogen of		
Sample	5%	10%	50%	5%	10%	50%
1	332	340	—	400	424	486
2	335	345	—	400	423	498
3	338	346	373	356	374	435
4	300	315	378	346	364	426

TABLE 6.5. *Activation Energies, Ea, for Breakdown of the Polysiloxanes in Air and Nitrogen over the Range 10–15% Weight Loss*

Polymer vulcanizates	In air Ea, kJ/mole	In nitrogen Ea, kJ/mole
Polydimethylsiloxane	201	152
Polymethylvinylsiloxane	237	137
Polymethylphenylsiloxane	217	196
Polymethyltrifluoropropylsiloxane	133	212

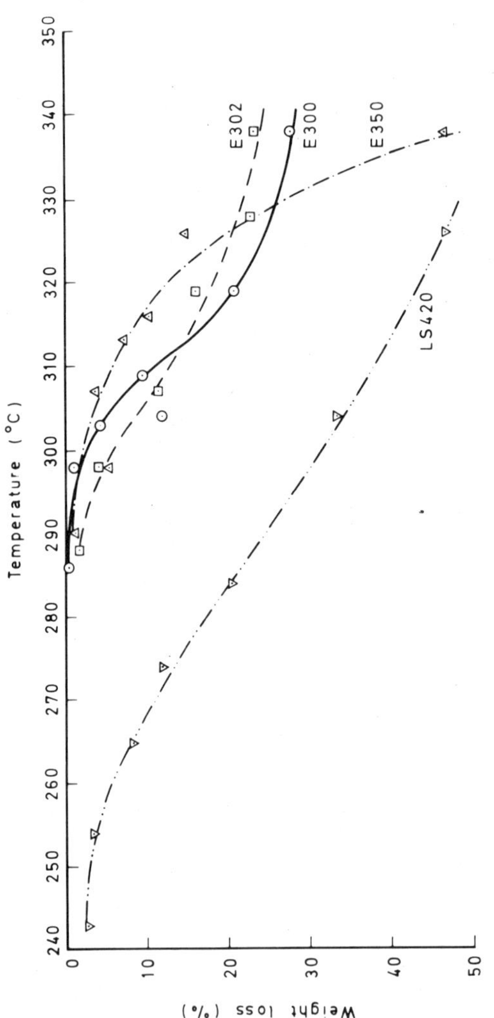

Figure 6.5. Isothermal weight loss of polysiloxane vulcanizates in air. Weight loss after two hours at temperature.

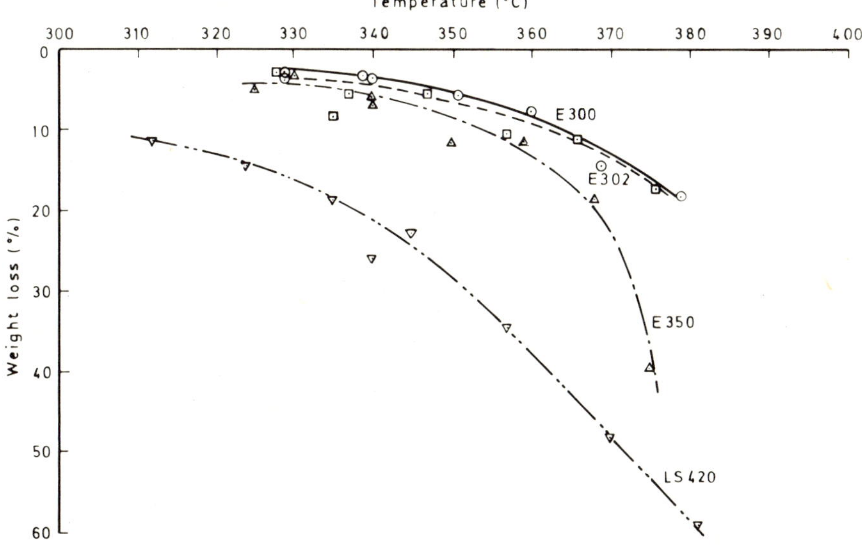

Figure 6.6. *Isothermal weight loss of polysiloxane vulcanizates in nitrogen. Weight loss after two hours at temperature.*

TABLE 6.6. *Bond Energies of Some Carbon and Silicon Bonds*

Bond	Bond energy (kJ)
Si—O	444
C—H	414
C—O	360
C—C	348
Si—C	318
Si—H	318
Si—Si	222

TABLE 6.7. *Yield of Cyclic Oligomers from a Polydimethylsiloxane of Molecular Weight 111,500 (Reference 21)*

Cyclic oligomer	(%)
Trimer	73
Tetramer	13
Pentamer	4
Hexamer	6
Heptamer	1
Octamer	0.9
Nonamer	0.3
Decamer	0.2
Undecamer	
Duodecamer	} Traces
Tridecamer	

Benzene is also formed in significant quantities, which increases with decreasing molecular weight and increasing temperature.

The thermo-oxidative degradation of the polysiloxanes is a free radical chain reaction similar to that which takes place in hydrocarbon polymers. Attack takes place at the methyl groups, and among the degradation products observed are methanol, formaldehyde, formic acid, carbon monoxide, carbon dioxide, and water. It is thought that silanol groups are formed in place of the methyl groups.[27] Cross-linking of the chains also results from the oxidation of the side groups of the macromolecules. It is reported that the incorporation of as little as 5 mole % of methylphenylsiloxane units into the polydimethylsiloxane chain significantly decreases the rate of Si-C cleavage and the formation of cyclosiloxanes.[28] Conversely, the incorporation of only 1% of vinyl groups increases the degradation process at 280°C by as much as 100%.[28] Of the other common modifications to the basic polydimethylsiloxane structure, the trifluoropropyl group lowers the thermal and thermo-oxidative stability, whereas the γ-cyanopropyl substituent is said to have relatively little effect.[29]

In practice, however, the polymers are used as vulcanizates containing fillers and stabilizers. The vulcanization process itself alters the stability of the base polymer, and the fillers and stabilizers incorporated have further effects on the degradation of the compounded material. For example, the stability of vulcanizates based on polymer containing from 0.1 to 0.5 mole % of vinyl units has been reported to be considerably higher than that of vulcanizates from unmodified rubber.[19] The reason suggested for this is that the dimethylsiloxane rubber was vulcanized using acyl peroxides, the products of conversion of these facilitating the degradation of the polymer chains. Vulcanization of the rubbers with vinyl groups was carried out using alkyl peroxides. Smaller quantities of these peroxides are needed and they give rise to products that are volatile and hence lost from the polymer during post-cure.

Similarly, although the incorporation of phenyl groups into the siloxane molecule has been shown to improve greatly the oxidative stability of the polymer compared to polydimethylsiloxane, in order to produce a vulcanizate with reasonable properties large quantities of peroxide are required. This causes reactions that are detrimental to the heat stability, and the resulting vulcanizates offer little or no advantage over peroxide-cured methyl vinyl polymers.

It has been shown that in filled vulcanized methyl vinyl silicone rubber, network scission at temperatures below 250°C is due largely to hydrolytic reactions in the main chain. At temperatures of 250°C and above, there are indications that a significant amount of scission arises from oxidative reaction in the cross-links, and it is this reaction that is catalyzed by acidic residues in the rubber.[30] In conventional heat aging tests in which the rubber remains in an unstrained state, the effects of hydrolysis will only be observed if the concentration of water in the system is allowed to rise. On aging the material under well-ventilated conditions, the effects of hydrolysis are not seen and the silicone rubber becomes brittle after long exposure at high temperature. This embrittlement results from additional cross-linking caused by oxidative reactions in the methyl groups of the main chain polymer. When the rubber is employed in compression or tension, hydrolytic scission will affect performance, and in applications of this sort it is important to dry out the rubber before use and to prevent access of moisture to the component during service. Silica filler provides a further source of moisture and drying needs to be carried out at elevated temperatures immediately prior to its incorporation. The fumed silicas, being naturally drier, are therefore preferable as fillers for rubber to be used in closed systems, while the precipitated hydrated silicas enhance the hot-air stability of vulcanizates.[31]

Antioxidants and Stabilizers for Silicone Elastomers

As one mode of breakdown of the polymers in an oxidizing environment is the normal free radical chain mechanism observed in hydrocarbon polymers, it should be possible to increase the lifetime of the siloxanes at elevated temperatures by the use of antioxidants. As the polymers have a high intrinsic thermal stability, the antioxidants themselves must be stable at high temperatures.

The standard free radical acceptors have been shown to interfere with the cure reaction brought about by organic peroxides. Polymeric aryl amine antioxidants, however, provide good protection to polysiloxanes without interfering with the vulcanization process. Such antioxidants have been prepared by the copolycondensation of hydroquinone with various aromatic diamines in the presence of tributyl phosphate.[32] In particular, the polymer obtained from 3,8-diaminopyrene gives a significant improvement in compression set resistance of silicones in air at 200°C[33]; this is illustrated in Figure 6.7.[34] Other workers have prepared poly-*p*-phenyleneamines of the general formula

by the condensation of *p*-halogenoaniline in the melt in argon at 260°C in the presence of catalytic amounts of copper iodide, or by the condensation of *p*-phenylenediamine with hydroquinone. Poly-*m*-phenyleneamine has been synthesized by condensation of *m*-phenylenediamine with resorcinol. These materials inhibit the oxidation of polymethylvinylsiloxane at 380°C.[35]

The other mechanism of polysiloxane degradation is the reversion reaction, the breakdown of the main siloxane chain with the formation of cyclic products.

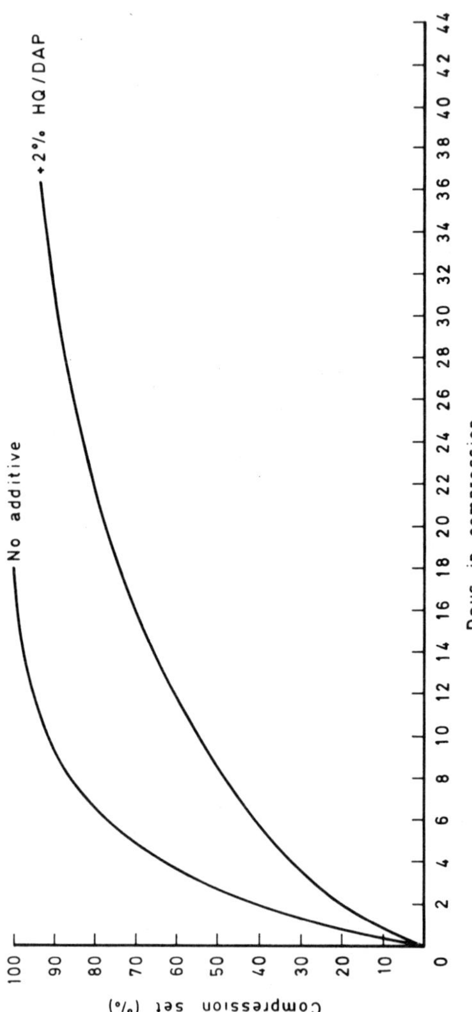

Figure 6.7. Effect of polymeric antioxidant on compression set resistance of silicone rubber in air at 200°C. Reference 34.

This hydrolysis-based reaction can be reduced by drying the rubber and additives.[30] The scission of the cross-links in a vulcanizate catalyzed by acidic or basic residues can be alleviated by careful purification of the polymer, or by the addition of materials that can neutralize the catalysts.[36] The silica commonly used as a reinforcing additive can also improve the thermal stability of a vulcanizate, possibly by reaction with the free Si-OH groups formed by the oxidation of the organic side groups.[22]

The most common stabilizing additives are metals and their derivatives, the best known of these being ferric oxide.[6,37] This material gives minimal inhibition of peroxide cure and can be included in formulations as masterbatch.[38] Numerous other metal derivatives have been evaluated, especially transition-metal and rare-earth derivatives,[39] though they seldom improve on the effectiveness of ferric oxide.[40,41] Synergistic effects have been observed between rare-earth metal oxides and ferric oxide, the combination with cerium oxide being the most efficient and giving protection at 300°C for more than 36 hours.[42] Examples of other effective compounds are ferric chloride-amine adducts that are reported to improve the compression set at 160°C,[43] and silylated ferrocene derivatives that are described as being very effective at 250 to 300°C.[44]

The mechanism of the protective action is not fully understood, but is probably a combination of several effects, including the ability of the transition metal ions to take part in redox reactions, eliminating free radical intermediates;

$$SiO\cdot + Fe^{3+} \longrightarrow SiO^- + Fe^{2+}$$

to act as special agents that modify the activity of the filler surface; or to function as amphoteric species which neutralize traces of acid or base that catalyze siloxane degradation.

The action of ferric oxide is both concentration- and temperature-dependent. The iron can catalyze the oxidation as well as inhibit it; an optimum iron level for a minimum oxidation rate exists for each temperature.[45] It has been shown that the metal oxides can act as catalysts in the decomposition of the free radical species formed during the oxidation of the polymer,[46] in that the stabilizer is not used up by the inhibitory process. Soluble iron compounds act in a manner similar to Fe_2O_3, but the optimum dose of the latter is nearly 60 times higher than that of the former.

Elevated Temperature Properties of Silicone Vulcanizates

As mentioned above, the cure, filler, and other additives all have an effect on the thermal stability of the base polymer. The cure of polydimethylsiloxane by benzoyl or 2,4-dichlorobenzoyl peroxide is not as efficient as the cure of polymer containing vinyl substituents. The vinyl substituted material may also be cured using smaller quantities of dicumyl or di-tertiary butyl peroxide, which do not leave residues to initiate degradation of the polymer chain. Hence a rubber prepared from a gum containing about 0.1% of vinyl groups is more stable than one made from pure polydimethylsiloxane, but if higher concentrations of vinyl groups are used then the vulcanizate's resistance to thermal aging is decreased.

γ-Cyanopropyl and trifluoropropyl groups are incorporated into silicones to increase the fluid resistance of the elastomer. For example, it has been shown possible to use cyanosiloxane polymers as channel sealants for integral fuel tanks in military aircraft.[47, 48] These polar substituents increase the thermo-oxidative stability of the propyl radical in the polymer, but the vulcanizates based on them are inferior in thermal aging resistance to those containing solely methyl groups.

The scission of the Si-O bonds at elevated temperatures and the cross-linking reactions that occur in oxidizing atmospheres lead to an interesting phenomenon, in that a rubber can appear to be stable when aged in air but show a drop in molecular weight when heated in inert atmosphere. This is because in oxidizing conditions the competing reactions cancel each other out and help maintain the original properties. Consideration must therefore be given in the design of a component to its ventilation. The oxidative cross-linking reaction also causes compression set in silicone rubbers when they are heated for prolonged periods. Nevertheless, these materials are still amongst the most resistant of elastomers to compression set as is illustrated in Figures 6.8 and 6.9.[49, 50] On the basis of oven aging tests, the following estimates of heat stability have been quoted for fluorosilicone and methylvinylsilicone vulcanizates[51] (Table 6.8).

Figure 6.10 compares the effect of long-term oven aging at 150°C on two methylvinyl silicones, a fluorosilicone, and Viton A.[52] Stress relaxation testing, however, shows that these times are overestimates.[53]

Some examples of typical properties obtained from silicone vulcanizates are shown in Table 6.9.

It will be noted that the values for tensile strength are comparatively low when compared, for example, with natural rubber (13–27 MPa), but the advantage of the silicones is that they show relatively small variation in properties with change in temperature. This is illustrated in Figure 6.11, which shows the effect of temperature on the tensile strength of various rubbers,[54] and also in Figure 6.12, which plots the percentage retention of tensile strength against temperature for a number of rubbers, including the fluoroelastomer Viton, a material noted for its high-temperature properties.[53]

Silicones are also used, because of their good electrical properties, for potting electrical components and for wire and cable insulation. They show excellent electrical insulating qualities at elevated temperature, very good resistance to ozone and corona, good voltage endurance, flame retardance, and ability to burn to a nonconducting ash that continues to insulate. Examples of some electrical

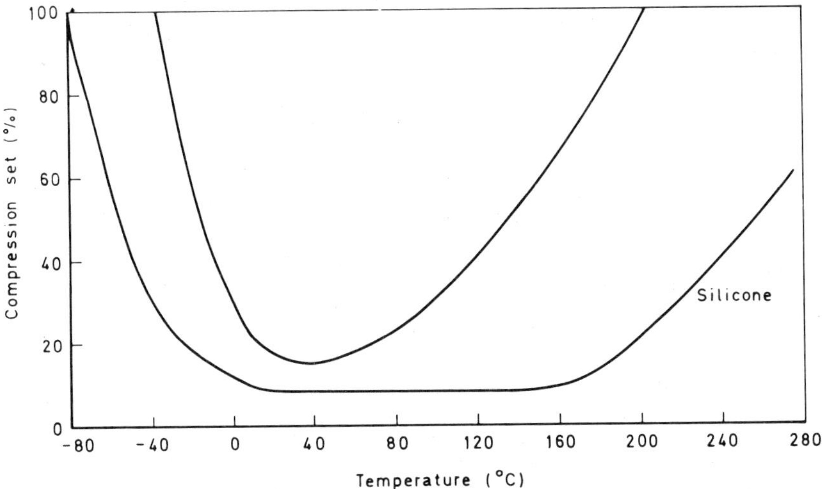

Figure 6.8. Compression set of silicone rubber compared to that of an organic elastomer. Reference 49.

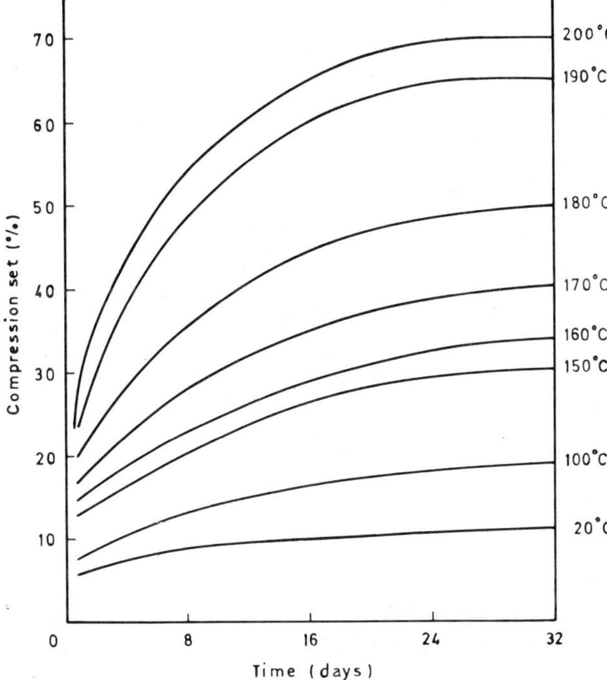

Figure 6.9. Compression set of methylvinyl silicone rubber as a function of time at different temperatures. Reference 50.

TABLE 6.8. Useful Life in Hours at Temperature of Silicone Rubbers (Reference 51)

Rubber	Life in hours at temperatures of		
	150°C	175°C	200°C
Fluorosilicone	20000	5000	4000
Methylvinylsilicone	30000	15000	10000

properties are given in Table 6.10, and Figure 6.13 illustrates the effect of temperature on the volume resistivity of various plastics, including a silicone rubber and two silicone resins.[55]

SILICONE RESINS

Synthesis

Silicone resins differ from fluids and gums in that they are synthesized from mixtures containing trichloro- and tetrachloro-silane as well as mono- and dichloro-silane. This results in the cured resins having a cross-linked structure. The Tgs of cured commercial silicone resins range up to 200°C, in contrast to typical silicone rubber Tgs of about −60°C. The resins are manufactured by first

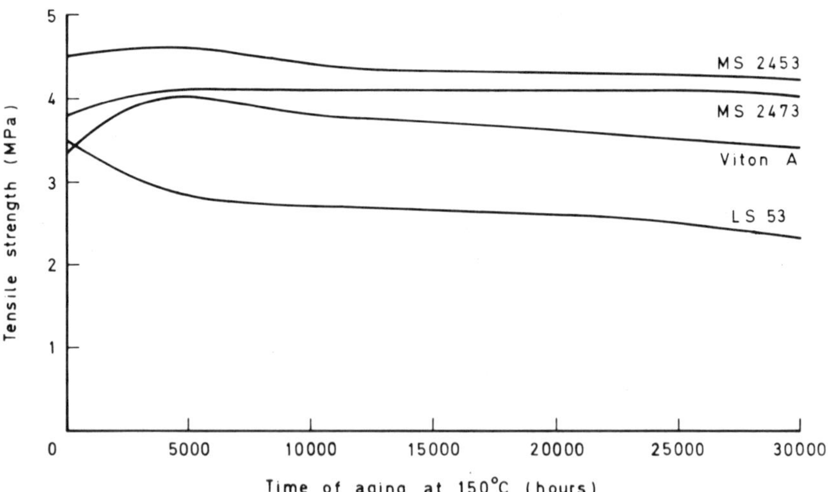

Figure 6.10. Effect of oven aging at 150°C on the tensile strength retention of two methylvinyl silicones, a fluorosilicone, and a hydrofluoro elastomer. Reference 52.

TABLE 6.9. Some Typical Properties of Silicone Rubber Vulcanizates (Reference 6)

Property	Low compression set	Heat resistant	Low temperature	High strength	Solvent resistant
Formulation	Vinyl silicone Fume silica Ground silica Vinyl specific peroxide	Vinyl silicone Fume silica Ground silica Vinyl specific peroxide	Phenylvinylsilicone Ground silica Vinyl specific peroxide	Phenylvinylsilicone Fume silica Vinyl specific peroxide	Fluorovinylsilicone Fume silica Red iron oxide Vinyl specific peroxide
Hardness, Shore A	60	68	50	63	50
Tensile strength (MPa)	6.4	5.5	6.2	10.3	6.9
Elongation (%)	200	200	400	700	220
Compression set					
(% after 22 hours at 177°C)	14	30	20	42	50
Brittle point (°C)	−65	−65	−101	−101	−68
Oil resistance, Volume change (%)					
After 7 days at 71°C in Skydrol 500A	—	—	+25	—	+30
After 70 hours at 149°C in ASTM #1	+7	+5	+10	+10	+1
After 70 hours at 149°C in ASTM #3	+40	—	+90	+90	+5
Heat aging					
After 48 hours at 316°C					
Hardness, Shore A	—	78	—	—	—
Tensile strength (MPa)	—	3.6	—	—	—
Elongation (%)	—	100	—	—	—

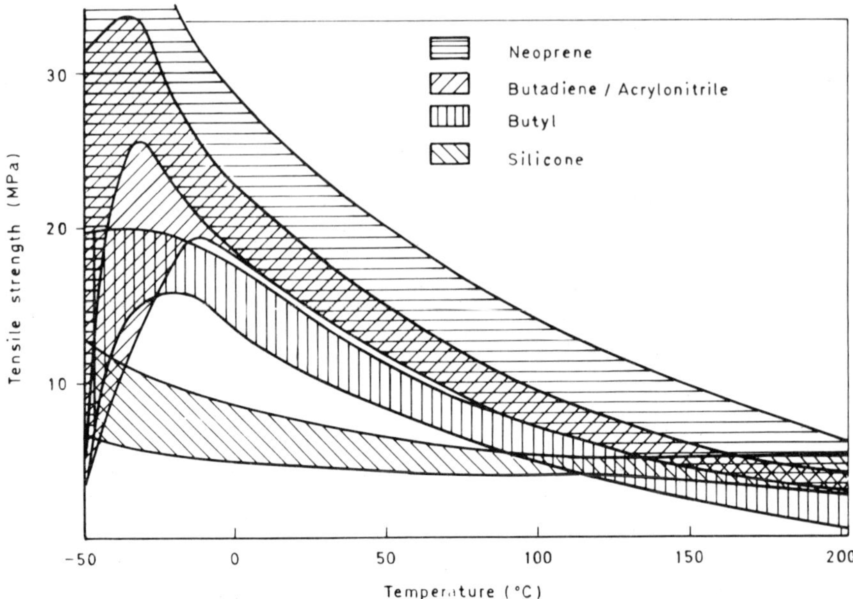

Figure 6.11. The effect of temperature on the tensile strength of various rubber vulcanizates. Reference 54.

formulating an appropriate blend of monomethyl, dimethyl, monophenyl, diphenyl, methylphenyl, monovinyl, and methylvinylchlorosilanes, together with silicon tetrachloride. The final properties of the resin depend as much on the processing and cure conditions as on the original composition, but it is possible to generalize and say that trichlorosilanes produce harder resins immiscible in other organic polymers; dichlorosilanes increase softness and flexibility; and phenylsilanes give resins that are more miscible with organic polymers, less brittle, and of superior thermal resistance. The chlorosilane blend, commonly dissolved in inert solvents that modify the rate of reaction, is then mixed with water. Because of the difference in the rates of hydrolysis of the various chlorosilanes, problems can arise in producing polymers incorporating a proper balance of all the constituents; it is possible to control this by the choice of reaction solvent, or if necessary by sequential addition of the chlorosilanes. After hydrolysis the resin incorporates a certain proportion of silanol groups; these serve to initiate the final cure of the resin to a fully cross-linked structure. This final cure is achieved by processing at elevated temperatures, either with small amounts of residual acid left in the resin as a catalyst, or by incorporation of metal soaps, such as tin or zinc octoate, cobalt naphthenate, or amines such as triethanolamine.

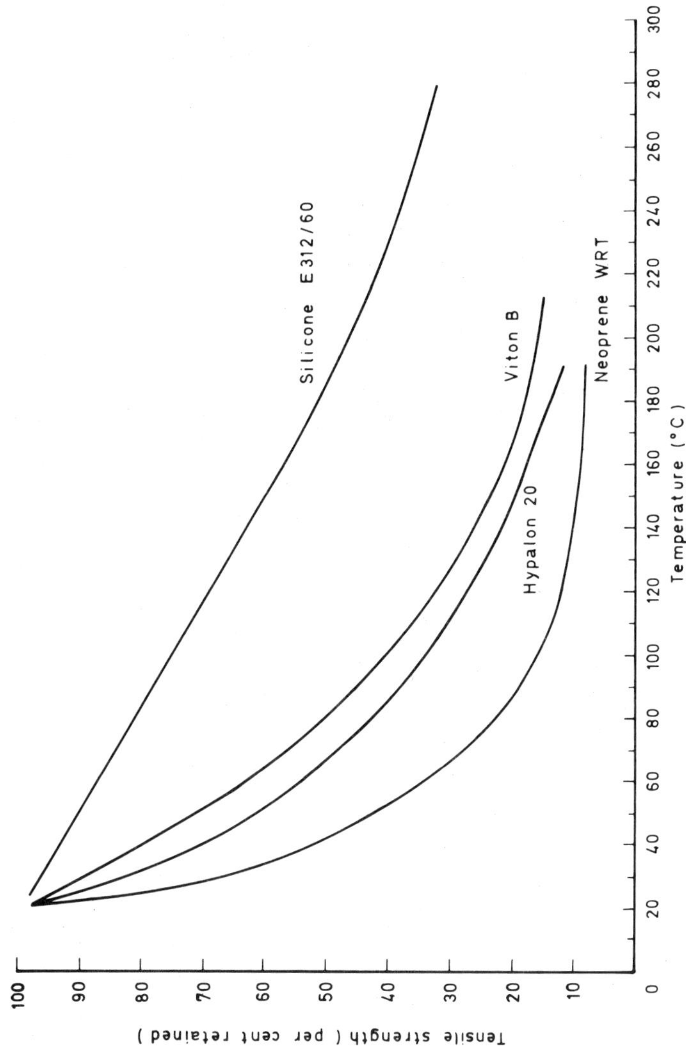

Figure 6.12. Percentage tensile strength retention of various rubber vulcanizates with increase in temperature. Reference 53.

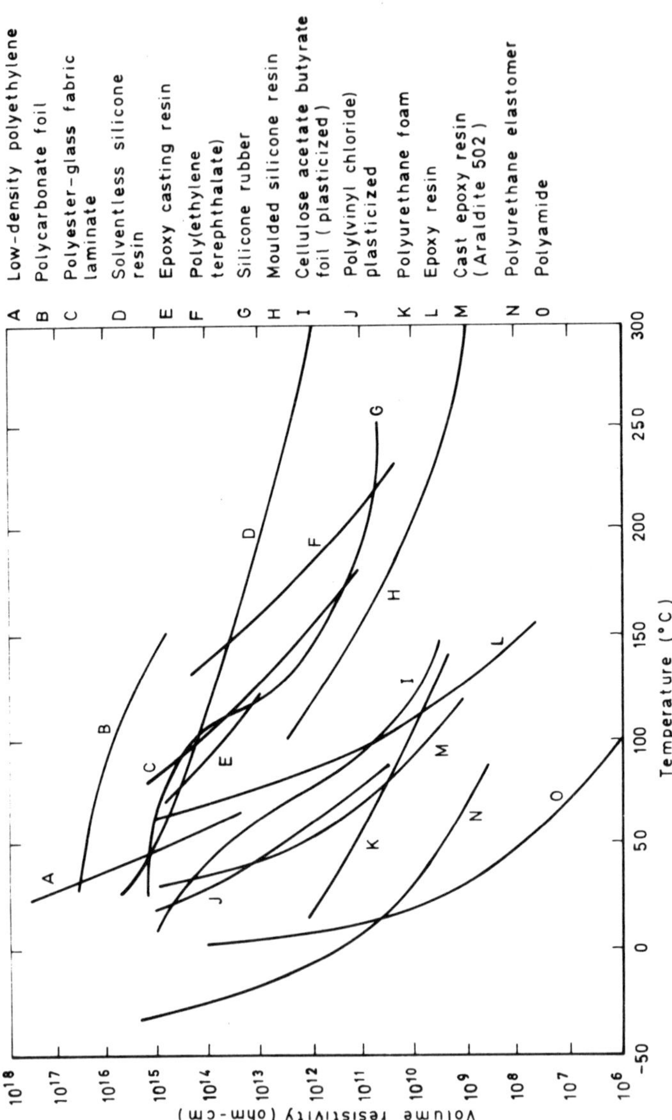

Figure 6.13. Effect of temperature on volume resistivity of various plastics. Reference 55.

TABLE 6.10. Some Electrical Properties of Silicone and Natural Rubber Vulcanizates (Reference 55)

Property	Silicone	Natural rubber
Volume resistivity (ohm-cm)	$1 \times 10^{15} - 5 \times 10^{15}$	10^{15}
Dielectric strength (V/mil)	650	1000
Dielectric constant 60 Hz	3.15–3.4	2.37
Power factor 60 Hz	0.001–0.005	0.15-0.2

Thermal Stability of Silicone Resins

The high-temperature stability of silicone resins is illustrated in Figure 6.14, which shows that weight loss starts to occur between 350 and 400°C in both air and nitrogen (rate of temperature rise 2°C/min). The silicone resins decompose in a similar manner to the elastomers; in inert atmosphere they give rise to cyclic siloxanes, and in oxidizing conditions attack takes place at the methyl groups. The thermal stability of cross-linked polymethylsiloxanes of known network structure has been investigated by Andrianov *et al.*,[56] who have shown that increasing the distance between cross-link points decreases the thermal and thermo-oxidative stability of the polymers, and that the structure of the elemental network unit has a greater influence on thermal stability than on oxidative thermal stability. Figures 6.15 and 6.16 show the weight loss curves resulting from heating in helium and air, respectively.

The products of thermal degradation were primarily hydrocarbons (methane, ethane, propane) and water at temperatures from 300 to 500°C; above 500°C these products were accompanied by increasing amounts of the cyclic siloxane oligomers, principally the trimer and tetramer. In air, DTA shows a large exotherm between 400 and 500°C related to the oxidation of the methyl groups. The same volatile products were observed as in thermal degradation, but there was a concomitant cross-linking reaction giving rise to a high yield of residual silica.

Elevated Temperature Properties of Silicone Resins

The resins are used as coatings, for laminating, and to a lesser extent as molding compounds.

As laminating resins the silicone-glass fiber prepregs, although lacking in drape and tack, can be processed into the complicated shapes normally associa-

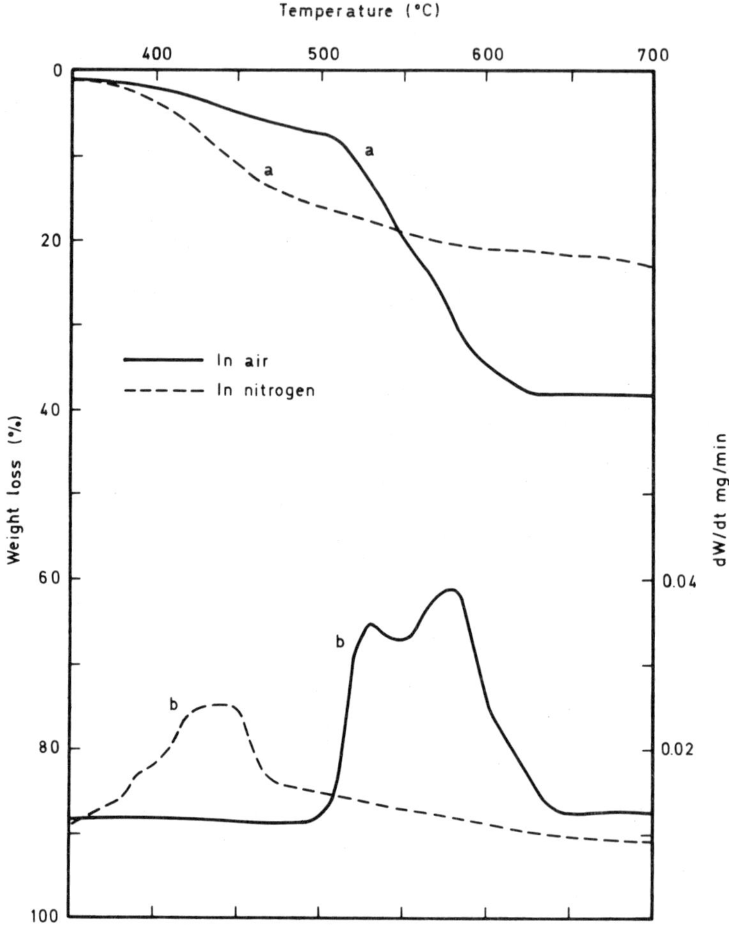

Figure 6.14. Silicone resin MS 840; (a) weight loss and (b) rate of weight loss dW/dt. (Heating rate 2°C/min)

ted with the more pliable polyester and epoxy prepregs. The shelf life of silicone prepregs is said to be outstanding, material stored one year at room temperature being capable of producing high quality structural parts.[57]

The silicone prepreg is processed at approximately 180°C under pressures of 0.1 to 0.7 MPa. After press cure, a post-cure at 250°C is used to remove any residual volatiles from the laminate. Such laminates are used in structural applications that have operating temperatures in the range 250 to 500°C, for specific electrical purposes, and for applications that require superior resistance to oxidation and fire retardance, the initial mechanical properties of the laminates being inferior to those of epoxy or polyester resins (Table 6.11). Some typical electri-

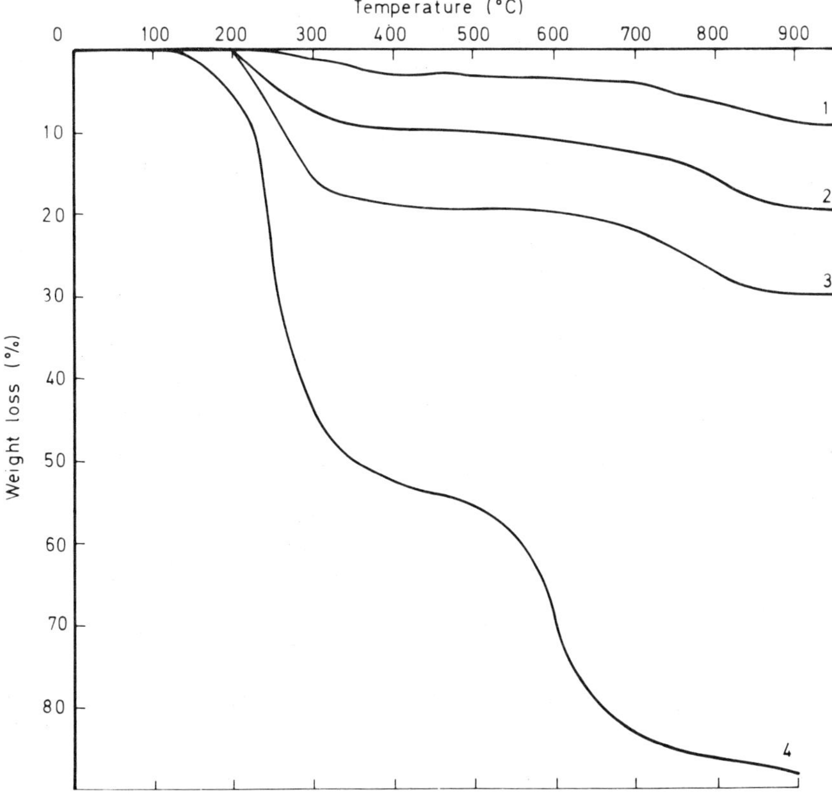

Figure 6.15. *Weight loss of polymethylsiloxane resins in helium. (Heating rate 5°C/min) Molecular weight between cross-links; 1 = 66, 2 = 77.5, 3 = 156, 4 = no cross-links. Reference 56.*

TABLE 6.11. *Properties of Silicone Resin/Glass Cloth Laminates Compared with Other Systems (Reference 59)*

Resin	Specific gravity	Tensile strength (MPa)	Tensile modulus (GPa)	Compressive strength (MPa)	Flexural strength (MPa)	Heat resistance (continuous) (°C)
Polyester	1.5–2.1	207–483	6.9–20.7	172–345	276–621	149–177
Epoxy	1.9–2.0	138–414	13.0–27.6	345–482	482–690	166–260
Phenolic	1.8–2.0	276–414	6.9–20.7	241–276	448–655	177–260
Silicone	1.6–1.9	69–241	6.9–13.8	172–317	69–262	204–371

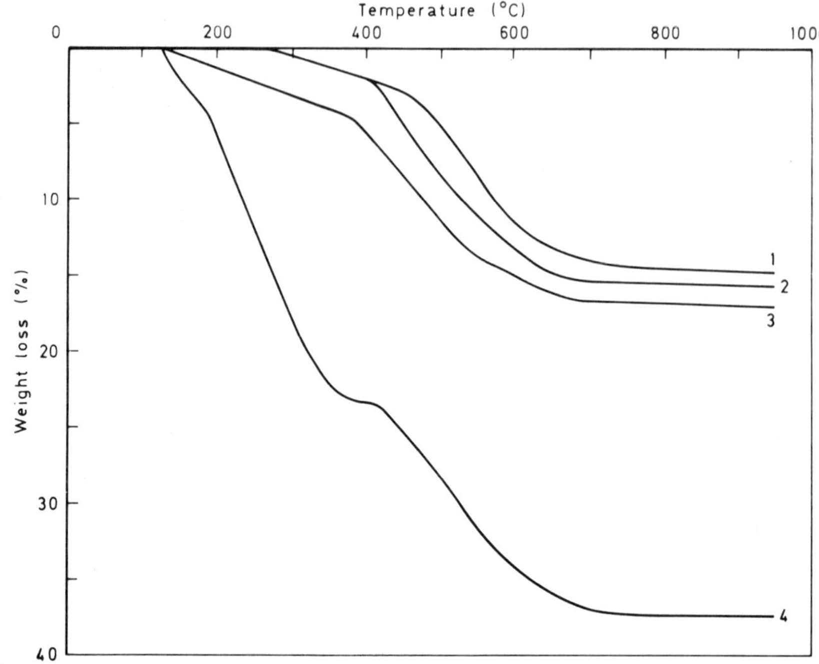

Figure 6.16.   Weight loss of polymethylsiloxane resins in air. (Heating rate 5°c/min) Molecular weight between cross-links; 1 = 66, 2 = 77.5, 3 = 156, 4 = no cross-links. Reference 56.

cal properties are given in Table 6.12. Mechanical properties as a function of temperature are shown in Figures 6.17 and 6.18.[58] The stability of the silicones on thermal aging is illustrated in Figures 6.19 and 6.20.[58] The effect of temperature on the volume resistivity of various plastics including two silicone resins has already been depicted in Figure 6.13, and the effect of aging at 260°C on dielectric constant and loss tangent is illustrated in Figure 6.21.[57]

For coating applications the resins can be used on their own, where high-temperature resistance is of importance, or mixed with various resins such as alkyd, esterified epoxy, phenol-, urea-, and melamine-formaldehyde. The film properties of silicones, such as flow and toughness, are inferior to those of conventional coating materials, hence the use of mixtures, but the heat resistance decreases rapidly with the addition of other resins. The cross-linked resin films are resistant to acids and aqueous solutions of chemicals, but they are attacked by bases and steam.

When the combined temperature and electrical requirements call for the use of these relatively weak resins as moldings, they can be reinforced with fibrous or particulate fillers to improve toughness.

TABLE 6.12. Some Typical Electrical Properties of Silicone and Epoxy Resins (Reference 55)

Property	Silicone		Epoxy			Silicone	Epoxy		
	Molding compounds		Molding compounds			Flexible casting resin	Cast resin		
	Glass fiber-filled	Mineral-filled	Unfilled	Mineral-filled	Mineral filler glass-reinforced		Unfilled	Silica-filled	Flexible
Volume resistivity									
(ohm-cm)	1×10^{14}	5×10^{13}	10.5×10^{15}	10^{15}	1.72–4.70×10^{15}	2×10^{15}	10^{12}–10^{17}	10^{13}–10^{16}	10^{12}–10^{14}
Dielectric strength									
(V/mil) short time	280	390	400–433	345–387	304–364	550	400–450	400–450	235–400
step-by-step	270		384	400	340–350		380		
Dielectric constant									
60 Hz	4.50	3.65	4.35–5.0	4.4–5.56	5.82–7.17	2.70	3.5–5.0	3.2–4.5	3.0–6.0
1 KHz	4.40	3.65	4.36–5.0	4.2–4.91	4.99–5.55	—	3.5–4.5	3.2–4.0	3.0–5.0
1 MHz	4.25	3.60	4.37–4.8	4.1–4.56	4.75–5.18	2.70	3.3–4.0	3.0–3.8	3.0–6.0
Dissipation factor									
60 Hz	0.010	0.0040	0.38	0.011–0.083	0.122–0.183	0.001	0.002–0.010	0.008–0.03	0.010–0.040
1 KHz	0.010	0.0035	0.02–0.24	0.019–0.140	0.117–0.15	—	0.002–0.020	0.008–0.03	0.012–0.050
1 MHz	0.004	0.0030	0.022–0.26	0.013–0.137	0.13–0.143	0.001	0.030–0.050	0.02–0.04	0.018–0.090
Arc resistance									
(secs)	240	230	154–180	128–180	123–133	—	45–120	150–300	50–180

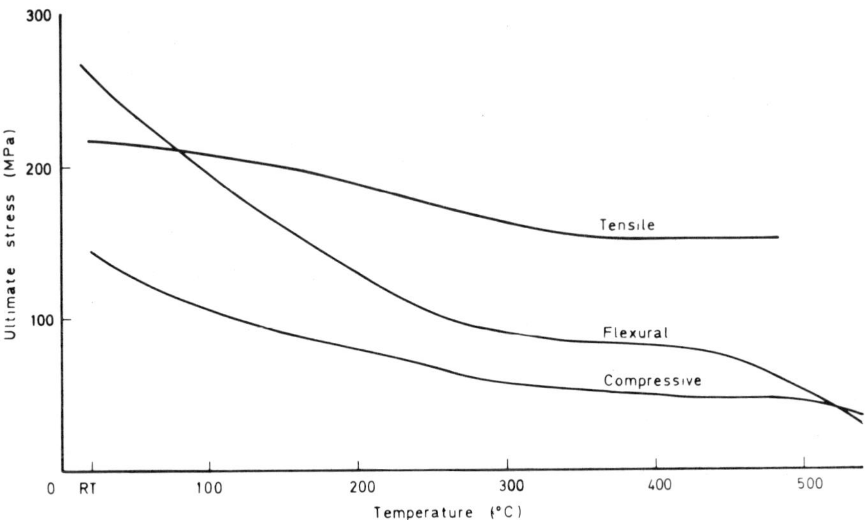

Figure 6.17. Mechanical properties vs. temperature for a structural grade silicone resin/ glass cloth laminate. Reference 58.

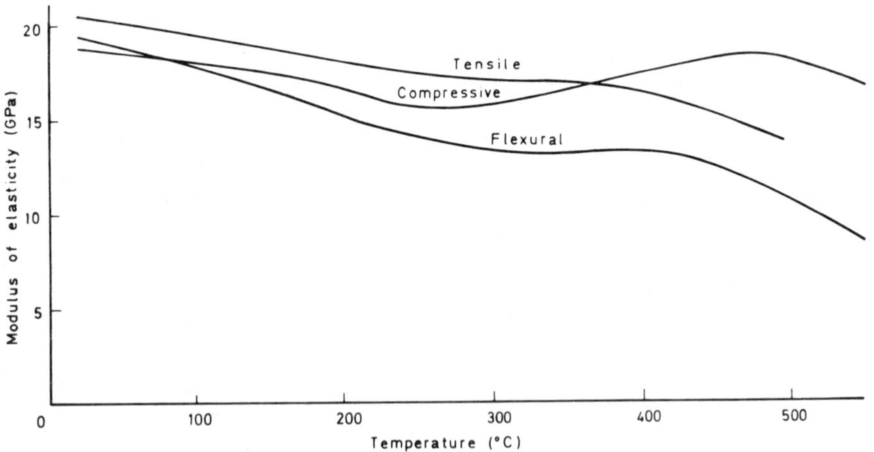

Figure 6.18. Stiffness properties vs. temperature for a structural grade silicone resin/glass cloth laminate. Reference 58.

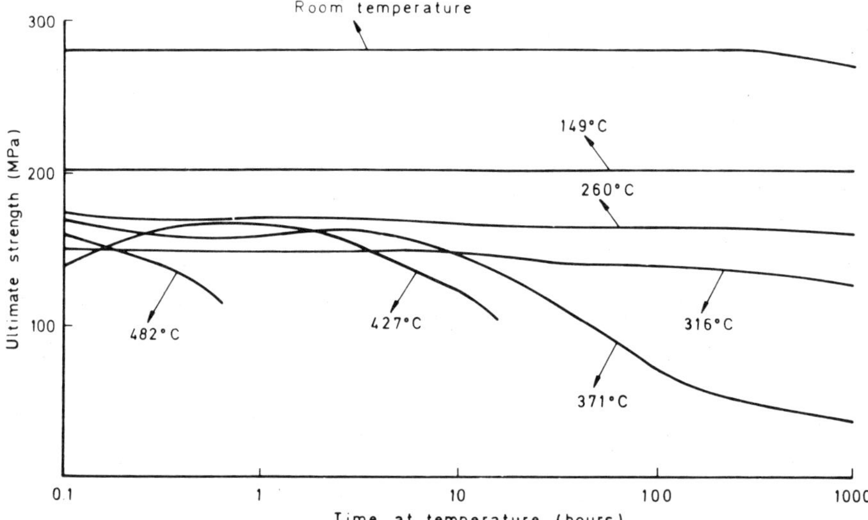

Figure 6.19. Long-term aging properties in tension for a structural grade silicone resin/glass cloth laminate. Reference 58.

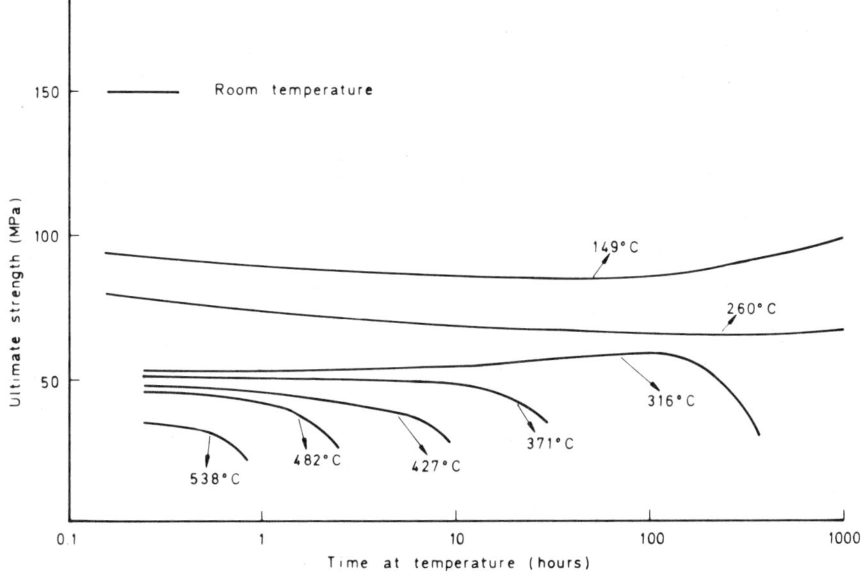

Figure 6.20. Long-term aging properties in compression for a structural grade silicone resin/glass fabric laminate. Reference 58.

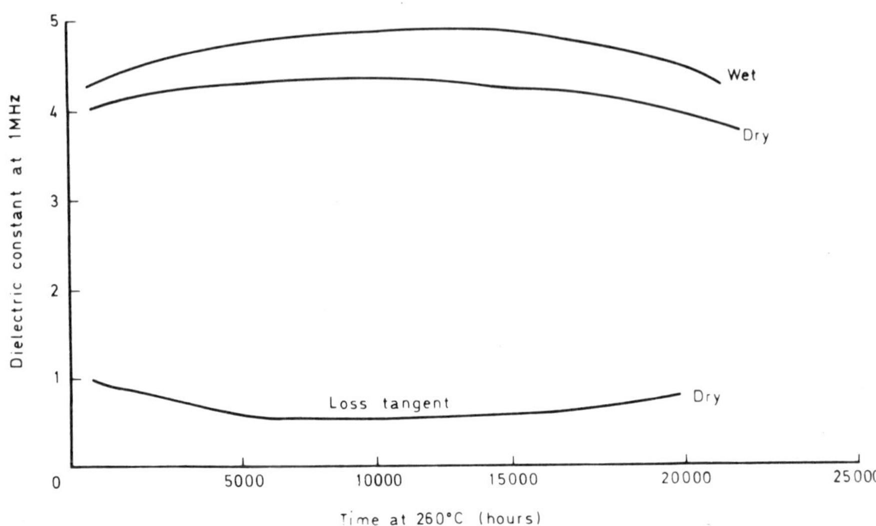

Figure 6.21.   Heat aging stability characteristics of electrical properties for a silicone resin/ glass fabric laminate. Reference 57.

SUMMARY

1. The silicones provide materials with a unique combination of both low- and high-temperature resistance.

2. The early problems of low physical strength and toughness have been to a large extent overcome by a combination of better fillers and better cross-linking agents. Research is continuing on the relationships between silicone rubber, fume silica, and the final properties of the vulcanizate,[14] and it is possible that such work will result in even stronger materials.

3. Stabilization improves high-temperature performance, and the search for better antioxidants promises extensions to 300–350°C as the upper limit of useful life.[37,56,60]

4. Modifications have been made to the structure of the main polysiloxane chain, as distinct from changing substituent groups as described above. Examples of such modifications include silphenylene[61,62]:

$$\left[\begin{array}{c} R_1 \\ | \\ Si \\ | \\ R_3 \end{array} \quad\bigcirc\quad \left(\begin{array}{c} R_2 \\ | \\ Si-O \\ | \\ R_4 \end{array}\right)_x \right]_n$$

and cyclodisilazane groups[63,64]:

$$\left[\begin{array}{c} \overset{\displaystyle R \diagdown \quad \diagup R}{\underset{\displaystyle R \diagup \quad \diagdown}{Si}} \\[4pt] \text{—}\!\!-\overset{R}{\underset{R}{Si}}\text{—N}\overset{}{\underset{}{\diagup\diagdown}}N\!\!-\!\!\left(\overset{R}{\underset{R}{Si}}\text{—O}\right)_{\!x} \\[4pt] \overset{}{\underset{\displaystyle R \diagup \quad \diagdown R}{Si}} \end{array}\right]_{n}$$

The aim of such structural changes has been to increase the thermal stability by reducing the tendency to reversion and by inhibiting the oxidation of the polymer.

The only structural change to date to bring about a worthwhile improvement in properties has been the incorporation of carborane linkages, which is dealt with in the following chapter.

REFERENCES

1. C. Friedel and J. M. Crafts, *Ann.* **136**, 203 (1865).
2. F. S. Kipping, *J. Soc. Chem. Ind.* **55**, 229T (1937).
3. E. G. Rochow, U.S. Patent No. 2,380,995 (1945).
4. J. A. C. Watt, *Chem. Brit.* **6**, 519 (1970).
5. B. A. Bluestein, U.S. Patent No. 3,435,000 (1969).
6. W. J. Bobear, Silicone rubber, in *Rubber Technology,* 2nd ed., M. Morton, (ed.), Van Nostrand-Reinhold, New York (1973), Chap. 15, p. 368.
7. G. G. Karelina and I. A. Pliner, *Kauch. Rezina* **28**, 3 (1969).
8. E. L. Warrick, *Rubber Chem. Technol.* **49**(4), 909 (1976).
9. W. Kniege, in *Chemistry and Technology of Silicones,* W. Noll, (ed.), Academic Press, New York (1968), p. 402.
10. L. E. St. Pierre and R. S. Chahal, *J. Polym. Sci. C* **30**, 429 (1970).
11. D. W. Southwart and T. Hunt, *J. Inst. Rubber Ind.* **2**(2), 77 (1968).
12. D. W. Southwart and T. Hunt, *J. Inst. Rubber Ind.* **2**(2), 79 (1968).
13. D. W. Southwart and T. Hunt, *J. Inst. Rubber Ind.* **3**(6), 249 (1969).
14. D. W. Southwart and T. Hunt, *J. Inst. Rubber Ind.* **4**(2), 74 (1970).
15. M. Schatz and K. Svehla, *Sb. Vys. Sk. Chem-Technol. Praze Org. Chem. Technol. C* **20**, 157 (1973); *Chem. Abs.* **81**, 122261 (1974).
16. H. A. Vaughn, U.S. Patent No. 3,328,340 (1967).
17. Deutsche Gold-und-Silber-Scheideanstalt Vormals Roessler, Brit. Patent No. 1,087,584 (1967).
18. W. W. Wright and W. A. Lee, The search for thermally stable polymers, in *Progress in High Polymers,* Vol. 2, J. C. Robb and F. Peaker, (eds.), Iliffe Books Ltd., London (1968).
19. E. A. Goldovskii and A. S. Kuzminskii, *Polym. Sci. Tech.* **6**(4), 75 (1979).
20. T. H. Thomas and T. C. Kendrick, *J. Polym. Sci. A 2* **7**, 537 (1969).
21. N. Grassie and I. G. MacFarlane, *Eur. Polym. J.* **14**, 875 (1978).

22. R. M. Aseyeva, S. M. Mezhikovskii, A. A. Kholodovskaya, O. G. Selskaya, and A. A. Berlin, *Polym. Sci. U.S.S.R. A* **15**(8), 2104 (1973).
23. M. A. Verkhotin, V. V. Rode, and S. R. Rafikov, *Vysok. Soed. B* **9**(11), 847 (1967).
24. N. Grassie, I. G. MacFarlane, and K. F. Francey, *Eur. Polym. J.* **15**, 415 (1979).
25. N. Grassie and K. F. Francey, *Polym. Degrdn. Stab.* **2**, 53 (1980).
26. N. Grassie, K. F. Francey, and I. G. MacFarlane, *Polym. Degrdn. Stab.* **2**, 67 (1980).
27. K. A. Andrianov and N. N. Sokolov, *Khim. Prom.* 329 (1955).
28. E. A. Goldovskii and A. S. Kuzminskii, *Rev. Gen. Caout. Plast.* **45** (3), 321, (4), 457 (1968).
29. T. C. Williams, R. A. Pike, and F. Fekete, *Ind. Eng. Chem.* **51**, 939 (1959).
30. D. K. Thomas, *Polymer* **7**, 99 (1966).
31. Reference 9, p. 401.
32. L. N. Phillips, D. K. Thomas, and W. W. Wright, Brit. Patent No. 1,144,641 (1969).
33. L. N. Phillips, D. K. Thomas, and W. W. Wright, Brit. Patent No. 1,161,909 (1969).
34. E. Kay, D. K. Thomas, and W. W. Wright, *Int. Rubber Conf. (Brighton)* (1972), F1.
35. V. V. Kopylov, N. D. Baikina, and A. N. Pravednikov, *Int. Polym. Sci. Tech.* **4**(9), 103 (1977).
36. D. K. Thomas, *Polymer* **7**, 243 (1966).
37. G. S. Tubjanskaya, R. I. Kobzova, E. M. Oparina, V. A. Zaitsev, A. A. Egorova, and B. K. Kabanov, *Plast. Massy* (10), 21 (1969).
38. W. J. Bobear, U.S. Patent No. 3,261,802 (1966).
39. K. Z. Gumargaliera, E. V. Kamzolkina, D. Kh. Kitaeva, and G. P. Gladyshev, *Vysok. Soed. B* **16**, 310 (1974).
40. M. P. Grinblat, N. I. Rozova, I. A. Kats, and Yu. V. Kokhanov, *Int. Polym. Sci. Tech.* **3**(12), 59 (1976).
41. V. P. Kolomytsyn, A. Ya. Borzenkova, B. G. Zoegintseva, and S. A. Spiridova, *Khim. Khim. Technol. (Minsk)* **9**, 104 (1975).
42. I. N. Kasyanova, F. A. Galil-Ogly, L. V. Kireeva, and A. S. Shaptin, *Int. Polym. Sci. Tech.* **2**(1), 5 (1975).
43. L. N. Phillips and D. K. Thomas, French Patent No. 1,440,466 (1966).
44. R. I. Kobzova, G. S. Tubjanskaya, E. M. Oparina, and N. K. Lerkina, *Tr. Vses. Nauch. Issled. Inst. Pererab. Nefti.* (11), 30 (1969).
45. J. M. Nielson, *J. Polym. Sci. Symp.* **40**, 189 (1973).
46. E. A. Goldovskii, R. F. Fatkulina, A. S. Kuzminskii, and A. A. Dontsov, *Int. Polym. Sci. Tech.* **5**(12), 22 (1978).
47. L. Morris, *2nd Nat. SAMPE Tech. Conf.* **2**, 183 (1970).
48. H. Singh, *2nd Nat. SAMPE Nat. Symp. and Exhib.* **22**, 227 (1977).
49. I.C.I. Data Sheet, Designers Guide to I.C.I. Silicone Elastomers (Nov. 1973).
50. W. Kniege, *Rubber India* **25** (10), 14 (1973).
51. D. K. Thomas, *Polymer* **13**, 479 (1972).
52. R. Sinnot (unpublished results).
53. D. K. Thomas and R. Sinnot, *J. Inst. Rubber Ind.* **3**, 163 (1969).
54. Reference 9, p. 497.
55. K. N. Mathes, Electrical properties, in *Encyclopedia of Polymer Science and Technology*, H. F. Mark, N. G. Gaylord, and N. M. Bikales, (eds.), Interscience Publishers, New York (1966), Vol. 5, p. 575.
56. Ya. V. Kovalenko, M. L. Pustyl'nik, T. S. Bebchuk, M. N. Muzafarova, Ya. V. Zherdev, V. D. Nederosol, and K. A. Andrianov, *Polym. Sci. U.S.S.R.* **20**(7), 1824 (1978).
57. H. J. Doyle and S. C. Harrier, Phenolics and silicones, in *Handbook of Fiberglass and Advanced Plastics Composites*, G. Lubin, (ed.), Van Nostrand Reinhold Co., New York (1969), p. 85.
58. K. H. Boller and K. E. Kimball, Wright Air Development Center Report, WADC TR-59-229 (1959).

59. K. Hattori, Reinforced plastics, in *Encyclopedia of Polymer Science and Technology*, H. F. Mark, N. G. Gaylord, N. M. Bikales, (eds.), Interscience Publishers, New York (1970), Vol. 12, p. 1.
60. A. A. Berlin, *Polym. Sci. U.S.S.R.* **13** (2), 309 (1971).
61. R. L. Merker and M. J. Scott, *J. Polym. Sci. A* **2,** 15 (1964).
62. L. W. Breed, R. L. Elliot, and M. E. Whitehead, *J. Polym. Sci. A 1* **5,** 2745 (1967).
63. L. W. Breed, R. L. Elliot, and H. Rosenberg, U.S. Patent No. 3,702,317 (1972).
64. L. W. Breed and J. C. Wiley, U.S. Patent No. 3,803,086 (1974).

SUPPLEMENTARY BIBLIOGRAPHY

R. R. McGregor, *Silicones and Their Uses,* McGraw-Hill, London (1954).

G. G. Freeman, *Silicones, An Introduction to Their Chemistry and Applications,* Iliffe Books Ltd., London (1962).

W. Noll, *Chemistry and Technology of Silicones,* Academic Press, London (1968).

F. M. Lewis, Elastomers by condensation polymerisation. B. The chemistry of silicone elastomers, *High Polymers* **23**(2), 767 (1968).

T. N. Balykova and V. V. Rhode, Progress in the study of the degradation and stabilization of siloxane polymers, *Russ. Chem. Rev.* **38** (4), 306 (1969).

H. K. Lichtenwalner and M. N. Sprung, Silicones, in *Encyclopedia of Polymer Science and Technology,* H. F. Mark, N. G. Gaylord, and N. M. Bikales, (eds.), Interscience, New York (1970), Vol. 12, p. 464.

C. M. Blow, Silicone rubbers, in *Rubber Technology and Manufacture,* Butterworths, London (1971).

K. J. Saunders, Silicones, in *Organic Polymer Chemistry,* Chapman and Hall, London (1973), p. 347.

E. L. Warrick, O. R. Pierce, K. E. Polmanteer, and J. C. Saam, Silicone elastomer developments, 1967–1977, *Rubber Chem. and Technol.* **52** (3), 437 (1979).

D. J. Bannister and J. A. Semlyen, Studies of cyclic and linear poly(dimethylsiloxanes). 6. Effect of heat, *Polymer* **22,** 377 (1981).

7

BORON-CONTAINING POLYMERS—THE CARBORANESILOXANES

INTRODUCTION

During the 1950s much research effort was devoted to boron hydride chemistry in a search for high-energy liquid fuels. This led to the discovery of polyhedral organoborane compounds, which were designated carboranes. Of all the boron hydrides, decaborane received the most attention because of its availability and inherent stability.

Decaborane skeleton

Under certain conditions decaborane reacts with acetylene in the presence of a Lewis base to form the molecule $B_{10}H_{10}C_2H_2$, commonly called o-carborane.[1,2] On heating this compound in a closed system at approximately 500°C, or in a flow system at about 600°C, it isomerizes to m-carborane in quantitative yield.[3,4] Heating to still higher temperatures, 630–700°C, results in further isomerization to p-carborane.[5]

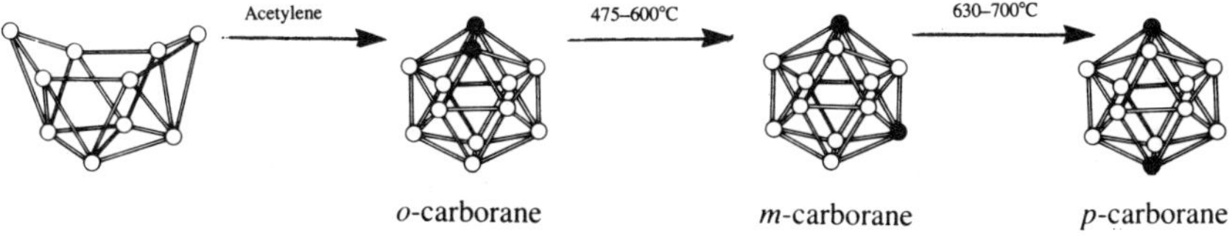

These four compounds have been utilized as building blocks in synthesizing many types of polymers. For example, 1-vinyl- or 1-allyl-*o*-carborane may be polymerized[6,7]; *m*-carborane units may be incorporated into polyamides, polyesters, or polyurethanes[8,9,10]; *m*- or *p*-carborane moieties may be linked by single atoms,[11, 12] such as S, Hg, $Sn(CH_3)_2$; decaborane will form a polymer with P–O–P or P–N=P linkages.[13] Greatest success has been achieved, however, with carboranesiloxanes and it is with these that the remainder of this chapter is concerned.

Preparation of Carboranesiloxanes

The first mention of polymers of this type, in which *o*-carborane nuclei formed pendant groups to a siloxane chain, was in a note that appeared in Polymer Letters in January, 1964.[14] This briefly stated that condensation of a carboranesilane with a dimethylsilane yielded products of molecular weight up to 13,000 that were tough, flexible, and had high thermal stability.

$$R-\underset{\underset{CH_3}{|}}{\overset{\overset{CH_3}{|}}{Si}}-R \quad + \quad R'-\underset{\underset{[CH_2]_4}{|}}{\overset{\overset{CH_3}{|}}{Si}}-R' \longrightarrow \left[\underset{\underset{CH_3}{|}}{\overset{\overset{CH_3}{|}}{Si}}-O-\underset{\underset{[CH_2]_4}{|}}{\overset{\overset{CH_3}{|}}{Si}}-O\right]_n$$

where $R,R' = Cl$ or OC_2H_5

A more detailed account of the preparation of these polymers was not published until 1967.[15] Also described in this paper was the synthesis of carboranesiloxanes containing *m*-carborane units in the main chain separated from the silicon atoms by straight-chain alkyl groups.

$$\left[\underset{\underset{CH_3}{|}}{\overset{\overset{CH_3}{|}}{Si}}-[CH_2]_3-CB_{10}H_{10}C-[CH_2]_3-\underset{\underset{CH_3}{|}}{\overset{\overset{CH_3}{|}}{Si}}-O\left[\underset{\underset{CH_3}{|}}{\overset{\overset{CH_3}{|}}{Si}}-O\right]_x\right]_n$$

Since neither of these two polymer types were marketed commercially, they will not be considered further here.

Meanwhile the preparation of poly-*m*-carboranylenesiloxanes without alkyl groups in the chain had been described at the 150th American Chemical Society meeting in 1965.[16] Polymers above 1,000–2,000 in molecular weight could not be prepared by the conventional route of hydrolysis of 1,7-bis(chlorodimethylsilyl)*m*-carborane, this failure being ascribed to the extremely electronegative nature of the carborane nucleus. Success was achieved by ferric chloride-catalyzed elimination of an alkyl halide from equimolar mixtures of dichloro- and dialkoxysilanes at temperatures of 140–165°C.

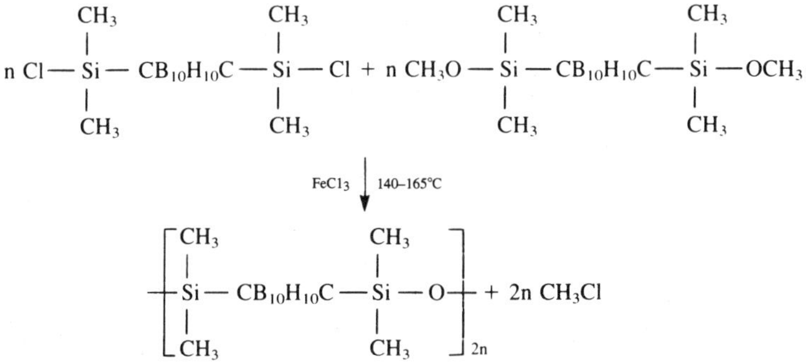

$$n\ Cl-\underset{\underset{CH_3}{|}}{\overset{\overset{CH_3}{|}}{Si}}-CB_{10}H_{10}C-\underset{\underset{CH_3}{|}}{\overset{\overset{CH_3}{|}}{Si}}-Cl\ +\ n\ CH_3O-\underset{\underset{CH_3}{|}}{\overset{\overset{CH_3}{|}}{Si}}-CB_{10}H_{10}C-\underset{\underset{CH_3}{|}}{\overset{\overset{CH_3}{|}}{Si}}-OCH_3$$

$$\xrightarrow[\text{FeCl}_3]{140-165°C}$$

$$\left[-\underset{\underset{CH_3}{|}}{\overset{\overset{CH_3}{|}}{Si}}-CB_{10}H_{10}C-\underset{\underset{CH_3}{|}}{\overset{\overset{CH_3}{|}}{Si}}-O-\right]_{2n}\ +\ 2n\ CH_3Cl$$

SiB-1 or Dexsil 100

The product, a white crystalline solid with a melting range of 235–255°C, was soluble in a number of hot organic solvents and had a molecular weight of up to 16,000 as measured by vapor pressure osmometry. Higher molecular weight material was too insoluble for accurate measurements to be made. The product was first named SiB-1, but this was later changed to Dexsil 100.

It was possible by the above preparative route to introduce further dimethylsiloxyl groups between the carborane nuclei.

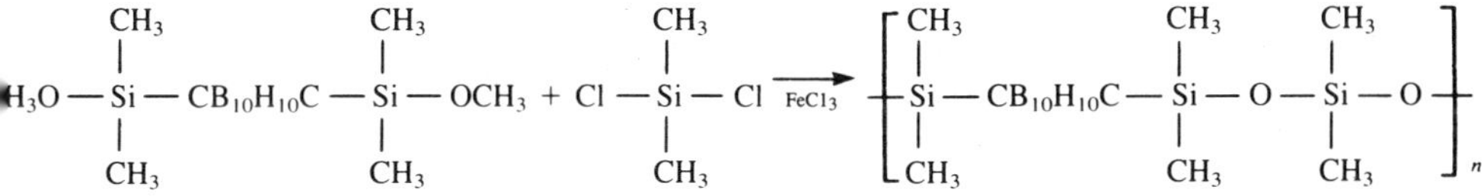

$$CH_3O-\underset{\underset{CH_3}{|}}{\overset{\overset{CH_3}{|}}{Si}}-CB_{10}H_{10}C-\underset{\underset{CH_3}{|}}{\overset{\overset{CH_3}{|}}{Si}}-OCH_3\ +\ Cl-\underset{\underset{CH_3}{|}}{\overset{\overset{CH_3}{|}}{Si}}-Cl\ \xrightarrow{\text{FeCl}_3}\ \left[-\underset{\underset{CH_3}{|}}{\overset{\overset{CH_3}{|}}{Si}}-CB_{10}H_{10}C-\underset{\underset{CH_3}{|}}{\overset{\overset{CH_3}{|}}{Si}}-O-\underset{\underset{CH_3}{|}}{\overset{\overset{CH_3}{|}}{Si}}-O-\right]_{n}$$

SiB-2 or Dexsil 200

The product SiB-2 (Dexsil 200) was highly elastomeric, although retaining some crystallinity. The next member of the series was prepared similarly.

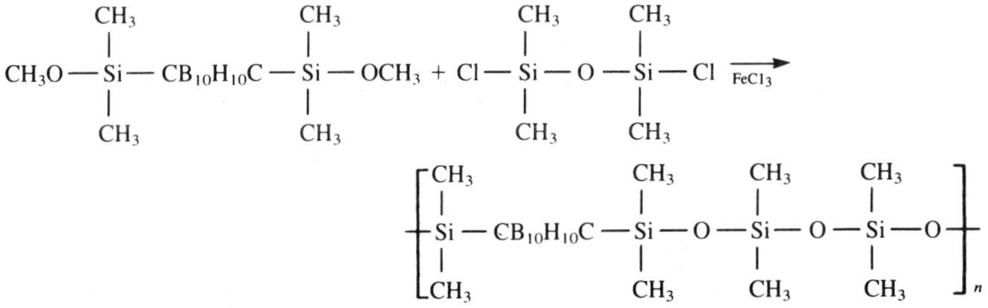

$$CH_3O-\underset{\underset{CH_3}{|}}{\overset{\overset{CH_3}{|}}{Si}}-CB_{10}H_{10}C-\underset{\underset{CH_3}{|}}{\overset{\overset{CH_3}{|}}{Si}}-OCH_3\ +\ Cl-\underset{\underset{CH_3}{|}}{\overset{\overset{CH_3}{|}}{Si}}-O-\underset{\underset{CH_3}{|}}{\overset{\overset{CH_3}{|}}{Si}}-Cl\ \xrightarrow{\text{FeCl}_3}$$

$$\left[-\underset{\underset{CH_3}{|}}{\overset{\overset{CH_3}{|}}{Si}}-CB_{10}H_{10}C-\underset{\underset{CH_3}{|}}{\overset{\overset{CH_3}{|}}{Si}}-O-\underset{\underset{CH_3}{|}}{\overset{\overset{CH_3}{|}}{Si}}-O-\underset{\underset{CH_3}{|}}{\overset{\overset{CH_3}{|}}{Si}}-O-\right]_{n}$$

SiB-3 or Dexsil 300

In this polymer crystallinity was completely absent.

For the next higher homologue a slightly different preparative route was necessary.

$$\text{Li} - \text{CB}_{10}\text{H}_{10}\text{C} - \text{Li} \; + \; 2\,\text{Cl} - \underset{\underset{\text{CH}_3}{|}}{\overset{\overset{\text{CH}_3}{|}}{\text{Si}}} - \text{O} - \underset{\underset{\text{CH}_3}{|}}{\overset{\overset{\text{CH}_3}{|}}{\text{Si}}} - \text{Cl} \longrightarrow$$

$$\text{Cl} - \underset{\underset{\text{CH}_3}{|}}{\overset{\overset{\text{CH}_3}{|}}{\text{Si}}} - \text{O} - \underset{\underset{\text{CH}_3}{|}}{\overset{\overset{\text{CH}_3}{|}}{\text{Si}}} - \text{CB}_{10}\text{H}_{10}\text{C} - \underset{\underset{\text{CH}_3}{|}}{\overset{\overset{\text{CH}_3}{|}}{\text{Si}}} - \text{O} - \underset{\underset{\text{CH}_3}{|}}{\overset{\overset{\text{CH}_3}{|}}{\text{Si}}} - \text{Cl}$$

$$\xrightarrow[\text{FeCl}_3]{\quad \text{C}_2\text{H}_5\text{O} - \underset{\underset{\text{CH}_3}{|}}{\overset{\overset{\text{CH}_3}{|}}{\text{Si}}} - \text{OC}_2\text{H}_5 \quad}$$

$$\left[\underset{\underset{\text{CH}_3}{|}}{\overset{\overset{\text{CH}_3}{|}}{\text{Si}}} - \text{CB}_{10}\text{H}_{10}\text{C} - \underset{\underset{\text{CH}_3}{|}}{\overset{\overset{\text{CH}_3}{|}}{\text{Si}}} - \text{O} - \underset{\underset{\text{CH}_3}{|}}{\overset{\overset{\text{CH}_3}{|}}{\text{Si}}} - \text{O} - \underset{\underset{\text{CH}_3}{|}}{\overset{\overset{\text{CH}_3}{|}}{\text{Si}}} - \text{O} - \underset{\underset{\text{CH}_3}{|}}{\overset{\overset{\text{CH}_3}{|}}{\text{Si}}} - \text{O} \right]_n$$

SiB-4 or Dexsil 400

With increasing siloxane content, better low-temperature elastomeric behavior was observed. All the above equations show methyl substituents, but other alkyl or phenyl groups could readily be incorporated. The introduction of phenyl groups improved thermo-oxidative resistance but reduced the elastomeric character of the polymer. Products containing methyl and phenyl substituents were designated, for example, Dexsil 202 and Dexsil 402.[17]

$$\left[\underset{\underset{\text{CH}_3}{|}}{\overset{\overset{\text{CH}_3}{|}}{\text{Si}}} - \text{CB}_{10}\text{H}_{10}\text{C} - \underset{\underset{\text{CH}_3}{|}}{\overset{\overset{\text{CH}_3}{|}}{\text{Si}}} - \text{O} - \underset{\underset{\text{C}_6\text{H}_5}{|}}{\overset{\overset{\text{C}_6\text{H}_5}{|}}{\text{Si}}} - \text{O} \right]_n$$

Dexsil 202

$$\left[\underset{\underset{\text{CH}_3}{|}}{\overset{\overset{\text{CH}_3}{|}}{\text{Si}}} - \text{CB}_{10}\text{H}_{10}\text{C} - \underset{\underset{\text{CH}_3}{|}}{\overset{\overset{\text{CH}_3}{|}}{\text{Si}}} - \text{O} - \underset{\underset{\text{CH}_3}{|}}{\overset{\overset{\text{CH}_3}{|}}{\text{Si}}} - \text{O} - \underset{\underset{\text{C}_6\text{H}_5}{|}}{\overset{\overset{\text{C}_6\text{H}_5}{|}}{\text{Si}}} - \text{O} - \underset{\underset{\text{CH}_3}{|}}{\overset{\overset{\text{CH}_3}{|}}{\text{Si}}} - \text{O} \right]_n$$

Dexsil 402

Random copolymers were also prepared containing a mixture of Dexsil 100 and Dexsil 200 units.[18] These were called Dexsil 125, 150, 160, and 175, depending upon the relative proportions of the two units. They ranged from powdery to rubbery materials.

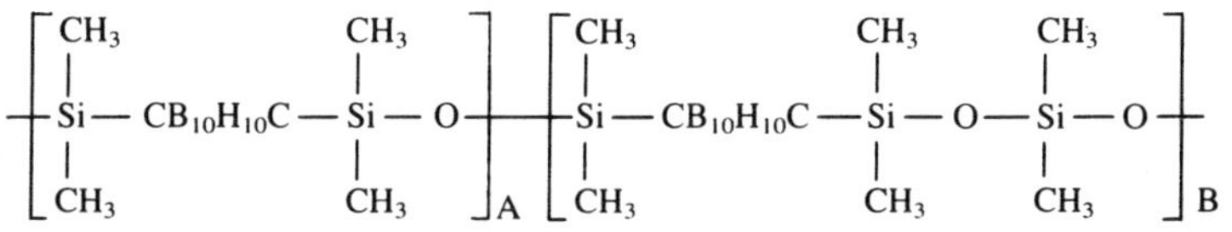

$$\left[\begin{array}{c} CH_3 \\ | \\ Si-CB_{10}H_{10}C-Si-O \\ | \\ CH_3 \end{array}\begin{array}{c} CH_3 \\ | \\ \\ | \\ CH_3 \end{array}\right]_A\left[\begin{array}{c} CH_3 \\ | \\ Si-CB_{10}H_{10}C-Si-O-Si-O \\ | \\ CH_3 \end{array}\begin{array}{c} CH_3 \\ | \\ \\ | \\ CH_3 \end{array}\begin{array}{c} CH_3 \\ | \\ \\ | \\ CH_3 \end{array}\right]_B$$

Dexsil 100 =	100% A,	0% B.
Dexsil 125 =	75% A,	25% B.
Dexsil 150 =	50% A,	50% B.
Dexsil 160 =	40% A,	60% B.
Dexsil 175 =	25% A,	75% B.
Dexsil 200 =	0% A,	100% B.

Although cross-linking of these polymers was possible via free-radical hydrogen abstraction reactions, it was greatly facilitated if vinylic groups were present, and this was achieved by the incorporation of a small number of 1-vinyl-o-carborane groups.[17]

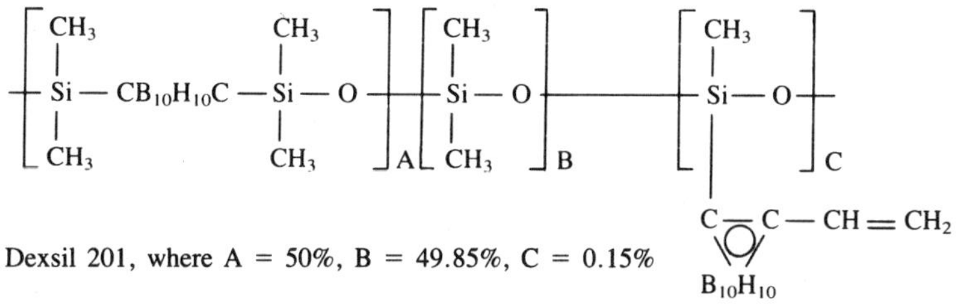

$$\left[\begin{array}{c} CH_3 \\ | \\ Si-CB_{10}H_{10}C-Si-O \\ | \\ CH_3 \end{array}\begin{array}{c} CH_3 \\ | \\ \\ | \\ CH_3 \end{array}\right]_A\left[\begin{array}{c} CH_3 \\ | \\ Si-O \\ | \\ CH_3 \end{array}\right]_B\left[\begin{array}{c} CH_3 \\ | \\ Si-O \\ | \end{array}\right]_C$$

Dexsil 201, where A = 50%, B = 49.85%, C = 0.15%

with the C side group: $C\!\!=\!\!C-CH=CH_2$ over $B_{10}H_{10}$

A more detailed study of the ferric chloride catalyzed reaction showed that in the final stage of reaction cross-linking occurred.[19] Efforts were therefore devoted to finding a route to linear polymers. This was accomplished by simple hydrolytic condensation reactions of chloroterminated monomers of the type shown[20]:

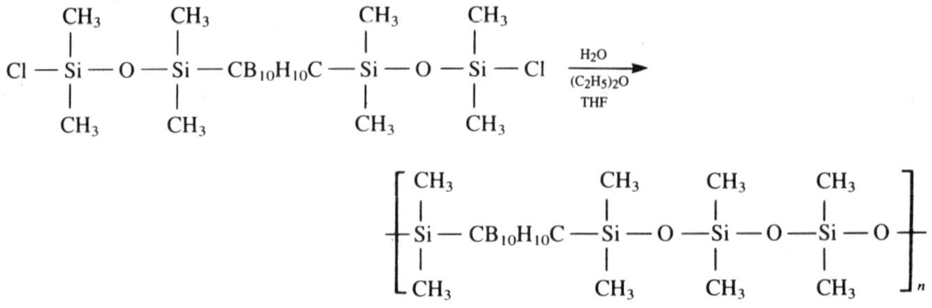

$$
\begin{array}{c}
\qquad\ CH_3 \quad\ CH_3 \qquad\qquad\quad CH_3 \quad\ CH_3 \\
\qquad\ | \qquad\quad | \qquad\qquad\qquad | \qquad\quad | \\
Cl-Si-O-Si-CB_{10}H_{10}C-Si-O-Si-Cl \xrightarrow[\substack{(C_2H_5)_2O \\ THF}]{H_2O} \\
\qquad\ | \qquad\quad | \qquad\qquad\qquad | \qquad\quad | \\
\qquad\ CH_3 \quad\ CH_3 \qquad\qquad\quad CH_3 \quad\ CH_3
\end{array}
$$

$$
\begin{bmatrix}
\ CH_3 \qquad\qquad\qquad\ CH_3 \quad\ CH_3 \quad\ CH_3 \\
| \qquad\qquad\qquad\qquad | \qquad\quad | \qquad\quad | \\
Si-CB_{10}H_{10}C-Si-O-Si-O-Si-O \\
| \qquad\qquad\qquad\qquad | \qquad\quad | \qquad\quad | \\
CH_3 \qquad\qquad\qquad\ CH_3 \quad\ CH_3 \quad\ CH_3
\end{bmatrix}_n
$$

Dexsil 300

$$
\begin{array}{c}
CH_3 \quad\ CH_3 \quad\ CH_3 \qquad\qquad\quad CH_3 \quad\ CH_3 \quad\ CH_3 \\
| \qquad\quad | \qquad\quad | \qquad\qquad\qquad | \qquad\quad | \qquad\quad | \\
Cl-Si-O-Si-O-Si-CB_{10}H_{10}C-Si-O-Si-O-Si-Cl \\
| \qquad\quad | \qquad\quad | \qquad\qquad\qquad | \qquad\quad | \qquad\quad | \\
CH_3 \quad\ CH_3 \quad\ CH_3 \qquad\qquad\quad CH_3 \quad\ CH_3 \quad\ CH_3
\end{array}
$$

$\xrightarrow[\substack{(C_2H_5)_2O \\ THF}]{\qquad H_2O}$

$$
\begin{bmatrix}
CH_3 \qquad\qquad\qquad\ CH_3 \quad\ CH_3 \quad\ CH_3 \quad\ CH_3 \quad\ CH_3 \\
| \qquad\qquad\qquad\qquad | \qquad\quad | \qquad\quad | \qquad\quad | \qquad\quad | \\
Si-CB_{10}H_{10}C-Si-O-Si-O-Si-O-Si-O-Si-O \\
| \qquad\qquad\qquad\qquad | \qquad\quad | \qquad\quad | \qquad\quad | \qquad\quad | \\
CH_3 \qquad\qquad\qquad\ CH_3 \quad\ CH_3 \quad\ CH_3 \quad\ CH_3 \quad\ CH_3
\end{bmatrix}_n
$$

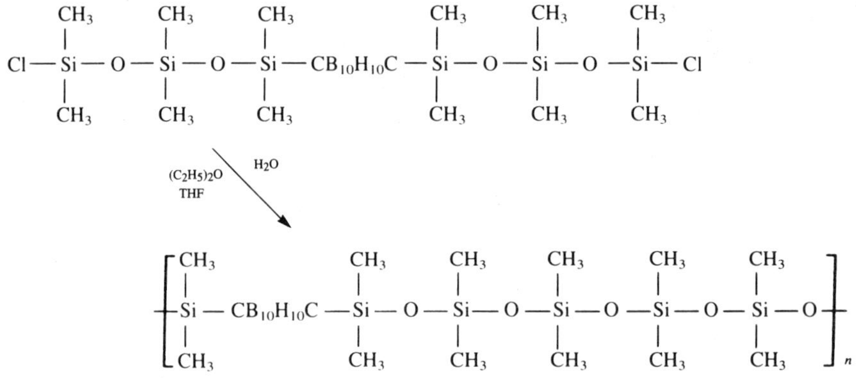

Dexsil 500

Advantages of these reactions were that they could be carried out at ice-bath temperatures and that the molecular weight distribution could be controlled by variation of the monomer/water ratio. Molecular weights up to 30,000 were obtained, but the products contained some relatively low-molecular-weight cyclic compounds.

An alternative route to linear polymers was the acid-catalyzed condensation of hydroxy-terminated monomers, e.g.,

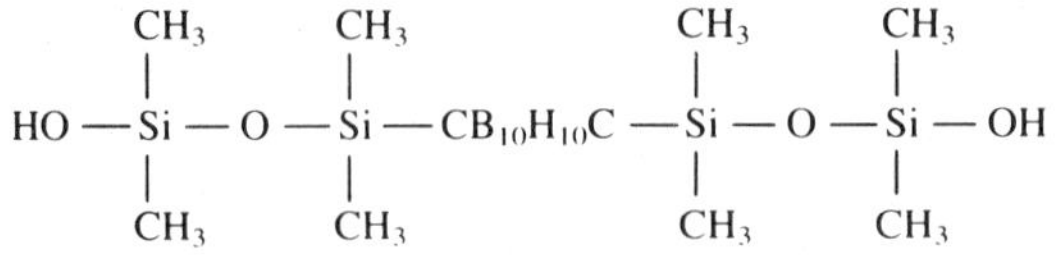

$$
\begin{array}{c}
\qquad\quad CH_3 \quad\ CH_3 \qquad\qquad\quad CH_3 \quad\ CH_3 \\
\qquad\quad | \qquad\quad | \qquad\qquad\qquad | \qquad\quad | \\
HO-Si-O-Si-CB_{10}H_{10}C-Si-O-Si-OH \\
\qquad\quad | \qquad\quad | \qquad\qquad\qquad | \qquad\quad | \\
\qquad\quad CH_3 \quad\ CH_3 \qquad\qquad\quad CH_3 \quad\ CH_3
\end{array}
$$

but this was in no way preferential to the hydrolytic condensation method.

With the development of the trifluoropropyl-substituted polysiloxanes and the demonstration of their greatly increased resistance towards swelling in various solvents, it was only natural that the same substituent should be incorporated into carboranesiloxanes. All the above preparative routes were investigated,[21] i.e., ferric chloride-catalyzed condensation of chloro- and methoxyterminated monomers, hydrolytic condensation of chloroterminated monomers, and condensation of hydroxyterminated monomers. In comparison with the completely methyl-substituted polymers, however, the molecular weights attained were comparatively low. The highest molecular weight of 13,000 was achieved via condensation of a hydroxyterminated monomer using a catalyst composed of tetramethylguanidine and sulphuric acid. The reaction scheme was as follows:

$$
\begin{array}{ccccc}
 & CH_2CH_2CF_3 & & CH_2CH_2CF_3 & & CH_3 \\
 & | & & | & & | \\
HO-\!\!\!\!\!\!\!\!\! & Si-CB_{10}H_{10}C-\!\!\!\!\!\!\!\!\! & Si-OH & + \ 2\,Cl-\!\!\!\!\!\!\!\!\! & Si-Cl & \xrightarrow{\ \Delta\ } \\
 & | & & | & & | \\
 & CH_3 & & CH_3 & & CH_3
\end{array}
$$

$$
\begin{array}{ccccccc}
 & CH_3 & & CH_2CH_2CF_3 & & CH_2CH_2CF_3 & CH_3 \\
 & | & & | & & | & | \\
Cl-\!Si-\!O-\!Si-\!CB_{10}H_{10}C-\!Si-\;O\;-\!Si-\!Cl \\
 & | & & | & & | & | \\
 & CH_3 & & CH_3 & & CH_3 & CH_3
\end{array}
$$

$$\Big\downarrow\ \ \mathrm{NaHCO_3} \quad \mathrm{Moist\ (C_2H_5)_2O}$$

$$
\xleftarrow[\ \mathrm{H_2SO_4}\ ]{\ \substack{\text{Tetramethyl}\\\text{guanidine}}}
\begin{array}{ccccccc}
 & CH_3 & & CH_2CH_2CF_3 & & CH_2CH_2CF_3 & CH_3 \\
 & | & & | & & | & | \\
HO-\!Si-\!O-\!Si-\!CB_{10}H_{10}C-\!Si-\!O-\!Si-\!OH \\
 & | & & | & & | & | \\
 & CH_3 & & CH_3 & & CH_3 & CH_3
\end{array}
$$

$$
\left[
\begin{array}{ccccccc}
CH_3 & & CH_2CH_2CF_3 & & CH_2CH_2CF_3 & CH_3 \\
| & & | & & | & | \\
Si-\!O-\!Si-\!CB_{10}H_{10}C-\!Si-\!O-\!Si-\!O \\
| & & | & & | & | \\
CH_3 & & CH_3 & & CH_3 & CH_3
\end{array}
\right]_n
$$

Dexsil 300-F$_2$

All the work on Dexsils described above was carried out by the Chemicals Division of the Olin Research Center. The Union Carbide Corp. then entered the field and, by condensing a carborane silanol with a reactive silane comonomer, was able to prepare linear *m*-carboranesiloxanes with molecular weights of 250,000 or more.[22]

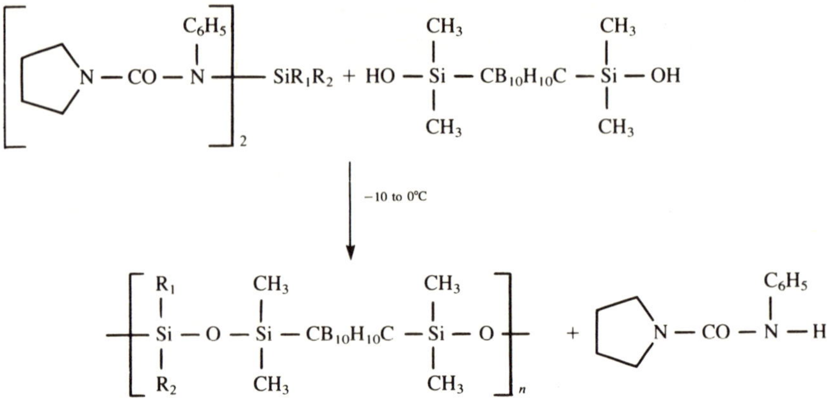

$$\left[\underset{\text{(pyrrolidine ring)}}{N-CO-N} \left| \underset{C_6H_5}{} \right. \right]_2 SiR_1R_2 \; + \; HO-\underset{CH_3}{\overset{CH_3}{Si}}-CB_{10}H_{10}C-\underset{CH_3}{\overset{CH_3}{Si}}-OH$$

$$\downarrow \quad {-10 \text{ to } 0°C}$$

$$\left\{ -\underset{R_2}{\overset{R_1}{Si}}-O-\underset{CH_3}{\overset{CH_3}{Si}}-CB_{10}H_{10}C-\underset{CH_3}{\overset{CH_3}{Si}}-O- \right\}_n \quad + \quad \underset{}{N-CO-\overset{C_6H_5}{N}-H}$$

where $R_1 R_2$ = alkyl or phenyl

The process is illustrated above and involves the condensation of, e.g., bis
(N-pyrrolidino-N^1-phenylureido)dialkylsilane with bis(hydroxydimethylsilyl)-
m-carborane. Nuclear magnetic resonance spectroscopy was used to monitor the
reaction stoichiometry[23] and to adjust for side reactions and monomer impurities. The products were initially referred to as D_2-carboranesiloxanes but were
later given the trade name of Ucarsils. In a similar manner, trifluoropropyl-
modified polymers were also prepared[24] with molecular weights between
100,000 and 220,000.

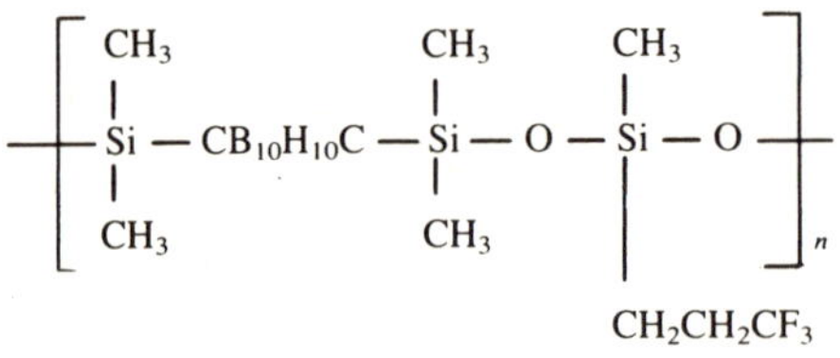

$$\left\{ -\underset{CH_3}{\overset{CH_3}{Si}}-CB_{10}H_{10}C-\underset{CH_3}{\overset{CH_3}{Si}}-O-\underset{CH_2CH_2CF_3}{\overset{CH_3}{Si}}-O- \right\}_n$$

Ucarsil F_1

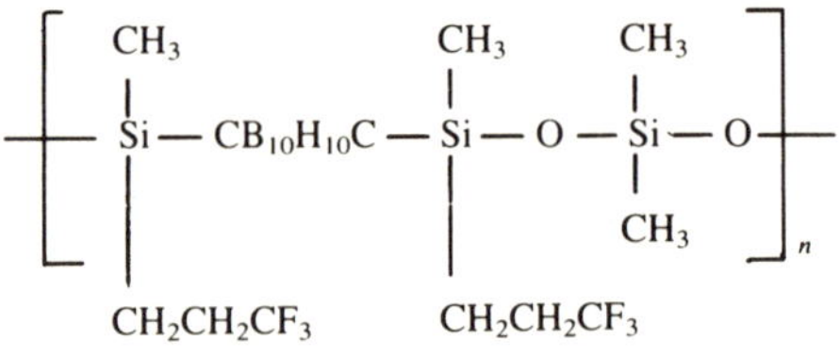

$$\left\{ -\underset{}{\overset{CH_3}{Si}}-CB_{10}H_{10}C-\underset{CH_2CH_2CF_3}{\overset{CH_3}{Si}}-O-\underset{CH_3}{\overset{CH_3}{Si}}-O- \right\}_n$$

Ucarsil F_2

$$\left\{ -\underset{CH_2CH_2CF_3}{\overset{CH_3}{Si}}-CB_{10}H_{10}C-\underset{CH_2CH_2CF_3}{\overset{CH_3}{Si}}-O-\underset{CH_2CH_2CF_3}{\overset{CH_3}{Si}}-O- \right\}_n$$

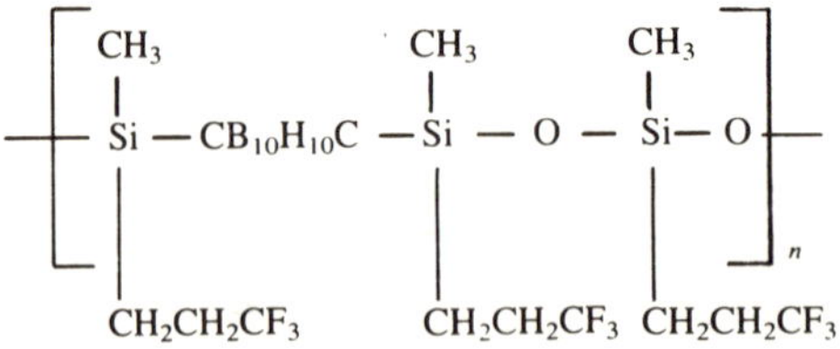

Ucarsil F_3

Meanwhile Chemical Systems Incorporated had been working on the synthesis of smaller carborane nuclei and, by direct reaction of C_2H_2, H_2, and B_5H_9, were able to produce a mixture of $C_2B_3H_5$, $C_2B_4H_6$, and $C_2B_5H_7$, together with a number of B-methyl derivatives.[25] The products were separated by gas chromatography. A poly[$C_2B_5H_5$-carboranylenesiloxane] analog of SiB-1 (Dexsil 100) was prepared via the ferric chloride-catalyzed elimination of alkylhalide route.[26] The material was a hard wax of molecular weight 12,500. If, however, as little as 5 mole % of the larger carborane moiety $C_2B_{10}H_{10}$ were incorporated into the polymer chain, crystallinity was disrupted, and an elastomer with a glass transition temperature of $-59°C$ was obtained.[27] Polymers based mainly on $C_2B_5H_5$ were called Pentasils.

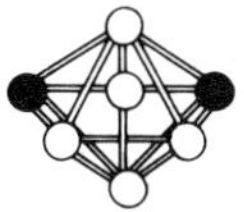

A new synthetic route was later discovered,[28] which involved the alcoholysis of the bischlorodimethylsilylcarborane monomers and the generation of the HCl catalyst in situ. A simplified reaction scheme is as follows:

The method could be used for the synthesis of most of the carboranesiloxane polymers, with the probable exception of the SiB-1 homopolymers of the larger carboranes. Molecular weights attained, however, were only of the order of 1,000–3,000.

Thermal Stability

The thermal stabilities of uncompounded poly(carboranesiloxanes) have been assessed by differential scanning calorimetry, differential thermal analysis, and thermogravimetry. The earliest results using differential scanning calorimetry[16] showed that the polymers Dexsil 100, Dexsil 200, and Dexsil 300 were stable up to 500°C in an inert atmosphere. In air, however, only Dexsil 100 remained unchanged up to this temperature. The highest members of the series underwent cross-linking reactions. Replacement of methyl groups by phenyl groups in the siloxane part of the molecules improved the thermo-oxidative stability. Most later investigations used thermogravimetry and the results of these studies are summarized here.[20,21,24,29–31] Figure 7.1 shows the weight

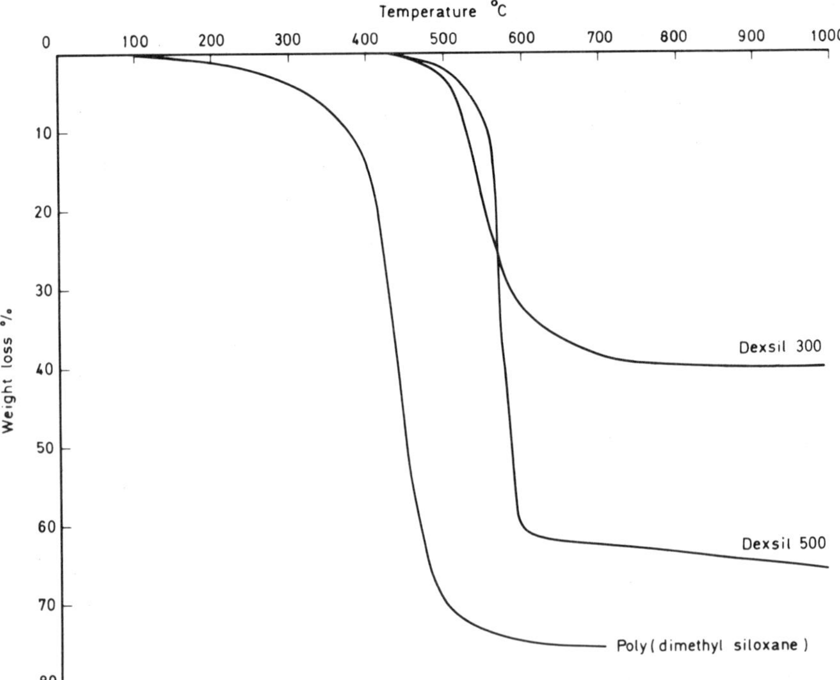

Figure 7.1. Thermogravimetry of Dexsil polymers in argon. (Heating rate 2.5°C/min) Reference 20.

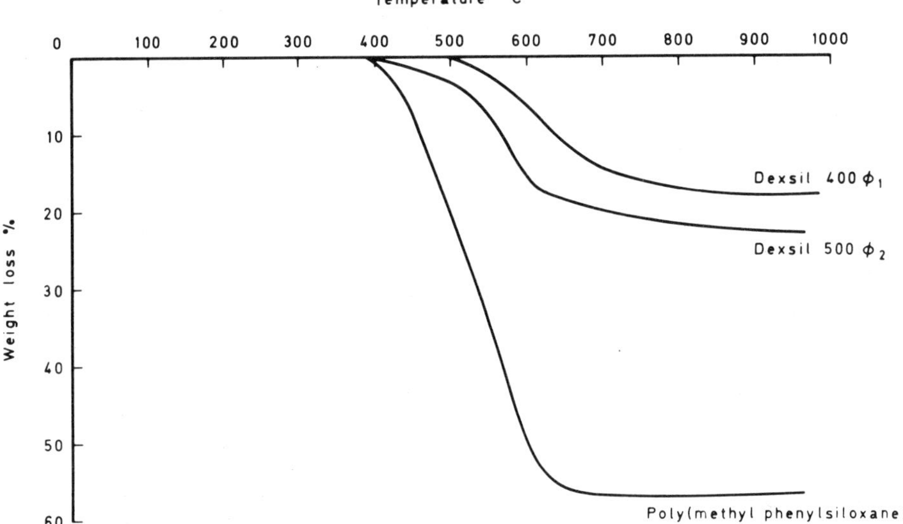

Figure 7.2. Thermogravimetry of Dexsil polymers in argon. (Heating rate 2.5°C/min) Reference 20.

losses for Dexsil 300 and 500 and poly(dimethylsiloxane) when heated in argon at a rate of 2.5°C/minute. The greatly improved thermal stability of the poly(carborane siloxanes) compared with poly(dimethylsiloxane) is immediately evident. There is little difference in the temperatures for initial weight loss of the two Dexsils, but with increasing proportion of carborane nuclei in the polymer the total weight lost at the highest temperatures is markedly reduced. Figure 7.2 gives similar data for phenyl-substituted polymers. The introduction of phenyl groups into poly(dimethylsiloxane) results in a significant increase in stability, but with the Dexsils the effect is not to increase the temperatures for initial breakdown but to reduce the total weight loss over the range 500–1,000°C. In the case of trifluoropropyl substituents (Figure 7.3), the stability of the carborane-containing polymers is *reduced* and Dexsil 300 F4 appears to behave very similarly to poly(methyltrifluoropropylsiloxane). That this is misleading, however, is shown by the isothermal results of Figure 7.4, which point to considerably better long-term thermal stability for the Dexsils. Similar dynamic weight loss curves have been obtained for Ucarsil polymers (Figure 7.5), but the results are not strictly comparable with those for the Dexsils because a heating rate of 10°C/minute was used instead of 2.5°C/minute. In air, the shape of the weight loss curves is very much influenced by the formation of inorganic residues—the oxides of silicon and boron (Figures 7.6 and 7.7)—and hence is very difficult to interpret. The trend is, however, the same as in an inert atmosphere, i.e., improved stability with phenyl substitution and decreased stability with

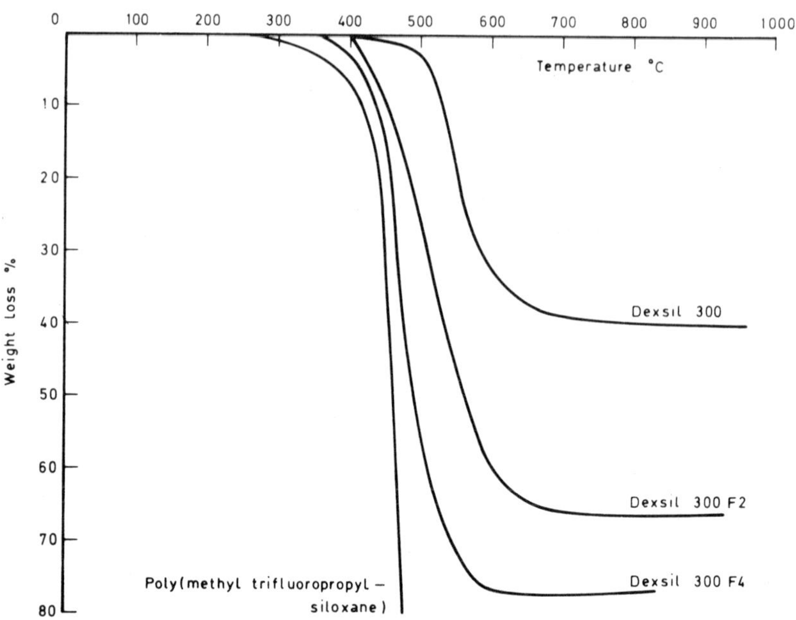

Figure 7.3. *Thermogravimetry of trifluoropropyl-substituted Dexsil polymers in nitrogen. (Heating rate 2.5°C/min) Reference 21.*

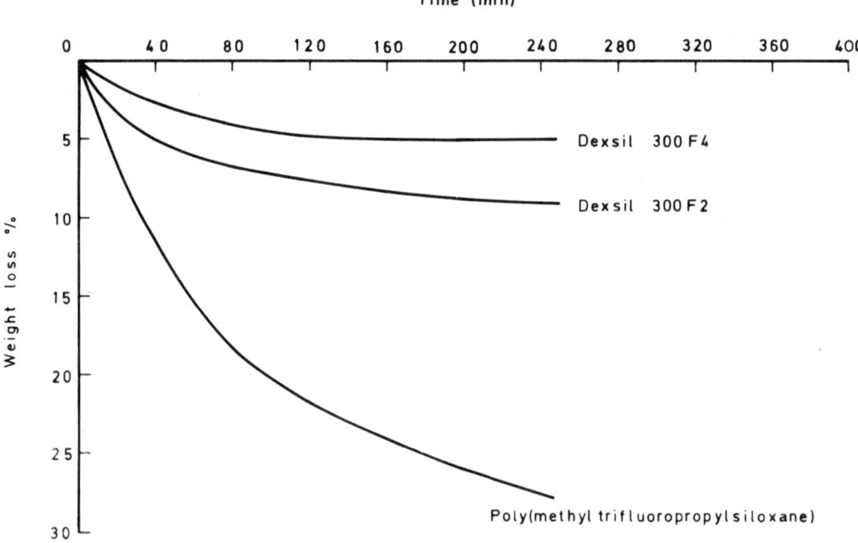

Figure 7.4. *Isothermal weight loss of trifluoropropyl-substituted Dexsil polymers in nitrogen at 350°C. Reference 21.*

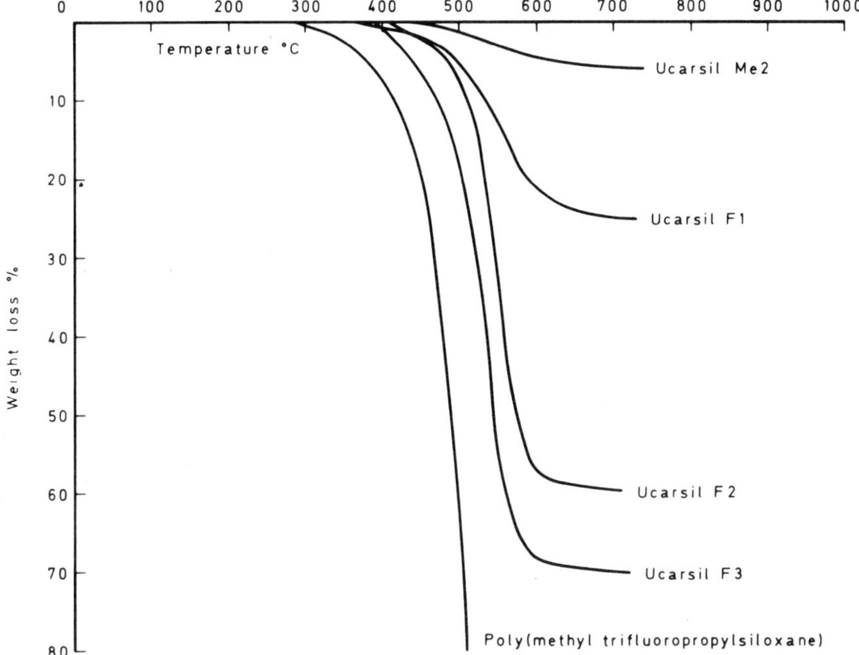

Figure 7.5. Thermogravimetry of Ucarsil polymers in nitrogen. (Heating rate 10°C/min) Reference 24.

trifluoropropyl substitution. This can perhaps best be seen in Table 7.1, which quotes the temperatures for 10% weight loss of the various polymers together with the total weight lost at 700°C, arranging the polymers in order of increasing stability.

Roller and Gillham[30] have correlated the total weight loss on heating to 800°C in argon with the number of $-Si(CH_3)_2O-$ linkages in the polymer repeat unit, showing that there is an inverse linear relationship between the two.

Data are much sparser on the $-CB_5H_5C-$ polymers, but work by Augl[32] indicates that the temperatures for initial weight loss and maximum rate of decomposition are about 100°C lower than for the corresponding $-CB_{10}H_{10}C-$ materials.

The improved thermal stability of the poly(carboranesiloxanes) has been attributed to a stabilizing action of the carborane nuclei on the neighboring methyl groups, this protection becoming progressively smaller as the distance between the two increases.

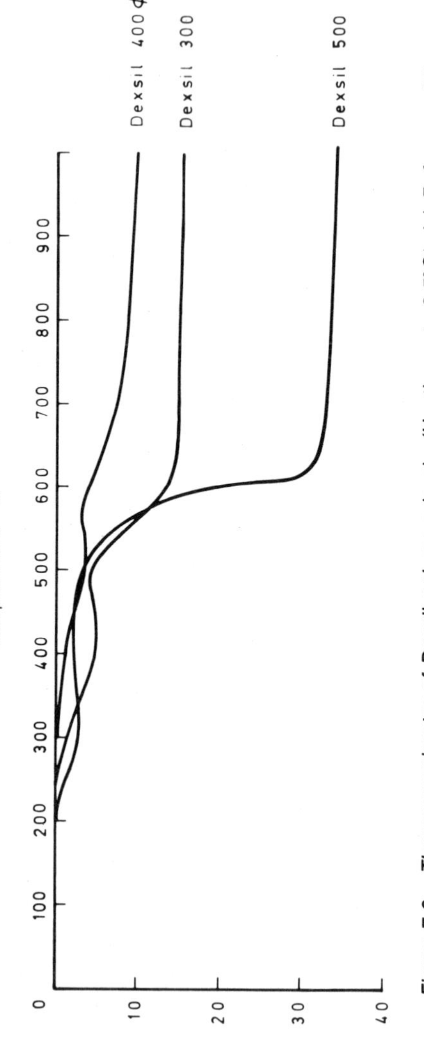

Figure 7.6. Thermogravimetry of Dexsil polymers in air. (Heating rate 2.5°C/min) Reference 20.

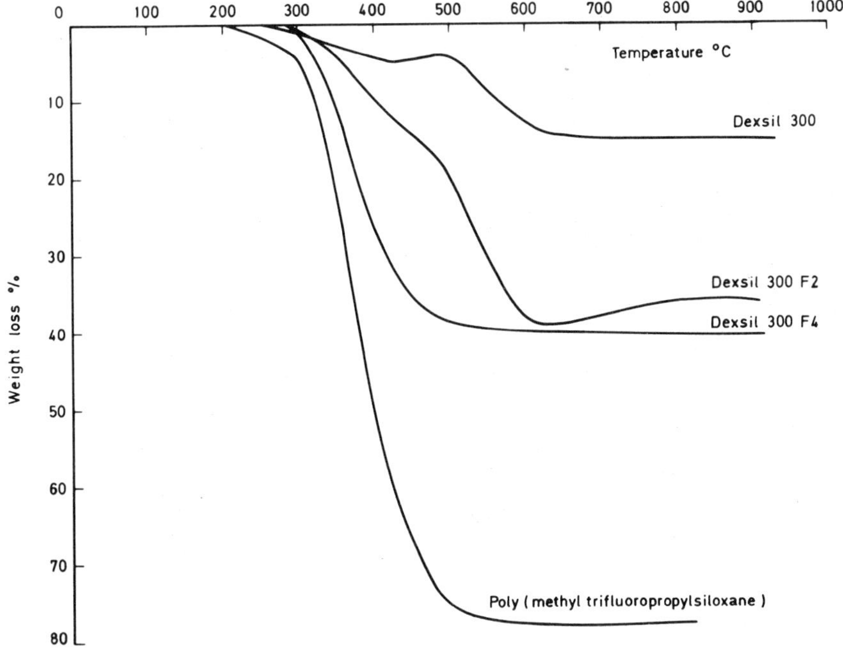

Figure 7.7. Thermogravimetry of trifluoropropyl-substituted Dexsil polymers in air. (Heating rate 2.5°C/min) Reference 21.

TABLE 7.1. Comparison of Thermal Stability from Weight Loss Data (Derived at a heating rate of 2.5°C/min)

Polymer	Temperature (°C) for 10% weight loss	Total weight loss (%) at 700°C	Temperature (°C) for 10% weight loss	Total weight loss (%) at 700°C
	Inert atmosphere		Oxidizing atmosphere	
Poly(dimethylsiloxane)	380	76		
Poly(methyltrifluoro-propylsiloxane)	420	> 90	320	78
Dexsil 300 F4	440	77	350	40
Dexsil 300 F2	450	66	400	48
Poly(methylphenyl-siloxane)	460	57		
Dexsil 300	530	39	560	15
Dexsil 500	560	63	570	33
Dexsil 500 ϕ2	565	20		
Dexsil 400 ϕ1	635	15	~ 1000	7

Physical Characterization

An extensive study of the physical and mechanical properties of the poly(carboranesiloxanes) has been made over the years at Princeton University. The first series of experiments by Tobolsky and coworkers utilized stress relaxation and torsional modulus versus temperature measurements.[33-37] The whole range of Dexsils from 100 to 400 were investigated in inert and oxidizing atmospheres, mainly in the form of peroxide cross-linked materials but also in some cases with silica and iron oxide fillers. In air the unfilled polymers showed an abrupt rise in modulus above about 300°C as oxidative cross-linking occurred. The material became progressively more brittle and then failed. A temperature for onset of oxidative cross-linking may be derived from the modulus-temperature curves and the values for a number of Dexsils are listed in Table 7.2. The most important feature of this table is the relatively low value of the oxidation temperature for Dexsil 300, and the improvement which occurs by the combined use of silica and Fe_2O_3.

More marked differences between samples were observed in the stress relaxation experiments. Figure 7.8 shows curves obtained at 350°C in vacuum. The Dexsil 201 with no filler or added peroxide clearly had the best stability. Some cross-linking of this material occurred during the polymerization process as stated earlier. A more detailed study of the peroxide cross-linked polymer[35] showed in confirmation of this that there were two types of cross-link, one of which broke down at a rate of about ten times that of the other. The data of Figure 7.8 also show the decrease in stability as the spacing between the carborane groups was increased (compare Dexsil 201 and 402). The poly(dimethylsiloxanes) behaved in a different manner, the stress relaxing linearly over the time scale of the experiment. More detailed evidence of this is given in Reference 34. In air the stress relaxation curves were completely different (Figure 7.9) because of the competition between chain scission and oxidative cross-linking. As the temperature was increased, the cross-linking occurred in shorter time in-

TABLE 7.2. *Temperatures for Onset of Oxidative Cross-Linking*

Polymer sample	Oxidation temperature (°C)
Dexsil 100—peroxide cure	340
Dexsil 125—peroxide cure	320
Dexsil 150—peroxide cure	330
Dexsil 160—peroxide cure	355
Dexsil 175—peroxide cure	310
Dexsil 200—peroxide cure	340
Dexsil 300—peroxide cure	275
Dexsil 300—peroxide cure + silica filler	300
Dexsil 300—peroxide cure + Fe_2O_3	305
Dexsil 300—peroxide cure + silica filler + Fe_2O_3	320

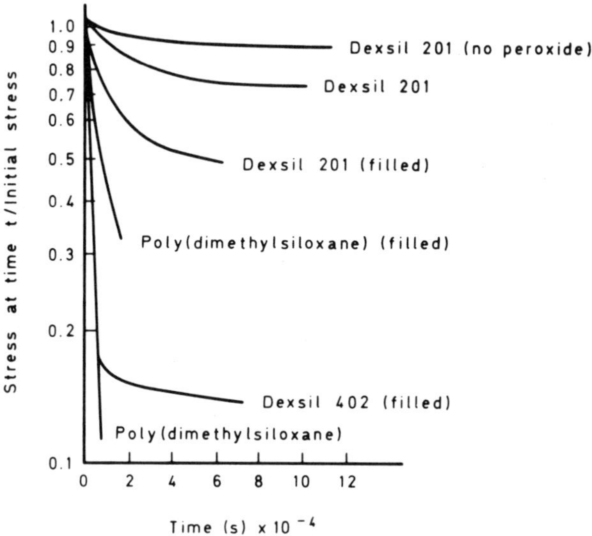

Figure 7.8. *Stress relaxation of poly(carboranesiloxanes) at 350°C in vacuum. Reference 36.*

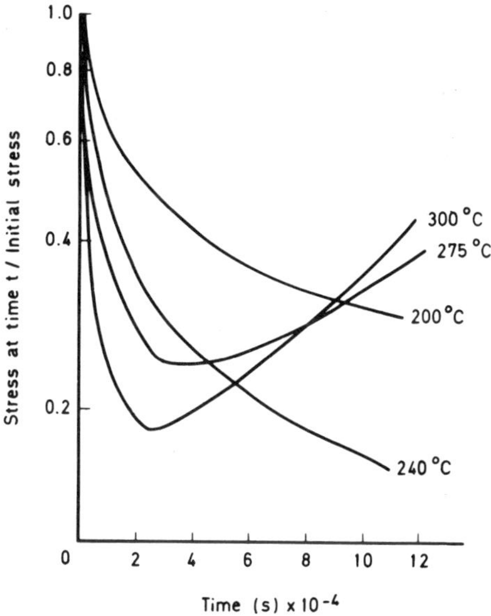

Figure 7.9. *Stress relaxation of filled Dexsil 300 in air at different temperatures. Reference 37.*

TABLE 7.3. Maximum Use Temperatures
for 24 Hour Exposures (Reference 34)

Polymer sample	Temperature (°C)	
	Air	Nitrogen
Dexsil 200	260	325
Dexsil 150	275	—
Dexsil 100	285	—
Dexsil 200 (filled)	300	> 416
Poly(dimethylsiloxane)	230	280

tervals and was more pronounced in its effect. A maximum use temperature can be derived from the stress relaxation curves based on the retention of $1/e$ (0.368) of the initial modulus over a period of 24 hours. Some of these derived values are given in Table 7.3.

The superiority of the poly(carboranesiloxanes) over poly(dimethylsiloxane) is again evident.

The more recent work at Princeton University by Gillham and Roller used the technique of torsional braid analysis (TBA).[38–44] All the Dexsils from 100–500 were examined, including samples of different molecular weight, with different end-groups and with phenyl- and trifluoropropyl-substituents. Pentasils from 100–500 were studied similarly. The TBA curves obtained were complex (see Figure 7.10 for an example), and for a detailed analysis of them recourse must be had to the original papers. The results may, however, be summarized as follows. In air the $CB_{10}H_{10}C$-based polymers may be divided into two classes. All those with only methyl substituents started to stiffen (showing the onset of cross-linking) at 280–300°C. The polymers with phenyl substituents were more stable and began to stiffen at 350°C, as did poly(dimethylsiloxane). The CB_5H_5C-based materials were less stable, cross-linking being observed at 250°C. Incorporation of 20 mole % $CB_{10}H_{10}C$ nuclei into the chain raised the onset temperature to 300°C. In inert atmospheres differences between the various types of polymer were less marked, no stiffening occurring below 400°C. Overall the Dexsil 300 structure was the most stable.

The TBA results also showed that blends of the different poly(carboranesiloxanes) were incompatible.

Elastomer Formulation and Elevated Temperature Properties

Since the poly(carboranesiloxanes) are structurally similar to the polysiloxanes, it is not surprising that compounding, molding, and vulcanization processes are basically alike.[17,22,24,45–47] Compounded poly(carborane-

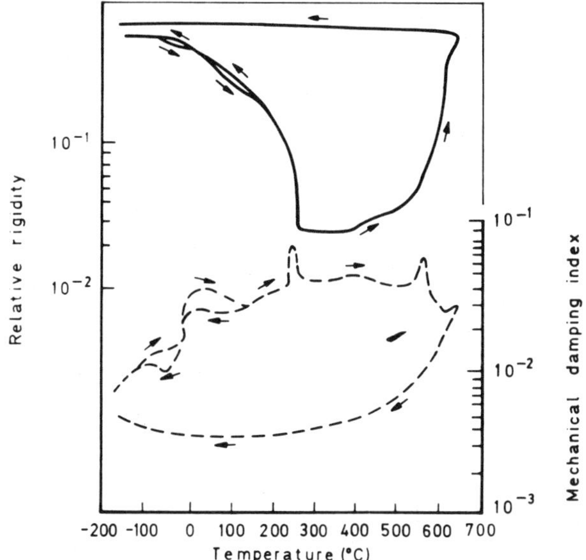

Figure 7.10. Thermochemical behavior of Dexsil 100 in nitrogen. [Heating rate 3.6°/min above 25°C; 2.0°C/min below 25°C; (130°C→−180°C→625°C→−180°C→25°C)] Reference 41.

siloxanes) contain four basic ingredients: the polymer itself, reinforcing filler, antioxidant, and curing agent. The typical range of formulations is polymer 100 parts, filler (silicas or silicates) 10–100 parts, antioxidant (metal oxide) 1–10 parts, and curing agent (organic peroxides) 0.5–3 parts. The major part of Olin Mathieson's evaluation was devoted to the Dexsil 201 polymer.[17,45,46] The cured, but uncompounded, material had a tensile strength in the range 0.7–1.0 MPa, and an elongation at break of 120–250%. With fillers such as silicas, clays, and talcs, tensile strengths of 1.4–3.1 MPa and elongations of 100–200% could be achieved. By using filler loadings of 100 parts or more, and by applying a high initial pressure (∼ 10 MPa) for a short time followed by greatly reduced pressure during the main curing period, the tensile strength could be raised to 4.2 MPa with an elongation at break approaching 300%.

Red iron oxides proved to be the best antioxidants, the quantity to be added depending upon the service temperature. For 260–315°C, more than 5 parts per hundred of polymer were needed, but at higher temperatures, 425–540°C, 1–5 parts were sufficient.

Curing was accomplished using organic peroxides such as benzoyl peroxide, dichlorobenzoyl peroxide, and dicumyl peroxide. The free radicals produced by their thermal decomposition were believed to cause the formation of carbon-carbon bonds between the organic groups on the polymer chain. The vulcanization temperature depended upon the particular peroxide used, but the

TABLE 7.4. Properties of Various Dexsil 201 Formulations (Reference 45)

Additive (parts per hundred)		Shore A hardness	Elongation at break (%)		Tensile strength (MPa)		% Tensile strength retention at 315°C
			RT	315°C	RT	315°C	
SiO_2	10	45	145	—	1.00	—	—
SiO_2	25	63	145	102	2.57	0.58	23
SiO_2	37.5	64	145	91	2.79	0.72	26
SiO_2	50	84	111	53	2.94	0.75	26
SiO_2	60	70	84	37	2.95	0.68	23
SiO_2	75	85	16	13	2.95	0.68	23
SiO_2 } Fe_2O_3	50 / 1	66	100	62	2.65	0.79	30
SiO_2 } Fe_2O_3	50 / 3	70	69	75	2.21	1.01	46

vulcanizates were post-cured for 16 hours at 100°C, followed by 8 hours at 150°C and 24 hours at 260°C. The properties of a number of formulations tested at ambient temperature and after 15 minutes at 315°C are given in Table 7.4.

One particular formulation, 100 parts Dexsil 201, 35 parts SiO_2, 15 parts Fe_2O_3, and 1 part dicumylperoxide, was heated at different temperatures for different lengths of time and then tested at ambient temperature. The results are summarized in Table 7.5.

Resistance to compression set is an important property of an elastomer when used as a seal, and Figure 7.11 compares the compression set of Dexsil 201, a polysiloxane, and a fluorocarbon elastomer when heated at 205°C. The superiority of the poly(carboranesiloxane) after long time periods at temperature is evident. Figure 7.12 shows the weight loss to be expected for a compounded Dexsil 201 with time of aging at 315°C. The total weight loss is less than 10% even after 500 hours at temperature.

A Dexsil 402 formulation (Dexsil 402, 100 parts; SiO_2, 37.5 parts; Fe_2O_3, 5 parts; benzoyl peroxide, 2 parts) was also studied to a certain extent. After aging for 24 hours at 315°C, this showed complete retention of tensile strength when tested at ambient temperature and a 35% retention when tested at 315°C.

TABLE 7.5. Properties of Compounded Dexsil 201 after Various Heat Aging Treatments (Reference 46)

Temperature (°C)	Time (hours)	Shore A hardness	Elongation at break (%)	Tensile strength (MPa)	Tensile strength retention (%)
260	1000	70	60	2.70	87
285	200	75	58	3.14	100
425	24	68	87	1.66	61
480	2	52	156	0.70	26

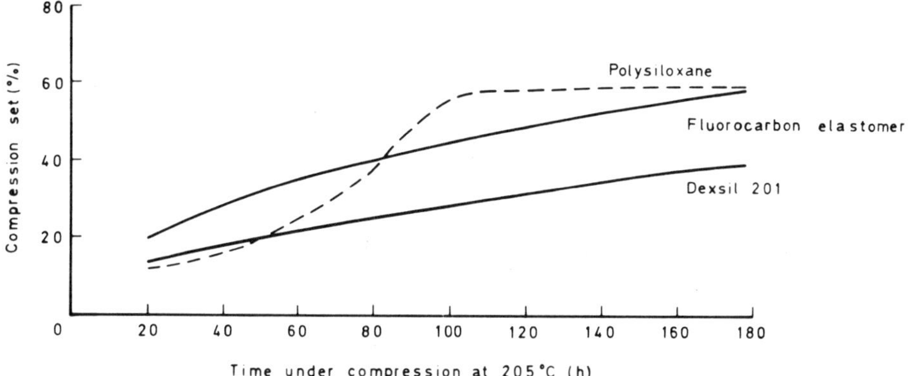

Figure 7.11. Comparison of compression set at 205°C of different elastomers. Reference 46.

The effect of variation in formulation parameters upon the properties of the Ucarsil materials was studied in some detail.[47] With these higher molecular weight gumstocks, it was found that substantial improvements in retention of properties after high-temperature aging were achieved by using a hydrophobic silica as reinforcing filler instead of a fumed silica, and by employing low (1.0 to 1.5 parts) peroxide levels for curing. Also, the higher the phenyl content of the vulcanizate, the less were the changes in mechanical properties during heat aging. Table 7.6 compares the mechanical properties as a function of temperature of a number of Ucarsils.

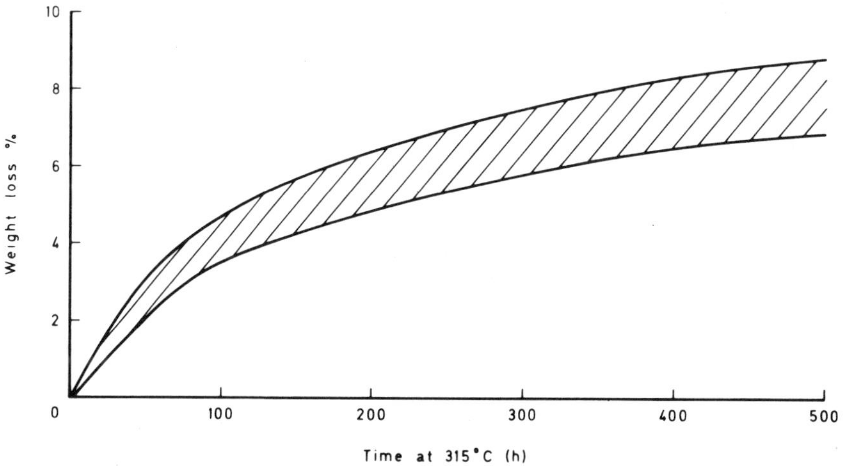

Figure 7.12. Weight loss of compounded Dexsil 200 at 315°C. Reference 46.

Table 7.7 compares property retention at ambient temperature after aging for different periods at temperatures close to 300°C.

The increase in Young's modulus and the decrease in the elongation at break showed that all the polymers were becoming progressively more brittle with time at elevated temperature. The diphenyl-substituted polymer had the best thermo-oxidative stability, and the Ucarsil F3, with the greatest number of trifluoropropyl substituents, the worst.

SUMMARY

From the available data on the poly(carboranesiloxanes) the following conclusions can be drawn:

1. Their thermal and thermo-oxidative stability is superior to that of the polysiloxanes.

TABLE 7.6. *Mechanical Properties as a Function of Temperature for Ucarsil Elastomers (References 22 and 24)*

Elastomer	Temperature (°C)	Elongation at break (%)	Tensile strength (MPa)	Youngs modulus (MPa)
Methylphenyl	−30	55	10.42	212.52
substituted[a]	−20	180	3.85	17.25
	25	130	3.58	3.45
	200	40	1.08	3.97
	260	30	0.80	3.73
	300	15	0.41	3.45
Ucarsil F1	−30	2	17.04	1524.90
	−20	15	7.80	147.66
	25	100	3.73	5.40
	200	40	1.03	3.37
	260	30	0.63	2.90
	290	15	0.36	2.82
Ucarsil F3	−30	—	—	1918.20
	−20	1	1.26	1787.10
	25	200	1.36	15.49
	200	11	0.50	7.04
	260	5	0.18	6.76
	290	3	0.11	4.37

[a]Polymer (33% methylphenyl) 100, SiO_2 30, Fe_2O_3 2.5, and dicumyl peroxide 2.5 parts. Similar proportions for Ucarsil F1 and F3.

TABLE 7.7. *Mechanical Properties of Ucarsil Elastomers Tested at Ambient Temperature after Aging for Different Times at Elevated Temperatures (References 24 and 47)*

Elastomer	Temperature ($^\circ$C)	Time (hours)	Elongation at break (%)	Tensile strength (MPa)	Youngs modulus (MPa)
Diphenyl	315	0	100	4.23	3.68
modified		25	60	3.05	6.36
		50	55	2.95	7.71
		75	50	3.35	9.05
		150	45	3.68	12.51
		300	30	3.93	23.46
Methylphenyl	315	0	110	4.44	4.53
modified		25	55	3.70	7.56
		50	40	3.58	8.96
		75	30	3.02	11.76
		150	20	6.04	42.78
		300	10	8.42	186.30
Ucarsil F1	315	0	100	3.73	5.40
		25	2	1.38	51.06
		50		Too brittle to test	
Ucarsil F1	290	0	100	3.73	5.40
		25	40	3.09	3.51
		150	10	2.62	38.64
Ucarsil F2	290	0	60	1.35	9.80
		50	20	2.19	29.19
		150	20	3.48	51.82
Ucarsil F3	290	0	200	1.36	15.49
		50	90	2.92	17.87
		150	30	3.27	26.15

Formulation: Polymer 100, SiO_2 30, Fe_2O_3 2.5, and dicumyl peroxide 2.5 parts.

2. As with many other polymer systems, the *p*-carborane conveys better thermal stability than the *m*-isomer.

3. Thermal stability is inversely related to the siloxane content.

4. Phenyl substituents improve thermo-oxidative stability. Methyl groups are susceptible to attack by oxygen, this becoming easier the further the groups are from a carborane nucleus.

5. There is little difference in the thermal stability of $CB_{10}H_{10}C$- and CB_5H_5C-based polymers, but the former have superior thermo-oxidative stability.

6. Oxidation resistance can be improved tremendously by proper compounding and cure.

7. Possible uses are as sealants, gaskets, O-rings, coatings, and diaphragms. The polymers have already been extensively used in high-temperature gas chromatography.[48, 49]

REFERENCES

1. T. L. Heying, J. W. Ager, S. L. Clark, D. J. Mangold, H. L. Goldstein, M. Hillmann, R. J. Polak, and J. W. Szymanski, *Inorg. Chem.* **2,** 1089 (1963).
2. D. Grafstein, J. Bobinski, J. Dvorak, H. Smith, N. Schwartz, M. S. Cohen, and M. M. Fein, *Inorg. Chem.* **2,** 1120 (1963).
3. D. Grafstein and J. Dvorak, *Inorg. Chem.* **2,** 1128 (1963).
4. H. A. Schroeder and G. D. Vickers, *Inorg. Chem.* **2,** 1317 (1963).
5. S. Papetti and T. L. Heying, *J. Am. Chem. Soc.* **86,** 2295 (1964).
6. T. L. Heying and H. L. Goldstein, U.S. Patent No. 3,109,031.
7. T. L. Heying and E. W. Cox, U.S. Patent No. 3,137,734.
8. V. V. Korshak, N. I. Bekasova, and L. G. Komarova, *J. Polym. Sci.* **8,** 2351 (1970).
9. T. L. Heying, R. P. Alexander, and H. L. Goldstein, U.S. Patent No. 3,258,479.
10. T. L. Heying, J. A. Reid, and S. I. Trotz, U.S. Patent No. 3,311,593.
11. N. S. Semenuk, S. Papetti, and H. A. Schroeder, *Inorg. Chem.* **8,** 2441 (1969).
12. H. A. Schroeder, S. Papetti, R. P. Alexander, J. F. Sieckhaus, and T. L. Heying, *Inorg Chem.* **8,** 2444 (1969).
13. H. A. Schroeder, J. R. Reiner, and T. A. Knowles, *Inorg. Chem.* **2,** 393 (1963).
14. J. Green, N. Mayes, A. P. Kotloby, M. Fein, E. L. O'Brien, and M. S. Cohen, *J. Polym. Sci. B* **2,** 109 (1964).
15. N. Mayes, J. Green, and M. S. Cohen, *J. Polym. Sci. A 1* **5,** 365 (1967).
16. S. Papetti, B. B. Schaeffer, A. P. Gray, and T. L. Heying, *J. Polym. Sci. A 1* **4,** 1623 (1966).
17. H. A. Schroeder, *Inorg. Macromol. Rev.* **1,** 45 (1970).
18. S. Papetti and H. A. Schroeder, U.S. Patent No. 3,463,801 (1969).
19. H. J. Dietrich, R. P. Alexander, T. L. Heying, H. Kwasnik, C. O. Obenland, and H. A. Schroeder, *Makromol. Chem.* **175,** 425 (1974).
20. K. O. Knollmueller, R. N. Scott, H. Kwasnik, and J. F. Sieckhaus, *J. Polym. Sci. A 1* **9,** 1071 (1971).
21. R. N.Scott, K. O. Knollmueller, H. Hooks, and J. F. Sieckhaus, *J. Polym. Sci. A 1* **10,** 2303 (1972).
22. E. N. Peters, E. Hedaya, J. H. Kawakami, G. T. Kwiatkowski, D. W. McNeil, and R. W. Tulis, *Rubber Chem. Technol.* **48,** 14 (1975).
23. E. Hedaya, J. H. Kawakami, P. W. Kopf, G. T. Kwiatkowski, D. W. McNeil, D. A. Owen, E. N. Peters, and R. W. Tulis, *J. Polym. Sci., Polym. Chem. Ed.* **15,** 2229 (1977).
24. E. N. Peters, D. D. Stewart, J. J. Bohan, R. Moffitt, C. D. Beard, G. T. Kwiatkowski, and E. Hedaya, *J. Polym. Sci., Polym. Chem. Ed.* **15,** 973 (1977).
25. J. Ditter, J. Oakes, E. Klusmann, and R. Williams, *J. Inorg. Chem.* **9,** 889 (1970).
26. R. E. Kesting, K. F. Jackson, E. B. Klusmann, and F. J. Gehart, *J. Appl. Polym. Sci.* **14,** 2525 (1970).
27. R. E. Kesting, K. F. Jackson, and J. M. Newman, *J. Appl. Polym. Sci.* **15,** 1527 (1971).
28. R. E. Kesting, K. F. Jackson, and J. M. Newman, *J. Appl. Polym. Sci.* **15,** 2645 (1971).
29. A. D. Delman, A. A. Stein, J. J. Kelly, and B. B. Simms, *J. Appl. Polym. Sci.* **11,** 1979 (1967).
30. M. B. Roller and J. K. Gilham, *J. Appl. Polym. Sci.* **17,** 2141 (1973).
31. E. N. Peters, J. H. Kawakami, G. T. Kwiatkowski, E. Hedaya, B. L. Joesten, D. W. McNeil, and D. A. Owens, *J. Polym. Sci., Polym. Phys. Ed.* **15,** 723 (1977).
32. J. M. Augl, NOLTR 70-240 (1970).
33. L. H. Sperling, S. L. Cooper, and A. V. Tobolsky, *J. Appl. Polym. Sci.* **10,** 1725 (1966).
34. E. J. Zaganiaris, L. H. Sperling, and A. V. Tobolsky, *J. Macromol. Sci. Chem. A* **1,** 1111 (1967).
35. M. T. Shaw and A. V. Tobolsky, *Macromolecules* **3,** 552 (1970).

36. M. T. Shaw and A. V. Tobolsky, *Polym. Eng. Sci.* **10,** 225 (1970).
37. M. Mininni and A. V. Tobolsky, *J. Appl. Polym. Sci.* **16,** 2555 (1972).
38. M. B. Roller and J. K. Gillham, *J. Appl. Polym. Sci.* **16,** 3073 (1972).
39. M. B. Roller and J. K. Gillham, *J. Appl. Polym. Sci.* **16,** 3095 (1972).
40. M. B. Roller and J. K. Gillham, *J. Appl. Polym. Sci.* **16,** 3105 (1972).
41. M. B. Roller and J. K. Gillham, *J. Appl. Polym. Sci.* **17,** 2141 (1973).
42. M. B. Roller and J. K. Gillham, *J. Appl. Polym. Sci.* **17,** 2623 (1973).
43. Y. Mohadger, M. B. Roller, and J. K. Gillham, *J. Appl. Polym. Sci.* **17,** 2635 (1973).
44. M. B. Roller and J. K. Gillham, *Polym. Eng. Sci.* **14,** 567 (1974).
45. H. Schroeder, O. G. Schaffling, T. B. Larchar, F. F. Frulla, and T. L. Heying, *Rubber Chem. Technol.* **39,** 1184 (1966).
46. D. J. Mangold, *Appl. Polym. Symp.* **11,** 157 (1969).
47. E. N. Peters, D. D. Stewart, J. J. Bohan, G. T. Kwiatkowski, C. D. Beard, R. Moffitt, and E. Hedaya, *J. Elast. Plast.* **9,** 177 (1977).
48. M. N. Novotny, R. Segura, and A. Zlatkis, *Anal. Chem.* **44,** 9 (1972).
49. G. E. Pollock, *Anal. Chem.* **44,** 634 (1972).

SUPPLEMENTARY BIBLIOGRAPHY

Polycarboranes, V. V. Korshak, I. G. Sarishvila, A. F. Zhigach, and M. V. Sobolevskii, *Russ. Chem. Rev.* **36,** 903 (1967).
On the thermal behavior of 4-(o-carboranyl)-1-butylmethylsiloxane-dimethylsiloxane copolymers, A. D. Delman, J. J. Kelly, A. A. Stein, and B. B. Simms, *J. Polym. Sci. A 1* **5,** 2119 (1967).
Carborane polymers, R. E. Williams, *Pure and Appl. Chem.* **29,** 569 (1972).
Carborane polymers, H. A. Schroeder, *Polym. Preprints* **13,** 764 (1972).
Carborane-siloxane copolymers, J. F. Ditter and A. J. Gotcher, AD-770 625 (November 1973).
Polycarboranes, NTIS/PS-77/0705 (9/12/77).
Poly(dodecacarborane siloxanes), E. N. Peters, *J. Macromol. Sci. Rev. Macromol. Chem. C* **17** (2), 173 (1979).

8

PHOSPHORUS-CONTAINING POLYMERS—THE PHOSPHAZENES

INTRODUCTION

Rubber-like networks made from polyphosphazene chains were reported as early as 1897 by Stokes.[1] One particular compound, poly(dichlorophosphazene), came to be known as "inorganic rubber" because of the similarity of many of its mechanical properties to those of natural rubber. Thermal decomposition of the polymer did not commence until about 300°C, but rapid hydrolysis to inorganic salts in moist air precluded any practical applications. Much effort was devoted to overcoming this problem by the synthesis of analogous polymers without halogen groups, or by attempting total replacement of the halogen groups in the polymer itself. The early attempts were frustrated by the insoluble, intractable nature of the materials produced and the consequent difficulties of chemical modification and processing. The first report of the preparation of a linear, high-molecular-weight poly(dichlorophosphazene), which was completely soluble in benzene, was made by Allcock and Kugel in 1965.[2] The chloro groups of this polymer could be completely replaced by others to yield products of good hydrolytic stability.

Preparation of Polyphosphazenes

Purified hexachlorocyclotriphosphazene (phosphonitrilic chloride trimer $(NPCl_2)_3$) was heated in an evacuated, sealed tube for 24 to 48 hours at 250°C. The polymer formed initially was soluble, but prolonged reaction times led to the formation of a cross-linked material:

389

where n = 3 to 15,000.

The product formed contained about 70% high-molecular-weight polymer and 30% cyclic oligomers. This mixture, or the separated high-molecular-weight portion, was then treated with alkoxides or aryloxides in refluxing solvents such as tetrahydrofuran, benzene, or toluene to give polyalkoxy- or polyaryloxy-phosphazenes.

Using this technique, polymers have been prepared with many different substituent groups, e.g., OCH_3, OC_2H_5, OCH_2CF_3, $OCH_2C_2F_5$, $OCH_2C_3F_7$, OC_6H_5, OC_6H_4F, $OC_6H_4CF_3$, $OC_6H_3Cl_2$, and $OC_6H_4C_6H_5$.[3,4]

A similar reaction with primary or secondary amines yields poly(bisaminophosphazenes),

where R is methyl, ethyl, n-propyl, n-butyl, and phenyl.[3,4]

The products range from glasses through flexible and rubbery thermoplastics to elastomers with glass transition temperatures varying from below $-90°C$ to $+90°C$ (Table 8.1). Most industrial interest has been shown in polymers containing two or more different substituent groups. The reason for this is that the presence of different substituents serves to inhibit crystallization and enhance elastomeric characteristics. The preparation of a polymer containing roughly equal numbers of trifluoroethoxy and heptafluorobutoxy side-groups randomly distributed was described by Rose in 1968.[5] This polymer had a glass transition temperature of $-77°C$, a decomposition temperature in air of $300°C$ (heating rate $2.5°C$ per minute), and it was unaffected by prolonged immersion in boiling water. In addition, when exposed to a direct flame, it softened and vaporized but did not burn. Polymers containing approximately equal quantities of $-OCH_2CF_3$ and $-OCH_2C_4F_8H$ groups and $-OCH_2C_2F_4H$ and $-OCH_2C_6F_{12}H$ groups have also been reported.[6] The Firestone Tire and Rubber Company's PNF

TABLE 8.1. Properties of Selected Poly(Organophosphazenes) (Reference 3)

Polymer	Glass transition temperature Tg (°C)	Microcrystalline melting point Tm (°C)	Physical form at 25°C
$(NP(OCH_3)_2)_n$	−76	—	Elastomer
$(NP(OC_2H_5)_2)_n$	−84	—	Elastomer
$(NP(OCH_2CF_3)_2)_n$	−66	242	Flexible thermoplastic
$(NP(OCH_2C_3F_7)_2)_n$	−65	100	Thermoplastic
$(NP(OC_6H_5)_2)_n$	− 8	390	Flexible thermoplastic
$(NP(OC_6H_4F)_2)_n$	−14	—	Flexible thermoplastic
$(NP(OC_6H_4CF_3)_2)_n$	−35	330	Flexible thermoplastic
$(NP(OC_6H_3Cl_2)_2)_n$	2	210	—
$(NP(OC_6H_4C_6H_5)_2)_n$	43	> 350	—
$(NP(NHCH_3)_2)_n$	14	—	Flexible thermoplastic
$(NP(NHC_2H_5)_2)_n$	30	—	Thermoplastic
$(NP(NHC_3H_7)_2)_n$	−92	—	Flexible thermoplastic
$(NP(NHC_4H_9)_2)_n$	—	—	Flexible thermoplastic
$(NP(NHC_6H_5)_2)_n$	91	—	Brittle glass
$(NP(OCH_2CF_3)(OCH_2C_3F_7))_n$	−77	—	Elastomer
$(NP(OCH_2C_2F_4H)(OCH_2C_6F_{12}H))_n$	−60	—	Elastomer

fluoroelastomer is based on polyphosphazenes containing mixed fluoroalkoxy substituents, and it is with these that the remainder of this chapter is concerned. Most data have been reported on the material containing mixed $-OCH_2CF_3$ and $-OCH_2C_4F_8H$ substituents together with a small amount of a third component to facilitate cross-linking.

Thermal Stability

Two main techniques have been used in the assessment of thermal stability: weight loss studies and measurement of the change in molecular weight, and molecular weight distribution as a function of time at temperature. Figure 8.1 shows the weight losses for the polymers $(NP(OCH_2CF_3)_2)_n$, $(NP(OCH_2C_4F_8H)_2)_n$, and $(NP(OCH_2CF_3)(OCH_2C_4F_8H))_n$ when heated in nitrogen at a rate of 2°C/minute.[7] The polymer with the mixed substituents is, by this criterion, considerably more stable than the others, the temperature for initial weight loss being almost 100°C higher. The enhanced stability of this structure was confirmed when measured at higher heating rates.[8] Figures 8.2 and 8.3 show isothermal weight loss results for this polymer in nitrogen and air.[9] As would be expected, degradation is more rapid in air than nitrogen. Using these data to calculate the mean rates of weight loss as a percentage of the residual weight, overall activation energies of 84 kJ/mole and 113 kJ/mole were derived for breakdown in nitrogen and air, respectively. These weight loss data indicate

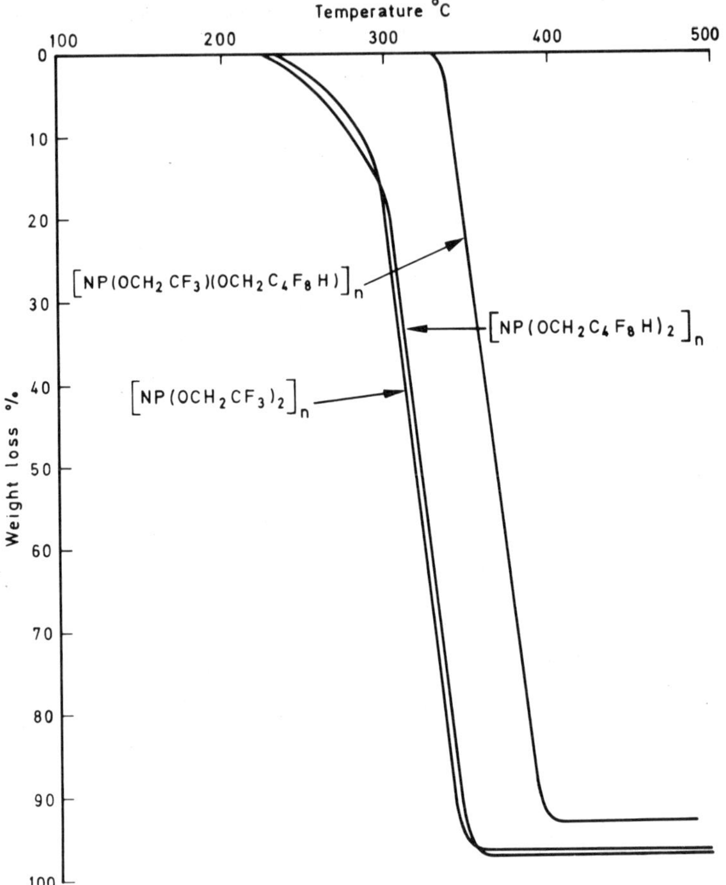

Figure 8.1. *Thermogravimetry of polyphosphazenes in nitrogen. (Heating rate 2°C/min)* Reference 7.

that a use temperature approaching 300°C might be expected for the polymer. A study of the change in molecular weight caused by aging in air for different times at different temperatures, however, tells a different story.[10] Figure 8.4 illustrates the change in solution viscosity as a function of time of heating at 135, 149, 177, and 200°C. Even at the lower temperatures, there was an initial rapid decrease in viscosity followed by a leveling off and approach towards a constant value. This type of viscosity-time behavior is consistent with a random chain scission process occurring at weak links in the polymer backbone. Similar curves have been reported for other polyphosphazenes.[11–14] Gel permeation chromatograms have also been obtained[10] on samples aged at 135, 149, 177, and 200°C, and the

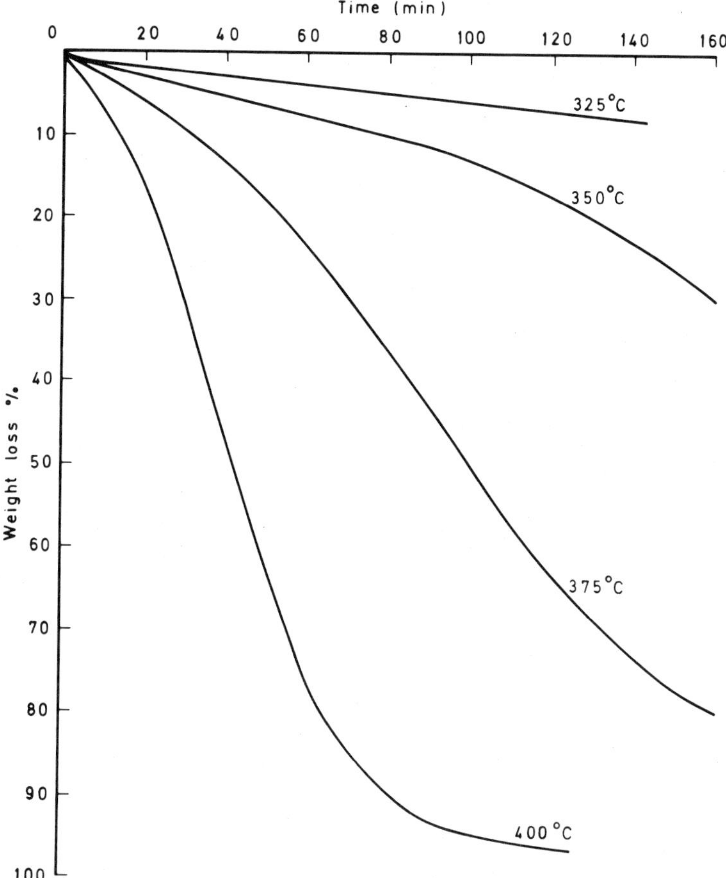

Figure 8.2. Isothermal weight loss of [NP(OCH₂CF₃) (OCH₂C₄F₈H)]ₙ in nitrogen. Reference 9.

results are given in Figure 8.5. The lowering of the molecular weight and the narrowing of the molecular weight distribution with increasing time at temperature are again characteristic of a random chain scission mechanism. Analysis of the GPC data indicated that several types of weak links may be spaced randomly along the chain. These may be of the following type:

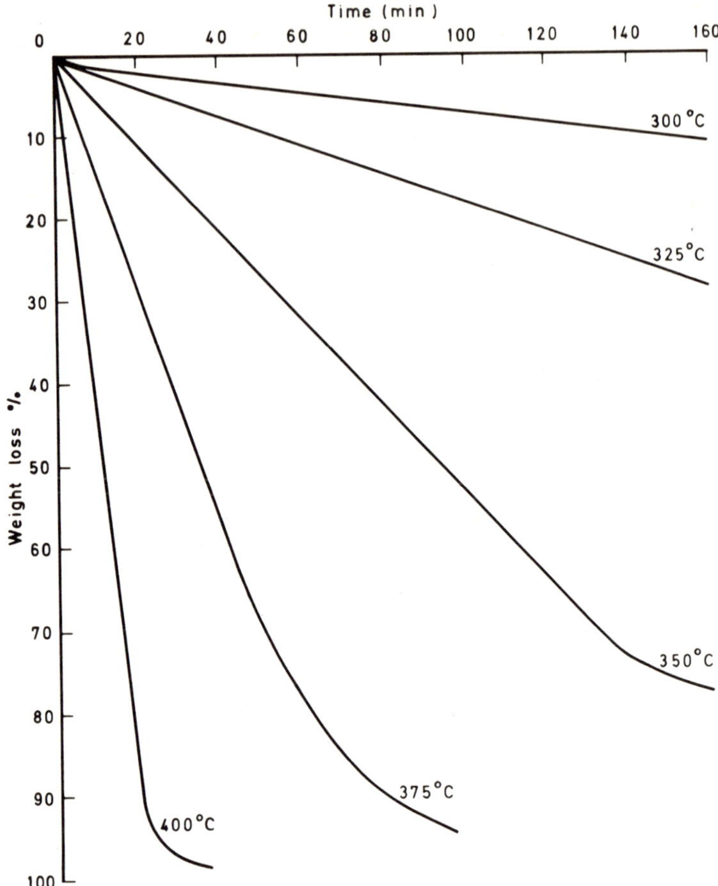

Figure 8.3.   Isothermal weight loss of $[NP(OCH_2CF_3)(OCH_2C_4F_8H)]_n$ in air. Reference 9.

Structure A results from incomplete substitution of the initial poly(di-chlorophosphazene) and structures B and C from the hydrolysis of A. Structure C is thought to be the most likely weak link. One approach to improving the thermal stability of the polymer would therefore be to deactivate these moieties. This has been attempted using metal dialkyldithiocarbamates, Group IA and IIA oxides, hydroxides, and carbonates and bis(8-oxyquinolate) zinc (II) and magnesium (II) as stabilizers. Figure 8.6 shows the change in solution viscosity with time at 177°C in the presence of differing amounts of bis(8-oxyquinolate) zinc (II) and Figure 8.7 shows the gel permeation chromatograms for similar expo-sures. There is a marked increase in thermal stability. A possible mechanism for the stabilization involves complexing of the zinc compound with POH groups to form thermally and hydrolytically stable compounds.

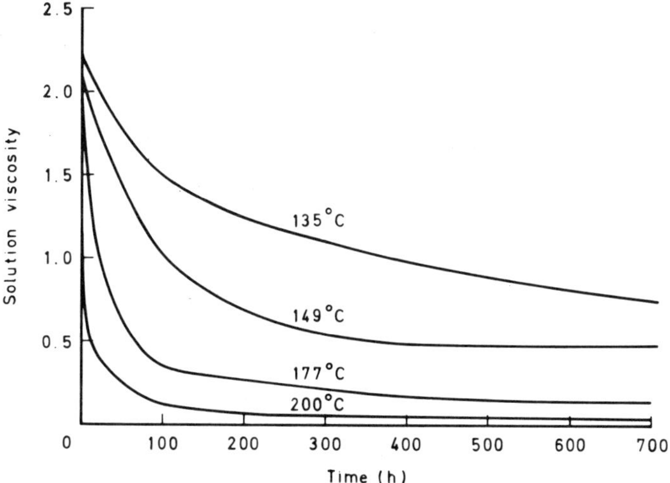

Figure 8.4. *Change in solution viscosity of [NP(OCH$_2$CF$_3$)(OCH$_2$C$_4$F$_8$H)]$_n$ after heating at different temperatures. Reference 10.*

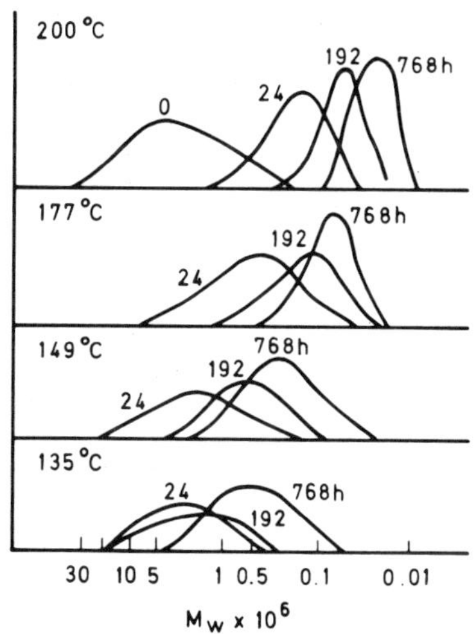

Figure 8.5. *Gel permeation chromatograms of [NP(OCH$_2$CF$_3$)(OCH$_2$C$_4$F$_8$H)]$_n$ aged at different temperatures. Reference 10.*

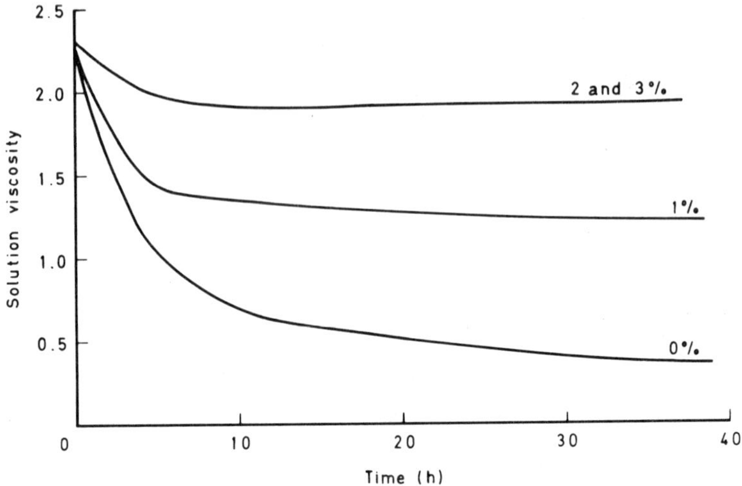

Figure 8.6. Change in solution viscosity of $[NP(OCH_2CF_3)(OCH_2C_4F_8H)]_n$ after heating at 177°C in the presence of bis[8-oxyquinolate] zinc (II). Reference 10.

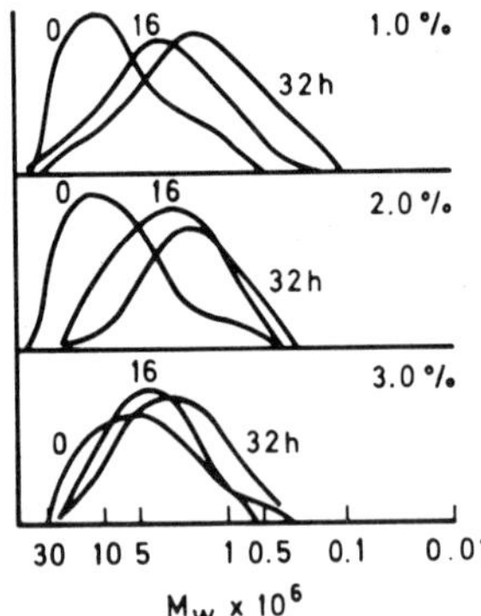

Figure 8.7. Gel permeation chromatograms of $[NP(OCH_2CF_3)(OCH_2C_4F_8H)]_n$ aged at 177°C in the presence of bis[8-oxyquinolate] zinc (II). Reference 10.

The toxicity of the thermo-oxidative decomposition products of polyaryloxyphosphazenes and polyfluoroalkoxyphosphazenes has been evaluated in terms of the sensory irritation stress index, using mice to ascertain adverse physiological effects.[8] This study shows that polyphosphazenes have lower hazard ratings than either polytetrafluoroethylene or poly(vinylchloride).

Physical Characterization

Thermomechanical transitions in polyphosphazenes have been measured as a function of temperature using torsional braid analysis.[7,15] Figure 8.8 gives the results for a continuous experiment running from 150° to $-180°$ to 250° to $-180°$ to 250°C. The copolymer appeared to be completely amorphous, no crystal melt transition region being observed. The loss spectra showed a glass transition peak ($-64°$C), a higher temperature loss peak ($-49°$C), and an intense glassy-state doublet (-160 and $-172°$C). The glass transition temperature lay close to that of the longer chain substituent homopolymer. The higher temperature peak or shoulder to the Tg loss peak may indicate a degree of heterogeneity (block character) in the copolymer structure. The glassy-state transitions were tentatively attributed to motion of the fluoroalkoxy substituents.

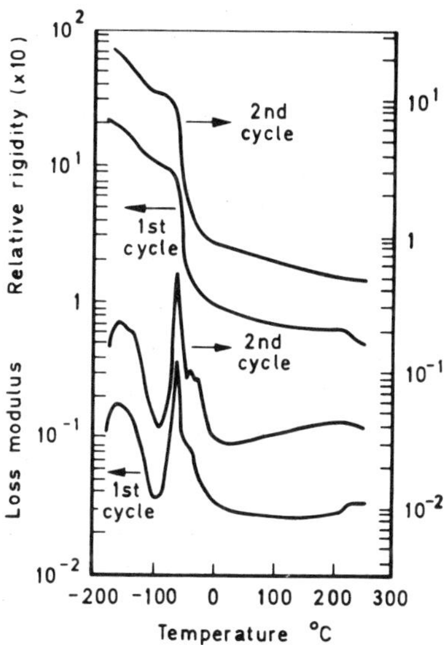

Figure 8.8. Thermomechanical spectra of [NP(OCH₂CF₃) (OCH₂C₄F₈H)]ₙ:
Helium atmosphere   1st cycle 150°C → −180°C → 250°C ⎫*Whole is a*
                        2nd cycle 250°C → −180°C → 250°C ⎭*continuous sequence*
Reference 7.

Elastomer Formulation and Elevated Temperature Properties

The $(NP(OCH_2CF_3)(OCH_2C_4F_9H))$ polymer is a soft gum that can be handled on conventional dry rubber mixing and processing equipment. It responds to variations in compounding in much the same way as do conventional hydrocarbon elastomers, although in some features it is unique. It can be vulcanized with organic peroxides, sulphur plus accelerators, and by high-energy radiation.[16–18] A peroxide cure is recommended for the best high-temperature properties. Reinforcing fillers are needed and coated silicas and clays give the best thermal stability. Carbon black fillers yield useful vulcanizates, but result in reduced retention of mechanical properties at elevated temperatures. Magnesia is added, as in the fluoroelastomers, to act as an acid acceptor and to promote the curing reaction. Bis(8-hydroxyquinolate) zinc (II) may be included as an antioxidant, but it also serves as a cure promoter. A typical formulation could therefore be

Polyphosphazene	100 parts,
Coated silica	30 parts,
Magnesium oxide	6 parts,
Antioxidant	2 parts,
Peroxide	1 part,

and a press cure of 20–60 minutes at 165–170°C suffices, followed by a post-cure at 100°C for 24 hours. Initial properties of such a formulation would be of the order: tensile strength 14 MPa, 100% tensile modulus 3 MPa, elongation at break 170%, and Shore A hardness 60. The changes in these properties with time of aging at 150, 175, and 200°C are shown in Figure 8.9. There is relatively little change in tensile strength at 150°C, but a definite fall off at 175°C and higher temperatures. The other properties are not as greatly affected. The improvement in thermo-oxidative stability which results from the incorporation of bis(8-hydroxyquinolate) zinc (II) into the formulation is illustrated in Figure 8.10. The stabilizer was added either during mixing of the formulation on a rubber mill, or by adding a solution to the gumstock prior to mixing. The latter procedure appears to give superior retention of modulus on aging. Table 8.2 lists properties as a function of temperature for the standard 100 parts polymer, 30 parts silica, 6 parts magnesium oxide, and 1 part peroxide mix, and Table 8.3 shows the changes in these properties with time of aging at temperature.

These data indicate a service life of at least 1,000 hours at 150°C, 300 hours at 175°C, and 120 hours at 200°C. The polyfluorophosphazenes also possess good solvent resistance (because of their fluorine content) and good low-

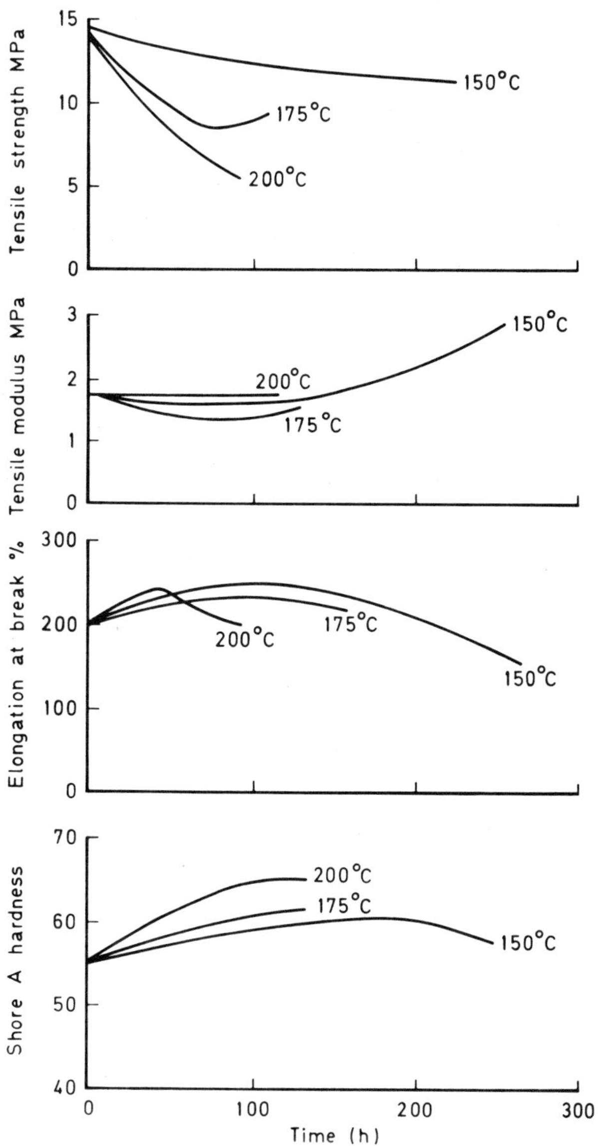

Figure 8.9. Change in properties of compounded $[NP(OCH_2CF_3)(OCH_2C_4F_8H)]_n$ on aging at various temperatures. Reference 16.

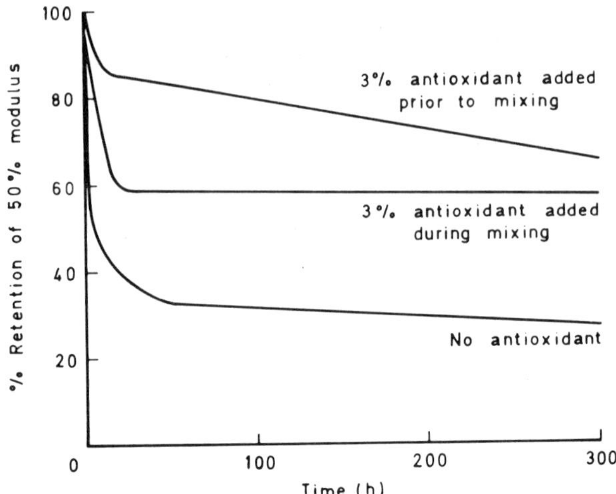

Figure 8.10. Effect of antioxidant on the retention of the modulus of polyphosphazene on aging in air at 200°C. Reference 10.

temperature flexibility. Figure 8.11 compares the approximate service temperature range for a number of oil-resistant rubbers, and it is concluded from this that the polyphosphazenes are only matched overall by the fluorosilicones. The *low*-temperature capabilities of the polyphosphazenes have undoubtedly attracted most attention. The polymers have been extensively investigated for fuel-resistant components for use in arctic conditions.[19] The conclusion from this exercise was that for critical applications below −46°C requiring a good balance of properties, in particular compression set resistance, the fluorophosphazene elastomers were the only acceptable candidates. O-ring seals have successfully completed a 1,000-hour dynamic qualification test in a rod-seal test apparatus over the temperature range −54 to −163°C and at pressures up to 21 MPa.[20]

TABLE 8.2. Mechanical Properties as a Function of Temperature for Polyphosphazene Elastomer (Reference 17)

Temperature (°C)	Tensile strength (MPa)	100% Youngs modulus (MPa)	Elongation at break (%)
20	14.9	5.9	160
100	5.5	1.7	160
150	5.5	1.7	150
175	4.0	1.1	170
200	4.6	1.7	165
Compression set after 70 hours at 150°C: 62%			
after 70 hours at 175°C: 90%			

TABLE 8.3. Mechanical Properties of Polyphosphazene Elastomer Tested at Ambient Temperature after Aging for Different Times at Elevated Temperatures (Reference 17)

Temperature (°C)	Time (hours)	Tensile strength (MPa)	100% Youngs modulus (MPa)	Elongation at break (%)
20	—	13.2	3.7	165
135	72	13.8	5.6	160
135	120	12.2	4.5	150
135	240	13.6	5.3	160
135	336	12.9	3.5	180
135	672	11.0	3.1	190
135	1000	10.3	4.4	170
150	72	13.1	3.8	170
150	120	12.2	3.3	185
150	240	11.6	3.7	175
150	336	10.1	3.0	175
150	672	10.5	3.5	190
150	1000	10.0	3.3	200
175	72	9.9	2.8	200
175	120	10.4	2.5	215
175	240	7.3	2.5	190
200	24	8.4	1.9	240
200	48	6.9	1.7	220
200	72	5.8	1.9	210
200	120	4.8	1.8	215
200	240	2.8	2.1	155

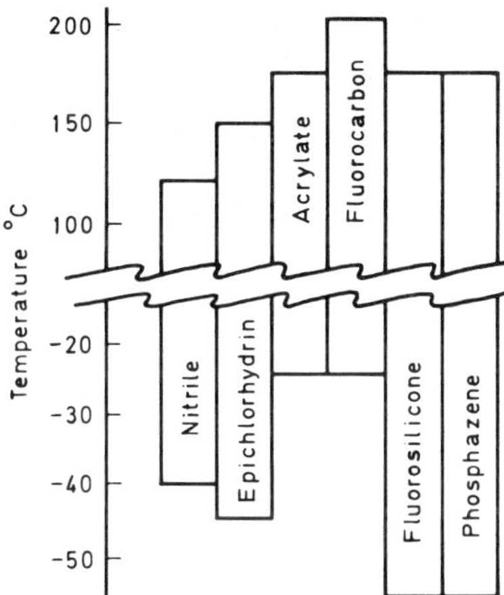

Figure 8.11. Approximate service temperature range for a number of oil-resistant rubbers. Reference 18.

SUMMARY

1. The polyphosphazene elastomers containing fluorine provide a combination of oil resistance with a wide operating temperature range ($+150°C$ to below $-25°C$).

2. They are flexible at low temperatures without recourse to plasticizers and are only matched in this respect by the fluorosilicones.

3. Their operating temperature range plus oil resistance is also only matched by the fluorosilicones and the latter have poorer flex and tear resistance.

4. Uses would appear to be as O-rings and oil seals in dynamic applications; as gaskets, diaphragms, hose, and protective fire-resistant coatings.

REFERENCES

1. H. N. Stokes, *Am. Chem. J.* **19,** 782 (1897)
2. H. R. Allcock and R. L. Kugel, *J. Amer. Chem. Soc.* **87,** 4216 (1965).
3. H. R. Allcock, *Chem. Rev.* **72,** 315 (1972).
4. H. R. Allcock, *Polym. Preprints* **13** (2), 774 (1972).
5. S. H. Rose, *J. Polym. Sci. B* **6,** 837 (1968).
6. S. H. Rose and K. A. Reynard, *Polym. Preprints* **13**(2), 778 (1972).
7. T. M. Connelly and J. K. Gillham, *J. Appl. Polym. Sci.* **20,** 473 (1976).
8. K. Sebata, J. H. Magill, and Y. C. Alarie, *J. Fire and Flammability* **9,** 50 (1978).
9. J. K. Valaitis and G. S. Kyker, *J. Appl. Polym. Sci.* **23,** 765 (1979).
10. G. S. Kyker and J. K. Valaitis, *Polym. Preprints* **18**(1), 488 (1977).
11. G. Allen, C. J. Lewis, and S. M. Todd, *Polymer* **77,** 44 (1970).
12. H. R. Allcock and W. J. Cook, *Macromolecules* **7,** 284 (1974).
13. H. R. Allcock, G. Y. Moore, and W. J. Cook, *Macromolecules* **7,** 571 (1974).
14. G. L. Hagnauer and B. R. Laliberte, *J. Appl. Polym. Sci.* **20,** 3073 (1976).
15. T. M. Connelly and J. K. Gillham, *Polym. Preprints* **15**(2), 458 (1974).
16. D. P. Tate, *J. Polym. Sci. Symp.* **48,** 33 (1974).
17. G. S. Kyker and T. A. Antkowiak, *Rubber Chem. Technol.* **47,** 32 (1974).
18. D. P. Tate, *Rubber World* **172**(6), 41 (1975).
19. P. Touchet and P. E. Gatza, *J. Elast. Plast.* **9,** 3 (1977).
20. J. C. Vicic and K. A. Reynard, *J. Appl. Polym. Sci.* **21,** 3185 (1977).

SUPPLEMENTARY BIBLIOGRAPHY

Phosphorus-Nitrogen Compounds, H. R. Allcock, Academic Press, N.Y. (1972).

Polyphosphazenes-synthesis-properties-applications, R. E. Singler, N. S. Schneider, and G. L. Hagnauer, *Polym. Eng. Sci.* **15,** 321 (1975).

The thermal transition behavior of polyorganophosphazenes, N. S. Schneider, C. R. Desper, and R. E. Singler, *J. Appl. Polym. Sci.* **20,** 3087 (1976).

Poly(organophosphazenes)—unusual new high polymers, H. R. Allcock, *Angew. Chem. Int. Ed. Engl.* **16,** 147 (1977).

Polyphosphazenes: structure and applications, R. E. Singler and G. L. Hagnauer, in *Organometallic Polymers,* C. E. Carraher, J. E. Sheats, and C. U. Pittman, (eds.), Academic Press, N.Y. (1978), p. 257.

Transition to the mesomorphic state in polyphosphazenes, N. S. Schneider, C. R. Desper, R. E. Singler, M. N. Alexander, and P. L. Sagalyn, in *Organometallic Polymers,* C. E. Carraher, J. E. Sheats, and C. U. Pittman, (eds.), Academic Press, N.Y. (1978), p. 271.

Poly(organophosphazenes) designed for biomedical uses, H. R. Allcock, in *Organometallic Polymers,* C. E. Carraher, J. E. Sheats, and C. U. Pittman, (eds.), Academic Press, N.Y. (1978), p. 283.

Biocompatibility of eight poly(organophosphazenes), C. W. R. Wade, S. Gourlay, R. Rice, A. Hegyeli, R. Singler, and J. White, in *Organometallic Polymers,* C. E. Carraher, J. E. Sheats, and C. U. Pittman, (eds.), Academic Press, N.Y. (1978), p. 289.

Thermal degradation of polybis(*p*-isopropylphenoxy)phosphazene, I. Goldfarb, N. D. Hann, R. L. Dieck, and D. C. Messersmith, *J. Polym. Sci., Polym. Chem. Ed.* **16,** 1505 (1978).

Poly(organophosphazenes). Synthesis and applications of a new class of technologically important polymers, J. M. McAndless, Defense Research Establishment, Ottawa, Report No. 795 (January 1979).

FUTURE DEVELOPMENTS

Grouped together in this chapter are a number of systems on which a considerable amount of work has been done and on which research is still in progress, although this to date has not resulted in large-scale exploitation. These systems are without exception based upon heterocyclic ring structures. Also considered in this chapter are a number of concepts that may have an influence upon future progress towards more thermally stable polymeric materials. The concepts include those of ladder polymers, intramolecular cyclization, ordered polymers, stabilization of heat resistant polymers, and finally inorganic polymer systems.

HETEROCYCLIC SYSTEMS UNDER DEVELOPMENT

Polytriazines

Two review articles have been written dealing specifically with the symmetrical triazines[1] and the asymmetrical triazines.[2] The most promising early work led to the discovery of polyperfluoroalkylenetriazine elastomers with high thermal stability and good physical properties.[3] These were produced by the thermal cyclization of perfluoroalkylenediamidines. Perfluoroadipodiamidine and perfluoroglutaroimidine, when heated above their melting points, evolve ammonia and yield insoluble, cross-linked materials of the structure shown:

The amount of cross-linking is reduced by copolymerizing the above monomers with perfluoroalkylmonoamidines. Despite the initial promise of these polymers and a considerable amount of further development work, overall control of the reaction proved too difficult for commercial exploitation to follow. Attention latterly has been devoted to resin systems containing the triazine ring. The first of these is based upon Bisphenol-A and is produced by the reaction shown using zinc octanoate as catalyst.[4]

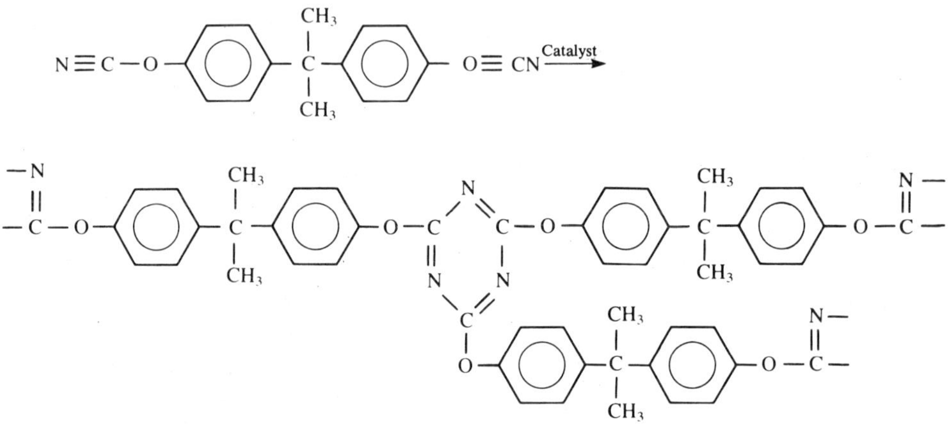

Triazine A resin (Bayer AG)

The resin is soluble in low-boiling-point solvents such as acetone and can be used for the fabrication of fiber-reinforced composites. The weight loss of a cast sample of Triazine A as a function of rising temperature is compared with those of an epoxy and a polyimide resin in Figure 9.1, and the retentions of flexural strength at elevated temperatures of copper-clad glass cloth laminates are similarly compared in Figure 9.2.

Triazine resins have also been prepared from phenolic novolacs or meta-cresol novolacs.

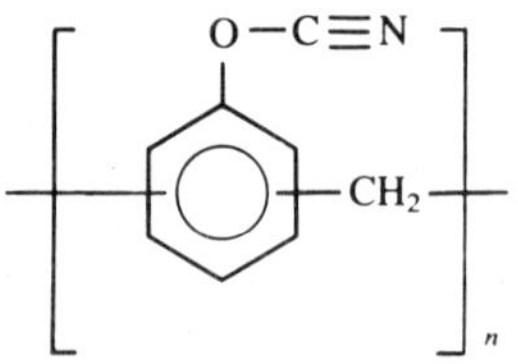

Phenolic novolac type 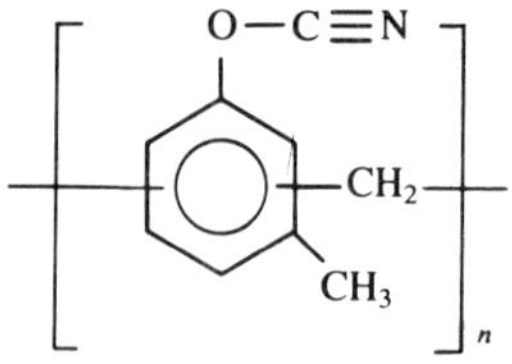   Meta-cresol novolac type

Although these have considerably higher thermal stability than the Triazine A resins, they have not been produced commercially because of poor shelf life,

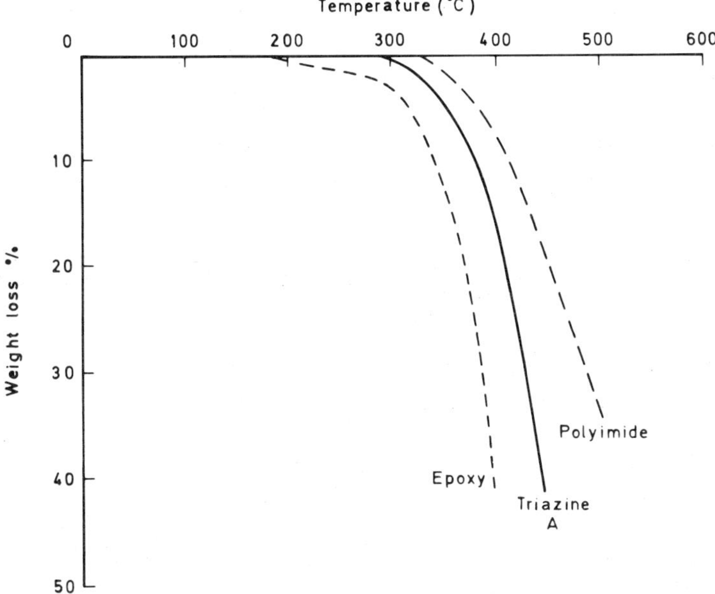

Figure 9.1. *Thermogravimetry in air of Triazine A resin compared with an epoxy and a polyimide resin. (Heating rate 10°C/min) Reference 4.*

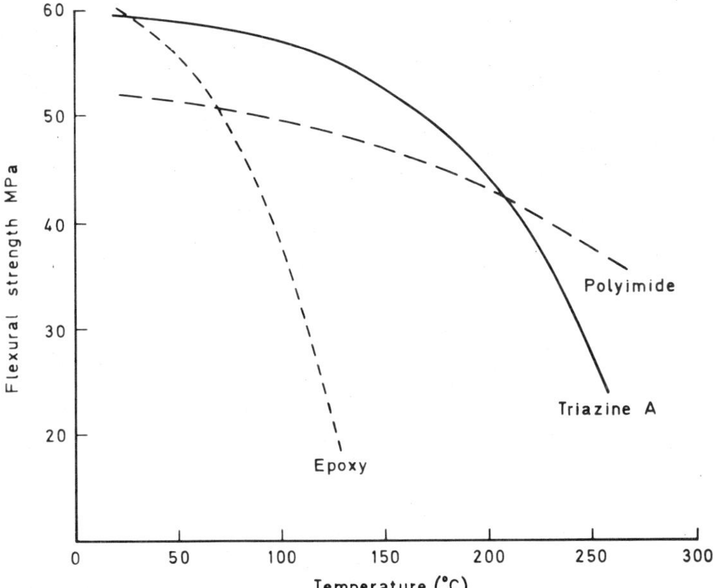

Figure 9.2. *Comparison of flexural strength retention of copper-clad glass cloth laminates. Reference 4.*

very short reaction times, high water absorption, and relatively poor adhesive strength.

Another triazine-containing material currently under development has been called[5,6] NCNS resin (Ciba–Geigy). This is composed of a primary bis-cyanamide and a secondary bis (N-sulphonamide) cyanamide, which under the influence of heat react by an addition process to form s-triazine rings and a three-dimensional cross-linked structure:

$$
\begin{array}{cc}
C\equiv N & C\equiv N \\
| & | \\
H-N-A-N-H
\end{array}
\qquad\qquad
\begin{array}{cc}
A' & A' \\
| & | \\
O=S=O\;\; O=S=O \\
| & | \\
N\equiv C-N-A-N-C\equiv N
\end{array}
$$

Primary bis-cyanamide   Secondary bis (N-sulphonamide) cyanamide

$$3\,R-C\equiv N \xrightarrow{\text{Trimerization}} \text{(s-triazine ring)} \qquad \text{where } A = -C_6H_4CH_2C_6H_4-$$

High-solids-content solutions for laminating purposes are prepared in alcohol/ethyl acetate or alcohol/chlorinated solvent mixtures. Prepregs made from these mixtures retain good processing characteristics after three months' storage at room temperature and can be fabricated by vacuum bag-autoclave or press-molding techniques. Some typical properties of glass cloth and carbon fiber composites at elevated temperatures are listed in Table 9.1.

Table 9.2 shows the retention of interlaminar shear strength as a function of time of aging at 205°C. There is very little change with time at temperature when tested hot.

If laminates are tested at 176°C after first boiling in water for 24 hours, or exposing at 95% RH and 50°C for 30 days, there are losses in ILSS of 37 and 58%, respectively. An NCNS resin based upon polymethylene-polyphenylamines

TABLE 9.1. *Properties of NCNS Resin Composites (Reference 6)*

Property	Glass cloth laminates [181 Fabric Volan A finish]	Carbon fiber laminates [HT-S fibers]
Flexural strength (MPa)		
At 25°C	548	1470
204°C	—	966
232°C	476	903
250°C	377	—
Flexural modulus (GPa)		
At 25°C	23	100
204°C	—	97
232°C	19	93
250°C	19	—
Interlaminar shear strength (MPa)		
At 25°C	46	81
232°C	43	41

TABLE 9.2. *Effect of Aging at 205°C on the ILSS of NCNS Resin/HT-S Carbon Fiber Laminates (Reference 5)*

Test temperature	Interlaminar shear strength (MPa) after time at 205°C of			
	0 Hours	500 Hours	1250 Hours	1500 Hours
Room temperature	81	110	116	117
205°C	57	50	61	49

rather than 4,4,′-diaminodiphenylmethane (DDM)

$$H_2N - C_6H_4 - CH_2 - C_6H_4 - NH_2$$

however, shows losses of only 6 and 24% when tested under the same conditions.

Polyphenylquinoxalines

The preparation of polyquinoxalines was first reported in 1964,[7,8] and since that time considerable effort has been devoted to the synthesis of variants of the structure and the evaluation of their properties. The basic reaction involves the condensation of aromatic tetraamines and aromatic tetracarbonyl compounds, either in the melt or in solution (hexamethylphosphoramide or dioxane).

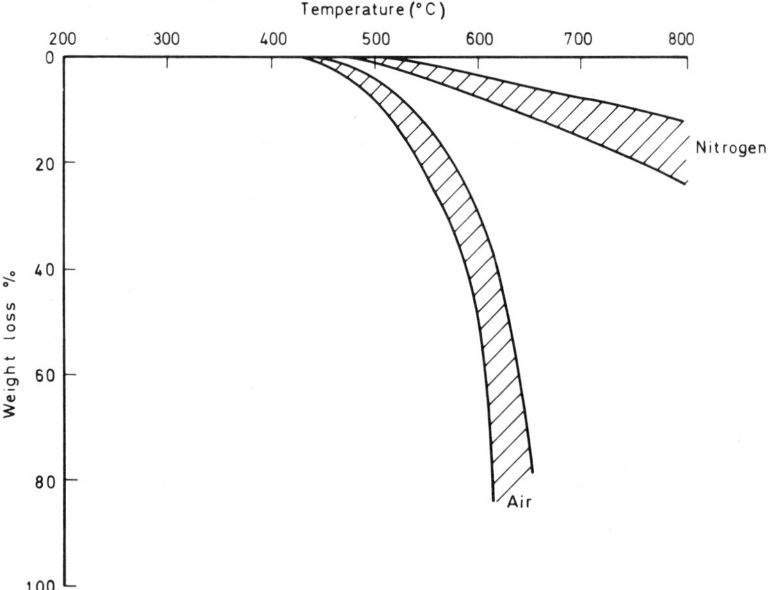

A range of aromatic bis (*o*-diamines) and aromatic bisglyoxals have been used, and ether,[9] sulphide,[10] and sulphone groups[10] have been incorporated in the polymer chain as flexibilizing units to improve solubility. Figure 9.3 shows weight loss bands in nitrogen and air for polyquinoxalines with and without ether linkages. The thermal stability in an oxidizing atmosphere is appreciably less than that of the polyimides or polybenzimidazoles. Thermo-oxidative stability was, however, greatly improved by the preparation of polyphenylquinoxalines (PPQs).[11, 12] These also had better solubility and were more amenable to proc-

Figure 9.3. Thermogravimetry of polyquinoxalines in air and nitrogen. (Heating rate 2°C/min) Reference 9.

essing. The synthetic route involved the reaction of aromatic bis (*o*-diamines) with aromatic dibenzils.

Figure 9.4 compares the isothermal weight loss in air at 400°C of a polyquinoxaline and the analogous phenyl-substituted polymer. The superiority of the phenyl-substituted material is very evident. Subsequent effort therefore concentrated upon the synthesis, thermal characterization, and mechanical evaluation of PPQs. Progress up to 1976 has been summarized in three review articles by Hergenrother.[13–15] Table 9.3 gives the glass transition temperatures determined by dielectric loss measurements of films of different PPQs dried in vacuum at 200°C. It is noteworthy that the highest figure attained is only 325°C, so some thermoplasticity and creep would be expected in PPQ samples exposed to temperatures of this order. Reduction in thermoplasticity requires some form of cross-linking and initial work concentrated upon the incorporation of cyano and cyanato groups in the polymer.[16,17] It was thought that these groups could be induced to trimerise, thus providing very stable *sym*-triazine cross-links. Cycling cyano- or cyanato-substituted PPQs to 400°C resulted in increases in the glass transition temperature of up to 100°C. Latterly, however, attention has been directed to acetylene-terminated quinoxalines (ATQs).[18–20] The synthetic route to the ATQ oligomers is as shown. These oligomers homopolymerize by both inter- and intramolecular addition reactions (no evolution of volatiles) to give fused aromatic ring systems of good thermal and thermo-oxidative stability. Figure 9.5 shows the weight loss range for a number of ATQ polymers at 316°C in air.

One further synthetic variation that deserves mention is the preparation of

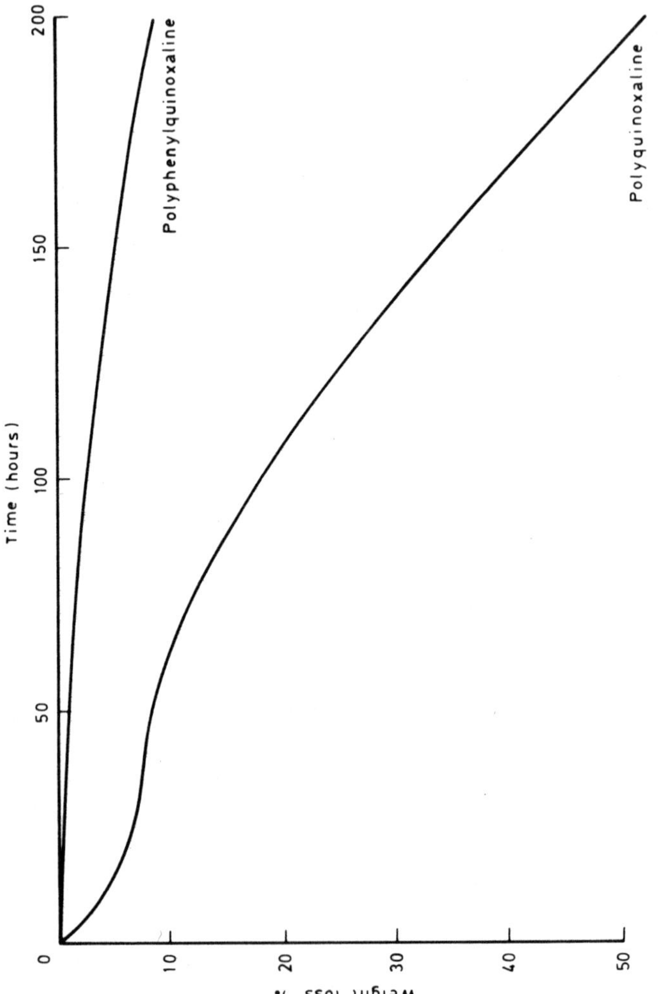

Figure 9.4. Weight losses of a polyquinoxaline and its phenyl-substituted analog in air at 400°C. Reference 13.

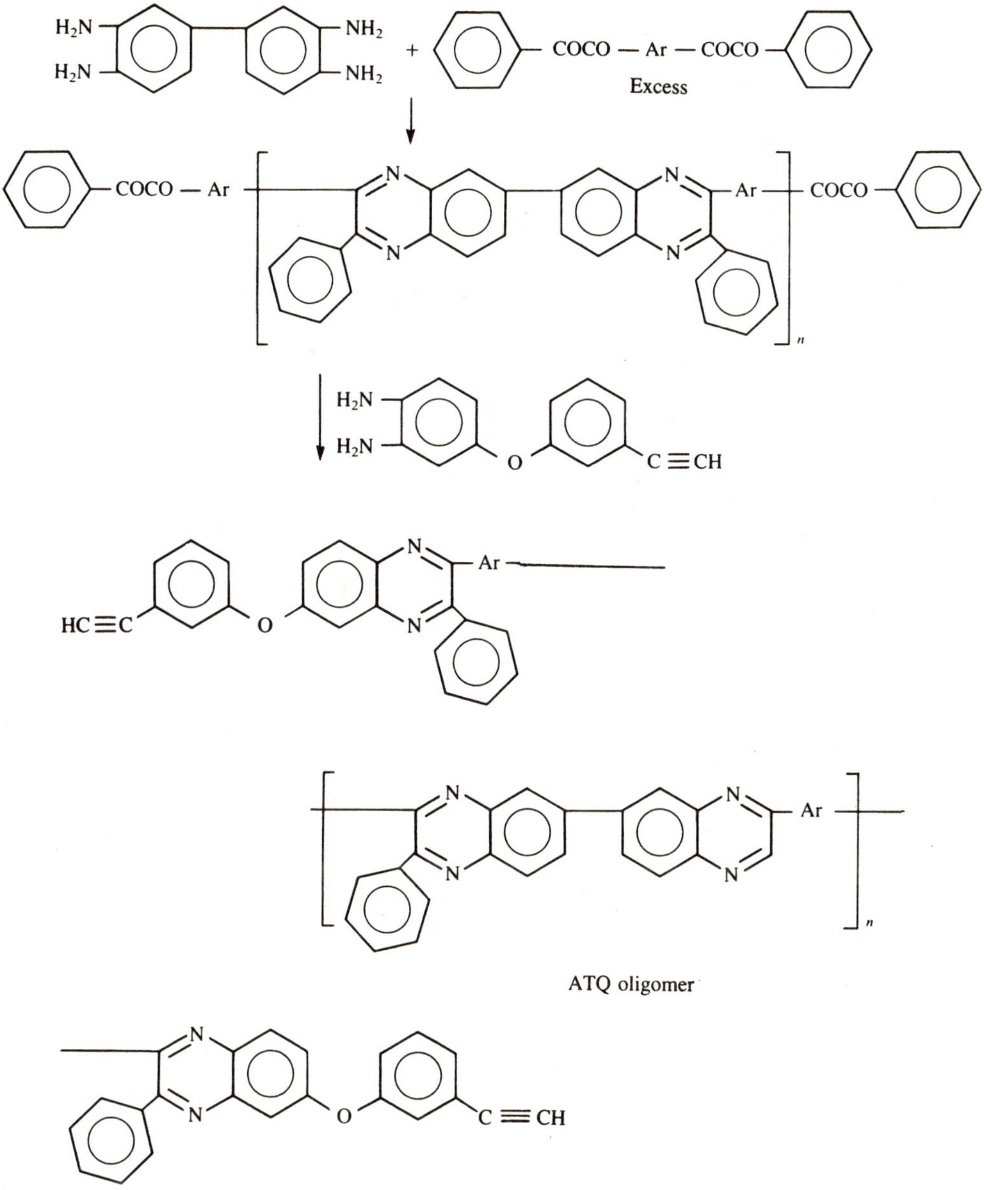

ATQ oligomer

a "ladder" polyquinoxaline.[21, 22] (See later for a discussion of the ladder polymer concept.) This, presumably because of incomplete cyclization, had a thermal stability no greater than that of a linear polyquinoxaline and so was not proceeded with further.

TABLE 9.3. *Glass Transition Temperatures of Polyphenylquinoxalines (Reference 13)*

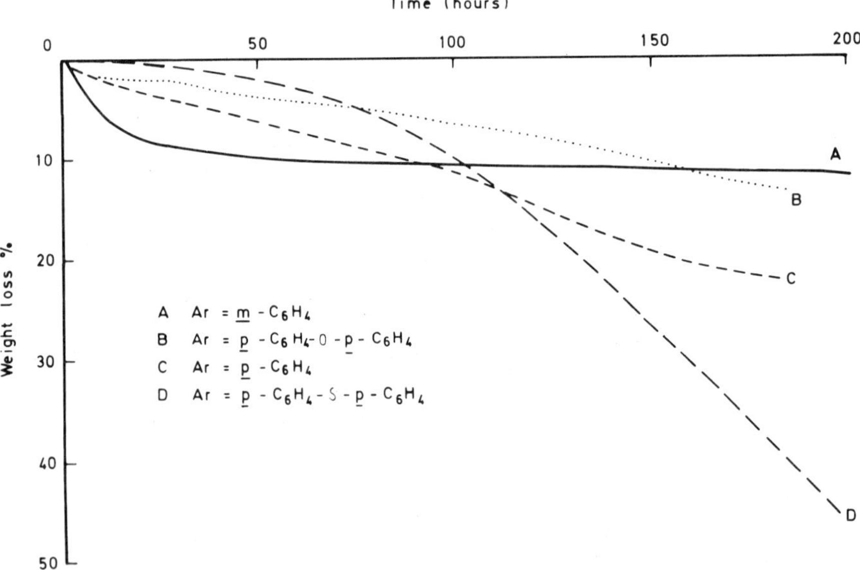

Polymer		
X	Y	Tg (°C)
Single bond	p-C_6H_4	325
Single bond	m-C_6H_4	320
Single bond	p,p'-$C_6H_4OC_6H_4$	285
Single bond	p,p'-$C_6H_4SC_6H_4$	260
—O—	p-C_6H_4	298
—O—	m-C_6H_4	253
—O—	p,p'-$C_6H_4OC_6H_4$	268
—SO_2—	p-C_6H_4	212
—SO_2—	p,p'-$C_6H_4OC_6H_4$	195
—CO—	p-C_6H_4	258
—CO—	p,p'-$C_6H_4OC_6H_4$	243

Figure 9.5. *Weight losses of ATQ polymers at 316°C in air. Reference 19.*

Ladder polyquinoxaline

Laminates were prepared from glass cloth (1581 style AF-994 glass plus HTS finish), boron fibers, and carbon fibers (Modmor II) using the polyquinoxaline shown as the matrix resin

A variety of press cures were used, but these were all followed by an extended post-cure under nitrogen of 4 hours at 205°C, 4 hours at 232°C, 10 hours at 260°C, 8 hours at 288°C, 4 hours at 316°C, 4 hours at 344°C, 4 hours at 371°C, and finally 4 hours at 398°C. The properties of these laminates are summarized in Tables 9.4, 9.5, and 9.6. The same polyquinoxaline was also used on a glass fabric carrier to form adhesive bonds between stainless steel panels. The cure cycle used was one hour each at 344, 426, and 455°C under 1.38 kPa pressure

TABLE 9.4. Properties of Polyquinoxaline/Glass Cloth Laminates (Reference 23)

Test condition	Flexural strength (MPa)	Flexural modulus (GPa)
RT	687	25
RT after 1 hour at 371°C	425	17
RT after 10 hours at 371°C	304	17
RT after 10 min at 538°C	117	15

TABLE 9.5. Properties of Polyquinoxaline/Boron Fiber Unidirectional Laminate (Reference 23)

Test temperature (°C)	Flexural strength (MPa)	Flexural modulus (GPa)
RT	1746	292
133	1359	250
177	1384	259
212	1242	255
316	1524	167
361	1666	83

TABLE 9.6. Properties of Polyquinoxaline/Carbon Fiber Unidirectional Laminate (Reference 23)

Test Condition	Flexural strength (MPa)	Flexural modulus (GPa)	ILSS (MPa)
RT	842	107	88
316°C after 1 hour at 316°C	816	105	59
316°C after 100 hours at 316°C	802	104	58
316°C after 200 hours at 316°C	651	101	48
371°C after 1 hour at 371°C	748	104	61
371°C after 50 hours at 371°C	713	100	61

TABLE 9.7. Tensile Shear Strengths of Stainless Steel/Polyquinoxaline Lap-Joints (Reference 23)

Test condition	Tensile shear strength (MPa)
RT	23
316°C after 1 hour at 316°C	20
316°C after 200 hours at 316°C	16
371°C after 1 hour at 371°C	13
371°C after 50 hours at 371°C	18
538°C after 10 min at 538°C	9

and in a nitrogen atmosphere. The tensile strengths of lap-shear joints are given in Table 9.7.

The flexural strength and interlaminar shear strength ranges for polyphenylquinoxaline/carbon fiber composites aged and tested at 343°C are shown in Figures 9.6 and 9.7.[24] For an ATQ/HT-S carbon fiber unidirectional composite, the room temperature flexural strength and modulus were 1055 MPa and 93 GPa, respectively. On testing at 260°C the corresponding figures were 931 MPa and 86 GPa, retentions of 88 and 92%.[20] This system shows very good resistance to hot-wet conditions, the flexural strength and modulus of dry and wet specimens being essentially the same when tested at temperatures up to 260°C. This contrasts vividly with the behavior of epoxy resin/carbon fiber composites (Figure 9.8).

Polyimidazoquinazolines

Synthetic routes to polyimidazoquinazolines are summarized in a review article by Yoda and Kurihara.[25] There are four methods involving the reaction of a tetraamine with a bisbenzoxazinone (A), with N,N'-bisbenzoylmethylene-dianthranilic acid (B), with 2,2'-dicarboxyphenylterephthalamide (C), and with o-aminobenzoic acid and aromatic dicarboxylic acid derivatives (D).

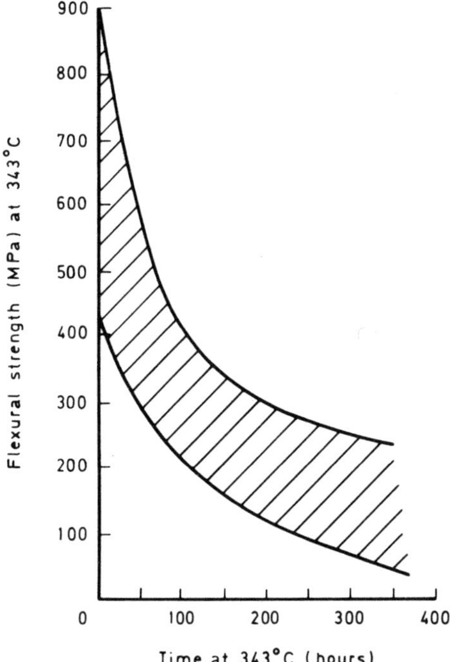

Figure 9.6. Flexural strength range on aging at 343°C of polyphenylquinoxaline/carbon fiber laminates. Reference 24.

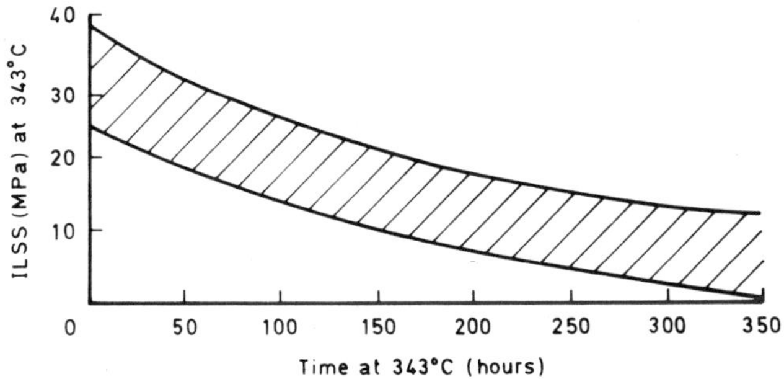

Figure 9.7. Interlaminar shear strength range on aging at 343°C of polyphenylquinoxaline/ carbon fiber laminates. Reference 24.

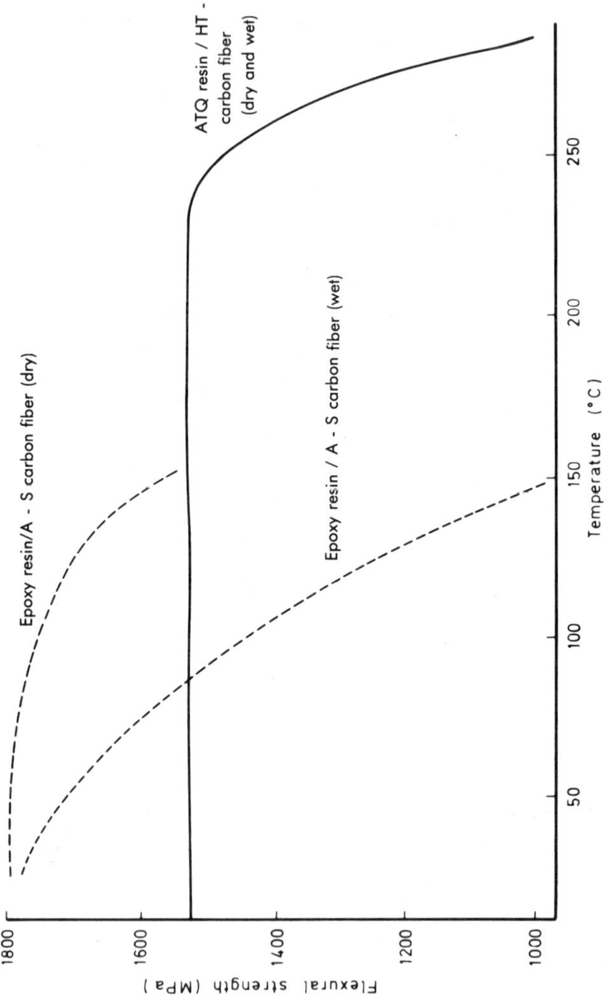

Figure 9.8. Flexural strength at elevated temperatures of dry and wet ATQ resin and epoxy resin–carbon fiber composites.

Route A.

Polyimidazoquinazoline

where R = CH$_3$, C$_6$H$_5$
X = O, CH$_2$

TABLE 9.8. *Flexural Properties of AF-R-530/Modmor II Carbon Fiber Laminates at Elevated Temperatures (Reference 26)*

Heat aging conditions	Flexural strength (MPa)	Flexural modulus (GPa)	Retention of RT strength (%)
RT	1540	119	—
200 hours at 316°C	1249	115	81
500 hours at 316°C	1252	126	81
1000 hours at 316°C	598	81	39
RT	1939	119	—
1 hour at 371°C	1181	126	61
50 hours at 371°C	1187	117	61
100 hours at 371°C	992	100	51
RT	1617	116	—
1 hour at 426°C	1497	113	93
5 hours at 426°C	987	102	61
10 hours at 426°C	331	98	20
1 hour at 481°C	1181	110	73
1 hour at 536°C	629	63	39

Route B.

where $X = CH_2$

Route C.

Reaction steps as in B.

Route D.

Resins designated AF-R-530 and AF-R-533 have been prepared essentially by route D and used in carbon fiber-reinforced laminates.[26-28] The systems require very high curing temperatures—up to 470°C in a nitrogen atmosphere. This does, however, result in exceptional retention of flexural properties, and measurements have been made at temperatures as high as 536°C (Table 9.8).

Polyphthalocyanines

The reaction of 4-amino-phthalonitrile with aliphatic diacid chlorides yields compounds of the type shown[29]:

$$2\ \ NC\text{-}C_6H_3(CN)\text{-}NH_2\ +\ Cl\,CO-R-CO\,Cl\ \longrightarrow$$

$$NC\text{-}C_6H_3(CN)-NH-CO-R-CO-NH-C_6H_3(CN)\text{-}CN$$

where $R = [CH_2]_n$ and $n = 4$ to 18.

On heating these diamides above their melting points (160 to 300°C, depending on the length of the aliphatic chain), addition reactions occur and very stable phthalocyanine nuclei are formed. This is shown schematically on the following page.

The polymers have been shown[30,31] to be thermally stable on a weight loss basis up to 260°C and their cure behavior has been studied using the torsional braid analysis method.[32] Recently a paper has been published citing the dynamic mechanical properties of the polymers with C_6, C_{10}, C_{22}, and C_{36} aliphatic chains.[33] The C_{10} resin has been used to prepare carbon fiber prepreg by a hot-melt technique, and the processability of this material has been assessed by differential scanning calorimetry, thermogravimetric analysis, and dynamic dielectric analysis. Fabrication of the prepreg has been accomplished using a conventional vacuum bag/hydraulic press system, and the mechanical properties of the laminates have been compared with those of similar samples made with an epoxy and a polyimide resin (Table 9.9).

The mechanical properties of the three different systems are very similar. Advantages claimed for the phthalocyanine-based material are:

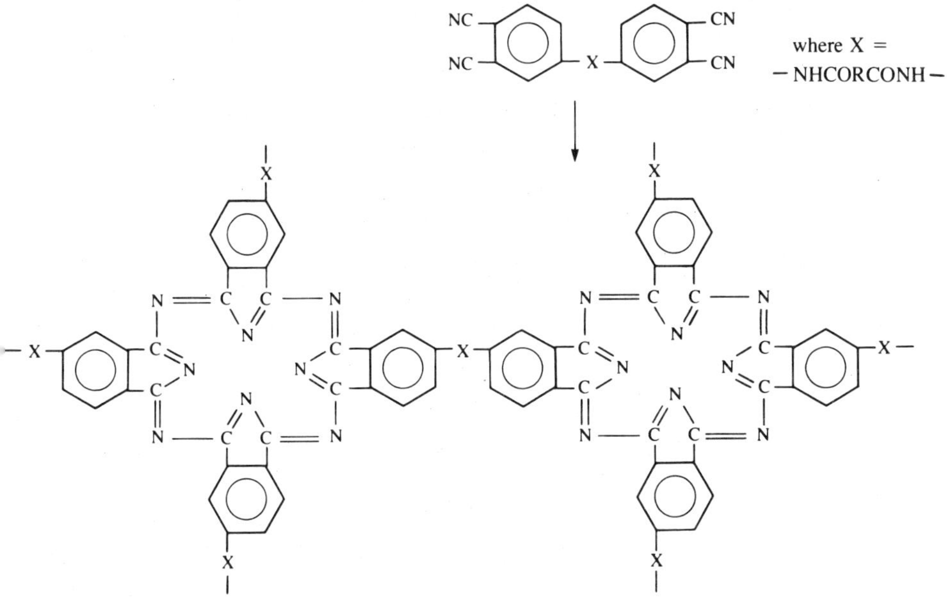

where X =
$-\mathrm{NHCORCONH}-$

TABLE 9.9. Mechanical Properties of Phthalocyanine Resin/Carbon Fiber Laminates Compared with an Epoxy and a Polyimide Resin System (Reference 33)

Laminate system	Fiber orientation	Tensile strength (GPa)	Tensile modulus (GPa)	Flexural strength (GPa)	Flexural modulus (GPa)
C_{10} phthalocyanine resin					
T300 carbon fiber	Quasi-isotropic	0.47	75.2	0.46	33.8
	±15°	0.56	93.2	0.66	75.2
	±45°	0.11	37.3	0.18	4.7
5208 epoxy resin					
T300 carbon fiber	Quasi-isotropic	0.48	48.3	—	—
	±15°	0.75	104.9	1.03	107.0
	±45°	0.16	26.9	—	—
F178 polyimide resin					
T300 carbon fiber	Quasi-isotropic	0.39	62.8	0.52	37.3
	±15°	0.61	97.3	0.81	91.1
	±45°	0.13	40.7	0.28	17.3

1. The resin monomer is a chemically simple and pure system, thus making for easy quality control.

2. The resin monomer and prepolymers are virtually inert at room temperatures, and hence cold storage is not necessary.

3. The moisture uptake is less than that of epoxy and polyimide resins.

4. The phthalocyanine reaction proceeds under less severe thermal and catalytic conditions than those required for reaction of terminal acetylenic groups.

During 1978 a new class of diether-linked polyphthalocyanines was synthesized.[34] These were based upon dodecanediol, Bisphenol-A, or Bisphenol- S. The preparation of a resin from Bisphenol-A is illustrated below:

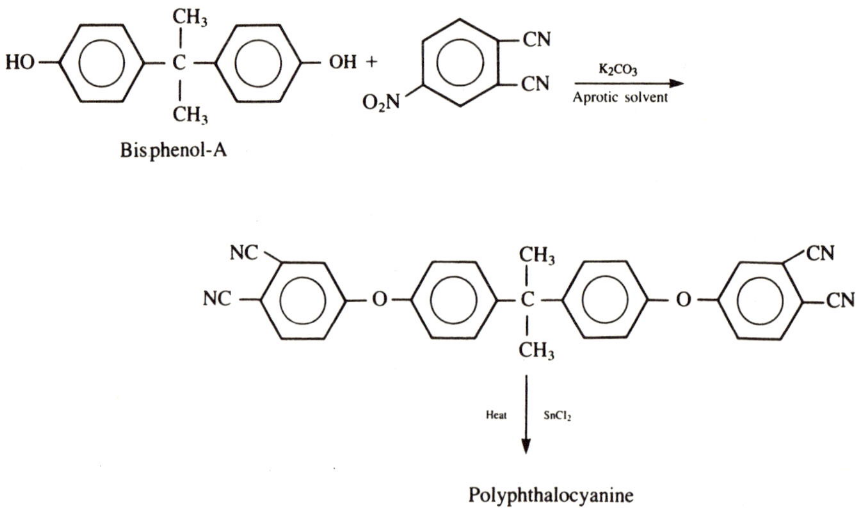

Bisphenol-A

Polyphthalocyanine

The nitro substituent is displaced from the 4-nitrophthalonitrile by an alkoxide- or a phenoxide-containing unit, the process being activated by the cyano groups. The thermal stability of the three systems is shown in Figure 9.9. As would be expected, the resins based on Bisphenols have much greater thermo-oxidative stability than that of the resin based on dodecanediol. The equilibrium water contents of the resins after long-term immersion in water at room temperature are Bisphenol-S resin 3.7%, dodecanediol resin 1.5%, and Bisphenol-A resin 1.3%. It is estimated that the bisphenol-linked polyphthalocyanines should have long-term operational capabilities at temperatures greater than 232°C and should be relatively insensitive to the effects of high humidity.

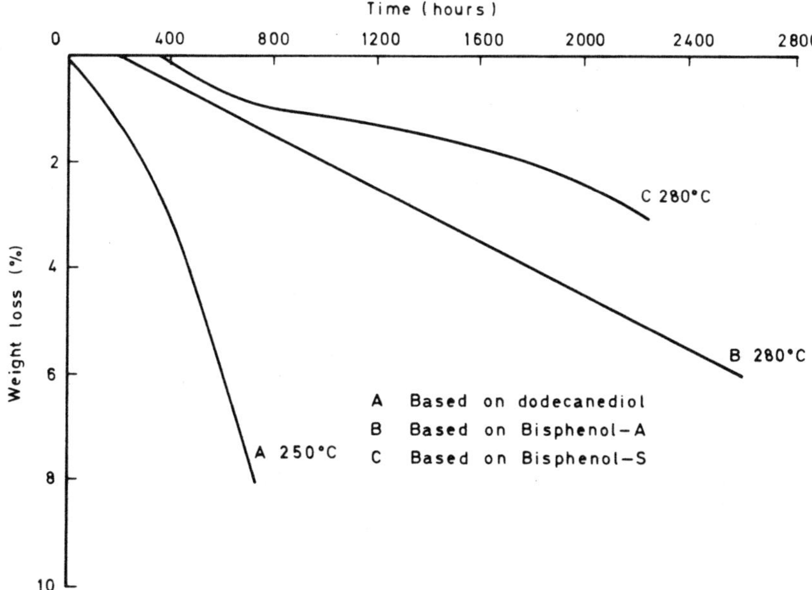

Figure 9.9.   Isothermal weight loss of polyphthalocyanine resins. Reference 34.

LADDER POLYMER CONCEPT

It has been postulated that greater thermal stability would be achieved with organic polymers if they were of a ''ladder'' or ''double-strand'' structure. The reasoning behind this is that two bonds must be broken instead of one in order to sever the polymer chain. Theoretical calculations have been made by Tessler on hypothetical structures of the following types[35]:

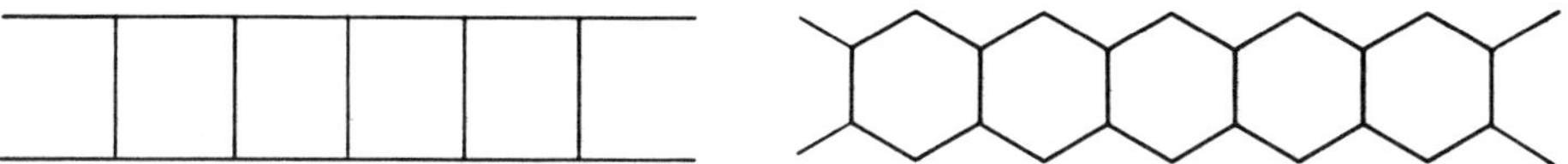

It was assumed that each bond was of equal stability and that scission occurred entirely at random at an arbitrary rate. Weight losses and changes in molecular weight were computated and compared with those for an analogous linear (single-strand) polymer. The weight loss data are illustrated in Figures 9.10 and 9.11. From these plots it can be seen that the improvement in stability based on weight loss of a ladder polymer compared with the single-strand analog

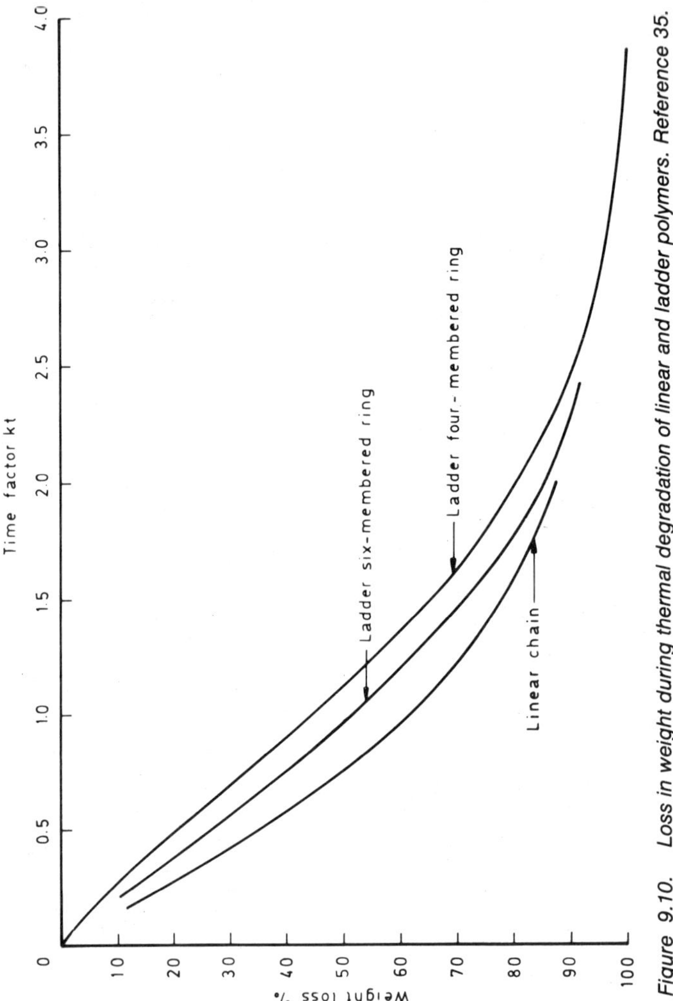

Figure 9.10. Loss in weight during thermal degradation of linear and ladder polymers. Reference 35.

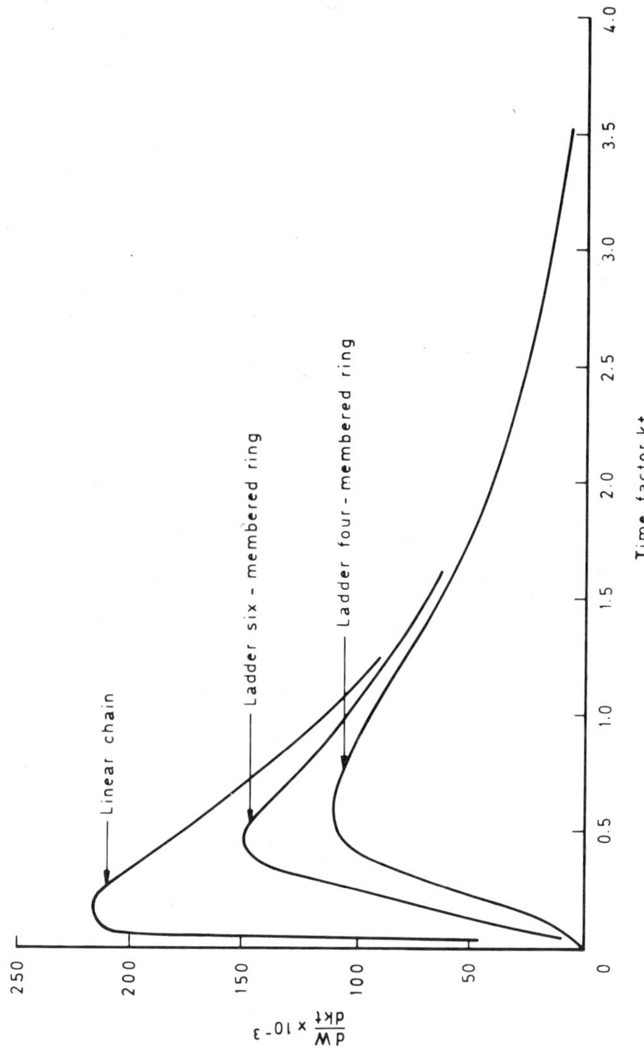

Figure 9.11. Rates of weight loss during thermal degradation of linear and ladder polymers. Reference 35.

is only marginal. In practice, of course, it would also be very difficult to prepare a perfect ladder polymer (i.e., one without any single bonds in the chain) and hence to achieve even the theoretical improvement in stability. It is also likely that such polymers would be difficult to fabricate.

The ladder system that has received the most intensive investigation is that based upon the reaction of a dianhydride with an aromatic tetramine and specifically of 1,4,5,8-naphthalenetetracarboxylic acid dianhydride with either 3,3'-diaminobenzidine or 1,2,4,5-tetraaminobenzene.[36–39]

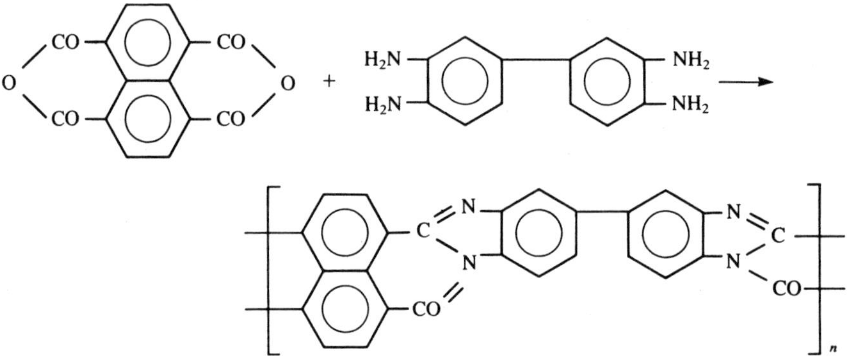

BBB [semi-ladder]

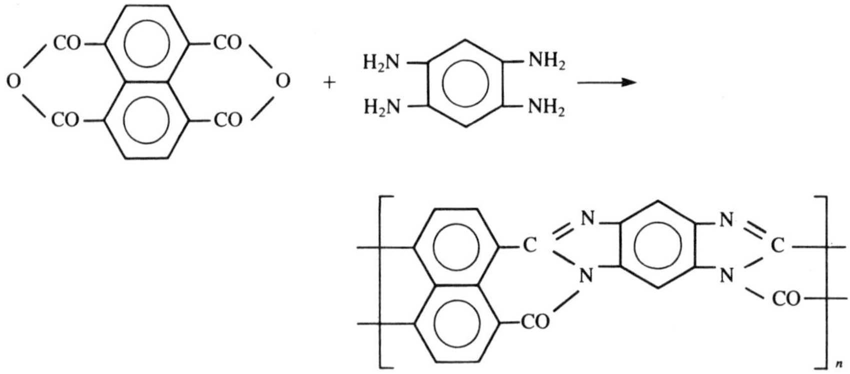

BBL [ladder]

These polymers were called poly-*bis*-(*benzimidazo benzo* phenanthrolines)— hence the BBB nomenclature—and together with the polymers derived from tetraamines and pyromellitic dianhydride [the poly(benzimidazo pyrrolones)] were given the trivial name pyrrones.

The synthesis is essentially a combination of the reactions used to produce polyimides and polybenzimidazoles. It takes place in two stages. In the first a polyaminoamic acid is formed that is soluble and fusible.

This reaction is carried out in organic aprotic solvents at moderate temperatures. Raising the temperature from 130 to 150°C results in partial cyclization to the polyaminoimide form.

Heating to still higher temperatures (250–325°C) causes further cyclization to the pyrrone structure. The final polymers are soluble only in concentrated sulphuric acid, but fibers have been successfully spun from this solvent. Several review articles have been written on the preparation of the pyrrones and related polymers.[40–42]

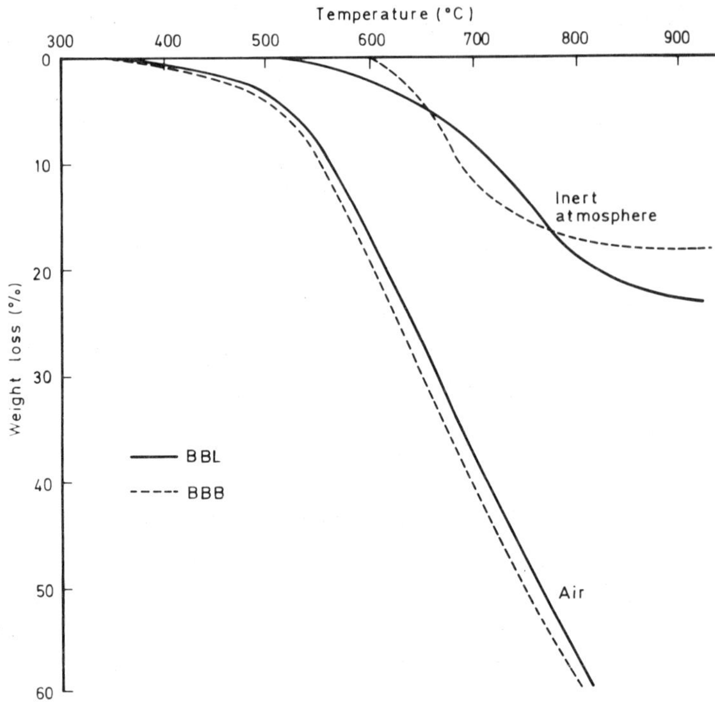

Figure 9.12.   Thermogravimetry of pyrrone polymers. (Heating rate 5°C/min) Reference 42.

Thermogravimetric curves for the BBB and BBL polymers in air and in an inert atmosphere are shown in Figure 9.12. From this it can be seen that the presence of the single linking bond in the BBB polymer appears to have little detrimental effect on thermal stability, especially in air. The curves differ only slightly, however, in the early stages (10–20% weight loss) from those obtained for linear polybenzimidazoles or polyimides. This comparatively poor stability of the ladder polyheteroarylenes has been attributed to incomplete cyclization in the chain, although as pointed out earlier purely theoretical considerations indicate that even with a perfect ladder structure the increase in thermal stability would not be great.

Although no commercial product is available, the greatest achievement in the practical application of ladder polymers has been in the development of fibers from the BBB material.[43–48] The fibers are spun from sulphuric acid solutions and, after being stretched 80–120% at 525–600°C, have at room temperature a tensile strength of 5 g/denier and an extension at break of 5%. In air 55–60% of the initial strength is retained after 30 hours at 360°C, or 1 minute at 660°C. The BBB fibers are currently the most thermally stable of those based upon organic

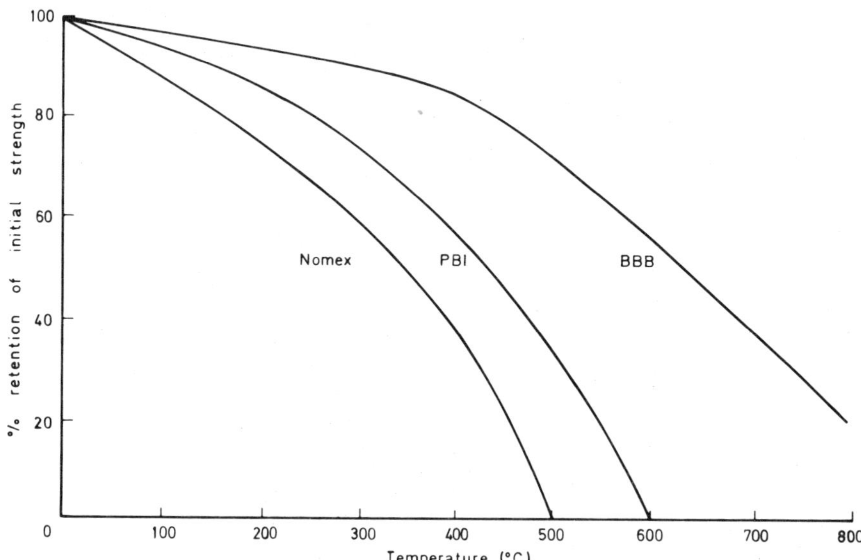

Figure 9.13 Strength retention of various heat resistant fibers at temperature.

polymers (Figure 9.13). The BBB system has also been used to produce films[49] and moldings.[50] The latter were made by a one-step method involving the simultaneous polymerization and molding of pyrrone monomers. The thermomechanical behavior was studied by the torsional braid technique.[51]

INTRAMOLECULAR CYCLIZATION

One of the problems in the use of thermally stable polymers lies in their processability. A high-temperature thermoplastic must perforce be fabricated above its softening point. A thermosetting material will flow initially, but as cure proceeds the mobility of the molecules is progressively arrested, and once the glass transition temperature of the system reaches the cure temperature, reaction effectively ceases. Raising the temperature increases molecular movement, frees reactive sites, and the curing process continues. This means in practice that the ultimate use temperature of a thermosetting resin is governed by the maximum temperature used in the curing cycle. The problem is especially acute with the thermally stable aromatic and heterocyclic materials because of the inherent rigidity of the polymer chains. This very rigidity is, however, put to good use in the principle of intramolecular cycloaddition, or IMC cure.[52] The concept is to synthesize a linear mobile polymer chain, which will flow and can be molded at relatively low temperatures, but which possesses pairs of pendant groups that can undergo an intramolecular cycloaddition reaction, thus producing rigidity. The pendant groups are held in close proximity along the chain and reaction depends only on rotational movement of the polymer backbone, not on the translational movement required for intermolecular coupling. As a result, reaction can continue long after the resin is in the glassy state, and hence the glass transition and use temperatures may be substantially higher than the "cure" temperature. Additional bonuses are that no volatile by-products are evolved in the cycloaddition process and that a linear polymer is obtained, thus avoiding the inherent brittleness of a three-dimensional structure. The validity of the intramolecular cycloaddition process has been demonstrated in both polyphenylquinoxaline[52,53] and polyimide systems.[54,55]

The route to the PPQ polymer is as shown on the following page. The intermediate polymer had a glass transition temperature (Tg) of 215°C. After an overnight cure at 245°C, the Tg was raised to 365°C, an increase of 120°C. The absence of volatile by-products during the reaction was confirmed by thermogravimetric-mass spectrometric analysis. Similar results were obtained with the polyimide system shown on p. 434. After a cure of 24 hours at 240–250°C, the measured Tg was 350–360°C, again an increase of 110–120°C. The problem with these systems is that the amount of flow of the prepolymer at Tg is insufficient for satisfactory processing.

Other potential chemical systems with intramolecular cycloaddition cure feasibility have recently been reviewed[56] and recommendations have been made as to which systems should be investigated.

ORDERED POLYMERS

Another new concept currently under investigation is aimed at producing ordered polymers, which have mechanical properties comparable with those obtained from fiber-reinforced composites, but *without* the use of a fiber reinforcement.[57] It is hoped to attain this goal through para-linked rigid rod-like polymers possessing significantly higher thermal stability and greater environmental resistance than those of present-day structural composites. Additional advantages which could accrue from these materials would be simpler processing and fabrication methods, the elimination of fiber-resin interfaces, and easier, and hence better, quality control. Three lines of study are being followed:

1. Completely para-linked aromatic/heterocyclic polymers. These are molecularly orientable, high stability polymers with rigid, rod-like, extended chain configurations.

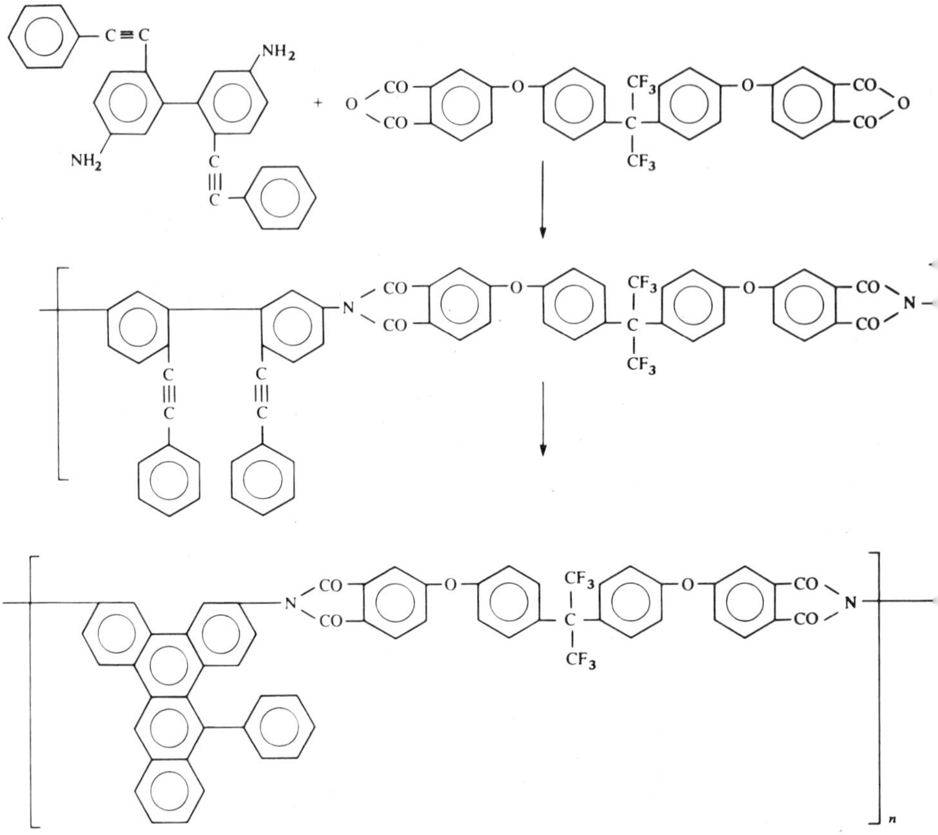

2. "Articulated" or "swivel" polymers. These are essentially the para-linked polymers of (1), structurally modified by the incorporation of flexibilizing "swivel" groups into the polymer backbone.

3. Molecular composites. Here polymers of type (1) are molecularly dispersed in flexible, coil-like variants of the *same* polymer.

The para-linked polymers upon which attention has been concentrated are polybenzimidazole, polybenzoxazole, and polybenzothiazole, and most progress has been made with the last of these.[58]

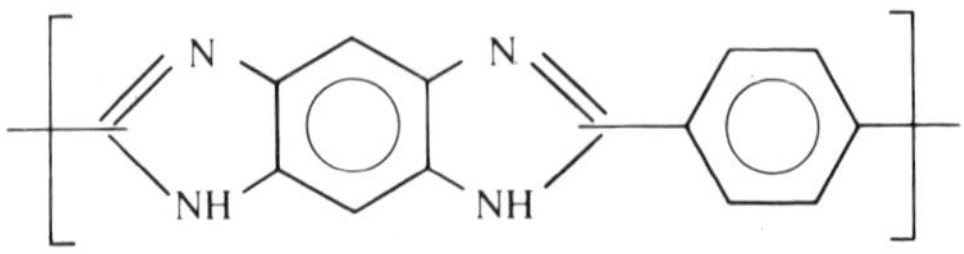

Polybenzimidazole

Polybenzoxazole

Polybenzothiazole

The synthesis involves the reaction of

where X is NH_2, OH, or SH, respectively, with

The excellent thermal stabilities of these three polymers in air and nitrogen are illustrated in Figure 9.14. On the basis of these results, obtained under rising temperature conditions, the oxidation stability of the polymers would appear to be identical. Isothermal weight loss studies (Figure 9.15), however, indicate that the polybenzothiazole has superior long-term oxidative stability compared with the polybenzoxazole. The polybenzothiazole is soluble in methane sulphonic acid, exhibiting intrinsic viscosities as high as 26. Fibers have been successfully spun from such solutions and their properties are compared with those of other materials in Table 9.10.

In the case of the articulated polymers, the emphasis is upon the production of rod-like polymers with a controlled number of flexibilizing segments, e.g.,

The behavior of such polymers has been treated theoretically by Flory.[59]

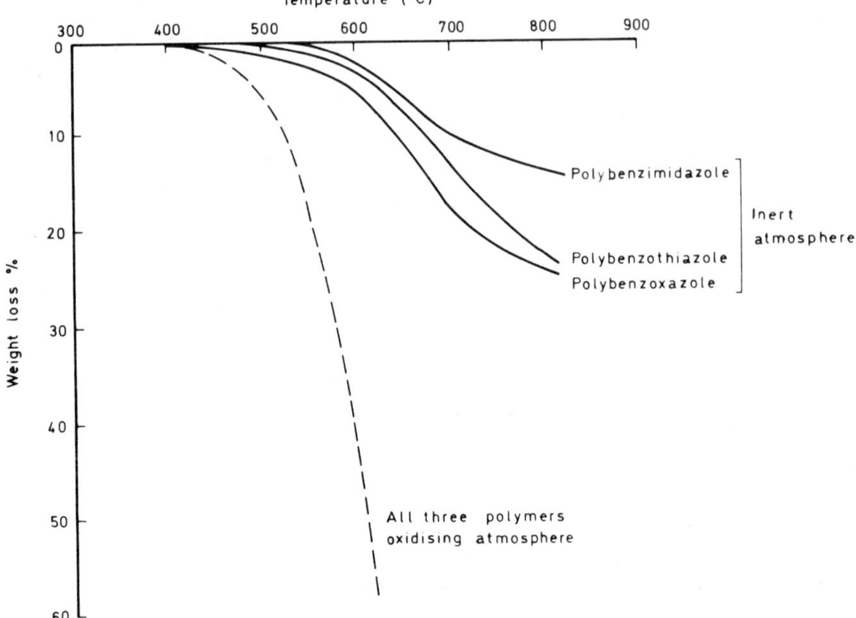

Figure 9.14. Comparison of thermal stabilities of a polybenzimidazole, a poly-benzothiazole, and a polybenzoxazole.

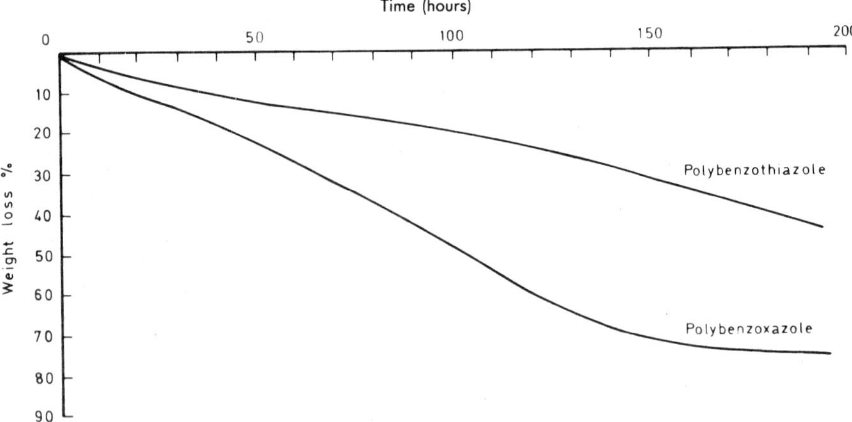

Figure 9.15. Isothermal weight loss of a polybenzothiazole and a polybenzoxazole at 371°C in air. Reference 58.

TABLE 9.10. Properties of Polybenzothiazole Fibers Compared with Those of Other Materials

Fiber material	Tensile strength (MPa)	Tensile modulus (GPa)	Specific gravity
Boron	3174	414	2.60
Alumina	1380–3450	380	3.60
Silicon carbide	~ 3450	290–331	3.20
Graphite	2346–2415	207–345	1.74
Polybenzamide	2760	131	1.45
Polyethylene	3001	90	0.95–0.98
Polybenzothiazole	1518	186	1.40
E Glass	1725	69	2.54

The molecular composite approach has been described in a recent paper[60] in which the preparation and testing of films was described. The morphology of these materials was studied by X-ray diffraction and scanning electron microscopy. Pronounced structuring and ordering of the rod-like polymer was observed, and significant increases in tensile strength and modulus were measured for films containing as little as 10% of rod-like polymer in the matrix. The improvement was especially marked if the films were stretched. Considering the quality and the quantity of effort currently being devoted to ordered polymers, rapid progress in this field should ensue.

STABILIZATION OF HEAT-RESISTANT POLYMERS

Many polymers when used even at modest temperatures require the addition of stabilizers of one type or another. The question may be posed—can additives be developed which would be effective with the higher temperature capability polymers? Some indications that this may be possible have been given earlier in the text. The major research on this topic, however, has been carried out by Russian workers and specifically by B. M. Kovarskaya *et al.* and G. P. Gladyshev *et al.* The approach of Kovarskaya is conventional in that compounds are added to the polymer which inhibit thermal oxidation by interacting with peroxide radicals.[61] It has been shown that phosphorus-containing compounds (ethers, amides, phosphites, phosphates, cyclophosphazene derivatives, and orthophosphoric acid), when added in concentrations of 0.5 to 5.0% by weight to polyheteroarylenes, specifically polyimides and polybenzoxazoles, considerably slow down the weight loss of the polymers, reduce the liberation of volatile breakdown products in the temperature range 300–450°C, and improve the retention of physical and mechanical properties on prolonged heat aging. (Other Rus-

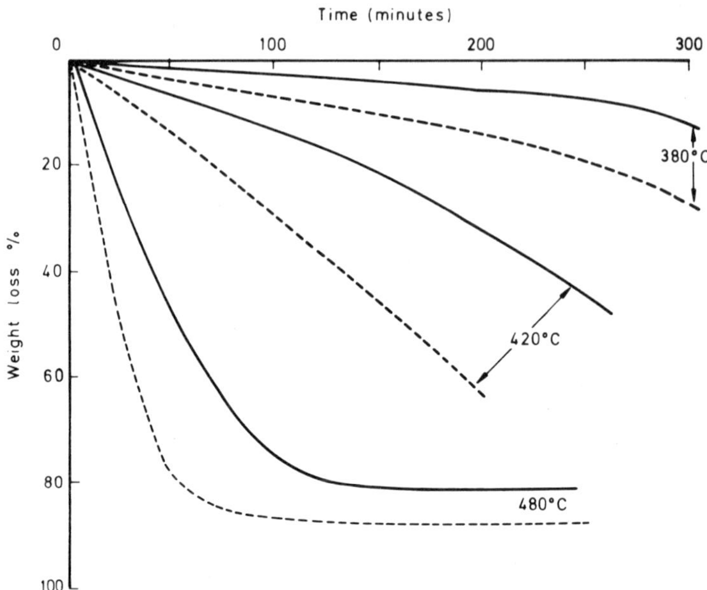

Figure 9.16. Isothermal weight-loss curves for polybenzoxazole (--------) and polybenzoxazole containing 2% β-naphthyl ester of phosphoric acid (________). Reference 61.

sian workers have studied the effect of inorganic acids, alkalies, and polyconjugated systems.[62]) Figure 9.16 shows the effect of 2% by weight of the β-naphthyl ester of phosphoric acid upon the weight loss of a polybenzoxazole in the temperature range 380–480°C. Some stabilization is evident; at the maximum there is about a 50% increase in the time to a specific weight loss. Analysis of the residues indicates that very little, if any, of the phosphorus is lost from the polymer, and that its presence promotes cross-linking and gel formation. In respect of the phosphorus-containing compounds studied, it was concluded that:

1. Phosphoric acid esters and amide-esters do not react with ROO radicals either in the liquid or solid phase and their stabilizing action therefore does not involve this reaction.

2. The conversion products of phosphorus acid amide-esters are, however, very reactive with ROO radicals.

3. Substituted phosphazene is also very reactive with peroxide radicals and may be recommended as an inhibitor for oxidation reactions occurring at 100–200°C.

The approach adopted by Gladyshev is quite different and involves the introduction of a stabilizer that has a very high reactivity with oxygen at elevated temperatures and combines with it to form inert products, i.e., the stabilizer acts as an oxygen scavenger.[63–67] In general, such a highly active material cannot be introduced directly into a polymer, although there are some compounds whose reaction with oxygen is governed by such a high-temperature coefficient that they form exceptions. The aim normally is to generate the stabilizer through thermal transformation or decomposition of a relatively inert compound mixed with the polymer previously. The stabilizer is, of course, used up during the oxidation process and its effectiveness is determined both by its concentration and the rate of diffusion of oxygen. The process of greatest interest is the formation of finely dispersed metals and lower valence metal oxides from organic salts of the metals, e.g., formates, oxalates. The theory has been reviewed in detail in several papers.[65,67] It is claimed that by using this method the lifetimes at 250–450°C of articles based upon polysiloxanes have been increased by a factor of ten or more. Figure 9.17 depicts the lifetime at 400 and 500°C of film samples of trimethylcyclopolysiloxane stabilized with finely dispersed iron.[67] The solid curves are theoretical plots and the points experimental determinations showing the very good agreement that is obtained between theory and practice.

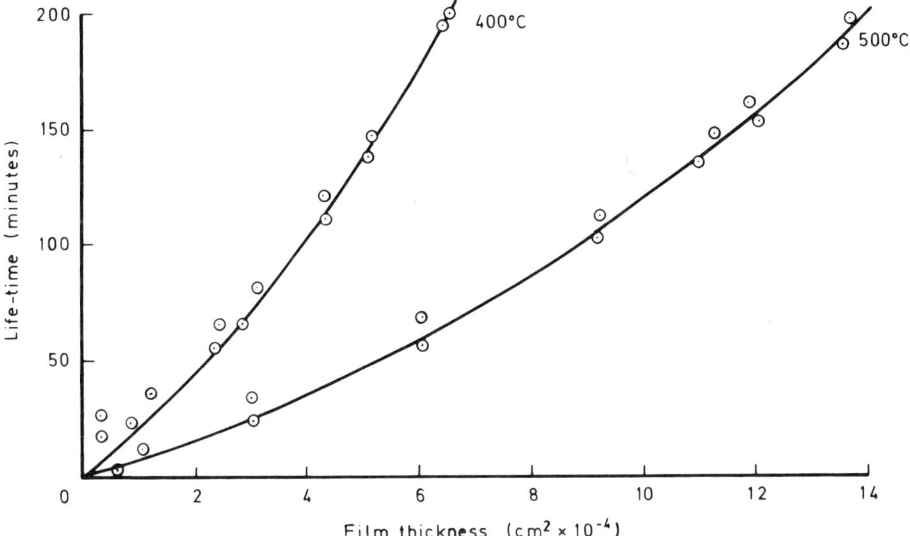

Figure 9.17. Lifetimes of trimethylcyclopolysiloxane films stabilized with finely dispersed iron (0.28 mole/Kg) as a function of sample thickness (lines are theoretical, points experimental). Reference 67.

INORGANIC POLYMERS

Based solely on bond energy considerations, inorganic polymers would appear to hold out good prospects of providing materials with higher thermal stability than carbon chain polymers, and in the 1950s and 1960s much effort was devoted to synthesizing a variety of such structures. The volume of work was such that several books were written reviewing progress at that time.[68–73] Three main areas of research may be delineated:

1. Polymers having linear chains consisting of repeating units such as –Si–N–, –B–N–, –P–N–. (The phosphazene polymers described earlier are one example.)

2. Polymers of the siloxane structure having the silicon partly replaced by elements such as aluminum, titanium, tin, etc.

3. Metal chelate or coordination polymers.

The types of linkage investigated are summarized in Table 9.11. The main routes to the chelate polymers were as follows (p. 441):

TABLE 9.11. Linkages Studied as the Basis of Inorganic Polymers

Type 1. Linear chain	Type 2. Modified siloxane structures
—B—N—	—Si—O—Al—
—B—N—O—	—Si—O—Ti—
—N—B—O—P—O— B—N—	—Si—O—Sn—
—B—N—C—	—Si—O—As—
—B—O—C—	—Si—O—B—
	—Si—O—S—
—P—N—	—Si—O—Sb—
—P—O—	—Si—O—P—
	—Si—O—Ge—
—Al—O—	—Si—O—Pb—
—Al—N—	—Si—O—Mg—
	—Si—O—Cu—
—Sn—O—	—Si—O—Zn—
—Sn—C—	—Si—O—Hg—
—Ge—O—	—Si—O—Co—
	—Si—O—Cr—
—C—As—	

Type 3. Chelate or coordination polymers
Chelated through O
Chelated through O and N
Chelated through N
Chelated through N and S
Chelated through S and O

1. Combination of metal ions and ligands. This is made possible by the fact that metal ions are polyfunctional, i.e., they are capable of accepting more than one pair of electrons. Consequently, in many cases reaction with polyfunctional donor molecules will result in linear or cross-linked polymeric materials.

2. Reaction of metal ions with preformed polymeric ligands having coordination sites.

3. Polymerization of suitable monomers already containing metal ions in their structure.

The information available up to 1970 on the thermal degradation of inorganic polymers was reviewed by Economy and Mason.[74] As stated above, from purely thermodynamic considerations inorganic bonds possessing greater strengths than carbon-carbon bonds would be expected to be more resistant to thermal degradation. The inorganic bonds are, however, ionic in nature and this, together with the availability of bonding orbitals in many of the elements, provides low-energy routes for three general types of degradation reaction— structural reorganization, oxidation, and hydrolysis. Small, unstrained ring compounds may be formed in many inorganic systems, and the production of these on thermal degradation of an inorganic polymer is often energetically favored, e.g., the formation of cyclic siloxanes from a polysiloxane. In the case of oxidative reactions, the organic groups pendant from the inorganic chain are generally preferentially attacked, but metal ions present in the system can also act as powerful oxidation catalysts; e.g., the oxidative stabilities of many chelate polymers are less than those of the organic ligands themselves. The majority of inorganic polymers are susceptible to nucleophilic or electrophilic attack and one of the most serious problems encountered has been the ease with which many of them hydrolyze in the presence of water; e.g., phosphonitrilic chloride polymers very rapidly lose their elastic properties on exposure to the atmosphere. Problems such as these have to date bedevilled the search for useful inorganic polymers and relatively few examples are known.

The chelate polymers are those in which the most attention has been paid to structural factors determining thermal stability, and the following conclusions have been drawn[75–77]:

1. The nature of the metal ion is the major factor governing thermal stability.

2. Chelates with a planar distribution of 5 or 6 atoms in the ring are the most stable.

3. Five-membered rings are more stable than six-membered rings unless conjugation occurs, when the reverse is true.

4. Chelate polymers in which the metal atoms are part of several rings are more thermally stable than analogous compounds with fewer rings.

5. The stability depends on the type of bond between the metal ion and the donor groups and on the structure of the ion.

6. Aromatic ligands confer higher stabilities than aliphatic ligands.

7. All ionic valences of the metal must be satisfied by the ligand for high thermal stability to be achieved.

8. Thermal stability appears to increase with increasing molecular weight.

N. H. Ray[78–79] has suggested that the principal reason for the failure of past work to produce useful new inorganic polymers has been the almost universal belief that only linear polymers would have the required combination of properties. This belief arises because of experience in the organic polymer field, where network polymers are insoluble and infusible, and hence incapable of fabrication except by very exotic techniques. Inorganic polymers, however, can undergo reversible dissociation of network bonds at elevated temperatures, and hence can be thermoplastic whether they have a chain or network structure. This postulate that the search for processable inorganic polymers need not be restricted to linear chain molecules overcomes one of the principal difficulties experienced to date—the inability to obtain thermal, oxidative, and hydrolytic stability at one and the same time in a single polymer. The difficulty is caused because all the polymer-forming elements, with the exception of linking atoms such as oxygen and sulphur, have valences of three or more. Hence, to form a linear polymer one or more unreactive substituents must be attached to each atom in order to produce bifunctionality. These univalent groups, when attached to a multivalent polymer-forming element other than carbon or silicon, tend to be susceptible to oxidation, hydrolysis, or straight thermal scission. For the production of a network polymer, however, a much smaller proportion of unreactive substituents is required, because in this case they are only needed to control the cross-link density of the network. The rate of hydrolysis of bonds in a highly cross-linked polymer can also be several orders of magnitude less than in the corresponding linear polymer because of steric and diffusional effects. Ray believes that by adopting this new approach to the production of inorganic polymers useful materials should be obtained. The approach will necessitate the concomitant development of new methods of processing and basic research into the principles underlying structure-property relationships in inorganic materials.

REFERENCES

1. V. A. Pankratov and S. V. Vinogradova, *Russ. Chem. Rev.* **41,** 66 (1972).
2. P. M. Hergenrother, *J. Macromol. Sci. Rev. Macromol. Chem. C* **13,** 189 (1975).
3. H. C. Brown, *J. Polym. Sci.* **44,** 9 (1960).
4. M. Gaku, K. Suzuki, and N. Ikeguchi, *Proc. Elec. and Electronic Insul. Conf.* **13,** 11 (1977).
5. R. J. Kray, *8th SAMPE Nat. Tech. Conf.* **8,** 66 (1976).
6. W. Collins and T. Villani, Paper presented at Conference on Thermosets versus High Performance Thermoplastics, Connecticut, 1977.

7. G. P. De Gaudemaris and B. J. Sillion, *J. Polym. Sci. B* **2**, 203 (1964).

8. J. K. Stille and J. R. Williamson, *J. Polym. Sci. B* **2**, 209 (1964).

9. J. K. Stille, J. R. Williamson, and F. E. Arnold, *J. Polym. Sci. A* **3**, 1013 (1965).

10. J. K. Stille and F. E. Arnold, *J. Polym. Sci. A 1* **4**, 551 (1966).

11. P. M. Hergenrother and H. H. Levine, *J. Polym. Sci. A 1* **5**, 1453 (1967).

12. P. M. Hergenrother, *J. Polym. Sci. A 1* **6**, 3170 (1968).

13. P. M. Hergenrother, *J. Macromol. Sci. Rev. Macromol. Chem. C* **6**, 1 (1971).

14. P. M. Hergenrother, Polyphenyl-*as*-triazines and polyphenylquinoxalines. Report prepared for a series entitled *The Synthesis of New Polymers: Modern Methods*, N. Yoda, (ed.), 1974.

15. P. M. Hergenrother, *Polym. Eng. Sci.* **16**, 303 (1976).

16. P. M. Hergenrother, *Polym. Preprints* **15**(1), 781 (1974). Also *Macromolecules* **7**, 575 (1974).

17. W. B. Alston, *8th SAMPE Nat. Tech. Conf.* **8**, 114 (1976).

18. R. F. Kovar and G. F. Ehlers, *Polym. Preprints* **16**(2), 246 (1975).

19. F. E. Arnold and R. F. Kovar, *8th SAMPE Nat. Tech. Conf.* **8**, 106 (1976).

20. C. E. Browning, A. Wereta, J. T. Hartness, and R. F. Kovar, *21st SAMPE Nat. Symp. Exhib.* **21**, 83 (1976).

21. J. K. Stille and E. L. Mainen, *J. Polym. Sci. B* **4**, 39 (1966).

22. J. K. Stille and E. L. Mainen, *Macromolecules* **1**, 36 (1968).

23. P. M. Hergenrother, Paper presented at High Temperature Polymer Reinforcements and Composites Conference, Utah, 1970.

24. M. L. Santelli, NOL TR 71-187 (1971).

25. N. Yoda and M. Kurihara, *Macromol. Rev.* **5**, 109 (1971).

26. R. B. Gosnell, W. P. Fitzgerald, and R. J. Milligan, *Appl. Polym. Symp.* **22**, 21 (1973).

27. C. B. Delano and R. J. Milligan, AFML TR-73-291 (December 1973).

28. C. B. Delano, J. D. Dodson, and R. J. Milligan, Aerotherm TR-77-254 (July 1977).

29. J. R. Griffith, J. G. O'Rear, and T. R. Walton, *Copolymers, Polyblends and Composites*, (Advances in Chemistry Series No. 142, N. A. J. Platzer, ed.), American Chemical Society, Washington (1975), p. 458.

30. T. R. Walton, J. R. Griffith, and J. G. O'Rear, in *Adhesive Science and Technology*, L. H. Lee, (ed.), Plenum Press, New York (1975), Vol. 9, p. 655.

31. W. D. Bascom, J. L. Bitner, and R. L. Cottington, *Org. Coat. and Plast. Chem. Preprints* **38**, 477 (1978).

32. J. L. Gillham, *Org. Coat. and Plast. Chem. Preprints* **40**, 866 (1979).

33. R. Y. Ting, H. C. Nash, R. L. Cottington, and C. J. N. Rall, in *Advances in Composite Materials*, A. R. Bunsell, C. Bathias, A. Martrenchar, D. Menkes, and G. Verchery, (eds.), Pergamon Press, Oxford (1980), Vol. 2, p. 1501.

34. T. M. Keller and J. R. Griffith, *Naval Research Laboratory 1978 Review*, p. 148.

35. M. M. Tessler, *Polym. Preprints* **8**(1), 152 (1967).

36. V. L. Bell and G. F. Pezdirtz, *J. Polym. Sci. B* **3**, 977 (1965).

37. F. Dawans and C. S. Marvel, *J. Polym. Sci. A* **3**, 3549 (1965).

38. J. G. Colson, R. H. Michel, and R. M. Paufler, *J. Polym. Sci. A 1* **4**, 59 (1966).

39. R. L. Van Deusen, *J. Polym. Sci. B* **4**, 211 (1966).

40. B. Nartsissov, *J. Macromol. Sci. Rev. Macromol. Chem. C* **11** (1), 143 (1974).

41. A. L. Rusanov, S. N. Leontova, and T. G. Iremashvili, *Russ. Chem. Rev.* **46**, 76 (1977).

42. A. L. Rusanov, *Russ. Chem. Rev.* **48**, 62 (1979).

43. W. H. Gloor, *Polym. Preprints* **7**, 819 (1966).

44. R. L. Van Deusen, O. K. Goins, and A. J. Sicree, *Polym. Preprints* **7**, 528 (1966).

45. W. H. Gloor, *Appl. Polym. Symp.* **6**, 151 (1967).

46. W. H. Gloor, *Appl. Polym. Symp.* **9**, 159 (1969).

47. G. I. Kudryavtsev and A. M. Shchetinin, *Khim. Volokna* (6), 2 (1968).

48. G. I. Kudryavtsev and A. M. Shchetinin, *Khim. Volokna* (3), 68 (1975).

49. F. E. Arnold and R. L. Van Deusen, *J. Appl. Polym. Sci.* **15**, 2035 (1971).
50. P. E. D. Morgan and H. J. Scott, *J. Appl. Polym. Sci.* **16**, 2029 (1972).
51. J. K. Gillham and K. C. Glazier, *J. Appl. Polym. Sci.* **16**, 2153 (1972).
52. F. L. Hedberg and F. E. Arnold, *J. Polym. Sci., Polym. Chem. Ed.* **14**, 2607 (1976).
53. F. L. Hedberg and F. E. Arnold, *Polym. Preprints* **16**, 677 (1975).
54. F. E. Arnold, U.S. Patent Application Nos. 678,324 and 678,325 (1976).
55. F. L. Hedberg and F. E. Arnold, AFML TR-76-198 March (1977).
56. K. S. Lau and F. E. Arnold, AFML TR-79-4065 May (1979).
57. T. E. Helminiak, F. E. Arnold, and C. L. Benner, *Polym. Preprints* **16**(2), 659 (1975).
58. J. F. Wolfe, B. H. Lau, and F. E. Arnold, *Polym. Preprints* **19**(2), 1 (1978).
59. P. J. Flory, *Macromolecules* **11**, 1141 (1978).
60. G. Husman, T. Helminiak, W. Adams, D. Wiff, and C. Benner, *Org. Coat. and Plast. Chem. Preprints* **40**, 797 (1979).
61. V. V. Guryanova, N. G. Annenkova, T. N. Novotortseva, A. B. Blyumenfeld, V. A. Baranova, F. S. Malyukova, L. A. Shesternina, and B. M. Kovarskaya, *Polym. Sci. U.S.S.R.* **20**, 238 (1978).
62. O. A. Shustova and G. P. Gladyshev, *Russ. Chem. Rev.* **45**, 865 (1976).
63. G. P. Gladyshev, *Dokl. Akad. Nauk S.S.R.* **216**, 585 (1974).
64. O. A. Shustova and G. P. Gladyshev, *Dokl. Akad. Nauk S.S.R.* **221**, 301 (1975).
65. G. P. Gladyshev, *Polym. Sci. U.S.S.R.* **17**, 1441 (1975).
66. G. P. Gladyshev, *J. Polym. Sci., Polym. Chem. Ed.* **14**, 1753 (1976).
67. E. B. Brun, O. A. Shustova, S. I. Kuchanov, and G. P. Gladyshev, *J. Polym. Sci., Polym. Chem. Ed.* **18**, 2461 (1980).
68. F. G. A. Stone and W. A. G. Graham (eds.), *Inorganic Polymers,* Academic Press, New York (1962).
69. M. F. Lappert and G. J. Leigh (eds.), *Developments in Inorganic Polymer Chemistry,* Elsevier, New York (1962).
70. D. N. Hunter, *Inorganic Polymers,* John Wiley & Sons, New York (1963).
71. F. G. R. Gimblett, *Inorganic Polymer Chemistry,* Butterworths, London (1963).
72. K. A. Andrianov, *Metal-Organic Polymers,* John Wiley & Sons, New York (1965).
73. H. R. Allcock, *Heteroatom Ring Systems and Polymers,* Academic Press, New York (1967).
74. J. Economy and J. H. Mason, Degradation of inorganic polymers, in *Thermal Stability of Polymers,* R. T. Conley, (ed.), Marcel Dekker, New York (1970), Vol. 1, Ch. 14, p. 577.
75. A. A. Berlin and N. G. Matveeva, *Adv. Chem. Moscow* **29**, 119 (1960).
76. E. A. Tomic, *J. Appl. Polym. Sci.* **9**, 3745 (1965).
77. S. V. Vinogradova and O. V. Vinogradova, *Russ. Chem. Rev.* **44**, 510 (1975).
78. N. H. Ray, *Endeavor* **34**, 9 (1975).
79. N. H. Ray, *Inorganic Polymers,* Academic Press, London (1978).

SUPPLEMENTARY BIBLIOGRAPHY

Polytriazines

R. J. Kray, NCNS resins for structural and electrical applications, *Proc. 30th Ann. Tech. Conf. Reinf. Plast./Comp. Inst.* **30** (1975), Sect. 19-C.
R. J. Kray, The NCNS resins for potential applications in advanced composites, *20th Nat. SAMPE Symp.* **20**, 227 (1975).

Polyphenylquinoxalines

W. Wrasidlo and J. M. Augl, Phenylated polypyrazinoquinoxalines, *J. Polym. Sci. B* **8,** 69 (1970).

J. M. Augl, Phenylated imide-quinoxaline copolymers, *J. Polym. Sci. A 1* **8,** 3145 (1970).

J. V. Duffy and J. M. Augl, Amide-quinoxaline copolymers, *J. Polym. Sci. A 1* **10,** 1123 (1972).

J. M. Augl, Thermal degradation of a polyphenylquinoxaline in air and vacuum, *J. Polym. Sci. A 1* **10,** 2403 (1972).

R. T. Rafter and W. P. Fitzgerald, PPQ resin systems for high-temperature composite materials, AMMRC CTR 72 (10 June 1972).

J. M. Augl and H. J. Booth, Thermomechanical behavior of a polyphenylquinoxaline, *J. Polym. Sci., Polym. Chem. Ed.* **11,** 2179 (1973).

J. T. Hoggatt, P. M. Hergenrother, and J. G. Shdo, Development of Polyphenylquinoxaline Graphite Composites, NASA CR-121109 (January 1973).

P. M. Hergenrother, Phenylquinoxaline copolymers, *Appl. Polym. Symp.* **22,** 57 (1973).

J. T. Hoggatt, S. G. Hill, and J. G. Shdo, Development of polyphenylquinoxaline graphite composites, NASA CR-134674 (July 1974).

T. T. Serafini, P. Delvigs, and R. D. Vannucci, In situ polymerization of monomers for polyphenylquinoxaline/graphite fiber composites, NASA TN-7793 (October 1974).

S. E. Wentworth and G. D. Mulligan, Preparation and preliminary evaluation of poly-*p*-tolylquinoxaline, a cross-linkable polyphenylquinoxaline, AMMRC TR 74-12 (May 1974).

J. M. Augl, J. V. Duffy, and S. E. Wentworth, Isothermal crosslinking studies on polyquinoxalines by dynamic mechanical methods, *J. Polym. Sci., Polym. Chem. Ed.* **12,** 1023 (1974).

P. M. Hergenrother, Polyphenylquinoxalines: Synthesis, characterization and mechanical properties, *J. Appl. Polym. Sci.* **18,** 1779 (1974).

J. G. Shdo, Development of polyphenylquinoxaline graphite composites, NASA CR-135041 (1976).

R. T. Rafter and E. S. Harrison, Tris-benzil crosslinked polyphenylquinoxalines, *Polym. Eng. Sci.* **16,** 318 (1976).

P. M. Hergenrother, Quinoxaline/phenylquinoxaline copolymers, *J. Appl. Polym. Sci.* **21,** 2157 (1977).

C. E. Browning, A. Wereta, J. T. Hartness, and R. F. Kovar, Initial development of ATQ matrix resin, AFML TR-75-214 (1975).

F. E. Arnold, R. F. Kovar, and G. F. Ehlers, Acetylene terminated quinoxalines, ATQ, AFML TR-76-71 (1976).

P. M. Hergenrother, Acetylene-containing precursor polymers, *J. Macromol. Sci. C* **19,** 1 (1980).

Ladder Polymers

Y. E. Doroshenko, Ladder and parquet polymers, *Russ. Chem. Rev.* **36,** 563 (1967).

R. A. Jewell, Thermal degradation studies of several pyrrone films, *J. Appl. Polym. Sci.* **12,** 1137 (1968).

A. A. Berlin, B. I. Liogonkii, and G. M. Shamraev, Thermostable polymers from dianhydrides of aromatic tetracarboxylic acids and tetra-amines, *Russ. Chem. Rev.* **40,** 284 (1971).

L. W. Frost and G. M. Bower, Polyimidazopyrrolones and related polymers. I. Dianhydrides and *o*-acetamidodiamines, *J. Polym. Sci. A 1* **9,** 1045 (1971).

G. M. Bower, L. W. Frost, Polyimidazopyrrolones and related polymers. II. Polyimidazopyrrolones from aromatic dianhydrides and 3,3'-dichloro-5,5'-diaminobenzidine, *J. Appl. Polym. Sci.* **16,** 345 (1972).

N. J. Johnston, Polymerization of pyromellitic dianhydride with 3,3'-diaminobenzidine in aprotic solvents. I. Reaction variables, *J. Polym. Sci. A 1* **10,** 2727 (1972).

N. J. Johnston and L. B. Epps, Polymerization of pyromellitic dianhydride with 3,3'-diaminobenzi-
dine in aprotic solvents. II. Dilute solution studies, *J. Polym. Sci. A 1* **10,** 2751 (1972).

A. H. Gerber, Thermally stable polymers derived from 2,3,5,6-tetraaminopyridine, *J. Polym. Sci.,
Polym. Chem. Ed.* **11,** 1703 (1973).

Ordered Polymers

P. J. Flory and A. Abe, Statistical thermodynamics of mixtures of rodlike particles. 1. Theory for
polydisperse systems, *Macromolecules* **11,** 1119 (1978).

A. Abe and P. J. Flory, Statistical thermodynamics of mixtures of rodlike particles. 2. Ternary systems, *Macromolecules* **11,** 1122 (1978).

P. J. Flory and R. S. Frost, Statistical thermodynamics of mixtures of rodlike particles. 3. The most
probable distribution, *Macromolecules* **11,** 1126 (1978).

R. S. Frost and P. J. Flory, Statistical thermodynamics of mixtures of rodlike particles. 4. The
Poisson distribution, *Macromolecules* **11,** 1134 (1978).

P. J. Flory, Statistical thermodynamics of mixtures of rodlike particles. 5. Mixtures with random
coils, *Macromolecules* **11,** 1138 (1978).

T. E. Helminiak, The Air Force ordered polymers research program—an overview, *Org. Coat. and
Plast. Chem. Preprints* **40,** 475 (1979).

J. F. Wolfe and B. H. Loo, Synthesis of all para-polybenzobisthiazole. Structural modification of
rod-like aromatic polymers, *Polym. Preprints* **20** (1), 82 (1979).

G. C. Berry, P. C. Metzger, S. VenKatraman, and D. B. Cotts, Properties of rodlike polymers in
solution, *Polym. Preprints* **20** (1), 42 (1979).

Inorganic Polymers

C. N. Kenney, Co-ordination polymers: a review, *Chem. and Ind. (London)* 880 (1960).

R. A. Shaw, Inorganic polymers, *SCI Monograph* **13,** 24 (1961).

S. Yolles, Review of inorganic polymers, *Official Digest* 1027 (1964).

J. R. MacCallum, Inorganic polymers, in *Kinetics and Mechanisms of Polycondensation*, D. H. Sol-
omon, (ed.), Marcel Dekker, New York (1972), Vol. 3, Ch. 7, p. 333.

M. Zeldin, A survey of inorganic polymers, *Polym. News* **3,** 65 (1976).

J. Hausmann, Ways to inorganic polymers, *Gummi Asbest Kunstst.* **29,** 440 (1976).

APPENDIX—SOME GENERAL REVIEW ARTICLES FROM 1970 ONWARDS

SUPPLEMENTARY BIBLIOGRAPHY

A. Lapierre, Super thermostable polymers, *Rev. Gen. Caout. Plast.* **47,** 1185 (1970).

R. T. Conley, High temperature polymers, *Org. Coat. and Plast. Chem. Preprints* **30** (2), 159 (1970).

R. T. Conley (ed.), *Thermal Stability of Polymers, Vol. 1,* Marcel Dekker, New York (1970).

W. G. Miller, High performance polymers, *Br. Plast.* 95 (June 1970).

W. B. Black, Structure-property relationships in high-temperature fibers, *Trans. NY Acad. Sci.* **32,** 765 (1970).

J. Preston, Fibers that resist high temperatures, *Chem. Tech.* 664 (November 1971).

S. Gonzalez-Babe and J. D. Abajo, Heat resistant polymers, *Rev. Plast. Mod.* **22,** 181, 1844 (1971).

B. Sillion, P. Guiliani, and J. Corbiere, Les Hautes Polymères Thermostables, Institut Français du Pétrole, Paris (1971).

V. V. Korshak, *The Chemical Structure and Thermal Characteristics of Polymers,* Israel Program for Scientific Translations, Jerusalam (1971).

G. F. Pezdirtz and N. J. Johnston, Thermally stable macromolecules, *Chemistry in Space Research,* R. Landel and A. Rembaum, (eds.), Elsevier, New York (1971), p. 155.

P. M. Hergenrother, High temperature organic adhesives, *SAMPE Q.* **3,** 1 (1971).

M. W. Pascoe, High temperature plastics, *Engineering* 230 (April 1973).

Anon., Applications prove the worth of high-heat resistant plastics, *Mater. Eng.* 41 (February 1973).

A. M. Houston, How heat resistant thermoplastics compare at elevated temperatures, *Mater. Eng.* **81**(6), 28 (1975).

J. Fontan and J. Abajo, Relationship between structure and properties of the new heat-resistant polymers, *Int. Polym. Sci. Technol.* **3**(11), T/75 (1976).

J. Preston, High performance fibers from aromatic polymers, *Polym. Eng. Sci.* **16,** 298 (1976).

M. L. Booth, High-temperature plastics, CME **23,** 71 (1976).

J. A. Vaccari, New heat resistant plastics cover 275–700°F range, *Mater. Eng.* **85**(5), 39 (1977).

J. A. Vaccari, New heat resistant elastomers are more versatile, *Mater. Eng.* **85**(5), 45 (1977).

J. A. Mock, Jumping the heat hurdle with organic, inorganic coatings, *Mater. Eng.* **85**(5), 51 (1977).

K. H. Miska, Adhesives and sealants take the heat up to 600°F, *Mater. Eng.* **85**(5), 57 (1977).

H. F. Mark, Polymers for extreme service conditions, *Macromolecules* **10**, 881 (1977).

J. L. Cotter, Heat resistant adhesives, *Rev. High Temp. Materials* **3**, 277 (1978).

S. Oldham, High temperature polymers, *SAMPE Q.* **10**(2), 1 (1979).

C. Arnold, Stability of high temperature polymers, *J. Polym. Sci. Macromol. Rev.* **14**, 265 (1979).

I. A. Gribova and O. V. Vinogradova, Advances in the production of thermostable antifriction plastics, *Russ. Chem. Rev.* **48**, 94 (1979).

P. E. Cassidy, *Thermally Stable Polymers. Synthesis and Properties,* Marcel Dekker, New York (1980).

J. Preston, Heat resistant polymers, in *Kirk-Othmer Encyclopedia Chem. Technol.,* 3rd ed., H. F. Mark, J. J. McKetta, and D. F. Othmer, (eds.), Interscience, New York (1980), Vol. 12, p. 203.

A. A. Askadskii and Y. S. Kochergin, Structure and properties of networks in heat-resistant polymers, *Russ. Chem. Rev.* **49**, 445 (1980).

P. E. Cassidy, History of heat-resistant polymers, *J. Macromol. Sci. Chem. A* **15**, 1435 (1981).

SUBJECT INDEX

TRADE NAMES INDEX